Sediment Cascades

Sediment Cascades

Sediment Cascades

An Integrated Approach

Edited by

Timothy P. Burt, *Durham University*

Robert J. Allison, *University of Sussex*

WILEY-BLACKWELL
A John Wiley & Sons, Ltd., Publication

This edition first published 2010

Wiley-Blackwell is an imprint of John Wiley & Sons, formed by the merger of Wiley's global Scientific, Technical and Medical business with Blackwell Publishing.

Registered office: John Wiley & Sons Ltd, The Atrium, Southern Gate, Chichester, West Sussex, PO19 8SQ, UK

Other Editorial Offices:
9600 Garsington Road, Oxford, OX4 2DQ, UK

111 River Street, Hoboken, NJ 07030-5774, USA

For details of our global editorial offices, for customer services and for information about how to apply for permission to reuse the copyright material in this book please see our website at www.wiley.com/wiley-blackwell

Library of Congress Cataloguing-in-Publication Data

Sediment cascades: an integrated approach / edited by Timothy Burt and Robert Allison.
p. cm.
Includes bibliographical references and index.
ISBN 978-0-470-84962-0 (cloth)
1. Sediment transport. 2. Sedimentation and deposition. 3. Sediments (Geology)
I. Burt, T. P. II. Allison, R. J. (Robert J.)
TC175.2.S33 2010
627′.122–dc22
2009026524

ISBN: 978-0-470-84962-0

A catalogue record for this book is available from the British Library.

Set in 10/12 pt Times by Thomson Digital, Noida, India.
Printed and bound in Singapore by Markono Print Media Pte Ltd.

Contents

Contributors

Professor Timothy P. Burt
Department of Geography
Durham University
Science Laboratories
South Road
Durham DH1 3LE UK
Tel: 0191-334-2611
Fax: 0191-334-3101
Email: T.P.Burt@durham.ac.uk

Professor Robert J. Allison
Department of Geography
University of Sussex
Arts Building C
Falmer Brighton
BN1 9SJ
Tel: 01273 678253
Email: R.J.Allison@sussex.ac.uk

Dr Louise J. Bracken
Department of Geography
Science Laboratories
South Road
Durham DH1 3LE UK
Tel: 0191 334 1846
Fax: 0191 334 1801
Email: L.J.Bracken@durham.ac.uk

Professor Michael Church
Department of Geography
University of British Columbia
Vancouver, BC
Canada, V6T 1Z2
Tel: (604) 822 5870
Fax: (604) 822 6150
Email: mchurch@geog.ubc.ca

Dr Tim R. H. Davies
Geological Sciences
University of Canterbury
Private Bag 4800
Christchurch 8140
New Zealand
Tel: +64 3 364 2700
Fax: +64 3 364 2769
Email: tim.davies@canterbury.ac.nz

Dr Martin G. Evans
Department of Geography
Mansfield Cooper Building
University of Manchester
Manchester M13 9PL
Tel: 0161 275 3640
Fax: 0161 275 7878
Email: mfvssmge@man.ac.uk

Professor Ian D. L. Foster
School of Life Sciences
University of Westminster
115 New Cavendish Street,
London W1W 6UW, UK
Tel: +44 (0) 20 7911 500
Email: ian.foster@westminster.ac.uk

Professor Basil Gomez
Geomorphology Laboratory
Indiana State University
Terre Haute
IN 47809 USA
Tel: +(812) 237-2249
Fax: +(812) 237-8231
Email: bgomez@indstate.edu

Professor Adrian M. Harvey
Department of Geography
Roxby Building
University of Liverpool
Liverpool, L69 7ZT UK
Tel: (0151) 794 2866
Fax: (0151) 794 2874
Email: amharvey@liv.ac.uk

Professor David L. Higgitt
Department of Geography
National University of Singapore
1 Arts Link Singapore 117570
Tel: +65 6516 6809
Fax: +65 6777 3091
Email: geodlh@nus.edu.sg

Robert H. Meade
28603 Meadow Drive
Evergreen, Colorado 80439 USA
Tel: (303) 674 7001
Email: potamundi@comcast.net

Dr Pamela S. Naden
CEH Wallingford Oxon OX10 8BB UK
Tel: 01491 838800
Fax: 01491 692424
Email: psn@ceh.ac.uk

Professor Michael J. Page
GNS Science
PO Box 30-368
Lower Hutt 6315 New Zealand
Tel: +64 (4) 570 1444
Fax: +64 (4) 570 4600
Email: M.Page @gns.cri.nz

Professor David N. Petley
Department of Geography
Science Laboratories South Road
Durham DH1 3LE UK
Tel: 0191 334 1909
Fax: 0191 334 1801
Email: d.n.petley@durham.ac.uk

Professor Denise J. Reed
Department of Earth and Environmental Sciences
University of New Orleans
New Orleans LA 70148 USA
Tel: (504) 280 7395
Fax: (504) 280 7396
Email: djreed@uno.edu

Dr Nick J. Rosser
Department of Geography
Science Laboratories
South Road
Durham DH1 3LE UK
Tel: 0191 334 1918
Fax: 0191 334 1801
Email: n.j.rosser@durham.ac.uk

Dr Tom Spencer
Cambridge Coastal Research Unit
Department of Geography
Cambridge University
Downing Place
Cambridge CB2 3EN UK
Tel: 01223 333350
Fax: 01223 333392
Email: ts111@cam.ac.uk

Professor Stanley W. Trimble
Department of Geography
1255 Bunche Hall
Los Angeles, CA 90095-1524
Tel: 310 825 1314
Fax: 310 206 5976
Email: trimble@geog.ucla.edu

Dr Noel A. Trustrum
GNS Science
PO Box 30-368
Lower Hutt 6315
New Zealand
Tel: +64 (4) 570 1444
Fax: +64 (4) 570 4600
Email: n.trustrum@gns.cri.nz

Preface

The aim of *Sediment Cascades: An Integrated Approach* is to provide an advanced text that addresses the transport of sediment through the fluvial landscape. We felt that there was a gap in the research literature in that no single text integrates the landscape components included here, covering the transfer of sediment and water from points of generation in upland environments, through transport pathways and intermediate stores to the coastal zone and oceanic sinks on the nearshore continental shelf and beyond. There is both a contemporary focus, with implications for catchment management, and consideration of larger-scale landform evolution.

We conceived this book quite a few years ago. Having set off with the best of intentions, we then stalled, mainly because both editors were too preoccupied with administrative duties at Durham University, and so distracted from the more important task of geomorphological research. No doubt several authors wondered whether the book would ever appear. However, with encouragement from Wiley, we resurrected the project and here we are! Fiona Woods at Wiley has been exceptional – applying firm pressure and a tight (revised) timescale, never letting the editors off the hook; which, in turn, meant firm pressure for the authors who, more or less, delivered their chapters to schedule. Thank you to them for coming through in the end. Special thanks to Michael Church who, at very short notice, wrote Chapter 9 when one set of authors decided they just could not produce their chapter. As would be expected, Michael's chapter is excellent, doubly impressive given how little time he had to write it.

We must repeat our thanks to Fiona Woods at Wiley, for dealing with the manuscript and pushing through the production in good time. Thanks too to Izzy Canning for her help with the mechanics of publication. We would also like to record our thanks to Janet Raine, Master's Secretary at Hatfield College, Durham for her continued support, especially whenever we faced a file-handling challenge.

For one reason or another, quite a number of the authors have a connection with Hatfield College, Durham, on the banks of the River Wear (normally rather tranquil but quite an impressive sight when there is a very big flood). So, we felt that the preface should be written here. The British Society for Geomorphology met for its annual conference at Hatfield College in 1994 (then BGRG, not BSG), and does so again in September 2009. We hope the book will be ready by then, or soon after.

Timothy Burt, Robert Allison
Hatfield College, Durham
1st June 2009

1

Sediment Cascades in the Environment: An Integrated Approach

Timothy P. Burt[1] and Robert J. Allison[2]

[1]Department of Geography, Durham University, Science Laboratories, South Road, Durham DH1 3LE, UK

[2]Department of Geography, University of Sussex, Arts Building C, Falmer, Brighton BN1 9SJ, UK

1.1 Statement of the Problem

> To make a sediment budget for a drainage basin, one must quantify and relate the major processes responsible for the generation and transport of sediment. This task is difficult because the processes are often slow and highly variable in space and time. Data from short-term monitoring at a few localities are not easily extrapolated to compute average values for large areas.
>
> *(Dietrich and Dunne, 1978, p.192)*

When Bill Dietrich and Tom Dunne wrote these words, the integrated study of sediment delivery systems was very much in its infancy. This is not to imply that nothing was known about these systems, far from it, but there had been a tendency for each part of the system to be studied by specialists from different disciplines (Hadley *et al.*, 1985). A great deal of attention had been focused on soil erosion, especially in the USA, where the Universal Soil Loss Equation (USLE) had been derived from more than 8000 plot-years of erosion research data from agricultural land of low slope angle, to investigate long-term patterns and trends in soil erosion (Mitchell and Bubenzer, 1980: see also Chapter 7, this volume). At the other end of the system, there had been steady improvement in the accuracy of sediment load calculations for major rivers. On-site (gross) rates of soil erosion within a drainage basin had been linked to the (net) sediment yield at the basin outlet via the

Sediment Cascades: An Integrated Approach Edited by Timothy Burt and Robert Allison

sediment delivery ratio but few studies were available that had unlocked the 'black box' of sediment delivery (Walling, 1983).

Maner and Barnes (1953) were amongst the first to use the term *sediment delivery*; since then, it has been widely used to represent the transport of eroded sediment to the basin outlet. Hadley and Schumm (1961) developed relationships between drainage basin characteristics and rates of sedimentation in small reservoirs, but they did not use the term sediment delivery ratio. Working on sediment design requirements for flood-retention reservoirs in the southeast United States, Roehl (1962) focused explicit attention on the sediment delivery ratio (SDR), quantifying the fraction of eroded material delivered to the basin outlet (both expressed in tonnes per square kilometre per year or similar units). He found that SDR could be explained using three morphological variables: drainage area, relief-length ratio and weighted mean bifurcation ratio. Sediment yield per unit area (*specific sediment yield*) decreased with increasing size of drainage area (Figure 1.1). This was a typical result for early studies of SDR, but as Church *et al.* (1999) point out (see also below), most attention was paid to regions with known high rates of soil loss where within-catchment 'loss' via deposition on floodplains was the norm. The approach was, therefore, to combine results from plot studies of soil erosion (agricultural science) with measurement of sediment transport in large rivers (fluvial geomorphology); few scientists were working *within* the drainage basin, examining transfer processes and transient storages.

From the 1980s onwards, the situation has been transformed. The plea by Walling (1983) to unlock the black box of sediment delivery has been answered in very many and different ways. This volume is an attempt to gauge progress as well as to look forward to future prospects. Relevant studies now range over long timescales and a wide variety of regions and spatial scales; studies in agricultural basins represent only one small part of the effort, and the results can be as relevant to global-scale questions about mountain building as they are to

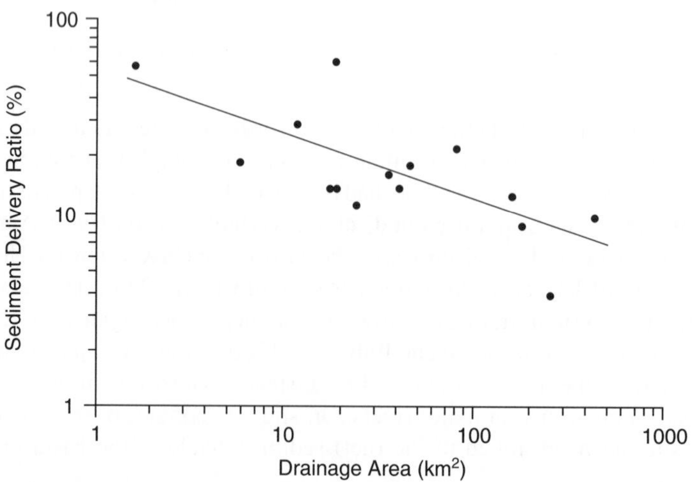

Figure 1.1 Relationship of sediment delivery ratio and drainage area for 15 basins in south eastern USA
(after Roehl, 1962 and Hadley *et al.*, 1985).

local management issues. However, the opening quotation still encapsulates the problem: the need in any situation to quantify the major processes responsible for the generation and transport of sediment through whatever geomorphological system is being studied.

1.2 Benchmark Papers

Individual chapters in this volume chart the development of research within their individual research areas. Here, we refer to three influential papers to show the way in which studies of the sediment cascade gradually developed in different research areas within the broad field of geomorphology.

1.2.1 The Degradation of a Coastal Slope, Dorset, England (Brunsden, 1974)

In part, the quantitative revolution in geomorphology arose from a need to establish rates of erosion for different processes, something which had been steadfastly sidestepped in traditional studies of landscape evolution (Chorley, 1978). Denys Brunsden spent five years monitoring a range of processes on a very active coastal landslide on the English Channel coast. He used several different techniques: aerial photographs, land surveying, geomorphological mapping and erosion pins, to study both large-scale (rotational landslides, mudslides) and small-scale processes (small slumps, wash erosion, frost action). His results emphasized the variability of erosion processes in time and space, and the difficulty of relying on short periods of observation to estimate long-term erosion rates. His results of monthly landslide volumes were some of the first to quantify landslide activity and the transfer of material downslope (in this case, into a complex system that eventually delivers material on to the beach).

1.2.2 Sediment Budget for a Small Catchment in Mountainous Terrain (Dietrich and Dunne, 1978)

This paper has already been referred to in the introduction. The research reported here lies somewhere between the geotechnical landslide study of Brunsden and the fluvial geomorphological work of Trimble (see below). Bill Dietrich and Tom Dunne worked in the coast range of Oregon. Their aim was to quantify all the processes of sediment transfer and all the various sediment stores in the study basin. Transfer of soil to stream channels was by two mechanisms: soil creep delivered soil to the slope base where it entered the channel, usually via small slumps; creep also infilled hillslope hollows which eventually filled and evacuated via debris flows. Half the soil discharged to channels left the basin as suspended sediment load whereas half was stored temporarily in tributaries, debris fans and the floodplain. The residence times of sediment in these storage elements increased down-valley from decades to 10 000 years (i.e. the entire postglacial period). The paper is significant for several reasons. They present a sediment budget model for Rock Creek basin as a systems diagram representing storages and transfers (Figure 1.2). This approach illustrates the popularity of a systems approach at the time (cf. Chorley and Kennedy, 1971); others

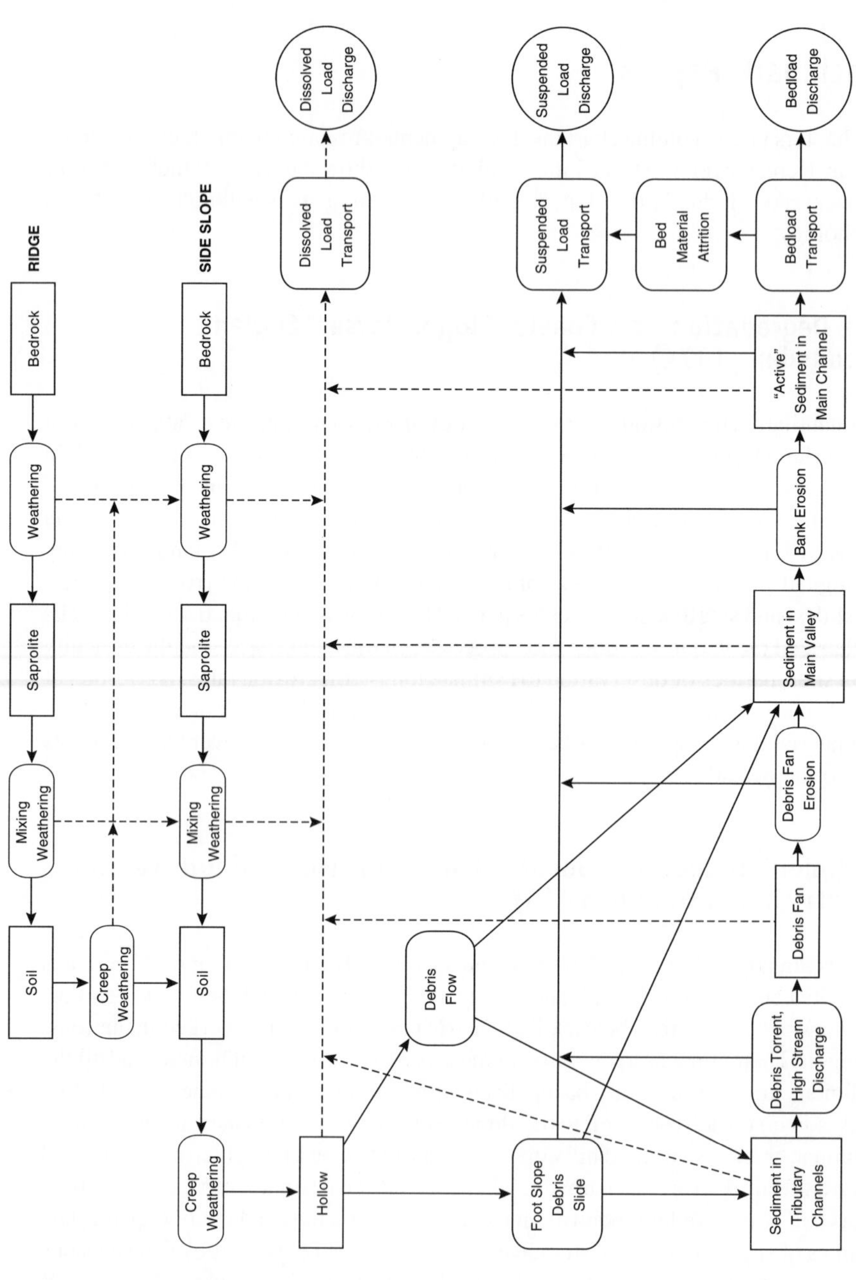

Figure 1.2 Sediment budget model for Rock Creek basin, Coast range, Oregon, USA (redrawn from Dietrich and Dunne, 1978).
Rectangles represent storages, round-cornered boxes indicate transfer processes, circles represent outputs, solid lines represent the transfer of sediment and dotted lines represent solute transport.

had defined the elements of the system and correlated different morphological variables (e.g. Kennedy, 1965; quoted in Chorley and Kennedy, 1971, p.157–8), but this was one of the first attempts to quantify the sediment cascade. It is also significant that Dietrich and Dunne tried to estimate residence times of sediment in the various storages. On the slopes, they paid particular attention to hillslope hollows from which debris flows originate and the rate at which they infill and then evacuate. Within the main channel, low-order tributaries, debris fans and 'active' sediment had residence times of the order of decades to a few centuries, whereas sediment within the floodplain was stored for an order of magnitude longer. They also compared the size of storages to the rates of suspended and bedload transport by the river. The variable timescale of sediment storage has become a recurrent theme, and in discussed in most chapters in this volume. For example, Davies and Korup (Chapter 4) show how mountain landscape evolution is modulated by sediment storages, between which sediment cascades in a manner determined by the sediment production rate and the characteristics of intermediate storage. As Brunsden (1974) also noted, these transient dynamics significantly complicate the long-term assumption of average erosion rates. In Bill Dietrich's case, later work continued to focus on the discontinuous occurrence of debris flows and how frequently hillslope hollows 'fail', delivering significant slugs of sediment to the drainage system (Dietrich, Wilson and Reneau, 1986; see also Chapter 2).

1.2.3 Changes in Sediment Storage in the Coon Creek Basin, Wisconsin, 1853–1975 (Trimble, 1981)

Stan Trimble's analysis of ten river basins in the southeastern United States (Trimble, 1975) indicated that of the material eroded from upland slopes since European settlement, only about 5% had been exported. The remainder, alluvium and colluvium, would probably not be exported because of extensive reservoir impoundments. He concluded that contemporary stream sediment loads were dubious indicators of regional denudation. He followed this up with a detailed study of the Coon Creek basin, a minor tributary of the Mississippi in the Driftless Area of southwestern Wisconsin (basin area: 360 km^2). The research neatly combines the use of the USLE to estimate source erosion rates and various survey work to establish sediment storage and transfer rates throughout the basin. Most storage is along streams, with alluvial storage accounting for about 75% of all measured sediment in the basin. Two sediment budgets are constructed: 1853–1938 and 1938–1975. At the time of European settlement, slope erosion and valley-floor accretion were both negligible, but erosion rates soon became very high as did accretion rates on the floodplain. Table 1.1 shows that erosion rates were 326 kt per year in this period, equivalent to just over 900 t km^{-2} yr^{-1}. Delivery to the upper main valley was almost as high (316 kt yr^{-1}) and most of this reached the lower main valley (245 t yr^{-1}). However, the sediment yield to the Mississippi was only 38 kt yr^{-1}, equivalent to just over 100 t km^{-2} yr^{-1}. Thus, most of the sediment transported out of the headwaters was being deposited within the valley, and floodplains at that time were aggrading at 150 mm yr^{-1} (Trimble, 1975). Soil conservation measures from the 1930s onwards severely curtailed upland erosion (Table 1.1). In his 1981 paper, Trimble speculated that the valley floor would become a net source of sediment in due course, although this was not the case as late as 1993 (Trimble, 1999) when the valley floor continued to be a small sediment sink, with the figures reflecting the much reduced inputs from farmed slopes. Remarkably, the sediment yield from the basin has stayed almost

Table 1.1 Sediment budget for Coon Creek (based on Trimble, 1999, Figure 1.1). All figures in 10^3 tonnes per year

	1853–1938	1938–1975	1975–1993
Sources			
Net upland sheet and rill erosion	326	114	76
Upland gullies	73	64	19
Tributaries	42	35	9
Upper main valley	0	27	13
Sinks			
Upland valleys	38	38	0
Tributary valleys	87	0	25
Upper main valley	71	27	4
Lower main valley	209	139	51
Total sources	441	240	117
Total soil erosion sources	399	178	95
Net export to upper main valley	316	175	79
Net export to lower main valley	245	175	88
Sediment yield to Mississippi	38	36	37

constant over a century and a half, not reflecting at all the changing patterns of erosion, transfer and sedimentation within the basin. Trimble (1975) concluded that the most significant pattern to emerge from this sediment budget is that sediment yield had been small compared with either erosion or change in sediment storage. He added that sediment yield must be a poor indicator of upland erosion in highly disturbed basins and that any established relationship between sediment yield per unit area and drainage area (Figure 1.1) would need to be rethought. Once again, the complexity of the sediment cascade in time and space is revealed once detailed measurements of storages and transfer rates are made.

These three papers illustrate the complexities of sediment delivery that began to emerge as researchers started to quantify the systems involved. These complexities are discussed in more detail in the following sections. In essence, the overly simple sediment-delivery-ratio–catchment-area relationship (e.g. Figure 1.1) indicates an average response, disguising complexity both in space and time (Walling, 1983).

1.3 Complexity of Sediment Delivery in Time and Space

Sediment cascades can be analysed over a wide range of timescales, from individual storms through to geological eras. In this volume, there is coverage ranging from single floods (Chapters 7 and 10) right through to events that are significant within the context of entire mountain-building episodes over millions of years (Chapter 4). Trimble (1999) notes that geomorphological effects ultimately caused by changes of vegetation cover and land management are not always directly linked in time and space. This makes predictions of erosion and sediment yield very uncertain. We have already seen from Table 1.1 how complicated changes in sediment transfer within a basin may not necessarily be reflected in sediment export from the basin. Trimble and Lund (1982) proposed a hysteresis model to explain the lag effects between land use and upland erosion at Coon Creek: note that this

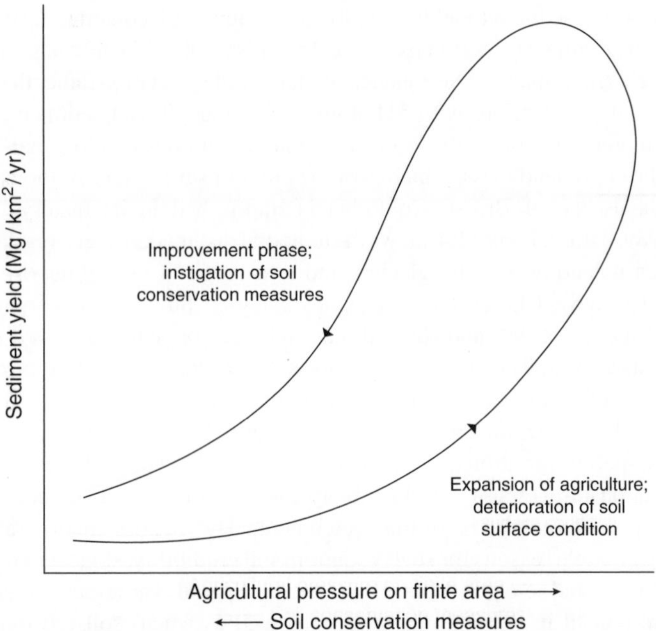

Figure 1.3 The hysteretic relationship between agricultural activity and sediment yield (based on Trimble and Lund, 1982).
Note that deterioration of soil surface condition lags behind the expansion of agriculture; similarly, improvement of soil surface condition lags behind implementation of soil conservation measures.

model only encompasses the farmed hillslopes; further lags would be evident within the floodplain–channel system, of course. Figure 1.3 shows how, as agriculture spread in the basin, it took years for soil condition to deteriorate; it was only in the later stages of soil misuse that off-site sediment transfer rose sharply, reflecting increased rilling and gullying, loss of soil organic matter and filling of transient storages, for example in riparian zones. Similarly, once soil conservation measures were introduced, it took time for soils to recover and for better management of upland riparian zones and channels to become effective. Trimble (1999) notes that, in highly disturbed basins where soil erosion is adding to alluvial storage, deposits from one period may be gradually released later, once headwater inputs decline (see also Chapter 11). Although, as noted above, this speculation proved unfounded in the case of Coon Creek, at least so far, the general point is a fair one, and might still occur at Coon Creek in due course, if soil erosion and sediment delivery rates were to fall back to pre-settlement levels.

At much longer timescales, the same sorts of lag effects can be seen in relation to the release of stored sediment. Church and Slaymaker (1989) presented data from British Columbian rivers that controverted the conventional model (Figure 1.1). They demonstrated increasing specific sediment yield at all basin scales up to $3 \times 10^4\,km^2$, resulting from the dominance of Quaternary sediments along river valleys. Sediment yields from the smallest basins studied were more than an order of magnitude lower than rates in medium to large basins, the headwaters having been cleaned out of sediments by glacial erosion. Working at a much longer timescale than Trimble, Church and Slaymaker drew much the same conclusion: there was no way in which sediment yields were in equilibrium with contemporary erosion of the land surface as determined by regional climate. In British Columbia, fluvial sediment yields at all scales above 1 km^2 remain a consequence of Pleistocene glaciation rather than of Holocene erosion. They agreed therefore with Schumm (1977) that 'fluvial sedimentation does not reflect contemporary erosion of the land surface at any timescale below that for significant evolution of the entire land mass, which is of the order of several tens of thousands of years. The natural landscape of British Columbia is imprisoned in its history'. (Church and Slaymaker, 1989, p.453). Figure 1.4 shows the temporal disjunction over a glacial–interglacial period between upland erosion by glaciers and evacuation of glacial deposits from valley floors. In a later study, Church and colleagues analysed fluvial clastic sediment yield in Canada (Church *et al.*, 1999) and showed that high rates of sediment loss downstream are typical only of agriculturally disturbed landscapes; for Canada at least, the type of relationship shown on Figure 1.1 is the exception rather than the rule.

Of course, such complexities in the temporal pattern of sediment delivery can be equally interpreted as spatial variations in the sediment cascade. The sediment-delivery-ratio–basin-area relationship (Figure 1.1) inevitably masks a good deal of spatial variation in erosion rates in different parts of the catchment. The factors incorporated in USLE summarize the possible reasons for such variation: soil erodibility, slope length and gradient, crop management and erosion control practices. Spatial variations in surface runoff production are crucial in determining where (as well as when) soil erosion occurs: both models of overland flow generation indicate that this is locally restricted to certain areas (Burt, 1989). In the partial area model, infiltration-excess overland flow is produced from areas of low infiltration capacity; these are often particular fields where tillage practices and subsequent heavy rainfall have compacted the soil surface (Imeson and Kwaad, 1990). Other fields nearby with different land use may generate no surface runoff at all. In the variable source area model, saturation of the soil profile generates saturation-excess overland flow;

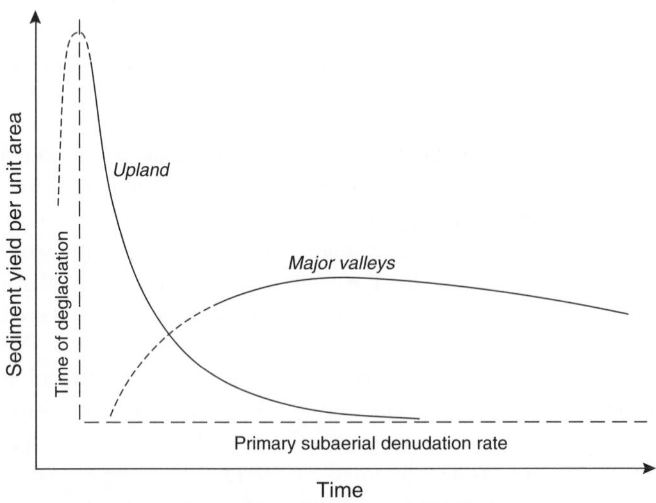

Figure 1.4 The paraglacial sedimentation cycle, modified to indicate the effect of spatial scale upon the temporal pattern of sediment yield. The time axis spans approximately 10 000 years (Church and Slaymaker, 1989).

this is most likely at the foot of slope, especially in hillslope hollows, or where the soil profile is thinner. For both models, there may be large areas of the catchment not producing any overland flow. The localized production of surface runoff and soil erosion, regardless of the exact runoff processes involved, has been emphasized in the term 'critical source area' (Heathwaite, Sharpley and Gburek, 2000).

At the larger scale, we have already noted that the sediment cascade in highly disturbed, agricultural catchments, with headwater erosion and floodplain deposition further downstream, may be very different to most other areas. The Canadian examples from Church and colleagues show that sediment delivery may well increase downstream, depending on the legacy of previous erosion episodes and the nature of the catchment area. In essence, Trimble's work at Coon Creek shows this too: although headwater sources remain the dominant source of sediment (just), remobilization of sediment from floodplain storages further downstream is now relatively much more important, and could become the dominant source in due course. The Coon Creek work may be regarded as a space–time substitution therefore, with changes through time equating with different spatial patterns under different circumstances.

One conclusion from this brief review is that detailed, distributed sediment budgets for catchment areas remain relatively scarce. Examples contained within this volume add to what is already known, and emphasize the differences to be found in different regions and at different periods of economic development. Such information may well inform government agencies responsible for land and water management. Such knowledge will be equally relevant to more traditional questions of geomorphology, questions given new impetus by our ability to quantify sediment cascades and model the implications. This is certainly true in relation to large-scale landform development, for example the relationship between erosion of mountains and feedback to the ongoing evolution of the mountains themselves (Chapter 4). The smaller-scale dynamics of landform evolution, for example, valley-side mass movements or terrace formation, will also be illuminated by better sediment budgeting.

1.4 This Book

Cascades in the natural environment transfer energy and material from source to sink (Chorley and Kennedy, 1971). The aim of *Sediment Cascades: An Integrated Approach* is to provide an advanced text that addresses the transport of sediment through the fluvial landscape. Sediments originate in a number of environments, from steep montane regions through to low-gradient farmland. The transfer from source to sink involves what Ferguson (1981) has called the 'jerky conveyor belt' of sediment delivery: 'temporary' storage of sediment *en route* in floodplains, alluvial fans and so on. Lakes, deltas, estuaries, coasts and the continental shelf provide the final sink for terrestrially derived material. There has been growing interest in the sediment cascade in recent years, across a range of disciplines, from geologists interested in analogues of ancient sedimentary basins, through Quaternary scientists reconstructing more recent sedimentary environments, to environmental scientists interested in sediment-bound pollutant transport. For geomorphologists, the main focus over several decades has been the study of the contemporary sediment cascade in small river basins over short timescales. However, increasingly, traditional, large-scale questions of landscape evolution have moved back centre stage, most notably in relation to mountain building and associated erosion. This book reflects that duality with coverage across a range of spatial scales. However, despite clear relevance for long-term landform evolution, the emphasis remains focused on contemporary sediment system dynamics.

Chapters 2–5 focus on upland areas. Church (Chapter 2) examines sediment transport in steep, montane channels where channel gradient is too large to permit unconstrained accumulation of alluvial sediment. Such channels exhibit close coupling to hillslope processes. Headmost hollows and hillside channels recruit sediment by mass wasting processes, and significant sediment transfers occur within these channels as debris slides and debris flows. Steep channels are dominated by episodic sediment supply and the channel may remain relatively stable until catastrophic evacuation of accumulated sediment occurs. In contrast to settled, lowland basins, the fluvial sediment yield often increases downstream, especially in glaciated landscapes. Rosser (Chapter 3) picks up on the sediment supply theme, examining the role that landslides and rockfalls can play in the generation and transfer of sediment into and through fluvial systems in areas of steep topography. Rosser argues that landslide mechanism determines the rate and position at which sediment enters the cascade, controls its character, and defines the inherited topography in which future failures develop and propagate. He demonstrates the need to marry together mechanistic understanding of the controls on slopes failure and their characteristics with both contemporary sediment dynamics and longer-term landscape evolution. Davies and Korup (Chapter 4) focus particularly on tectonically active mountain ranges. Interestingly, their conclusions very much echo those of Church (Chapter 2) in that they also conclude that the formation, development and failure of sediment storages in mountain belts are often rapid, if not catastrophic. At all but the shortest timescales, sediment accumulation and release exerts fundamental controls on the dynamics of mountain rivers. Whereas Rosser (Chapter 3) emphasizes links to landform evolution, Davies and Korup reflect on the way in which the dynamics of sediment cascades present severe problems to sustainable development in and close to mountain ranges.

Gomez, Page and Trustrum (Chapter 5) are also concerned with sediment dynamics in steep terrain, but generally less mountainous than those considered in previous chapters. They focus on shallow landslides and gully erosion, both of which are the outcome of land-use change and have the potential to impact the entire sediment cascade downstream, over timescales of decades to millennia. Given that erosion 'hot spots' can be especially problematic, consideration of landscape rehabilitation is therefore an important adjunct to their coverage of erosion and sediment delivery. They conclude that steepland rivers contribute a disproportionately large amount of riverine sediment and, by association, carbon to the coastal ocean, an essential reminder that stream sediment load is not just mineral but can contain significant amounts organic matter too. Harvey (Chapter 6) too is concerned with steep terrain, specifically debris cones and alluvial fans. These zones of sediment storage occur at breaks in the sediment cascade, especially where two process domains meet, such as the boundary between the hillslope and the channel system or within the channel system at tributary junctions or where there are topographic changes, for example, in valley-floor gradient. Not only are debris cones and alluvial fans very distinctive landforms, but they can have important effects on the dynamics of the sediment cascade. Harvey's review is mainly concerned with process and form, since there have been few attempts to construct sediment budgets for these landforms. Whether related to timescales of a single flood or the entire Quaternary, fans are sensitive indicators of environmental change and the sediments stored may preserve a record of past variations in the sediment cascade.

The next few chapters have a more 'lowland' theme, although clearly the headwater catchments referred to can still have an 'upland' character, if not mountainous by global standards. Bracken (Chapter 7) reviews process-based research on the hydraulics of overland flow, the entrainment of particulate material, rilling and gullying. She then broadens the discussion to review the factors influencing soil erosion of agricultural land as well as covering erosion in badland systems. She ends with coverage of the movement of sediment off-site, the role of sediment buffer zones and sediment conveyance into low-order channels, and so is complementary with Trimble (Chapter 11) in taking an interest in sediment delivery. In covering the management of fine-sediment transport, there is also a strong link to Naden (Chapter 10). Perhaps the most important conclusion is the importance of disconnection between the point of erosion and the stream channel if sediment delivery is to be managed. The subject of hydrological connectivity continues to be an important theme in catchment research (e.g. Lane *et al.*, 2009) Bracken's review covers a topic where there has been much research in recent decades – see for example: Kirkby and Morgan (1980), Boardman, Foster and Dearing (1990) and Boardman and Poesen (2006). Evans and Burt (Chapter 8) examine erosional processes and sediment transport in upland mires. Given extensive cover of friable organic soil, the sediment flux from an eroded mire can have a much higher carbon content than where mires are not present. The significance of gully erosion in headwater peatlands for reservoir sedimentation downstream has been recognized for some years; only more recently have the implications of peat erosion for carbon storage been realized. The formation of peat is climatically determined; it is something of an irony that the degradation of peat has direct feedback to the climate system.

The next three chapters consider sediment conveyance in lowland basins where human occupation, over centuries or millennia, has had a significant impact.

Church (Chapter 9) looks at rivers where the sediment transport is dominated by movement of gravel-sized material, predominantly in the montane zone and the mountain foreland. This has been an important topic of research for some fluvial geomorphologists in recent decades and this chapter reviews progress from the point of view of sediment conveyance and in-channel storage. Church shows that channel sediments and channel morphology in gravel-bed rivers are the consequence of the sediment transport, which in turn determines some distinctive characteristics of gravel-bed channels and a distinctive morphodynamic style. Compared with gravel-bed rivers, the conveyance of fine sediment in river channels has been a relatively neglected topic of research recently. Yet, fine sediment is arguably the most important component of the sediment cascade – whether measured in terms of flux, its role in transporting nutrients and contaminants, or its importance to aquatic ecosystems. Naden (Chapter 10) redresses the balance and provides a complement to the previous chapter. Topics covered include sediment transport processes; channel geometry and bedforms in fine-grained channels; sediment supply, storage and output. She spans the three major disciplines of hydraulics, hydrology and geomorphology in order to bring out the key advances and identify some of the remaining issues. The fundamental influence of time and space scales on current thinking in these areas is well demonstrated through reach-scale and whole-catchment comparison. Naden's review of the sediment delivery ratio leads neatly into the next chapter. The review by Trimble (Chapter 10) reflects his career-long interest in sediment cascades in agricultural catchments, most notably his studies at Coon Creek. Indeed, Trimble's Figure 8 is original to his chapter (based on Trimble and Lund, 1982) and picks up on the connectivity theme mentioned earlier. The concept of 'fast in, slow out' is highlighted throughout the chapter and is argued to apply in many if not all instances. The idea is that a 'wave' of sediment can accumulate very rapidly by vertical accretion on floodplains but removal can be achieved only by channel and bank erosion. As noted above (Section 1.2.3 above), the complexity of the sediment cascade in time and space is revealed wherever detailed measurements of storages and transfer rates are made. In the examples discussed by Trimble, Ferguson's (1981) 'jerky conveyor belt' of sediment delivery comes frequently to mind, and there is, paradoxically, little linkage between soil erosion rates and sediment delivery at the basin outlet at any point in time. Another important conclusion is that there seems to be extremely important modal differences separating inherently stable basins from inherently unstable ones (Trimble, Figure 1.1). Foster (Chapter 12) examines the origins and history of lake and reservoir basins. He considers their development and longevity, and their possible impact on the transfer of sediment from river catchments to the oceans. He evaluates the impact of reservoirs on the fluvial system both upstream and downstream of the impoundment and discusses the value of sedimentary archives contained within these basins for providing information on geomorphological processes operating in river catchments. He also briefly considers the potential for palaeoenvironmental data to play a role in the broad field of policy and governance, benchmarking environmental quality standards, including nutrient and contaminant concentrations and sediment yields. In relation to the sediment cascade for fine sediment, Chapter 12 complements the preceding two chapters in particular.

Higgitt (Chapter 13) widens the scale of interest by examining the role of large rivers in conveying sediments from continents to oceans. The chapter contains four broad themes. Following on from the previous group of chapters, Higgitt upscales field studies of land degradation to ask questions about the global extent of soil erosion and the implications of the downstream transfer of eroded sediment. Given increasing emphasis on global environmental change, attention is also paid to the role of sediment in biogeochemical cycles (see also Chapter 10). Higgitt also reviews the quest to determine the global flux of continent-to-ocean sediment transfer and, by extension, contemporary denudation rates. This links nicely to the study of large rivers as analogues for interpreting strata in sedimentary facies and anticipates Chapter 15 (Petley). Alluvial stratigraphy provides an archive from which to examine the development of large river basins and their response to environmental change over timescales from decades to millions of years. Spencer and Reed (Chapter 14) emphasize the integrated nature of fluvial and coastal sediment systems. They show that coastal wetlands are an important sink for riverine sediments over millennial timescales. They examine three estuarine environments: the subtidal reaches of estuaries; the low to middle intertidal zone sand and mudflats; and the upper intertidal saltmarshes. They range across a changing set of interactions between sedimentological and biological processes, from the subtidal zone where physical (and physico-chemical) processes are dominant, through a complex set of interactions on tidal flats where the stability and erodibility and deposition of sediments is mediated by biological activity, to saltmarshes where substrate stability is strongly controlled by the presence of a permanent vegetation canopy. Finally, there is consideration of the consequences of global environmental change for estuaries; these systems are vulnerable both in relation to sea-level rise and to increased storminess. Petley (Chapter 15) examines the dynamics of the sediment cascade on the continental shelf and continental slope, both poorly understood environments. In particular, so little is known about the role of high-magnitude–low-frequency mass movements on the continental slope in moving sediment into the deep ocean that even those components of the sediment cascade that are well understood are difficult to analyse in terms of a long-term sediment budget. Notwithstanding the large increases in sediment delivery that have accompanied human development in many river basins, the majority of continental shelves trap most of incoming sediments, with relatively small amounts reaching the deep ocean floor. However, lack of knowledge about the number and magnitude of large submarine landslides, which transport vast amounts of sediment from the shelf and upper slope to the deep ocean, means that true rate of sediment accumulation in the deep marine environment is probably underestimated.

1.5 Perspective

Finally, it is important to emphasize the important themes emerging across the various chapters. Several themes are evident in more than one chapter.

- There is the continuing challenge of quantifying sediment budgets and modelling the sediment cascade. In only a very few cases, sediment budgets are very well constrained in time and space: see, for example, Trimble's examples from Coon Creek in Chapter 11 or some of the examples cited by Foster in Chapter 12. However, in many environments, from steep montane channels to the continental shelf, budgets are barely quantified. One ever-present theme is what Ferguson (1981) has called the 'jerky conveyor belt' of sediment delivery: 'temporary' storage of sediment within the sediment cascade. Trimble's slogan of 'fast in, slow out' seems to have much to commend it in relation to the gradual transfer of sediment downstream, and seems as relevant in glaciated valleys (Chapter 2) as it is in agricultural basins (Chapter 11). In terms of modelling and prediction, reasonable predictions of sediment transport are possible for reach lengths, but as Church's pun (Chapter 9) states, estimation of local variations in transport – the variations that determine the local morphology of river channels – remains beyond reach!

- Many chapters include coverage of management issues relating to human impact in catchment systems and the need to adopt an integrated approach to catchment management. This theme is perhaps best illustrated in lowland basins where human impact has been significant over decades and centuries. But the theme is equally relevant in other environments too, for example the discussion of sustainable development in and close to mountain ranges by Davies and Korup in Chapter 4.

- There has been an explosion of interest in steep, montane environments in the past decade, partly because of the need to quantify the risks associated with natural hazards, but also in relation to landform evolution in active tectonic zones. This exciting new development in megageomorphology has given new impetus to studies of large-scale uplift and erosion, and the evolution of large-scale landforms. Process studies will inform modelling studies, with developments in cosmogenic dating providing much improved constraint on process rates. We await calibration of general models of long-term slope evolution such as Ahnert (1987) or Kirkby (1987).

- Fluvial sediments provide an important archive of environmental change. They are used to identify and date periods of Quaternary environmental change, especially the study of sedimentary deposits in lakes, reservoirs and on floodplains, and to document very recent changes in the sediment cascade. There is also the study of contemporary fluvial sediments as analogues for interpreting sedimentary deposits in geological formations. These themes emerge particularly in Chapters 6, 12 and 13.

In many ways, Des Walling's plea to unlock the black box of sediment delivery has been well addressed, at least in general terms. Sediment budgets are much better quantified than they used to be. The main research challenges are now to better quantify budgets in time and space and then to turn this information into effective models. Much progress has been made in computer modelling, notably in computational fluid dynamics (see reviews in Bates, Lane and Ferguson, 2005) but such work is only the beginning. The interplay between field data collection and modelling will remain the focus for some time to come, but there is clearly considerable potential for modelling studies to take forward the study of sediment cascades in many varied and significant ways.

References

Ahnert, F. (1987) Process–response models of denudation at different spatial scales. *Catena*, **10**, 31–50.

Bates, P.D., Lane, S.N. and Ferguson, R.I. (2005) *Computational Fluid Dynamics*, John Wiley and Sons, Chichester.

Boardman, J. and Poesen, J. (eds) (2006) *Soil Erosion in Europe*, John Wiley and Sons, Chichester.

Boardman, J., Foster, I.D.L. and Dearing, J.A. (eds) (1990) *Soil Erosion on Agricultural Land*, John Wiley and Sons, Chichester.

Brunsden, D. (1974) The degradation of a coastal slope, Dorset, England, in *Progress in Geomorphology* (eds E.H. Brown and R.S. Waters), Institute of British Geographers, Special Publication No. 7, pp. 79–98.

Burt, T.P. (1989) Storm runoff generation in small catchments in relation to the flood response of large basins, in *Floods* (eds K.J. Beven and P.A. Carling), John Wiley and Sons, Chichester, pp. 11–36.

Chorley, R.J. (1978) Bases for theory in geomorphology, in *Geomorphology: Present Problems and Future Prospects* (eds C. Embleton, D. Brunsden and D.K.C. Jones), Oxford University Press, Oxford, pp. 1–13.

Chorley, R.J. and Kennedy, B.A. (1971) *Physical Geography: A Systems Approach*, Prentice-Hall, London.

Church, M. and Slaymaker, O. (1989) Disequilibrium of Holocene sediment yield in glaciated British Columbia. *Nature*, **337**, 452–454.

Church, M., Ham, D., Hassan, M. and Slaymaker, O. (1999) Fluvial clastic sediment yield in Canada: scaled analysis. *Canadian Journal of Earth Sciences*, **36** (8), 1267–1280.

Dietrich, W.E. and Dunne, T. (1978) Sediment budget for a small catchment in mountainous terrain. *Zeitschrift für Geomorphologie, Supplementeband*, **29**, 191–206.

Dietrich, W.E., Wilson, C.J. and Reneau, S.L. (1986) Hollows, colluvium, and landslides in soil-mantled landscapes, in *Hillslope Processes* (ed. A.D. Abrahams), Allen and Unwin, Boston, pp. 361–388.

Ferguson, R.I. (1981) Channel forms and channel changes, in *British Rivers* (ed. J. Lewin), Allen and Unwin, London, pp. 90–125.

Hadley, R.F. and Schumm, S.A. (1961) Sediment sources and drainage-basin characteristics in upper Cheyenne River basin. *U.S. Geological Survey Water-Supply Paper* 1531-B, Washington, pp. 137–197.

Hadley, R.F., Lal, R., Onstad, C.A. *et al.* (1985) *Recent Developments in Erosion and Sediment Yield Studies*. Technical Document in Hydrology, UNESCO, Paris, p. 127.

Kennedy, B.A. (1965) *An analysis of the factors influencing slope development on the Charmouthian Limestone of the Plateau de Bassingny, Haute-Marne, France*. Unpublished BA Dissertation, Department of Geography, Cambridge University, 99 pp.

Heathwaite, A.L., Sharpley, A.N. and Gburek, W.J. (2000) A conceptual approach for integrating phosphorus and nitrogen management at catchment scales. *Journal of Environmental Quality*, **29**, 158–166.

Imeson, A.C. and Kwaad, F.J.P.M. (1990) The response of tilled soils to wetting by rainfall and the dynamic character of soil erodibility, in *Soil Erosion on Agricultural Land* (eds J. Boardman, I.D.L. Foster and J. Dearing), John Wiley and Sons, Chichester, pp. 3–14.

Kirkby, M.J. (1987) General models of slope evolution through mass-movement, in *Slope Stability* (eds M.G. Anderson and K.I.S. Richards), John Wiley and Sons, Chichester, pp. 359–379.

Kirkby, M.J. and Morgan, R.P.C. (1980) *Soil Erosion*, John Wiley and Sons, Chichester.

Lane, S.N., Reaney, S.M. and Heathwaite A.L. (2009) Representation of landscape hydrological connectivity using a topographically driven surface flow index. *Water Resources Research*, **45**, W08423, doi: 10.1029/2008WR007336.

Maner, S.B. and Barnes, L.H. (1953) *Suggested Criteria for Estimating Gross Sheet Erosion and Sediment Delivery Rates for Blackland Prairie Problem Areas in Soil Conservation*. U.S. Department of Agriculture, Soil Conservation Service, Fort Worth, Texas.

Mitchell, J.K. and Bubenzer, G.D. (1980) Soil loss estimation, in *Soil Erosion* (eds M.J. Kirkby and R.P.C. Morgan), John Wiley and Sons, Chichester, pp. 20–35.

Roehl, J.W. (1962) Sediment source areas, delivery ratios, and influencing morphological factors. *International Association of Hydrological Sciences Publication*, **59**, 202–213.

Schumm, S.A. (1977) *The Fluvial System*, John Wiley and Sons, New York.

Trimble, S.W. (1975) Denudation studies: can we assume stream steady state. *Science*, **188**, 1207–1208.

Trimble, S.W. (1981) Changes in sediment storage in the Coon Creek basin, Driftless Area, Wisconsin, 1853 to 1975. *Science*, **214**, 181–183.

Trimble, S.W. (1999) Decreased rates of alluvial sediment storage in the Coon Creek basin, Wisconsin, 1975–93. *Science*, **285**, 1244–1246.

Trimble, S.W. and Lund, S.W. (1982) Soil conservation and the reduction of erosion and sedimentation in the Coon Creek basin, Wisconsin. *U.S. Geological Survey Professional Paper* 1234, Washington.

Walling, D.E. (1983) The sediment delivery problem. *Journal of Hydrology*, **65**, 209–237.

2

Mountains and Montane Channels

Michael Church

Department of Geography, University of British Columbia, Vancouver, British Columbia, V6T 1Z2, Canada

2.1 Introduction

Steep mountains are reckoned, in general, to experience the largest sediment fluxes of any terrestrial landscapes. The reason for this belief is simply that they exhibit the largest gravitational potential energy gradients on Earth's terrestrial surface. Hence, one expects to find the most active portion of the sediment cascade in such places. In this chapter, aspects of the sediment cascade in steep mountains are introduced, with particular attention to the relative roles of hillslope and channel processes. This entails making a number of definitions.

2.1.1 Steep Mountains in the Sediment Cascade

The first question is 'what is steep?' In the context of the geomorphological development of the landscape, a useful definition is a slope steeper than the limit for unconstrained accumulation of sediments. The morphology of such slopes is determined by rock structure and erosional history. In contrast, the morphology of 'mild' slopes is determined by the conditions for stability of sedimentary material, whether the product of local weathering or the accumulation of material that has arrived from upslope. Consideration of the angle of repose for unstructured, cohesionless material would place the division between steep and mild slopes at about 35°, but material properties or ambient conditions might alter that figure. In particular, persistent seepage of water through fine granular materials might reduce the limit stability angle to well below 30°.

A 'steep mountain', then, is a mountain that includes a length of steep slopes sufficient to determine its character; one that exhibits significant rock surfaces that have been exposed and eroded by mass wasting. This definition is consistent with the simple mechanical criterion given in the last paragraph, but its application in the landscape is

Sediment Cascades: An Integrated Approach Edited by Timothy Burt and Robert Allison

not straightforward. Many level or mildly sloping rock surfaces have been bared by other processes, such as glaciation, and are maintained bare by wind or running water. Conversely, soil is retained on remarkably steep slopes under the restraining influence of vegetation. Hence, neither 'steep' nor 'mild' slopes can be defined just by the presence or absence of a soil or regolith.

Steep mountain slopes are subject to erosion by mass wasting and to the rapid removal of surface-weathered material by wind or water. At sufficient altitudes, steep mountains are also subject to erosion by the action of ice and snow.

2.1.2 Dominant Geomorphological Processes

One way to categorize mass wasting processes is to divide them into those which entail bedrock displacement and those which are restricted to the superficial cover. The action of water is implicated to a greater or lesser degree in nearly all mass movements but, although its action in the case of shallow bedrock movements usually is restricted to such *in situ* effects as facilitation of weathering on planes of weakness and to freeze–thaw action, water in sedimentary materials may directly trigger failure by altering the weight, the effective stress state, and the cohesion. Mass wasting of superficial materials is, then, relatively intimately connected to the hillslope hydrological cycle.

Bedrock failures and superficial failures are crudely zoned in the landscape. Bedrock-dominated surfaces are mostly at relatively high elevation. Characteristic failure modes on bedrock slopes include rockfall, rock glides and slides, rock avalanches, toppling failures, and sag displacements. They mostly are relatively rare, relatively high-magnitude events that create episodic large increments to local sediment budgets. These processes are dealt with in Chapter 4 of this volume. Since mass wasting moves material downhill, the sedimentary products of bedrock failures – which join the superficial cover – collect on lower slopes. Failure modes in superficial materials include creep, slump, slide, avalanche and flow phenomena. Inasmuch as they are more directly associated with the hydrological cycle, they are more closely related to the normal recruitment and onward transfer of material in the subaerially dominant fluvial sediment cascade.

Mass wasting processes that dominate the transfer of superficial materials are outlined in Figure 2.1. Their relative order of importance varies amongst different landscapes. There are few regions where sufficient evidence has been gathered to permit general comparisons. One such region, however, is the Pacific Northwest coast of North America, where intensive forestry activity in steep mountains has directed significant attention to questions of slope stability on forested montane slopes. An order-of-magnitude summary of results is given in Table 2.1. It shows that, when account is taken of the limited area occupied by landslide failures in forested terrain (typically less than 1% of the land surface, which reduces the summary effect of surface erosion from slide scars), episodic shallow landslides are the most significant means of sediment delivery downslope. Lesser processes, including soil slumps, soil creep, tree throw and dry ravel, should not be ignored, however. They are important in recharging hillslope hollows over many decades or even centuries, which may lead eventually to a major failure. They also deliver material directly to stream banks (see also Chapters 3 and 4, this volume).

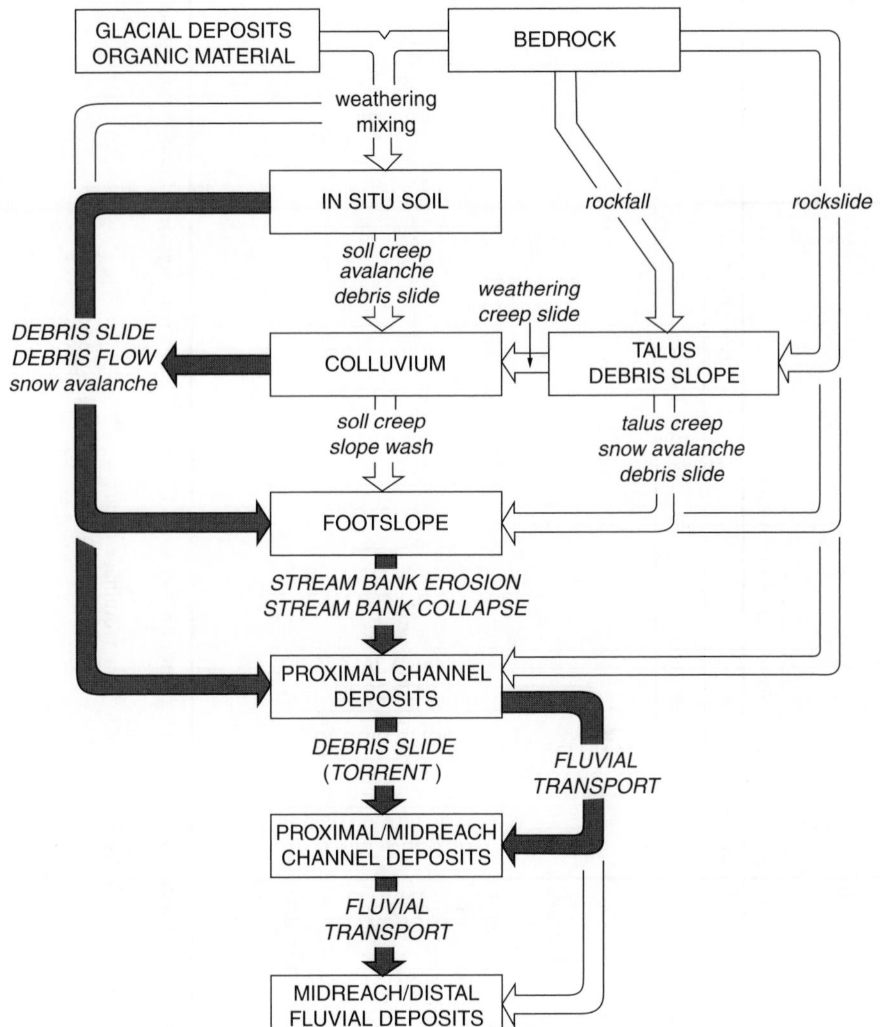

Figure 2.1 Conceptual representation of transfer processes for materials on steep hillslopes. The shaded pathways represent the dominant sediment transfer routes (after Roberts and Church, 1986, their figure 1, p.1093. Copyright 2008 NRC Canada or its licensors. Reproduced with permission).

In steep mountains, mass wasting processes normally deliver sedimentary material directly into water courses, which can thereby be said to be directly coupled to the hillslope sediment cascade.

2.1.3 What is a 'Steep Channel?'

By analogy with the definition of a steep slope, a steep channel is one in which channel gradient is too great to permit unconstrained alluvial sediment accumulation, implying the

Table 2.1 Sediment mobilization and yield from montane slopes in the Pacific Northwest of North America

Process	Mobilization rate		Yield rate to stream channels	
	Forested slopes	Cleared slopes	Forested slopes	Cleared slopes
Normal regime				
Soil creep (including animal effects)	$1\ m^3\ km^{-1}\ yr^{-1}$[a]	2×	$1\ m^3\ km^{-1}\ yr^{-1}$[a]	2×
Deep-seated creep	$10\ m^3\ km^{-1}\ yr^{-1}$[a]	1×	$10\ m^3\ km^{-1}\ yr^{-1}$[a]	1×
Tree throw	$1\ m^3\ km^{-2}\ yr^{-1}$	—	—	—
Surface erosion: forest floor	$<10\ m^3\ km^{-2}\ yr^{-1}$		$<1\ m^3\ km^{-2}\ yr^{-1}$	
Surface erosion: landslide scars, gully walls	$>10^3\ m^3\ km^{-2}\ yr^{-1}$ (slide area only)	1×	$>10^3\ m^3\ km^{-2}\ yr^{-1}$	1×
Surface erosion: active road surface	—	$10^4\ m^3\ km^{-2}\ yr^{-1}$ (road area only)	—	$10^4\ m^3\ km^{-2}\ yr^{-1}$ (road area only)
Episodic events				
Debris slides	$10^2\ m^3\ km^{-2}\ yr^{-1}$	2–10×	to $10^4\ m^3\ km^{-2}\ yr^{-1}$	to 10×
Rock failures (fall, slide)	No consistent data: not specifically associated with land use			

Results have been generalized to order of magnitude from more specific results given in Roberts and Church (1986), where sources of data which guided construction of the table are also given.

[a]These results reported as $m^3\ km^{-1}$ channel bank. All other results reported as $m^3\ km^{-2}$ drainage area.

condition that forces impressed by the water flow frequently exceed the threshold necessary to entrain material that is delivered to the channel. This condition can be investigated via the Shields criterion for clastic particle stability in a water flow, which can be expressed as

$$S = g(\rho_s - \rho)D\theta_c / g\rho d \tag{2.1}$$

wherein S is the channel gradient (strictly speaking, the energy gradient of the flowing water but, in steep channels and over some distance, there is no practical difference), g is the acceleration of gravity, ρ_s and ρ are the density of sediment and water respectively, D is sediment particle diameter, d is water depth, and θ_c is the limit value of the Shields number for particle stability. Substituting values for the physical constants, including $\rho_s = 2650$ kg m^{-3}, (2.1) becomes

$$S = 1.65\,\theta_c(D/d) \tag{2.1a}$$

To obtain explicit limiting gradients, we must make an assumption about relative roughness (D/d) in the channel and we must adopt a value for the Shields number. In montane channels, the largest clasts are customarily exposed above the water surface and typically have diameter similar to the depth of the channel. Hence, for exploratory arguments, it is reasonable to suppose that $D \sim d$ at high flow. For individual, well-exposed particles, a conventional value $\theta_c = 0.03$ has come to be accepted, whereas for widely graded mixtures $\theta_c = 0.045$ (cf. $\theta_c = 0.047$ by Meyer-Peter and Muller, 1948), in comparison with Shields' original specification of $\theta_c = 0.06$ for narrowly graded sediment beds. For $\theta_c = 0.03$, $S_c = 0.050$ (2.8°). At $\theta_c = 0.045$, $S_c = 0.074$, or 4.2°. At $\theta_c = 0.06$, $S_c = 0.099$ (5.7°). If $D/d < 1$, these values become smaller. It appears reasonable, then, to define channels steeper than about 3°, of which there are many in mountain regions, as steep channels. Such channels either are devoid of sediment accumulations (i.e. they are scoured to rock or other erosion-resistant substrate), or sediments are retained by virtue of conditions beyond particle inertia that constrain their mobility.

In fact, as $D/d \rightarrow 1.0$, an argument based on shear force is no longer strictly valid. Individual clasts form elements that experience fluid drag force $F = C_D \rho A u^2/2$, in which C_D is a drag coefficient, A is the projected area of the clast and u is flow velocity. Velocity fluctuations are, in general, significant as well, but are probably relatively unimportant in steep channels where mean velocity is comparably strong (except in boundary-induced eddies). Supposing that $C_D \approx 1$, that $A\ \alpha\ D/2$ and that grains are subequant, we obtain $\tau_D\ \alpha\ \rho u^2/2$. Recalling, from Chezy's formula, that $u^2\ \alpha\ RS$ (wherein R is the hydraulic radius), we recognize that we have an argument of the same form as that proceeding from Shields' formalism, except that the constant of proportion is apt to be different, as is in fact observed.

Of course, channels on lower gradients may also be devoid of mobilizable sediment: that depends on the balance of grain inertia (hence, grain size) and the force exerted by the flowing water. Therefore, sufficiently large streams with limited supplies of sufficiently small sediments may scour to a refractory boundary. Hence, absence of mobilizable sediment is not diagnostic for steep channels.

2.1.4 Relative Scales and Dominant Processes

The size (D) of clasts delivered to stream channels in steep terrane is apt to be large, whereas the size of headward channels, represented by channel depth (d), remains small because of the limited drainage area. In these circumstances, the channel scale converges toward the size of individual boundary elements. Downstream, channels become very much larger than the characteristic size of sediment grains. Divergence between channel scale and clast size is a major determinant of behaviour and morphology of river channels. The most fundamental division is between 'small channels', in which channel scale is comparable with the scale of individual sediment grains, and larger channels in which the boundary is made up of aggregate structures of many grains. In the former case, the sediment stability constraint explored above becomes critical for channel stability itself.

Channel gradient (S) determines the competence of the channel to move sediment. Absent constraints other than particle weight, channels steeper than about 4° do not accumulate sediments. On lower gradients, competence is governed by Equation 2.1, so that the limit size of transported material declines downstream as topographically delimited channel gradients decline and certain materials accumulate in the channel.

Sediment transport in steep channels is dominated by the episodic rearrangement of large clasts, which define the dominant morphological elements of the channel, typically a sequence of steps and pools. In headmost channels, such events may be limited to relatively rare mass-wasting events. Downstream, transport of more easily mobilized sediment is mediated by normal fluvial flows and sedimentary accumulations take the form of aggregated riffle–bar–pool units through which flows maintain the onward transport of material. Within forests, large wood debris has an important influence on channel morphology.

2.2 Hollows and Headmost Channels

2.2.1 Mobilization of Surficial Material on Steep Slopes

Slope stability is strongly influenced by slope drainage. Mountain slopes are rock dominated, with shallow surface materials consisting of frost-churned soil (on high mountains), glacial till (within glacial limits), or veneers of colluvium or residual soil. Slope declivities accumulate pockets of deeper soil. Where forest cover is well developed, a surface layer of highly permeable, humic forest soil and litter mantles the slope. Water is absorbed into the forest soil and moves downslope under the surface, usually on the bedrock, on cemented surfaces, or on the surface of unweathered till. The water exploits root channels, animal burrows, and soil discontinuities at the base of the plant rooting zone to move relatively quickly downslope (see, for example, McGlynn, McDonnell and Brammer, 2002). Slopes mantled with stony colluvium similarly experience rapid, subsurface drainage. Such drainage discourages surface erosion and slope instability.

The stability of surficial material on such slopes is governed by the Mohr–Coulomb equation, which gives material strength (s) as

$$s = c' + c_r + [(1-m)\rho_b g d_m + m(\rho_{sat} - \rho) g d_m] \cos^2\theta \tan\phi' \qquad (2.2)$$

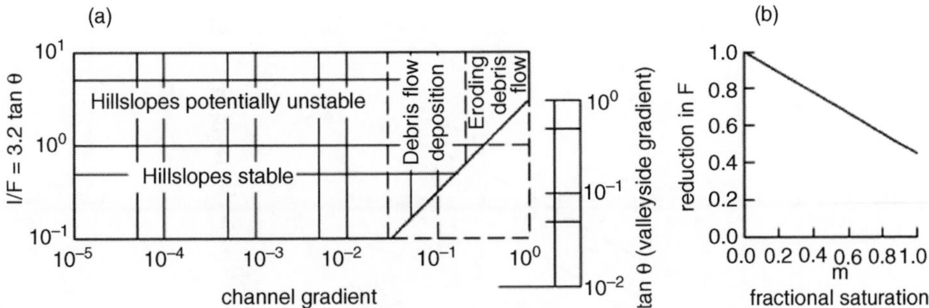

Figure 2.2 Hillslope and channel stability in headmost stream systems: (a) stability zoning according to gradient (modified after Whiting and Bradley, 1993; their Figure 1). The details of Whiting and Bradley's calculations differ from those presented here; (b) effect of water content on slope stability

in which c' is the effective material cohesion, c_r is pseudocohesion provided by root strength, ρ_b is material bulk density, ρ_{sat} is the density of saturated material, d_m is the depth of surface material above the potential failure plane, θ is the hillslope angle, m is fractional saturation (in effect, d/d_m, where d is the depth of the saturated zone), and ϕ' is the effective angle of the material's shear resistance. Material strength is compared with the shear stress (τ) on a candidate failure plane

$$\tau = [(1-m)\rho_b + m\rho_{sat}]gd_m \sin\theta\cos\theta \qquad (2.2a)$$

$F = s/\tau < 1.0$ indicates the likelihood of failure. Neglecting cohesion

$$F = \{(1-m)\rho_b + m(\rho_{sat}-\rho)/[(1-m)\rho_b + m\rho_{sat}]\}[\tan\phi'/\tan\theta] \qquad (2.2b)$$

For characteristic earth material properties ($\rho_b \approx 1650\,\text{kg m}^{-3}$, $\rho_{sat} \approx 1800\,\text{kg m}^{-3}$ and $\phi' \approx 35°$, and full saturation ($d = d_m$)), $F \approx 1/(3.2 \tan\theta)$. As $m \rightarrow 0$, indicating dry or rapidly drained material, $F \rightarrow 1/(1.4 \tan\theta)$. Cohesion and root strength would increase these values. If failure occurs, a debris slide or debris flow, or a slump or block slide happens, depending upon the water content of the material and the cohesion of the failed mass. On steep slopes and in hollows, debris slides or flows are most common.

Whiting and Bradley (1993) used the foregoing analysis to zone the stability of hillslopes and headmost channels (see Figure 2.2a). A more detailed analysis has been presented by Sidle and Ochiai (2006) that considers plant root strength and vegetative surcharge weight. Failures are most apt to occur in declivities where slope water becomes concentrated, since the summary effect of water in Equation 2.2b is to decrease the value of F (Figure 2.2b).

2.2.2 Hollows

Streams in mountains originate on steep slopes. Most hillslopes have on them a system of shallow declivities or hollows (Dietrich, Wilson and Reneau, 1986; cf. Hack and Goodlett, 1960) which may be obvious only after the slope is cleared of trees. In many

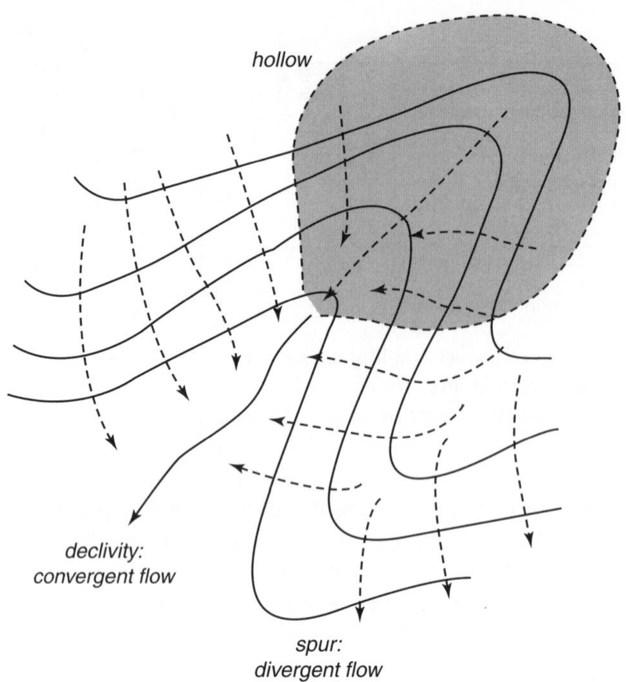

Figure 2.3 Conceptual representation of drainage lines on a hillside slope, emphasizing the convergence of flow into hollows

cases, these declivities are developed on lines of structural weakness or accelerated weathering determined by the geology; in others they are relict surface erosion lines left from an earlier time when forest cover was absent or incomplete. On steep slopes, they may be extended, downslope-oriented linear features. Water moves downslope, draining away from spurs and into hollows (Figure 2.3). The concentration of subsurface drainage in hollows creates seepage lines where the soil becomes completely saturated with water, and springs where larger subsurface flows emerge on to the surface. Hence, these features, not themselves channelled, are the origin of the surface drainage network. Hollows are also places where soil accumulates as the result of downslope creep, tree throw, and minor slippage. In depressions, subsurface water may create sufficiently high water pressures that the overlying accumulation of soil and debris is induced to fail. Therefore, shallow slope failures often occur at the head of drainage systems. This both extends the drainage system and delivers sediment into downstream channels. Hillslope declivities are the primary conduits of hillslope-derived water and sediment into the drainage network.

Dietrich, Wilson and Reneau (1986) speculate that the length of an unchannelled hollow might be controlled, in the long term, by the distance to cumulate sufficient water to prevent colluvium from accumulating in the hollow. A simple runoff model suffices to give some indication of the distance that might be involved and the principal controls. For shallow subsurface flow, Darcy's Law can be expressed in one dimension as

$$q = -K_s d \cos\theta(\mathrm{d}z/\mathrm{d}x) \tag{2.3}$$

wherein q is flow (per unit width), K_s is the saturated hydraulic conductivity, $d = md_m$ is the depth of the saturated section, θ is the slope gradient, and dz/dx is the gradient of the water plane. On a steep slope, $dz/dx \approx -\tan\theta$. Therefore

$$\theta = K_s m d_m \sin\theta \tag{2.3a}$$

At any distance, L, downslope, $q = iL$, wherein i is precipitation intensity. We may therefore rearrange (2.3a) as

$$K_s/i = (L_c/md_m)(\sin\theta)^{-1} \tag{2.3b}$$

which isolates three significant parameter groups, relative precipitation (or snowmelt) intensity; hollow length, L_c, scaled by the limit saturated depth for slope stability; and hillslope angle. Solution of this equation in association with Equation 2.2 (to get md_m) yields estimates of hollow length. For a real hollow of finite width, w, the geometry term becomes $w/\sin\theta$. Figure 2.4 shows the solution for a realistic range of values. For most soils on steep slopes there are preferred subsurface drainage lines of high conductivity (root

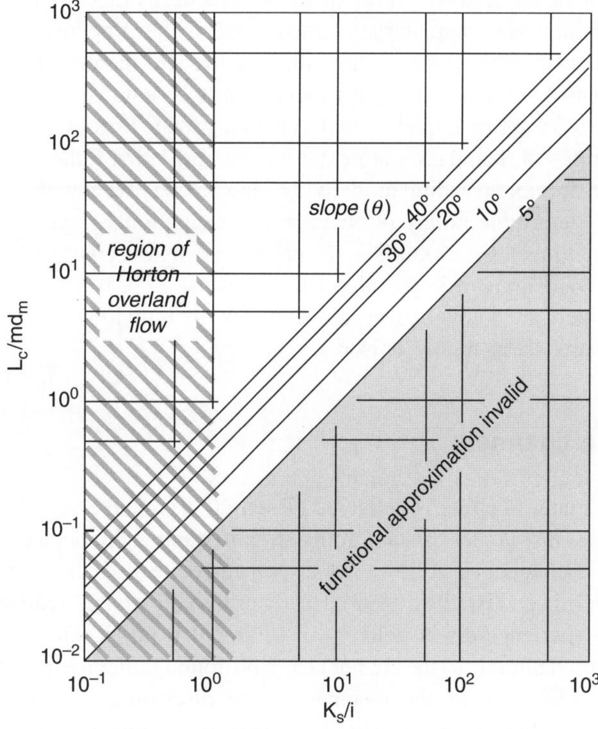

Figure 2.4 Hollow length in relation to slope geometry and relative precipitation intensity. See text for an explanation of the scales

channels; bedrock surface), so that typical values of K_s/i are in the range 10^1–10^3. If $m = 1$ for failure (soil column completely saturated), then a 1 m deep layer of surficial material on a 35° slope would fail at a distance 6–600 m. The significance of K_s in controlling hollow length is obvious. Finite width and fractional m would both act to shorten the distance, whereas surface vegetation (not parameterized here) would act to lengthen it. In humid mountains, hollows are often on the order of 100 m in length.

The question of how frequently hollows fail, delivering significant slugs of sediment to the drainage system, is an important one. Using a simple 'tipped trough' geometry to model the shape of hollows, Dietrich, Wilson and Reneau (1986) showed that hollow filling rate varies as $\sqrt{t}$. However, time to failure will also be influenced by the hollow gradient, hydrology, and material properties. Observations by Dietrich's group, and by Benda and Dunne (1987) indicate that times to failure in the Oregon and California Coast Ranges are of order 10^3–10^4 yr. These sites are beyond the glacial limit, where rock weathering controls material availability near drainage divides. Within the glacial border in British Columbia, times to failure of order 10^0 to $>10^3$ yr have been documented (Jakob and Bovis, 1996; Bovis, Millard and Oden, 1998). Similarly frequent debris flows are experienced in highly erodible mountains in Mediterranean regions, for example southern California (Johnson, McCuen and Hromadka, 1991). D'Odorico and Fagherazzi (2003) have made a theoretical analysis of the probability for shallow landslides to evacuate sediment from hillside hollows.

In most cases with frequent recurrence, specific, prolific sediment sources can be identified. Except near the lower limit of these temporal ranges, cumulative sediment storage and earth material characteristics must control the propensity for failure, since the times are considerably longer than the recurrence interval for suitably heavy precipitation inputs to create soil moisture conditions conducive to failure. But, on the poorly lithified rocks of earthquake- and typhoon-prone Taiwan, where some of the world's highest rates of denudation are experienced (Li, 1976; Fuller *et al.*, 2003), these specific triggers dominate sediment delivery downslope. Regionally, as well, major trigger events do determine failure occurrence. An exceptional regional record has been gathered from the Queen Charlotte Islands, off the northwest coast of British Columbia, by Schwab (1998) (Figure 2.5; see also Martin *et al.*, 2002). Six years dominate the record, all of which are known to be associated with individual major storms. Major sediment inputs are, then, highly episodic.

2.2.3 Hillside Channels

Once mobilized, material either is delivered directly to a stream channel, or it moves onto colluvial footslopes more or less distant from a stream channel. For open slope failures (ones not occurring in a channel or hollow), what happens depends mainly upon the width of valley floors (Whiting and Bradley, 1993). Headward, small streams tend to occupy narrow defiles, so they experience direct delivery of sediment from sideslopes. Whether or not sediment is delivered directly to the channel has a profound influence upon the organization and sedimentary character of the channel, hence discrimination of this condition is important. Channels that experience direct delivery of sediment from sideslopes can be said to be 'coupled' to the hillslopes (Figure 2.6). Ones which do not are 'buffered' or

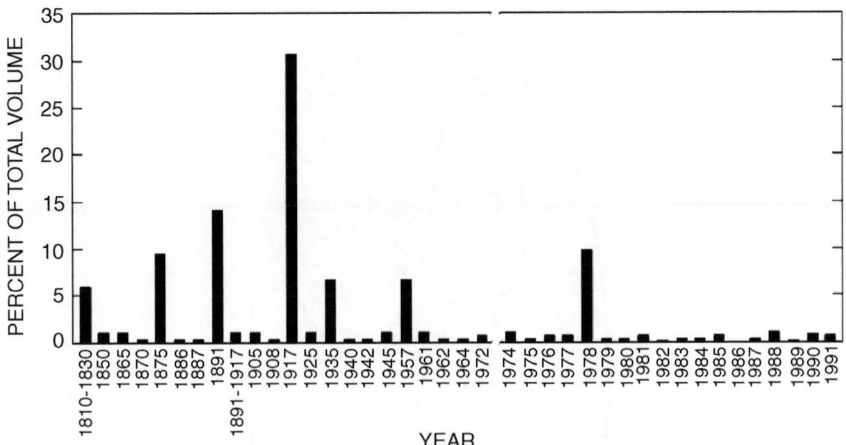

Figure 2.5 Frequency and relative magnitude of sediment delivery by debris slides and flows for study areas in the Queen Charlotte Islands and adjacent mainland coast of British Columbia. The study includes 970 events, of which 8% occurred in terrain disturbed by land use. Note that the abscissa is not uniformly scaled
(from Schwab 1998; reproduced by permission of British Columbia Ministry of Forests).

'uncoupled'. Buffered channels receive sediment inputs entirely by downstream sediment transport in the stream or by streambank erosion. They are flanked by floodplains or other valley-floor deposits. Between these two types is a class of 'intermittently coupled' channels, ones which flow against the valleyside in some places and are buffered by valley-bottom deposits in others. The occurrence of these types is correlated with position in the drainage network. Most small, headward streams are coupled.

A special class of slope failures in steep, headward channels is direct failure of the sedimentary fill in the channel or gully. Such failures commonly develop into debris flows that move rapidly down the channel (Innes, 1983; Costa, 1984; Iverson, 1997; Hungr *et al.*, 2001). Despite their in-channel occurrence, they are landslide-like phenomena rather than fluvial sediment transport by the stream. They are relatively rare in individual channels unless there is an unusually prolific source of sediment immediately adjacent to the upper channel, but they are by no means rare in upland and mountain landscapes. A debris flow – a moving mass of saturated material that deforms more or less continuously – is usually initiated in channel head declivities or hollows with gradients in excess of 0.35 (but, exceptionally, down to 0.25), but they may also begin when a landslide enters a steep channel from an adjacent slope, temporarily damming it, so that a significant weight of water collects behind the 'dam'.

The criterion for spontaneous mobilization of material in the channel is derived from Equation 2.2b with, after $m = 1$, the addition of the free surface fluid shear term (related to Equation 2.1)

$$(\rho_{sat} - \rho)\tan\phi' < [\rho_{sat} + (d/D)\rho]\tan\theta \qquad (2.3c)$$

wherein d is, here, water depth above the surface (channel bed) and D is mean bed material size (cf. Takahashi, 1981; Bovis, Dagg and Kaye, 1985). Bovis and Dagg (1988) make the

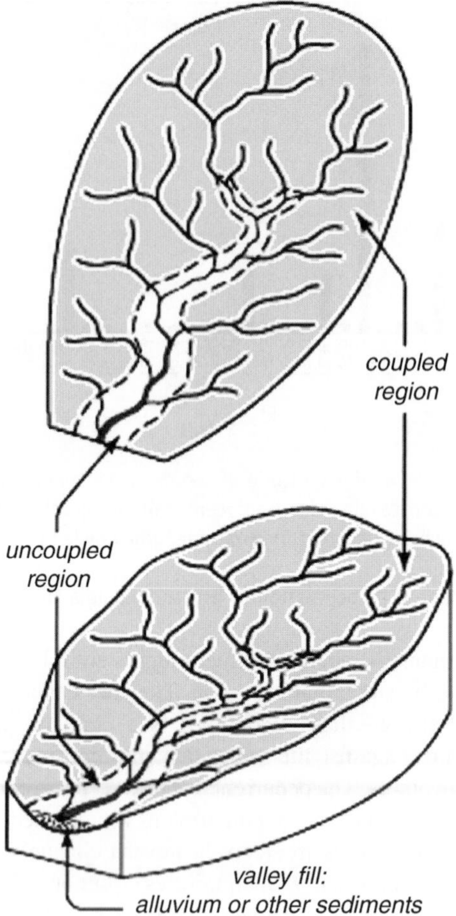

Figure 2.6 Drainage basin sketch to illustrate coupled and uncoupled reaches

interesting observation that, once colluvial rubble with a wide variety of grain sizes arrives in a declivity, modest water flows may wash out fine sediment by normal fluvial processes, so that the hydraulic conductivity of the remaining material may be increased by orders of magnitude. They have measured changes of order 10^{-5} m s^{-1} on hillside slopes to 10^{-1} m s^{-1} in gullies. This would create very rapid drainage of the declivity, so that the achievement of m values sufficient to cause failure would become rare. However, material would continue to accumulate, so that a failure, when it finally occurred, might be large. Conversely, if the range of material sizes is limited or matrix materials sufficiently abundant, trapping of fine material in the matrix of a deposit may progressively reduce hydraulic conductivity and increase the chance for failure.

Triggers for failure include prolonged heavy rain, intense rain induced by strong convection, or more moderate precipitation accompanied by impulsive destabilization (Bovis and Dagg, 1992), such as a debris slide or snow avalanche landing in the declivity, or seismic shaking. Although several precipitation threshold criteria have been published (e.g.

Caine, 1980) for the likelihood of debris flows to occur, there appears to be no universal criterion because antecedent conditions are complex (see Church and Miles, 1987).

Once mobilized, the flowing mass may reach speeds of 1 to 20 m s^{-1}, depending on the constitution of the material in the flow. Hydraulic characteristics are difficult to establish because of the variable constitution. A variety of rheological models has been suggested to describe the features (Costa, 1984; Iverson, 1997). On a practical level, granular and viscous (muddy) types can be distinguished. However, simple mechanical formulae or empirical relations generally have been adopted to describe the bulk behaviour of debris flows (Hungr, Morgan and Kellerhals, 1984; Mizuyama, Kobashi and Ou, 1992; Bovis and Jakob, 1999).

The total volume of material moved in a debris flow depends upon the length of the flowpath and the volume stored per unit length. Typically, only a few hundred cubic metres are involved in the initial mobilization, but debris flows commonly scour the flowpath to rock and grow to volumes of order 10^3–10^5 m^3 (van Dine, 1985; Aulitzky, 1989). The largest known debris flows exceed 10^6 m^3 volume. There is further discussion of channel 'dams' and associated debris flows in Chapter 4 (this volume).

2.2.4 Role of Hollows and Hillside Channels in Sediment Recruitment

Despite the relatively rare recurrence of hollow failures and subsequent debris flows, Benda and Dunne (1987) have calculated that they constitute the dominant means by which sediment is delivered into the low-order drainage system in the mountains of the Pacific Northwest of North America. Observations in British Columbia (cf. Table 2.1) imply the same situation (see also Slaymaker, 1993), while Wells and Harvey (1987) have made similar calculations for a site within the British uplands.

In the long term, sediment mobilized must be approximately equal to the rate of supply of material to a hollow or channel, except in glaciated regions where glacial deposits stored in declivities may still be subject to remobilization. However, there are so many contingencies associated with the accumulation of debris and triggering of a slide/flow that few generalizations are possible about individual events. Debris yield rate is defined as the volume of debris yielded by an event per unit length of channel traversed (Hungr, Morgan and Kellerhals, 1984). Inventories in British Columbia (Hungr, Morgan and Kellerhals, 1984; Fannin and Rollerson, 1993; Bovis, Millard and Oden, 1998) suggest that fully charged hollows and gullies deliver 5–25 m^3 m^{-1} of debris in failure. For a 100 m long hollow, this suggests initial failure volumes of order 500–2500 m^3, in good agreement with available observations. Where a debris flow follows, yield would be of order 5000–25 000 m^3 km^{-1} of scoured channel, which is also reasonable. By the implication of these results, it has proven possible to construct regional scale relations between drainage basin and soil characteristics, and potential maximum yield for a debris flow which are at least of indicative value for hazard studies (e.g. Bovis and Jakob, 1999).

2.3 Montane Channels

In Section 2.1.3, steep channels are defined as ones with gradients greater than 3°. Sediments resident in such channels must be constrained in some way. Constraint is achieved by wedging or jamming of grains against each other and against the channel bank. Sediments accumulate upstream of such barrages and, downstream, the channel is starved of sediment. These phenomena not only retain sediment in steep channels, but they may cause mild channels downstream to degrade to bedrock (Montgomery *et al.*, 1996; Massong and Montgomery, 2000). Wood, which can form high barriers, particularly acts to modify the distribution of bedrock-bound channels by establishing 'forced' alluvial channels on intermediate gradients (i.e. down to about 1° or 0.017).

2.3.1 Clastic Stepped Channels

The steepest sediment-retaining channels are characterized by discrete, channel-spanning structures, termed 'steps', composed of the larger clasts present in the channel, or of jammed accumulations of wood debris, and spaced approximately 1–4 channel widths (Montgomery *et al.*, 1995). On gradients that may range as high as 0.4 (22°), apparently regular sequences of steps and intervening pools are formed (Figure 2.7) (Chin, 1989; Grant, Swanson and Wolman, 1990). The main steps and pools are channel spanning, but they are often obliquely arranged, and smaller, non-spanning features may also occur. On still higher gradients 'cascades' occur, step-like boulder accumulations with micropools, but without intervening channel-spanning pools. These morphologies are dominant when the size of the largest clasts remains comparable with the size of the channel, as is found in headwaters (cf. Section 2.1.4). Chin and Wohl (2005) have reviewed the geometry of steep channels.

There have been three contending theories to describe the origin of step sequences. Judd and Peterson (1969) supposed that the phenomenon develops by wave propagation

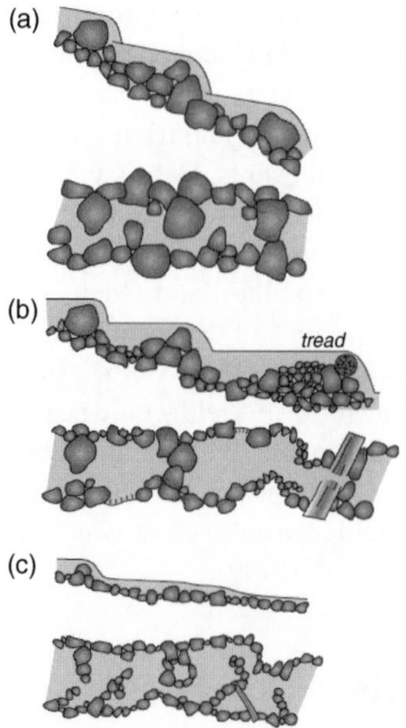

Figure 2.7 Definition sketch for the morphology of montane channels: (a) cascade; (b) step-pool; (c) rapid reach

downstream from an initial large, immobile clast. Whittaker and Jaeggi (1982) experimentally demonstrated the development of step-like features under antidunes in supercritical flow, whereas McDonald and Day (1978) experimentally induced an upstream stepping hydraulic jump that accumulated clasts to form steps. None of these mechanisms has been systematically observed in the field, but reformation of steps has been observed locally under the influence of quite moderate flows (cf. Hayward, 1980), and Lee and Ferguson (2002) created steps in a flume at near-critical flows but not at antidune spacing; it required complete bed mobilization to achieve this result.

The persistent structures, such as steps, that we observe in stream channels must, in some sense, be 'most stable forms'. If they were not, they would soon change and we would observe some different morphology. Hence, step–pool and cascade morphologies should be viewed as configurations that dissipate available stream energy on gradients that ordinarily would compel movement of any submerged material, without actually moving the step-forming material away. Structural reinforcement of the material's stability appears to be an essential feature of the morphology, so the presence of large clasts, immobile at most flows and able to become keystones, seems to be the necessary criterion for step formation (Wohl, Madsen and MacDonald (1997) forwarded a similar hypothesis with respect to log steps). It is not clear, however, whether there is any systematic propagation of step locations since wavelengths typically exhibit a wide variation (Figure 2.8, inset; and similar data in Judd and Peterson, 1969; Abrahams, Li and Atkinson, 1995; Chin, 1999; and Chartrand and Whiting, 2000). It is probable that the location of step keystones is largely random (Zimmermann and Church, 2001; Curran and Wilcock, 2005), although normally immobile stones may be locally shifted in the channel by undermining (see discussion by Mosley and Whittaker *in* Whittaker, 1987).

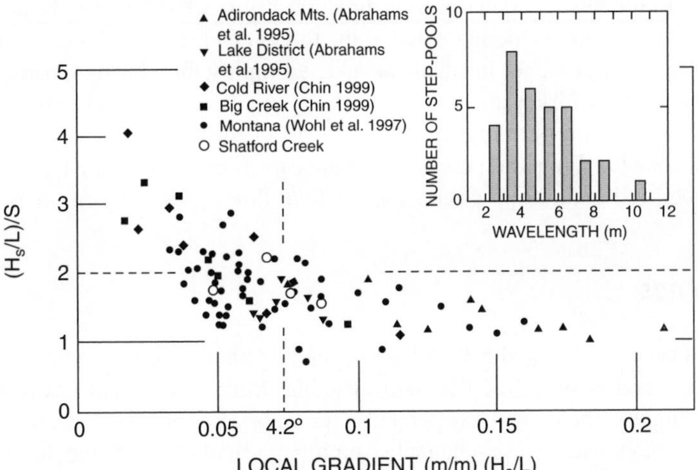

Figure 2.8 Relation between local gradient, H_s/L, and the ratio of step fall to reach gradient, S, in step-pool morphology. The data shown in this plot are reach means: individual steps may vary much more widely. Inset: distribution of individual step wavelengths in Shatford Creek, British Columbia: reach gradients vary from 0.07 to 0.09
(from Zimmermann and Church, 2001; reproduced by permission of Elsevier B.V.).

If this view is correct, then step–pool sequences maximize the proportion of flow energy that is dissipated without channel-forming work; that is, they are flow resistance maximizing structures. Abrahams, Li and Atkinson (1995) (see also Wohl, Madsen and MacDonald, 1997) pursued this possibility by comparing experimental channels designed to maximize energy dissipation with observed field morphologies. They found that experimental steps have the highest resistance number when local gradient, H/L, is no greater than $2S$ on average, where S is reach gradient, and they found similar morphology in several field examples. Figure 2.8 summarizes available data and shows that, above the 4.2° limit for unconstrained particle stability, Abrahams *et al.*'s criterion generally is observed. As gradients increase to approach 0.15 (8.5°), the constraint becomes more severe, so that $\langle H/L \rangle \rightarrow S$. At gradients below 4.2°, individual steps might locally isolate a large portion of the total reach drop, so that the ratio $\langle H/L \rangle /S$ may exceed the limit value specified by Abrahams *et al.*, but extended segments of plane bed may occur between steps. Restricted 'treads' remain present on gradients up to about 7 or 8°.

Energy dissipation in step-pools and cascades takes place largely in jet-like turbulence generated by the flow plunging off the step into the succeeding pool (Church and Zimmermann, 2007). This scours the upstream end of the pool to accentuate the step-pool morphology, but it achieves only limited sediment transporting work (see Section 2.3.4). Whittaker (1987) showed that the shape of the scour pool is constant under limited sediment feed (see, also, Comiti, Andreoli and Lenzi, 2005), and that departures from this shape reflect transient sediment storage in the pools. It appears, then, that stable step-pool morphology reflects low bed material transfer, consistent with the picture of persistence developed above.

This picture of step-pools and, by extension, cascades is consistent with an emerging picture of the physics of channelled granular systems that are subject to jamming (Cates *et al.*, 1998). A jammed state may develop due to mutual interference amongst individual large grains moving through a restricted opening. Once jammed, the grain structure may exhibit great strength against forces exerted in the same direction as that which created the jam (step). But the jam remains 'fragile' in the sense that, if the forces change their attack, the structure will quickly collapse. Large stones in small channels exemplify this situation, particularly when the channel is seeded with keystones that almost never move (Church and Zimmerman, 2007). But if bank configuration or the upstream channel are changed, the jam (step) may quickly be destroyed or changed. From this perspective, it appears more useful to think of steep channels as rarely mobilized grain-flow systems than as fluid flows with sediment transport.

2.3.2 Rapids

At gradients between 4.2° and 2.8° or less (depending upon relative roughness), individual unconstrained grains ($\theta_c \approx 0.03$) remain unstable. Such grains will easily move, once submerged, unless they are incorporated into imbricate or jammed structures. Such structures are more usually non-channel spanning, particularly as channels become larger and resident clasts smaller. At gradients down to 1°, these features take the form of usually submerged 'stone nets' and stone lines (Church, Hassan and Wolcott, 1998; see Figure 2.7) and form 'riffle-steps' in the terminology of Whittaker (1987). Relative roughness usually is in the range $0.3 < D/d < 1.0$, so that flow over the structures remains jet-like or wake-dominated. The overall morphology can be characterized as a 'rapid' (cf. 'plane bed' of Montgomery and Buffington, 1997).

Church, Hassan and Wolcott (1998) demonstrated experimentally that stone structures develop under conditions of low sediment transport, when the larger clasts on the bed are only very sporadically moved. They tend to roll into imbricate juxtaposition with other immobile clasts, to form the lines and nets. The keystones are the largest and least mobile stones of all, typically ones around D_{99}. Intermediate states include imbricate armour, stone clusters and transverse ribs.

Rapid-dominated channels typically have only subordinate pool development, usually as part of the non-spanning step structure, and are characteristically heavily armoured (the ratio of surface to subsurface D_{50} may be in the range 2 to 3). This state is typical of limited transport of bed material (Laronne and Carson, 1976). Because the surface material is locked in structures or imbricated, critical Shields numbers rise well above any specified for normally loose material. Church, Hassan and Wolcott (1998) reported experimental values as high as 0.075, and field observations have returned values in excess of 0.1. At sufficiently high flows, bed structures may be destroyed, but they tend to be re-established within one or two seasons as the result of flows near the threshold for competence.

Within the range of gradients for rapids to occur, major, discrete riffles and skeletal bar structures begin to appear. This represents the transition to the riffle-pool morphology characteristic of gravel-bed streams on mild gradients (see Chapter 9). Here, substantial storage of episodically mobile bed material may occur in pool bottoms and on bars superimposed upon the riffles.

2.3.3 Channels Dominated by Organic Debris

Many montane streams flow through forests where organic debris may enter the channel. Large organic debris (wood pieces greater than 10 cm in diameter, including limbs, roots and whole trees) may constitute significant morphological elements of the channel, and creates special morphologies (Montgomery *et al.*, 1995). Wood may also dominate sediment storage and transfer in the channel (Heede, 1972; Keller and Swanson, 1979).

In confined (coupled) headwaters, large pieces often span the channel above the water line. At the other extreme, when streams become much wider than the characteristic height of mature trees, wood pieces interfere in only a limited way with normal fluvial processes. Between these limits (typically, in channels of order 3–5), wood debris may lie in and block the channel (Bilby and Ward, (1989)). Individual pieces may form steps in much the same manner as jammed boulders do (Wohl, Madsen and MacDonald, 1997).

Wood debris up to some limit size may be moved by sufficiently high flows, and debris flow can move wood pieces of almost any size. These processes create a propensity for wood debris to collect into discrete jams along the channel, usually anchored by an immobile key piece. In headward, confined channels, jams are almost always the consequence of landslides and debris flow (Hogan, Bird and Hassan, 1998). They build vertically and can reach heights of up to 5 m or more. Upstream of wood jams, sediment is indiscriminately deposited into a wedge that becomes graded to the jam height. These 'forced alluvial reaches' are major storage zones for sediment in montane channels (Megahan, 1982; Nakamura and Swanson, 1993; Gomi *et al.*, 2001). Downstream, channels become degraded and heavily armoured for lack of sediment passage.

Wood decays relatively quickly. For most species, the time during which wood pieces retain some strength is 20 to 50 years. However, some species, particularly if continuously submerged, can retain strength for centuries. Because of wood decay, jams weaken with time and eventually are breached (Hogan, Bird and Hassan, 1998). The upstream sediment wedge then becomes a source for sediment transport downstream. Typically, degradation proceeds vertically through the sediment wedge, part of which is left behind as a terrace for a long period. If the upstream sediment wedge is debris flow deposits, the larger boulders may not be moved on in any case. Figure 2.9 illustrates a typical jam history.

Log jams serve as transient reservoirs, holding substantial amounts of potentially mobile sediment for periods of years to many decades (Bovis, Millard and Oden, 1998). They increase the total retention of sediment in the stream system (Megahan, 1982) and are, accordingly, a stability-promoting mechanism in steep channels. In certain circumstances, they may also disintegrate quickly, releasing a substantial store of sediment downstream and thereby reinforcing the episodic nature of sediment release into steep channels.

2.3.4 Sediment Transport in Montane Channels

There are two distinct regimes for sediment transport in montane channels and hollows. The initiation of mass movements is governed by Equation 2.2, while the initiation of grain movements in fluid shear flow is governed by Equation 2.1. Once mobilized, the magnitude of sediment transport in montane channels is generally limited by the sediment supply

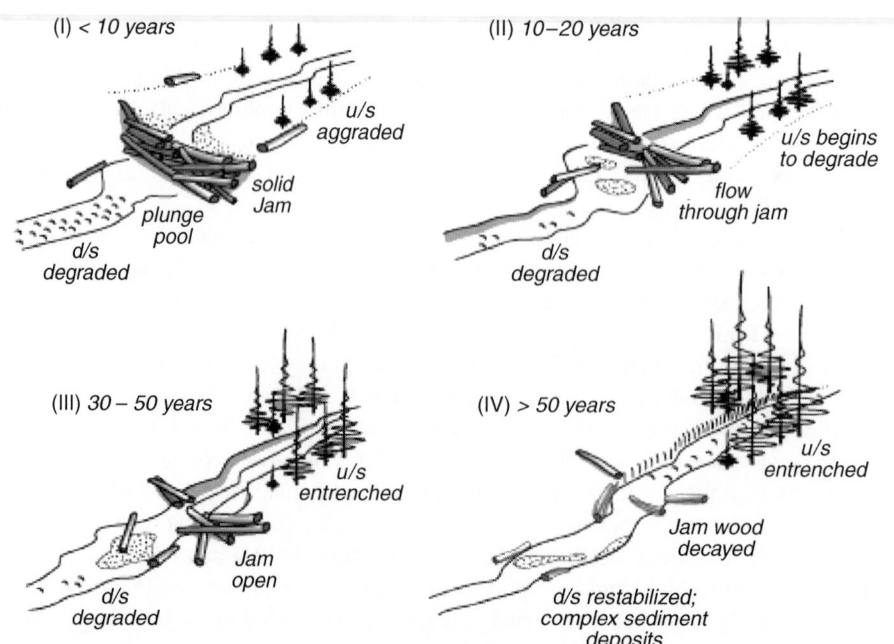

Figure 2.9 Morphology and characteristic history of development for vertically developed log jams: time scale may vary depending on wood type, initial jam configuration, and site

because, in steep channels, mobile sediment generally can be swept all the way to lower gradients, unless some special reservoir is encountered. Steep channels are stable only when the quantity of potentially mobile material in them is limited.

2.3.4.1 Mass Transport

Channelized mass transport events are commonly debris flows (Section 2.2.3). Flux rates are, in general, known only from post-event estimates of flow cross-section and velocity, the latter derived from flow superelevation observations in bends. Reported values range from order $10\,m^3\,s^{-1}$ to order $10^3\,m^3\,s^{-1}$ (cf. Bovis and Jakob, 1999), and are reasonably correlated with event total volume.

Debris flows deposit at the point where they begin to drain sufficiently rapidly to once again become frictional materials. The gradient at which this occurs depends upon debris texture, which controls pore-water movement, and upon channel topography, which influences the depth and velocity of the flow. Coarse materials (dominated by sand and coarser grains) may drain and stop on gradients as great as 14° (0.25), whereas muddy slurries may flow out on to gradients as low as 1.4° (0.025). The constitution of the material depends, in turn, on the geology of the source rocks or earth materials. Loss of confinement (see Fannin and Rollerson, 1993) and abrupt bends in the flow path (as at high-angle tributary junctions: Benda and Cundy, 1990) each may condition higher stopping angles. Debris fans at the base of mountainside gullies reveal, by their surface gradient, the characteristic stopping angle of local materials. These criteria generally maintain the flow within the range of channels considered in this chapter, so that debris flow does not usually export material beyond the limit of montane channels.

Debris flow deposits have been well described by many investigators (e.g. Costa, 1984). Material is generally unsorted and often diamict, consisting of boulders, cobbles, and substantially finer soil materials. Organic content, including large organic debris, may be high. In motion, large clasts and large organic debris are, in part, forced to the front, top and sides of the deposit. If they are sufficiently abundant, they may form a surface armour after deposition, protecting the mass of the deposit. Levées are commonly left along the track of the flow and may give rise to irregular, mounded deposits adjacent to montane channels. The distal lobe may represent a substantial sediment accumulation, through which normal runoff water eventually forms a channel. Often, debris flows deposit where a steep tributary enters a larger valley. Here, repeated debris flows may build a debris flow cone, the surface of which represents the characteristic stopping angle for material delivered from the tributary.

2.3.4.2 Fluvial Transport of Bed Material

The fluvial transport of bed material in mountain channels has been relatively little investigated. Whittaker (1987) emphasized the control of sediment supply over flux rates such that no reliable hydraulic relations are available for estimating transport. This circumstance is borne out by available measurements. Observers always report some or all of the following characteristics:

- there is no strict relation between hydraulics and sediment transport: the same transport may be observed for a range of different flows (Griffiths, 1980; Hayward, 1980; Lauffer and Sommer, 1982; Beschta, 1983; Whittaker, 1987; Warburton, 1992; Adenlof and Wohl, 1994; Rickenmann, 1997; Zimmermann and Church, 2001);

- the sizes transported are characteristically smaller than the full range of sizes present in the channel (Lauffer and Sommer, 1982; Adenlof and Wohl, 1994; Lenzi, D'Agostino and Billi, 1999; Whiting *et al.*, 1999; Zimmermann and Church, 2001; Church and Hassan, 2002; Brummer and Montgomery, 2006);

- the bed surface is armoured and usually remains stable (Hayward, 1980; Church, Wolcott and Fletcher, 1991; Warburton, 1992; Adenlof and Wohl, 1994);

- there are substantial variations in sediment transport along the channel as sediment goes into storage or is removed from storage in the pools (Beschta, 1983; Warburton, 1992; Adenlof and Wohl, 1994);

- bed material appears to travel through the channel in low-amplitude, irregular, long-period waves (Hayward, 1980; Whittaker, 1987; Ergenzinger, 1988) with apparently varying period and origin;

- bed material transport may be subject to seasonal exhaustion effects (Nanson, 1974; Moog and Whiting, 1998), much like suspended load;

- the overall tranport is small in comparison with classic concepts of 'hydraulic capacity' to transport sediment (Griffiths, 1980; Lauffer and Sommer, 1982; Bathurst, Graf and Cao, 1987; Rickenmann, 1997; D'Agostino and Lenzi, 1999; Yager, Kirchner and Dietrich, 2007).

In reaches dominated by rapids, investigators similarly report mobilization of only a portion of the bed material grain-size distribution (Church, Wolcott and Fletcher, 1991; Carling *et al.*, 1998), maintenance of the surface armour and low overall transport (Church, Hassan and Wolcott, 1998), and aperiodic pulses of sediment passage (Carling *et al.*, 1998).

A number of investigators have pointed out that these features are all consistent with supply limitation of mobile sediment. It is supposed that sediment originates from a limited number of slope, streambank and instream storage sites (Nanson, 1974; Bathurst, 1985; Whittaker, 1987; Carling *et al.*, 1998; Church and Hassan, 2005; Yager, Kirchner and Dietrich, 2007), and it is supposed that competence limitations affect the sediment transport, particularly in consequence of the structural locking of the larger grains and the dissipation of energy in the pools. Lenzi, Mao and Comiti (2006) have presented a comprehensive review of the inception of bedload movement in steep channels.

Table 2.2 presents an inventory of measurements of bed material transport in steep channels. Laboratory data are not included and, no doubt, the inventory is not complete. Most data are reported only graphically; in most cases it would be necessary to contact the authors in order to obtain the exact data. Two modes of measurement dominate the data: reports of measurements recovered from samplers or small capacity traps placed in the stream, and reports of synoptic or seasonal accumulations of sediment caught in large sediment retention structures. In a few cases (notably Rickenmann, [89, 90]; Rickenmann and McArdell, 2007), acoustic sensors or weighing devices have been used to recover a

Table 2.2 Sources of field data of bed material transport in steep channels

Stream	Gradient	Channel Width[a] (m)	D_{50}[b] (mm)	D_{large}[b] (mm)	Q[c] ($m^3 s^{-1}$)	Data[d]	Source
Bridge Cr., Alberta	0.067	~3	~32	~64	~2	18 meas., tab.	Nanson (1974)
Slate Cr, Idaho	0.026	6.4	43^{s}	74^{84s}	3.4^{bf}	4 meas., tab.	Emmett (1975)
Wickiup Cr., Idaho	0.204	2.1	34^{s}	98^{84s}	1.9^{bf}	2 meas., tab.	
Torlesse Stream, NZ	0.067	~3		$\sim 250^{s}$	~2	86 events, tab.	Hayward (1980)
Pitzbach, Austria	0.035	8	260^{s}		<10	480 meas., graph. Q-Q_b	Lauffer and Sommer (1982)
Carl Beck, UK	0.04	1.6	67^{s}	135^{90s}	3^{maf}	62 data, graph. ω_e-g_b	Carling (1989)
Pitzbach, Austria	0.040		98^{s}	155^{84s}		data not shown	Quoted in Bathurst ((2007), table 2; from Impasihardjo, 1999)
Bas Glacier d'Arolla, Switzerland	~0.10	10	35^{s}		>5	91 data, graph., ω-g_b	Warburton (1992)
E. St. Louis Cr., Colorado	0.058–0.197	3.5	40^{s}	72^{84s}	0.86^{bf}	111 data, 12 x-ss graph., $Q - G_b$	Adenlof and Wohl (1994)
Squaw Cr., Montana	0.02	20	125^{s}	$\sim 200^{84s}$	6^{bf}	39 meas., graph. $Q - g_b$	Bunte (1996)
7 streams, Switz.	0.05–0.17	3.3–5.5	$10–90^{t}$	$40–160^{90t}$		155 data, $Q_p - V_b$ graph. $V_e - V_b$	Rickenmann (1997; 2001)
Rio Cordon, Italy	0.13		90^{s}	330^{90s}	10.4^{P}	49 data, graph. $Q - G_b$	D'Agostino and Lenzi (1999)
10 streams, Idaho[e]	0.020–0.72		$27–207^{s}$	$74–450^{90s}$	$0.048–49^{bf}$	1031 data, on Web site graph. $Q - G_b$	Whiting *et al.* (1999), King *et al.* (2004)
Little Granite Cr., Wyoming	0.019	~30	100^{s}	220^{84s}	6.5^{bf}	280 meas., tab.	Ryan and Emmett (2002)
Toots Cr., Arkansas	0.088	4.2	68^{s}	190^{84s}	1.34^{f}	26 meas.(5 events) graph. $Q - g_b$	Marion and Weirich (2003)
13 streams, Colorado and Wyoming[g]	0.017–0.050	2.0–17.2	$31–146^{s}$	$71–543^{84s}$	$0.3–10.1^{bf}$	graph. $Q - G_b$	Ryan et al. (2005)Ryan, Porth and Troendle (2005)

(continued)

Table 2.2 (*Continued*)

Stream	Gradient	Channel Width[a] (m)	D_{50}[b] (mm)	D_{large}[b] (mm)	Q[c] ($\mathrm{m^3\,s^{-1}}$)	Data[d]	Source
Roaring R u/s, Colorado	0.036		104^{s}	211^{s}		22 meas., graph. $q - g_b$ (poorly resolved)	Bathurst (2007)
Torrent St. Pierre France	0.025	~11	16–48^{s}	58–300^{maxs}		graph. $q - g_b$[h]	Liu *et al.* (2008)
Urumqui He, China	0.025		21.5^{s}	158^{90s}		graph. $q - g_b$[h]	

[a]Variously reported as water surface width at time of measurements or as channel width; ~ indicates that the result is approximate or that the actual width varies with discharge over the range of the measurements.

[b]Superscript s indicates surface material; superscript b indicates subsurface material; t indicates transported material; no superscript indicates mixture or unknown. Under D_{large}, numbers indicate percent finer, max indicates maximum size.

[c]Superscript bf indicates bankfull; maf indicates mean annual flood; p indicates peak observed discharge.

[d]Meas. Indicates an individual measurement of sediment flux, events indicates integrated event total transport, data indicates flux measurements aggregated in some way (across channel; through event); tab. indicates a table of data; graph. indicates a graphical display of data; if both are present, only the tabulation is mentioned; letter notations indicate the form of graphical presentation, with the water flux term first; q indicates discharge per unit width; Q indicates total discharge; V indicates event water volume; ω indicates specific stream power; ω_e indicates specific stream power in excess of some threshold; g_b indicates sediment transport by mass or weight per unit width; G_b indicates total sediment transport by mass or weight; q_b, Q_b are corresponding volumetric terms; V_b indicates event sediment volume.

[e]There are a total of 33 streams reported in this compilation.

[f]Flows in this data set were controlled floodings.

[g]There are six additional streams reported in this compilation, with $S < 0.0175$.

[h]There are many measurements in these data sets, aggregated across channel in various ways in order to seek the best representation of total sediment flux.

continuous measure of sediment as well as the summary volume retained. Trap or sampler measurements are apt to be subject to very large errors but, given the large measure of real sample variance, errors are mainly unquantified.

As a framework to understand the phenomena discussed above, Warburton (1992) described a three-phase model for bed material transport in steep channels (cf. Jackson and Beschta, 1982). At moderate flows, fine material, dominantly sand, is mobilized from streambanks and the bed (phase 1). In many channels this constitutes the majority of all transported sediment (Whiting *et al.*, 1999). At high flows, gravel is entrained from pools, and from the bank (phase 2), and is episodically directly delivered from adjacent hillslopes. The mobile material remains smaller than the bed structure-forming clasts, and not all material in a given size class is mobilized. This is the regime of 'partial sediment transport' (Wilcock and McArdell, 1993) that characterizes transport near threshold. Rarely (more so as the channel becomes steeper), the bed structure is broken and all sizes move. This phase 3 phenomenon may occur moderately regularly in rapids, but is altogether exceptional in step–pool cascades, where it leads to a 'debris flood' or debris flow (cf. Batalla *et al.*, 1999).

The sequential mobilization of more and larger material as flow increases certainly induces a general scale correlation between flow and bed material transport (Figure 2.10: see

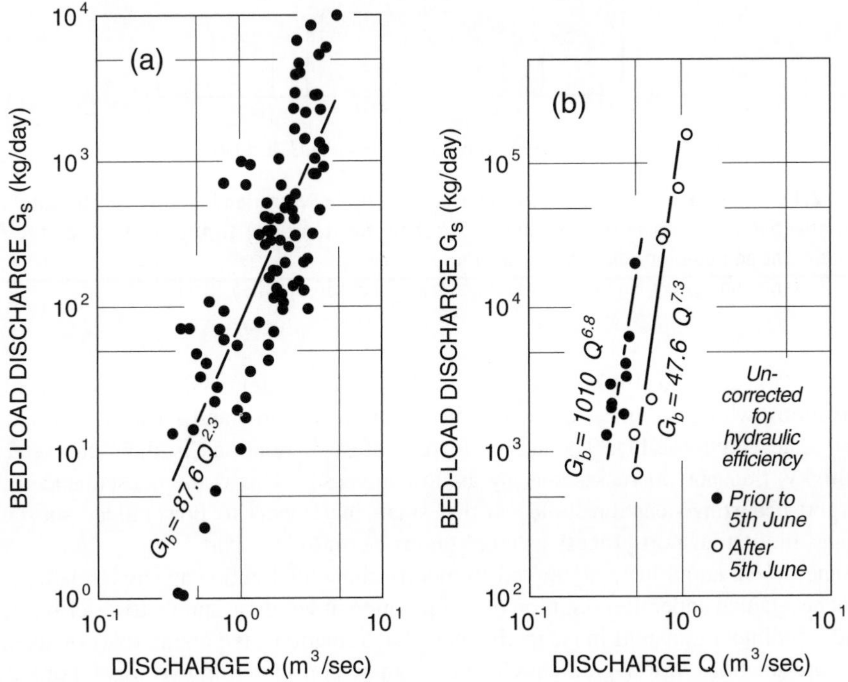

Figure 2.10 Bedload rating curves for steep channels: (a) Blackmare Creek, a rapid reach ($S = 0.03$; $A_d = 46.1\ \text{km}^2$) in Idaho, USA
(from Whiting *et al.*, 1999: Figure 8; reproduced by permission of the Geological Society of America).
(b) Bridge Creek, Alberta, Canada, a step-pool reach ($S = 0.07$; $A_d = 15.8\ \text{km}^2$); here, separate relations for rising and falling hydrographs reveal the supply limited nature of the transport. Data are relative, as no correction was made for the fractional efficiency of the basket sampler, with 6.4 mm screen. (from Nanson, 1974: Figure 5; reproduced by permission of American Journal of Science).

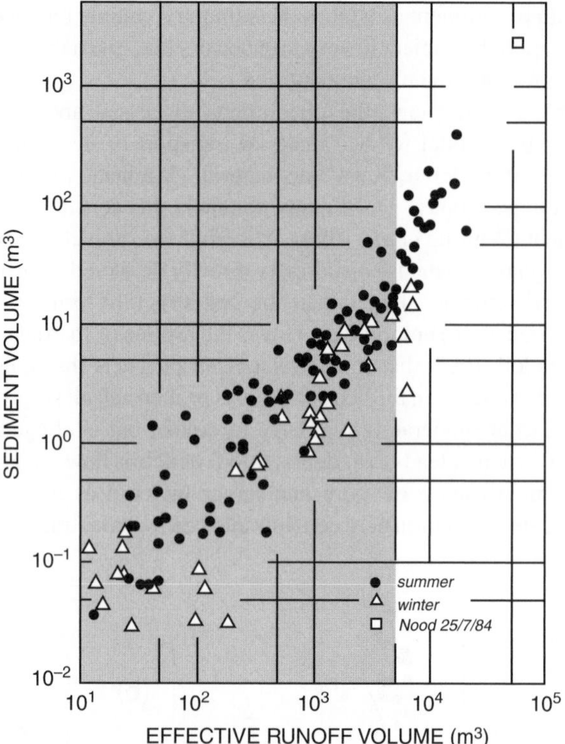

Figure 2.11 Erlenbach stream, a cascade reach ($S = 0.18$) in Switzerland: event load plotted against event effective flow volume (volume above threshold for transport) In the shaded area the data, despite event aggregation, display conspicuous scatter
(from Rickenmann, 1997: Figure 5; reproduced by permission of Wiley-Blackwell).

Bathurst, Graf and Cao (1987) and discussion therein for an attempted rationalization of this correlation), which becomes more clear if one integrates flow and sediment transport to event scale (Figure 2.11). The major feature of such relations is their characteristic sensitivity: transport increases sharply as flow increases. This is the consequence of the transport occurring near threshold, so that small increments to flow induce substantial changes in the transport process as more grains begin to take part.

Some investigators have attempted to modify limit 'hydraulic capacity' relations that might be attained under the condition of no limitation in sediment supply to describe phases 1 and 2 sediment transport in steep channels. Such relations, the counterpart of sediment transport functions in low-gradient channels, can be considered in two ways. For a given competent grain size, there is some maximum rate at which that material can be transported through a steep channel with some fixed framework geometry (such as steps and pools). This circumstance has been studied experimentally (Smart, 1984; Whittaker, 1987; Yager, Kirchner and Dietrich, 2007) and functional descriptions developed. Notably, Smart (see also Smart and Jaeggi, 1983) offered a modified Meyer-Peter and Muller function for gradients between 0.03 (1.7°) and 0.2 (11°), while Yager *et al.* presented a procedure for several classic bulk transport formulae modified to incorporate the following criteria:

(i) the median grain size of the mobile fraction is adopted as the representative bedload size for bulk transport estimates (this would not be necessary for fractional transport formulae);
(ii) the effective shear stress is the shear stress computed to act directly on the mobile sediment;
(iii) the computed transport is scaled according to the fraction of the bed covered by mobile grains (in an attempt to adapt the calculation to the available sediment supply).

Transport estimates are typically reduced, in comparison with unmodified calculations, by as much as an order of magnitude. However, Whittaker showed that the result depends closely upon channel geometry, particularly in step–pool systems, so that in many cases practical predictions will remain difficult to obtain, in conformity with the experience reported above.

Amongst bedload transport formulations, the classic Schoklitsch formula has been preferred by many investigators for application in steep channels because it is based on discharge. Accordingly, it is not necessary to directly measure or estimate either flow velocity or channel depth, one or both of which are required for calculations based on shear stress but very difficult to establish for the highly irregular geometry and high relative roughness typical of steep channels. The Schoklitsch (1962) equation is written in reduced form as

$$q_b = 0.93(q - q_c)S^{3/2} \tag{2.4}$$

wherein q_b is volumetric bedload transport per unit width, q is discharge per unit width, and q_c is the critical discharge at the inception of bedload movement. Rickenmann (2001) presented a similar equation based on experimental work in channels with gradient up to 0.2, but with coefficient 2.5, while D'Agostino and Lenzi (1999) have presented a number of closely related formulations, with and without explicit dependence on S. Equations in which the term $(q - q_c)$ is raised to some power $1.0 \leq b \leq 2.0$ have also been presented. There is no doubt that, at low rates of transport, this and similar equations greatly overestimate the transport (see Bathurst, Graf and Cao, 1987).

Rickenmann (2001) suggested that, near the threshold for motion, the large flow resistance presented by the high relative roughness in steep channels (typically, $3 > D_{max}/d > 0.7$ in high flow) acts to reduce the energy available for transport well below that expected in a channel with more subdued geometry, while Bathurst (2007) suggested that the effect may be related to the degree of surface armour development in steep channels. These factors are related through surface grain size and may be very difficult to separate functionally.

Noting the remarkable scatter in the measurements made at typically low transport rates (as in Figures 2.10 and 2.11, and further results in Whiting *et al.* (1999) and in Ryan, Porth and Troendle (2005)), Griffiths (1980) and Hegg and Rickenmann (1998) introduced stochastic approaches to estimate sediment flux. Griffiths' approach is based on a calibrated random variation of bedload yield in an event, while Hegg and Rickenmann based their approach on probabilistic assumptions about the distributions of driving and restraining forces acting on the grains in the streambed. The concept of variable restraining forces nicely summarizes the variable aggregate strength of structured arrangements of bed materials along a steep channel. But the generality of such assumptions remains to be investigated.

A second interpretation of a 'hydraulic capacity' relation would be one that accepts the criterion of equal mobility of all available material – the assumption implicit in classic bed material and bedload transport formulae (cf. Parker and Klingeman, 1982). This is phase 3 transport. In steep channels this would entail the mobilization of the boundary-forming elements. Such a circumstance can occur for some period in rapids, but it would destroy a step–pool cascade in which the structures form the channel. The ensuing situation would be entirely unstable and would propagate a debris flow – a catastrophic event in the history of the channel.

2.3.4.3 Wash Material

Wash material is material that, once entrained, is apt to travel a substantial distance in the stream, and is not found in significant quantities on the streambed. In most rivers this is material finer than about 180 μm, for which suspension velocity is lower than entrainment velocity. In steep channels, sand and granule gravel may also constitute wash material. The lower cut-off for bed material may fall within the range 250 μm to 4 mm or more. The occurrence of wash material is strictly supply-limited. Delivery to the channel depends upon the contingent effects of weather and recent geomorphological processes acting on the material sources in the drainage basin.

Fine sediments derive from mass wasting deposits delivered to the channel margins, from high stream banks in unconsolidated fine material, in particular from gully walls, from temporary in-stream storage, particularly behind log jams or debris dams, from weathering and attrition of larger materials along the channel, and in water draining into the channel from upslope. Debris flow deposits and landslide scars are important sources for chronic delivery of fine sediments into steep channels. Delivery can be divided into 'episodic' and 'normal regime' processes (Nistor and Church, 2005).

Episodic processes are high magnitude, low frequency events that deliver a substantial amount of fine sediment to the channel, such as debris flow, bank failure or the collapse of a log jam. In low-order streams, these are the cause of sudden, significant increases in sediment concentration which may persist for some time. In comparison, normal regime processes deliver modest amounts of sediment relatively frequently. They include dry ravel, frost heave, surface wash and splash on banks, creep, tree-throw and animal activity at bank tops, weathering processes and remobilization of fine material from the streambed. These processes set the 'background' level of fine sediment yield and represent the sediment characteristically washed out of the system in storm periods.

Because even normal fine sediment yield depends upon many contingent delivery processes, there is no simple relation between flow and sediment concentration or yield (Figure 2.12a). Storm-period sediment yield graphs tend to be complex (Figure 2.12b). Systemic sediment exhaustion during the event, leading to a clockwise hysteresis in the relation of sediment concentration to flow, is common. Material delivered to or prepared in the channel during the preceding low flow period is flushed away early in the event. Seasonal exhaustion also occurs (Beschta, 1981; Nistor and Church, 2005 Finally, a clearer correlation between flow and sediment concentration may emerge in the largest normal regime events when the stream is sufficiently high to access infrequently reached sources along the stream (Figure 2.12c).

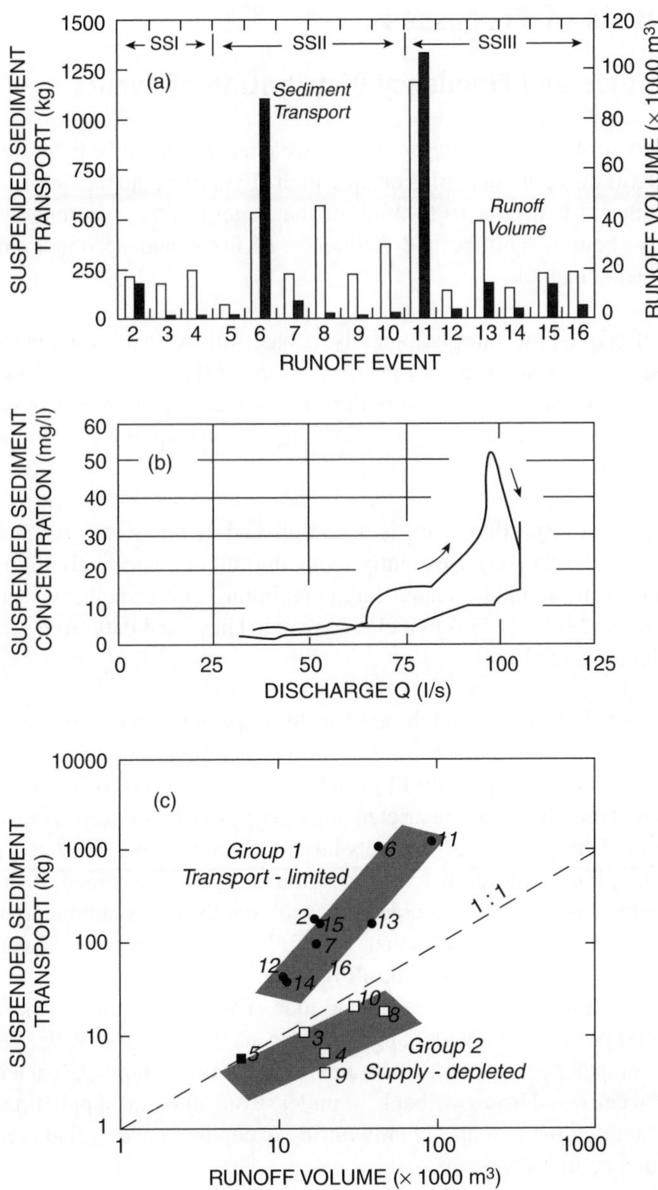

Figure 2.12 Fine sediment transport in steep channels. The example is a headmost gully in glacial till: (a) seasonal control through a winter season (October-May) by synoptic weather systems of normal regime fine sediment yield: notice the progressive reduction in sediment yield during each of three weather-delimited 'subseasons' (SSI-SSIII; (b) storm period sediment yield for event 6; (c) suspended sediment event ratings for a steep channel exhibiting sub-seasonal and synoptic control: the same gully illustrated in (a) and (b). Numbers indicate the events in (a). 1:1 indicates the slope of relations with constant sediment concentration
(after Nistor and Church, (2005) reproduced by permission of Wiley-Blackwell).

2.4 Synthesis of Processes

2.4.1 Magnitude and Frequency Distribution of Events

The processes described above are driven by weather, particularly by water influx into headward basins as the result of storms or snowmelt. To understand the implication of event forcing by weather it is necessary to think of the larger drainage system. There are two dominant ideas about magnitude and frequency of flows and accompanying sediment transport in stream channels.

(i) The most effective flow, morphologically, is bankfull. At that level, the forces exerted on the channel boundary are very nearly the largest that occur since, once overbank, flows may increase substantially without greatly increasing the flow depths or velocities further – water storage and conveyance are largely taken up by the dramatic increase of width overbank.

(ii) Most sediment transporting work is accomplished at flows near mean annual flood (MAF). This is a relatively frequently recurring but comparatively high flow, so the product of magnitude and frequency has a maximum. This principle was demonstrated by Wolman and Miller (1960) in a classic paper. They used data from relatively large rivers to demonstrate it.

Some researchers have combined these ideas to suppose that bankfull corresponds with MAF. However, it is not obvious why either of the statements should necessarily be true.

The first problem is defining 'bankfull'. Banks vary in elevation. When rivers flood, they initially flow out of their banks at a restricted number of places. On steep, headward streams, the variation of bank height often is relatively large, so bankfull stage may vary significantly along the channel. Furthermore, many headward channels are confined so that the actual elevation of bankfull is difficult to decide, and water depths may continue to increase with increasing flows. Nonetheless, the idea that 'bankfull' is an effective flow is plausible, so the question becomes 'how often does bankfull recur?'

There are greater problems with the second idea. The concept may work fairly well in rivers that are competent over a wide range of flows, as Wolman and Miller demonstrated. However, flow competence is a major constraint in headward channels. Channels dominated by large material delivered from overbank, or under strong material supply limitations, may move a significant volume of material only during exceptional events that recur much less frequently than once in a few years.

Two other phenomena underlie the occurrence of effective (meaning 'channel changing') events in hollows and stream channels. One is the distribution of significant water inputs in time and space in the landscape, and the way in which river systems combine those inputs from different source areas. The second is the geologically determined character of stream sediments and the forces necessary to move them. The combination of these factors determines the flows (and their frequency) that are capable of moving significant volumes of surficial material, hence of changing the morphology of the hollow or channel.

Extreme water inputs are quite local. The largest normal water inputs effective at catchment scale occur in severe convectional showers, perhaps embedded in frontal squall lines. The heaviest precipitation may last for less than an hour and may cover only a few

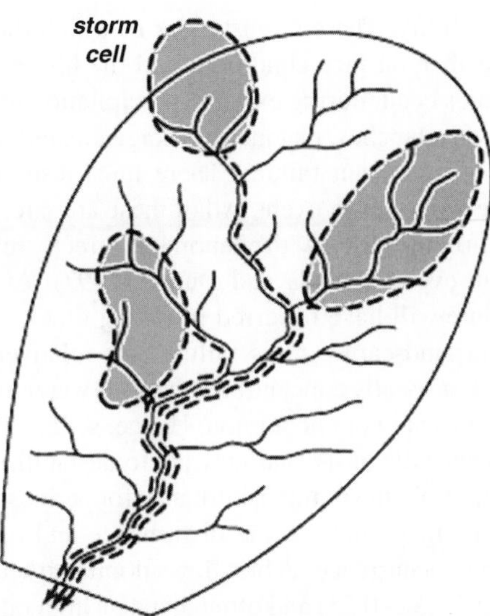

Figure 2.13 Conceptual diagram to illustrate the local nature of extreme forcing events in the landscape. Each storm cell represents a separate, extreme headwater event. Downstream, all events affect the higher order channels

hectares or square kilometres. These phenomena are not rare in the landscape, but they rarely recur at the same place within a few years (see Figure 2.13). Downstream, however, flows from extreme events and from regional heavy precipitation converge in higher order channels to yield more regularly recurring 'extreme flows'. Potentially mobile sediments in hollows, on the other hand, may possess substantial strength due to the effect of roots or of cohesion. Material on steep hillslopes may also possess significant drainage bypass routes in the form of subsurface macropores. Materials in steep headward channels are large or are constrained by strong structural arrangement. Lesser, loose material is quickly washed downstream. The combination of these conditions means that major sediment transfers and significant morphological change are rare events in steep channels. A channel-forming fluvial event may have a return period between decades and centuries. A debris flow may be initiated once in decades to once in a millennium, and major, hollow-clearing landslides may be even more infrequent (see Section 2.2.2).

As one moves to lower-gradient channels with greater confluence of upslope drainage and less formidable structural constraint, bed mobilization becomes more frequent. These temporal circumstances give rise to a complementary view of the steepland sediment transporting regime in space. As one moves headward, toward smaller channels with limited sediment supply and strong structural constraints on sediment movement, the recurrence of effective, channel-forming events becomes more and more infrequent.

2.4.2 The Sediment Transporting Regime in the Montane Landscape

Mass wasting phenomena around steep channels are subject to the constraint that mobilizing forces must overcome the frictional inertia of the material and constraints imposed by cohesion or pseudocohesion, either by addition of mass or by reduction in material strength to the point of shear instability. The circumstances in which these conditions occur are highly contingent on weather, surface condition, and the history of accumulation of the deposit. Most often, failures occur during extreme precipitation following an extended wet period. Sometimes, snow avalanches, channel blockage created by a small initial slide or seismic shaking trigger a significant failure. There must also be sufficient mobilizable material present to create a significant event. All of these are rare, contingent events. Mass wasting phenomena are, in comparison with phenomena directly related to channelled water flows, infrequent random events (Benda and Dunne, 1997). At most sites in mountain landscapes, no mass failure will have occurred in a long time.

But, in most mountain landscapes, mass failure is a relatively frequent event in the landscape as a whole because locally concentrated extreme water inputs or the occurrence of other potential trigger events are not uncommon. Hence, sediment is fed into some part of the stream system on an episodic basis, and may act to destabilize the channel system for some distance downstream from the initial disturbance for periods that range from weeks to years, depending upon the magnitude of the disturbance and the energy of the system. Despite their relatively rare occurrence, debris flows dominate sediment delivery from the steepest headwater channels ($S > 0.35$) and often dictate a high rate of sediment yield from headwater drainage lines (Figure 2.14). The dominance of debris flow imposes a characteristic regime of erosion on to steepland channels (Stock and Dietrich, 2006). Much of the sediment delivered from the steep headwaters may be sequestered at slope base in colluvial fans, so that onward movement of sediment in the fluvial system, typically beginning in

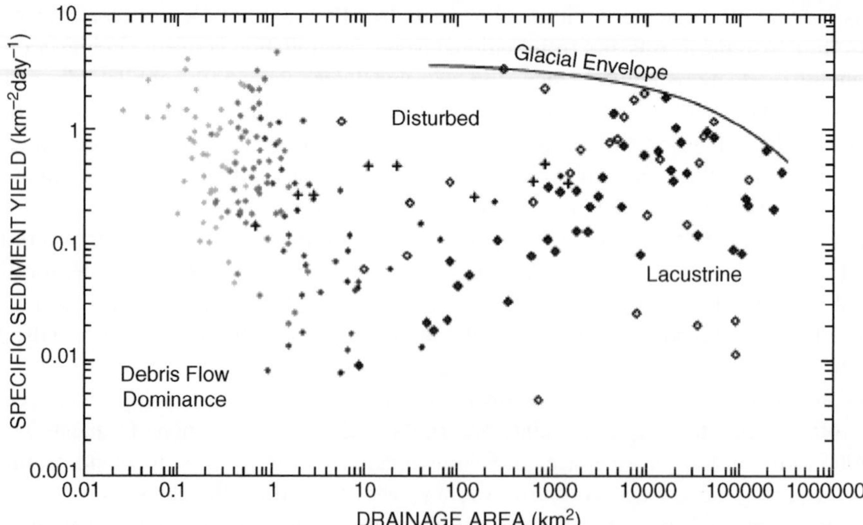

Figure 2.14 Distribution of sediment yield in the landscape. Drainage areas $<3\,km^2$ are hillside channels dominated by debris flow. Increasingly dark symbols indicate channels of increasingly high order; black symbols indicate order 5 and higher. Specific sediment yield declines through the low order channels as material is stored in hillslope-base fans and aprons. In higher order streams, specific sediment yield increases again as the streams erode Quaternary valley fill deposits. Open diamonds indicate channels off the main sequence that are either disturbed (by land use or contemporary glaciation), or are interrupted by lakes. + symbols indicate data of Kirchner *et al.* (2001) from Idaho (data compiled by J. Tunnicliffe).

second- or third-order tributaries, is often much reduced. In undisturbed montane landscapes, fluvial sediment yield often then increases downstream through the degrading fluvial system (Figure 2.14) as sediment is scoured from streambanks and the channel bed. The effect is particularly strong in formerly glaciated mountains where valley fills of glacially deposited materials are re-entrained by the streams. This signature is quite distinct from that of most settled landscapes, where land surface erosion consequent upon land use yields significant amounts of sediment to headwater streams, part of which is lost downstream in aggrading river systems.

In rivers on low gradient, we are used to thinking of the mobilization of bed material as a relatively frequent event, apt to happen for at least several days in each year even in gravels. In fact, the patterns of declining gradient, declining sediment calibre, and increasing integration of flow contributions down a drainage system produce systematic variation in the frequency of mobilization and magnitude of bed material (Figure 2.15). In steep channels, as defined in this paper, clastic materials must be structurally constrained to remain present at all. Accordingly, the Shields' number (cf. Equation 2.1) for entrainment is large ($\theta_c > 0.075$). Flows capable of delivering such a large number are rare. Consequently, the channel remains in a stable configuration for long periods. Smaller material is not moved efficiently, either. Plunge pools and shadow zones behind obstructions become reservoirs for limited quantities of potentially mobile pebbles. Only fine material moves through the system quickly; so quickly that the stream is usually starved of such material and headwater streams customarily run clear. The consequence is highly size-selective transport of a modest amount of material.

In this view, the entire fluvial system is a gigantic sorting machine, continuously selecting for transport those grains that are not stable in the ambient flow environment. What is left behind becomes part of the ordinarily stable channel that we observe. The effects are most extreme in steep channels. The converse side of this concept is the increasingly supply dominated nature of sediment transport in headward streams. Since only highly selected materials are stable there, most of the rest quickly moves down the system toward its position of 'customary stability'. Hence, the supply rate of material to the channel comes to govern the rate of sediment transport, and simple hydraulic correlations are not to be had. Sediment transport is a stochastic process governed by the delivery of sediment to the channel from adjacent slopes, and by the history of development and destruction of

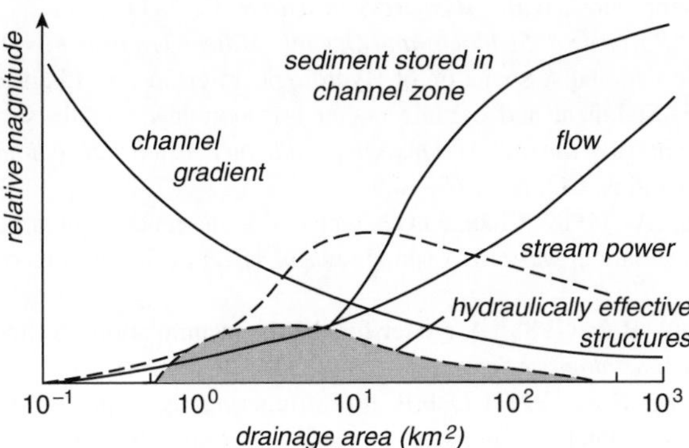

Figure 2.15 Diagram to illustrate the controls of spatially varying entrainment forces and frequencies in headward fluvial systems

sediment reservoirs along the channel, both processes being forced by the occurrence of extreme water supply to the channels. The system adapts to normal flows by adopting a state that is stable under the stresses they impose. Steep channels are self-organized, contingently stable systems. In a sense, they are near-critical systems since the boundary strength presumably does not become much greater than that required to withstand the range of flows to which the channel normally is subject. The classic Shields' criterion can be interpreted as the critical parameter defining the stability limit. On adjacent hillside slopes, a similar concept of constrained stability holds, but the system there is not so obviously near-critical, since processes such as vegetation development can, in favourable circumstances, locally move the system well away from the stability limit for the local gradient.

References

Abrahams, A.D., Li, G. and Atkinson, J.F. (1995) Step–pool streams: adjustment ato maximum flow resistance. *Water Resources Research*, **31**, 2593–2602.

Adenlof, K.A. and Wohl, E.E. (1994) Controls on bedload movement in a subalpine stream of the Colorado Rocky Mountains, USA. *Arctic and Alpine Research*, **26**, 77–85.

Aulitzky, H. (1989) The debris flows of Austria. *Bulletin of the International Association of Engineering Geology*, **40**, 5–13.

Batalla, R.J., deJong, C., Ergenzinger, P. and Sala, M. (1999) Field observations on hyperconcentrated flows in mountain torrents. *Earth Surface Processes and Landforms*, **24**, 247–253.

Bathurst, J.C. (1985) Flow resistance estimation in mountain rivers. *Journal of Hydraulic Engineering*, **111**, 625–643.

Bathurst, J.C. (2007) Effect of coarse surface layer on bedload transport. *Journal of Hydraulic Engineering*, **133**, 1192–1205.

Bathurst, J.C., Graf, W.H. and Cao, H.H. (1987) Bed load discharge equations for steep mountain rivers, in *Sediment Transport in Gravel-Bed Rivers* (eds C.R. Thorne, J.C. Bathurst and R.D. Hey), John Wiley and Sons, Chichester, pp. 453–477.

Benda, L.E. and Cundy, T.W. (1990) Predicting deposition of debris flows in mountain channels. *Canadian Geotechnical Journal*, **27**, 409–417.

Benda, L. and Dunne, T. (1987) Sediment routing by debris flow, *Erosion and Sedimentation in the Pacific Rim*, **165**, International Association of Hydrological Science, Wallingford, pp. 213–223.

Benda, L. and Dunne, T. (1997) Stochastic forcing of sediment supply to channel networks from landsliding and debris flow. *Water Resources Research*, **33**, 2849–2863.

Beschta, R.L. (1981) *Patterns of Sediment and Organic Matter Transport in Oregon Coast Range streams*, **132**, International Association of Hydrological Science, Wallingford, pp. 179–185.

Beschta, R.L. (1983) Sediment and organic matter transport in mountain streams of the Pacific Northwest. *Proceedings of the D.B. Simons Symposium on Erosion and Sedimentation*. Simons, Li and Assoc., Fort Collins, CO, pp. 1.69–1.89.

Bilby, R.E. and Ward, J.W. (1989) Change in characteristics and function of large woody debris with increasing size of stream in western Washington. *American Fisheries Society Transactions*, **118**, 368–378.

Bovis, M.J. and Dagg, B.R. (1988) A model for debris accumulation and mobilization in steep mountain stream. *Hydrological Sciences Journal*, **33**, 589–604.

Bovis, M.J. and Dagg, B.R. (1992) Debris flow triggering by impulsive loading: mechanical modellling and case studies. *Canadian Geotechnical Journal*, **29**, 345–352.

Bovis, M.J. and Jakob, M. (1999) The role of debris supply conditions in predicting debris flow activity. *Earth Surface Processes and Landforms*, **24**, 1039–1054.

Bovis, M.J., Dagg, B.R. and Kaye, D. (1985) Debris flow and debris torrents in the southern Canadian Cordillera: discussion. *Canadian Geotechnical Journal*, **22**, 608.

Bovis, M.J., Millard, T.H. and Oden, M.E. (1998) Gully processes in coastal British Columbia: the role of woody debris, *Carnation Creek and Queen Charlotte Islands Fish/Forestry Workshop. Land Management Handbook*, **41** (eds D.L. Hogan, P.J. Tschaplinski and S. Chatwin), British Columbia Ministry of Forests, pp. 49–75.

Brummer, C.J. and Montgomery, D.R. (2006) Influence of coarse lag formation on the mechanics of sediment pulse dispersion in a mountain stream, Squire Creek, North Cascades, Washington, United States. *Water Resources Research*, **42**, W07412, 16. doi: 10.1029/2005WR004776

Bunte, K. (1996) *Analyses of the Temporal Variation of Coarse Bedlaod Transport and its Grain Size Distribution, Squaw Creek, Montana, U.S.A.* United States Department of Agriculture, Forest Service, Rocky Mountain Forest and Range Experiment Station, General Technical Report RM-GTR-288, p. 123.

Caine, N. (1980) The rainfall intensity–duration control of shallow landslides and debris flows. *Geografiska Annaler*, **62A**, 23–27.

Carling, P.A. (1989) Bedload transport in two gravel-bedded streams. *Earth Surface Processes and Landforms*, **14**, 27–39.

Carling, P.A., Williams, J.J., Kelsey, A. *et al.* (1998) Coars bedload transport in a mountain river. *Earth Surface Processes and Landforms*, **23**, 141–157.

Cates, M.E., Wittmer, J.P., Bouchard, J.-P. and Claudin, P. (1998) Jamming, force chains and fragile matter. *Physical Review Letters*, **81**, 1841–1844.

Chartrand, S.M. and Whiting, P.J. (2000) Alluvial architecture in headwater streams with special emphasis on step-pool topography. *Earth Surface Processes and Landforms*, **25**, 583–600.

Chin, A. (1989) Step pools in stream channels. *Progress in Physical Geography*, **13**, 391–407.

Chin, A. (1999) The morphological structure of step-pools in mountain streams. *Geomorphology*, **27**, 191–204.

Chin, A. and Wohl, E. (2005) Toward a theory for step-pools in stream channels. *Progress in Physical Geography*, **29**, 275–296.

Church, M. and Hassan, M. (2002) Mobility of bed material in Harris Creek. *Water Resources Research*, **38, 1237**.

Church, M. and Hassan, M. (2005) Upland gravel-bed rivers with low sediment transport, in *Catchment Dynamics and River Processes: Mediterranean and other Climate Regions,* Developments in Earth Surface Processes, **7** (eds C. Garcia and R.J. Batalla), Elsevier, Amsterdam, pp. 141–168.

Church, M. and Miles, M.J. (1987) Meteorological antecedents to debris flow in southwestern British Columbia: some case studies. *Reviews in Engineering Geology*, **7**, 63–79.

Church, M. and Zimmerman, A. (2007) Form and stability of step–pool channels: research progress. *Water Resources Research*, **43**, W03415, 21. doi: 10.1029/WR005037

Church, M., Hassan, M.A. and Wolcott, J.F. (1998) Stabilizing self-organized structures in gravel-bed stream channels: field and experimental observations. *Water Resources Research*, **34**, 3169–3179.

Church, M., Wolcott, J.F. and Fletcher, K. (1991) A test of equal mobility in fluvial sediment transport: behavior of the sand fraction. *Water Resources Research*, **27**, 2941–2951.

Comiti, F., Andreoli, A. and Lenzi, M.A. (2005) Morphological effects of local scouring in step–pool streams. *Earth Surface Processes and Landforms*, **30**, 1567–1581.

Costa, J.E. (1984) Physical geomorphology of debris flows, in *Developments and Applications of Geomorphology* (eds J.E. Costa and P.J. Fleisher), Springer-Verlag, Berlin, pp. 268–317.

Curran, J.C. and Wilcock, P.R. (2005) Characteristic dimensions of the step-pool configuration: an experimental study. *Water Resources Research*, **41**, W02030. doi: 10.1029/WR003568

D'Agostino, V. and Lenzi, M.A. (1999) Bedload transport in the instrumented catchment of the Rio Cordon. Part II: Analysis of the bedload rate. *Catena*, **36**, 191–204.

D'Odorico, P. and Fagherazzi, S. (2003) A probabilistic model of rainfall-triggered shallow landslides in hollows: a long-term analysis. *Water Resources Research*, **39**, 1262, 14. doi: 10.1029/WR001595

Dietrich, W.E., Wilson, C.J. and Reneau, S.L. (1986) Hollows, colluvium, and landslides in soil-mantled landscapes, in *Hillslope Processes* (ed. A.D. Abrahams), Allen and Unwin, Boston, pp. 361–388.

Emmett, W.W. (1975) The channels and waters of the Upper Salmon River area, Idaho. *U.S. Geological Survey Professional Paper* 870-A, Washington, p. 116.

Ergenzinger, P. (1988) *The Nature of Coarse Material Bed Load Transport*, International Association of Hydrological Science, **174**, Wallingford, pp. 207–216.

Fannin, R.J. and Rollerson, T.P. (1993) Debris flows: some physical characteristics and behaviour. *Canadian Geotechnical Journal*, **30**, 71–81.

Fuller, C.W., Willett, S.D., Hovius, N. and Slingerland, R. (2003) Erosion rates for Taiwan mountain basins: new determinations from suspended sediment records and a stochastic model of their temporal variation. *Journal of Geology*, **111**, 71–87.

Gomi, T., Sidle, R.C., Bryant, M.D. and Woodsmith, R.D. (2001) The characteristics of woody debris and sediment distribution in headwater streams, southeastern Alaska. *Canadian Journal of Forest Research*, **31**, 1386–1399.

Grant, G.E., Swanson, F.J. and Wolman, M.G. (1990) Pattern and origin of stepped-bed morphology in high-gradient streams, western Cascades, *Oregon. Geological Society of America Bulletin*, **102**, 340–352.

Griffiths, G.A. (1980) Stochastic estimation of bedload yield in pool-and-riffle mountain streams. *Water Resources Research*, **16**, 931–937.

Hack, J.T. and Goodlett, J.C. (1960) Geomorphology and forest ecology of a mountain region in the central Appalachians. *U.S. Geological Survey Professional Paper* 347, Washington.

Hayward, J.A. (1980) Hydrology and Stream Sediment from Torlesse Stream Catchment. Tussock Grasslands and Mountain Lands Institute, Lincoln College, Canterbury, N.Z. Special Publication 17, p. 236.

Heede, B.H. (1972) Influence of a forest on the hydraulic geometry of two mountain streams. *Water Resources Bulletin*, **8**, 523–530.

Hegg, C. and Rickenmann, D. (1998) Short-time relations between runoff and bed load transport in a steep mountain torrent. *Modelling Soil Erosion, Sediment Transport and Closely Related Hydrological Processes (Proceedings of the Vienna Symposium, July, 1998).* International Association of Hydrological Sciences, 249, Wallingford, pp. 317–324.

Hogan, D.L., Bird, S.A. and Hassan, M.A. (1998) Spatial and temporal evolution of small coastal gravel-bed streams: influence of forest management on channel morphology and fish habitat, in *Gravel-Bed Rivers in the Environment* (eds P.C. Klingeman, R.L. Beschta, P.D. Komar and J.B. Bradley), Water Resources Publication, Highlands Ranch, CO, pp. 365–392.

Hungr, O., Morgan, G.C. and Kellerhals, R. (1984) Quantitative analysis of debris flow hazards for design of remedial measures. *Canadian Geotechnical Journal*, **22**, 663–677.

Hungr, O., Evans, S.G., Bovis, M.J. and Hutchinson, J.N. (2001) A review of the classification of landslides of the flow type. *Environmental and Engineering Geoscience*, **7**, 221–238.

Impasihardjo, K. (1991). *Bed load transport of nonuniform size sediment in mountain rivers*. PhD thesis, University of Newcastle upon Tyne.

Innes, J.L. (1983) Debris flows. *Progress in Physical Geography*, **7**, 469–501.

Iverson, R.M. (1997) The physics of debris flows. *Reviews of Geophysics*, **35**, 245–296.

Jackson, W.L. and Beschta, R.L. (1982) A model of two-phase bedload transport in an Oregon Coast Range stream. *Earth Surface Processes and Landforms*, **7**, 517–527.

Jakob, M. and Bovis, M.J. (1996) Morphometric and geotechnical controls of debris flow activity, southern Coast Mountains, British Columbia, Canada. *Zeitschrift fur geomorphologie*, **104** (Supp.), 13–26.

Johnson, P.A., McCuen, R.H. and Hromadka, T.V. (1991) Magnitude and frequency of debris flows. *Journal of Hydrology*, **123**, 69–82.

Judd, H.E. and Peterson, D.F. (1969) *Hydraulics Of Large Bed Element Channels*. College of Engineering, Utah State University, Utah Water Research Laboratory, Report PRWG 17-6.

Keller, E.A. and Swanson, F.J. (1979) Effects of large organic material on channel form and fluvial processes. *Earth Surface Processes and Landforms*, **4**, 361–380.

King, J.G., Emmett, W.W., Whiting, P.J. *et al.* (2004) *Sediment Transport Data and Related Information for Selected Coarse-Bed Streams and Rivers in Idaho*. United States Department of Agriculture, Forest Service, Rocky Mountain Research Station, General Technical Report RMRS-GTR-131, 26.

Kirchner, J., Finkel, R., Riebe, C. *et al.* (2001) Mountain erosion over 10 years, 10 kyr and 10 myr time scales. *Geology*, **29**, 591–594.

Laronne, J.B. and Carson, M.A. (1976) Interrelationships between bed morphology and bed-material transport for a small, gravel-bed channel. *Sedimentology*, **23**, 67–85.

Lauffer, H. and Sommer, N. (1982) Studies on sediment transport in mountain streams of the Eastern Alps. International Commission on Large Dams, 14th Congress, Proceedings, pp. 431–453.

Lee, A.J. and Ferguson, R.I. (2002) Velocity and flow resistance in step-pool streams. *Geomorphology*, **46**, 59–71.

Lenzi, M.A., D'Agostino, V. and Billi, P. (1999) Bedload transport in the instrumented catchment of the Rio Cordon. Part I: Analysis of bedload records, conditions and threshold of bedload entrainment. *Catena*, **36**, 171–190.

Lenzi, M.A., Mao, L. and Comiti, F. (2006) When does bedload transport begin in steep boulder-bed streams? *Hydrological Processes*, **20**, 3517–3533. doi: 10.1002/hyp.6168

Li, U.-H. (1976) Denudation of Taiwan Island since the Pliocene Epoch. *Geology*, **5**, 105–107.

Liu, Y., Métivier, F., Lajeuness, É. *et al.* (2008) Measuring bedload in gravel-bed mountain rivers: averaging methods and sampling strategies. *Geodinamica Acta*, **21**, 81–92. doi: 10.3166/ga.21.81-92

Marion, D.A. and Weirich, F. (2003) Equal-mobility bed load transport in a small step–pool channel in the Ouachita Mountains. *Geomorphology*, **55**, 139–154. doi: 10.1016/S0169-555X(03) 00137-5

Martin, Y., Rood, K., Schwab, J.W. and Church, M. (2002) Sediment transfer by shallow landsliding in the Queen Charlotte Islands, British Columbia. *Canadian Journal of Earth Sciences*, **39**, 189–205.

Massong, T.M. and Montgomery, D.R. (2000) Influence of sediment supply, lithology, and wood debris on the distribution of bedrock and alluvial channels. *Geological Society of America Bulletin*, **112**, 591–599.

McDonald, B.C. and Day, T.J. (1978) An experimental flume study on the formation of transverse ribs. *Geological Survey of Canada, Current Research, Part A, Paper*, 78-1A, 441–451.

McGlynn, B.L., McDonnell, J.J. and Brammer, D.D. (2002) A review of the evolving perceptual model of hillslope flowpaths at the Maimai catchments, New Zealand. *Journal of Hydrology*, **257**, 1–26.

Megahan, W.F. (1982) Channel sediment storage behind obstructions in forested drainage basins draining the granitic bedrock of the Idaho batholith, in *Sediment Budgets and Routing in Forested Drainage Basins* (eds F.J. Swanson, R.J. Janda, T. Dunne and D.N. Swanston), United States Department of Agriculture, Forest Service, Pacific Northwest Research Station, General Technical Report PNW-141, pp. 114–121.

Meyer-Peter, E. and Muller, R. (1948) Forumulas for bed-load transport. *Proceedings of the 2nd Meeting of the International Association for Hydraulic Structures Research*, Stockholm, Appendix 2, p. 26.

Mizuyama, T., Kobashi, S. and Ou, G. (1992) Prediction of debris flow peak discharge. *Interpraevent*, **4**, 99–108.

Montgomery, D.R. and Buffington, J.M. (1997) Channel-reach morphology in mountain drainage basins. *Geological Society of America Bulletin*, **109**, 596–611.

Montgomery, D.R., Buffington, J.M., Smith, R.D. *et al.* (1995) Pool spacing in forest channels. *Water Resources Research*, **31**, 1097–1105.

Montgomery, D.R., Abbe, T.B., Buffington, J.M. *et al.* (1996) Distribution of bedrock and alluvial channels in forested mountain drainage basins. *Nature*, **381**, 587–589.

Moog, D.B. and Whiting, P.J. (1998) Annual hysteresis in bed load rating curves. *Water Resources Research*, **34**, 2393–2399.

Nakamura, F. and Swanson, F.J. (1993) Effects of coarse woody debris on morphology and sediment storage in mountain stream systems in western Oregon. *Earth Surface Processes and Landforms*, **18**, 43–61.

Nanson, G.C. (1974) Bedload and suspended load transport in a small, steep mountain stream. *American Journal of Science*, **274**, 471–486.

Nistor, C. and Church, M. (2005) Fluvial suspended sediment transport regime in a steepland gully: Vancouver Island, British Columbia. *Hydrological Processes*, **19**: 861–885.

Parker, G. and Klingeman, P.C. (1982) On why gravel-bed streams are paved. *Water Resources Research*, **18**, 1409–1423.

Rickenmann, D. (1997) Sediment transport in Swiss torrents. *Earth Surface Processes and Landforms*, **22**, 937–951.

Rickenmann, D. (2001) Comparison of bed load transport in torrents and gravel bed streams. *Water Resources Research*, **37**, 3295–3305.

Rickenmann, D. and McArdell, B.W. (2007) Continuous measurement of sediment transport in the Erlenbach stream using piezoelectric bedload impact sensor. *Earth Surface Processes and Landforms*, **32**, 1362–1378. doi: 10.1002/esp.1478

Roberts, R.G. and Church, M. (1986) The sediment budget in severely disturbed watersheds, Queen Charlotte Ranges, British Columbia. *Canadian Journal of Forest Research*, **16**, 1092–1106.

Ryan, S.E. and Emmett, W.W. (2002) *The Nature of Flow and Sediment Movement in Little Granite Creek near Bondurant, Wyoming*. U.S. Department of Agriculture, Forest Service, Rocky Mountain Research Station, General Technical Report RMRS-GTR-90, p. 48.

Ryan, S.E., Porth, L.S. and Troendle, C.A. (2005) Coarse sediment transport in mountain streams in Colorado and Wyoming. *Earth Surface Processes and Landforms*, **30**, 269–288. doi: 10.1002/esp.1128

Schoklitsch, A. (1962) *Handbuch des Wasserbaues*, 3rd edn, Springer, Vienna.

Schwab, J.W. (1998) Landslides on the Queen Charlotte Islands: processes, rates and climatic events, in *Carnation Creek and Queen Charlotte Islands Fish/Forestry Workshop. Land Management Handbook*, **41** (eds D.L. Hogan, P.J. Tschaplinski and S. Chatwin), British Columbia Ministry of Forests, pp. 41–47.

Sidle, R.C. and Ochiai, H. (2006) Landslides and slope processes: prediction and land use. *American Geophysical Union, Water Resources Monograph*, **18**, 260.

Slaymaker, O. (1993) The sediment budget of the Lillooet River basin, British Columbia. *Physical Geography*, **14**, 304–320.

Smart, G.M. (1984) Sediment transport formula for steep channels. *Journal of Hydraulic Engineering*, **110**, 267–276.

Smart, G.M. and Jaeggi, M. (1983) Sediment transport on steep slopes. Eidgenossischen Technichen Hochschule, Zurich. *Versuchsanstalt fur Wasserbau, Hydrologie und Glaziologie, Mitteilungen*, 64.

Stock, J.D. and Dietrich, W.E. (2006) Erosion of steepland valleys by debris flows. *Geological Society of America Bulletin*, **118**, 1125–1148.

Takahashi, T. (1981) Debris flow. *Annual Review of Fluid Mechanics*, **13**, 57–77.

van Dine, D.F. (1985) Debris flows and debris torrents in the southern Canadian Cordillera. *Canadian Geotechnical Journal*, **22**, 44–67.

Warburton, J. (1992) Observations of bed load transport and channel bed changes in a proglacial mountain stream. *Arctic and Alpine Research*, **24**, 195–203.

Wells, S.G. and Harvey, A.M. (1987) Sedimentologic and geomorphic variations in storm generated alluvial fans, Howgill Fells, northwest England. *Geological Society of America Bulletin*, **98**, 192–198.

Whiting, P.J. and Bradley, J.B. (1993) A process-based classification system for headwater streams. *Earth Surface Processes and Landforms*, **18**, 603–612.

Whiting, P.J., Stamm, J.F., Moog, D.B. and Orndorff, R.L. (1999) Sediment-transporting flows in headwater streams. *Geological Society of America Bulletin*, **111**, 450–466.

Whittaker, J.G. (1987) Sediment transport in step-pool streams, in *Sediment Transport in Gravel-Bed Rivers* (eds C.R. Thorne, J.C. Bathurst and R.D. Hey), Wiley, Chichester, pp. 545–579, (includes discussion).

Whittaker, J.G. and Jaeggi, M.N.R. (1982) Origin of step pool systems in mountain streams. *American Society of Civil Engineers, Proceedings: Journal of the Hydraulics Division*, **108**, 758–773.

Wilcock, P.R. and McArdell, B.W. (1993) Surface-based fractional transport rates: mobilization thresholds and partial transport of a sand-gravel sediment. *Water Resources Research*, **29**, 1297–1312.

Wohl, E., Madsen, S. and MacDonald, L. (1997) Characteristics of log and clast bed-steps in step-pool streams of northwestern Montana, USA. *Geomorphology*, **20**, 1–10.

Wolman, M.G. and Miller, J.P. (1960) Magnitude and frequency of forces in geomorphic processes. *Journal of Geology*, **68**, 54–74.

Yager, E.M., Kirchner, J.W. and Dietrich, W.E. (2007) Calculating bed load transport in steep boulder bed channels. *Water Resources Research*, **43**, W07418, 24. doi: 10.1029/2006WR005432

Zimmermann, A. and Church, M. (2001) Channel morphology, gradient profiles and bed stresses during flood in a step-pool channel. *Geomorphology*, **40**, 311–327.

3

Landslides and Rockfalls

Nick J. Rosser

IHRR, Department of Geography, Durham University, South Road, Durham, UK

3.1 Introduction: Evidence for the Role of Landsliding in Sediment Cascades and Landscape Evolution

The sediment efflux from hillslopes acts as a boundary condition on the far-field fluvial system (Allen, 2008). The study of the contribution of landsliding to bulk sediment yield falls into two broad areas of interest: first, those studies concerned with the interaction of tectonics and climate on orogen-scale landscape evolution, which by definition have subsumed fine-scale process into wide-scale geomorphological indicators; and second, those studies that focus on the derivation of sediment budgets, which concentrate on material release and transport over (geologically) short timescales. As a central component in the uppermost stages of sediment cascades, knowledge of the role of landslides is restricted by an often weak understanding of the coupling between hillslope and channel, particularly in the case of rock-slope failures (Schlunegger, Detzner and Olsson, 2002). Given the dominance of interfluves in the overall landscape, the apparently simplistic approaches with which the processes that act upon them are predicted when compared to the fluvial system is startling (Korup, 2006a). There is a strong reliance upon inferring failed landslide volume from simple plan-area (Hovius, Stark and Allen, 1997; Guzzetti *et al.*, 2009) and upon *indirect* rather than *direct* measurement of change (Krautblatter and Dikau, 2007). Dynamics of failure are interpreted from commonly relict and reworked deposits (Bertran and Texier, 1999). Increasingly, arguments become entwined in tautological disparities. There are only sparse data on the relative volume of failed material transfer from the landslide deposit (hence, what is 'area'?). There is over-reliance on often poorly constrained fluvial sediment yields down-catchment. Molnar, Anderson and Anderson (2007) also suggest that there has been an overemphasis on the role of tectonics, relative to a consideration of the role of rock fracturing in the process of erosion. Each of these factors become increasingly problematic with increasing scale, particularly catchment size and landslide number (Slaymaker and Spencer, 1998; Otto, Goetz and Schrott, 2008; Guzzetti *et al.*, 2009). This has led to a tendency to rely upon empirical

Sediment Cascades: An Integrated Approach Edited by Timothy Burt and Robert Allison

models for estimates of denudation and sediment delivery, while paying less attention to storage (Otto, Goetz and Schrott, 2008); linking sediment delivery to sediment storage remains problematic, especially at scales relevant to landscape evolution.

As the primary geomorphological process and mechanism of sediment release into the landscape (Densmore *et al.*, 1997; Brardinoni and Church, 2004), an understanding of the controls on the spatial and temporal scales over which a slope is prepared for, and then undergoes, failure is central to constraining landscape-scale cascades of sediment. This is central to all aspects of the process of erosion, which is long established, as illustrated by Dutton (1882; cited in Molnar, Anderson and Anderson, 2007):

> Erosion is the result of two complex process. The first group comprises those which accomplish the disintegration of the rocks, reducing them to fragments, pebbles, sand, and clay. The second comprises those process which remove the debris and carry it away to another part of the world.
> *(Dutton, 1882, p. 64)*

The dominant controls on the rates of denudation due to landslides and rockfalls, and hence the generation of sediment, are manifest via the interplay of changing climate, evolving geology and variable processes in action (Bull, 2009). As such, landslides are not only the result of this interplay, but also their occurrence provides an insight into the relative function of both tectonics and climatic forcing on hillslope adjustment. As geomorphological agents, landslides also present a series of relatively simple-to-constrain characteristics that can and have been exploited to inform our understanding of the spatial and temporal nature of landscape development over extensive timescales.

Burbank (2002) reviews global rates of erosion and identifies the difference between denudation rates determined from lithospheric pressure–temperature–time studies (1–30 mm yr^{-1}) and that of weathering (0.005–0.02 mm yr^{-1}); weathering alone is therefore incapable of sustaining the rapid rates of unloading, suggesting a significant role for bedrock landsliding, rock avalanches and glacial erosion. Montgomery and Brandon (2002) argue that although the suggestion that greater relief and steeper topography intuitively leads to accelerated rates of erosion is an inherently logical assumption, few studies have been able to quantify this at spatial and temporal scales relevant to the widespread evolution of landscapes. Rates of rock uplift and river incision are commonly faster than that of sediment production derived by weathering of intact bedrock (Heimsath *et al.*, 1997), so the majority of mountain slopes are bedrock with only a thin or discontinuous regolith cover; hence the dominant mechanism of sediment production must reflect this. More recently this has been supported by the observation of a strong degree of feedback between the processes of erosion and tectonic forcing, and the view that erosion and topographical relief may become decoupled when (bedrock) landsliding allows hillslope lowering to act at a rate equivalent to fluvial incision (Burbank *et al.*, 1996). Hillslopes are therefore suggested to adjust to uplift and bedrock incision via increased rates of *relief-limiting landsliding*, rather than gradual slope steepening (Montgomery and Brandon, 2002). Burbank *et al.* (1996) identified the consistency in slope angle histograms for the northwest Himalayas, across a region of variable uplift and erosion (1–12 mm yr^{-1}). They inferred that landsliding, assumed to be constant on a geological timescale, responded to rapid river incision, deriving hillslopes closely aligned to their angle of repose. As such, the role of cohesion at this scale is considered negligible such that relative relief, and hence hillslope height, is limited by bulk rock mass strength (Carson and Petley, 1970; Montgomery and Brandon, 2002).

Landslide-dominated sediment yield increases non-linearly with slope, tailing off to an asymptotic level that maintains relief at a geologically determined angle of repose (Howard, 1997; Korup, 2006a). Authors have identified such threshold hillslopes in a range of mountainous landscapes, including the Tien Shan (Sobel *et al.*, 2006), the Himalayas (Burbank *et al.*, 1996), the Andes (Safran *et al.*, 2005), Taiwan (Lin *et al.*, 2008), New Zealand (Bull, 2009), the Tibetan Plateau (Whipple, Kirby and Brocklehurst, 1999) and the Olypmic Mountains, Washington (Montgomery, 2001) (Figure 3.1). Despite this work, as Korup (2006a) identifies, there are no definitive or unambiguous morphometric descriptors of threshold hillslopes, and hence a reliance of indirect measurement such as invariance of mean slope angles with respect to postulated rates of river incision (Safran *et al.*, 2005) or slope angle histograms (Burbank *et al.*, 1996). As analysis techniques evolve, the requirement to subsume site-specific heterogeneous, discontinuous and anisotropic behaviour of rock into geomorphological surrogates will reduce. Furthermore, perhaps as a function of infrequent occurrence, few data are available on landslides themselves that allow us to constrain the denudation rates, and efforts to upscale and link the mechanical controls on patterns of landsliding to geology remain to be addressed (Montgomery, 2001).

The highest sediment yields occur in tectonically active areas, which by definition are notable for regular earthquakes and extreme precipitation; in geomorphologically instantaneous time these trigger landslides and rockfalls, which is the focus here. Firstly, the evidence for the role of landslides in shaping the landscape and attempts to quantify rates of sediment generation is reviewed. Then, the controls upon this process that determine the significance of magnitude, frequency, recurrence, timing, geometry and dynamics of failure on sediment generation of transfer are considered. Critically, these properties define

Figure 3.1 Landslide-limited relief due to bedrock landsliding, Karakorum Highway, Pakistan (photograph courtesy of M. Duhnforth).

when, *where*, *what*, *why* and *how* sediment is produced, conditioned and enters the cascade, which are examined at global, orogen and local scales. The material covered in this chapter is constrained at the slope-toe interface with the drainage network, and at the ridge crest. The focus is particularly on rock slopes in mountainous terrain, as the major global sediment source, where regolith is effectively negligible. Given the explosion in literature dealing with this subject in recent years, the treatment is necessarily selective.

3.2 The Role of Landslides in Sediment Cascades

3.2.1 Landslide Type, Behaviour and Mechanism

3.2.1.1 Landslide Behaviour

In both the geomorphological and geotechnical literature, *landslide behaviour* is used to describe the characteristics of all types of ground failure and mass movement as controlled and driven by gravity (Cruden and Varnes, 1996; Dikau *et al.*, 1996; van Asch *et al.*, 2007). Behaviour dictates the rate of material flux upon and from a hillslope and hence defines sediment production at the point of entry into the fluvial system. A wide variety of observed landslide behaviour derives an equally diverse character of landslide sediment generation. Failure and deformation of landslide material controls the rate of downslope release, the total volume, the sediment–water mix pre-, during and post-failure, and the degree of fragmentation. This complexity is reflected in the diversity of published theories on landslide behaviour suggested previously (e.g. Heim, 1932; Sassa, 1988; Kilburn and Sorensen, 1998; Erismann and Abele, 2001). Interestingly, on actively failing slopes, the average long-term velocity of downslope movement can often be in excess of both weathering and tectonic uplift rates, yet this material transfer is only considered in landscape evolution post-catastrophic failure.

3.2.1.2 Landslide Types

In this context, the most pertinent approach to classification is based upon behaviour, dynamics and material. A plethora of classification schemes is available, but perhaps the most simplistic and convenient in this context is that initially proposed by Varnes (1978), later refined by Leroueil (2001). This is based upon slope movements according to type, material, state and change in state of activity. Types of movement considered were distinguished by Cruden and Varnes (1996) and Dikau *et al.* (1996), who identify five divisions: falls, topples, lateral spreading, slides and flows.

The principal types of movement may act alone, or be superimposed in time and space, leading to a description as 'complex' or 'composite' (Cruden and Varnes, 1996) (Figure 3.2). In such instances, at one instance or location within the failing mass, one mechanism often dominates. Hence, excluding rockfall and topples, many slope failures are not first-time. A wide range of dominant failure types exhibit complex characteristics, including rock and debris avalanches, debris flows, rockfalls and slides, earth flows and shallow bedrock landslides. The majority of failures are mixtures of material including rock and soil, or a mix of the two plus water, ice and air, promoting high levels of post-failure mobility. The

Figure 3.2 Illustration of the local heterogeneity of landslide types, connection and distributions – Arolla, Switzerland
(photo courtesy of A Densmore).

principal consequence of variability of these factors is the diversity of observed mechanical behaviours, the occurrence of large runout distances (Hunter and Fell, 2003) and the extensive lateral spread of deposits, each of which conditions the landslide-generated sediments prior to expulsion (Bertran and Texier, 1999). Landslides are closely influenced by, and feedback into fluvial, hydrological and glacial systems. Uniquely in *landslide-limited relief*, the scale and rigidity of the failing mass has the potential to directly couple the fluvial system and the altitude extremes of local relief (Bigi *et al.*, 2006). Slides are commonly characterized by an ability to erode and deposit sediment when in motion, deriving a complex and poorly understood pattern of topographic interaction, commonly termed *lag-rate* (Crosta, Imposimato and Roddeman, 2003; Stock and Dietrich, 2006), making estimates of *landslide volume* or *area* difficult to define, and harder to measure. The requirement for a theory of landslide sediment transport or a more mechanistic representation of the role of landslides in contributing to sediment budgets is clear. An understanding of the inherited character of the initial failure is therefore paramount to comprehending present movement dynamics.

3.2.1.3 Slope Movement

Pre-failure slope movement is controlled by five broad factors: intact rock properties, *in situ* stress and history, pore fluid pressure, structure and discontinuities, and loading. Slope materials are invariably discontinuous, porous and fractured, anisotropic, heterogeneous and have complex deformation characteristics under high stress and dynamic loading from tectonics, seismicity, and various forms of crustal flexure. The complexity of the constituent parts of landslide materials combined with the long history of formation makes analysis challenging (van Asch *et al.*, 2007). Landslides display a range of styles of behaviour including long-term creep, catastrophic movement preceded by long-term creep and sudden catastrophic failure with no apparent creep phase or trigger (Petley, Bulmer and Murphy, 2002). These observations have been distilled into the concept of *progressive failure*, as first described by Bjerrum (1967), who noted that failure in cohesive materials requires the shear zone to undergo a transition or progression from peak to residual strength, requiring the formation and growth of a shear surface across the footprint of the incipient landslide. Only when this shear surface is fully formed can failure occur. Reported landslide movement patterns are primarily the propagation to the surface of basal deformation of the (developing) shear zone, which may be a discrete band, or the coalescence of nucleated fractures in a zone of shear. The pattern of movement on the slope surface reflects, at least in part, the processes occurring on the basal shear (Terzaghi, 1950), and hence failure mechanism dictates the rate of downslope sediment flux. Note, however, that zero topographic change does not necessarily equate to zero sediment flux if the slope is undergoing planar/translational failure. Mechanism and dynamics hold influence over magnitude and duration of forces to which failing and failed sediments are exposed, which ultimately condition this sediment.

The behaviour of slope movements is time-dependent (Bjerrum, 1967). It is now established that the model of three distinct phases of movement for first time failures is valid (Petley, Bulmer and Murphy, 2002). This pattern of behaviour is found in many natural systems, including the propagation of fracture in rock, deep mine roof failures, ice collapse and the evolution of some volcanic systems (Main, 2000; Kilburn and Sammonds, 2005). The three phases are: *primary movement*, representing the initiation of displacement, characterized by an initially high but decreasing rate of displacement; *secondary movement*, during which the rate of displacement is low but may fluctuate in response to the stress state of the landslide; and, *tertiary movement*, in which the displacement rapidly increases as the landslide accelerates to final failure. Linear behaviour in $1/v$ space, termed 'Satio linearity', is observed during this third and final phase of movement for first time failures in brittle materials (Petley, Petley and Allison, 2008). This behaviour is associated on the microscale with the interrelationship between strain hardening and strain weakening processes (Petley, Petley and Allison, 2008). This behaviour can be mathematically modelled from two mechanisms. First, from state- and rate-dependent creep, derived from the Dieterich–Ruina friction laws (Helmstetter and Sornette, 2004), and second, through a reinforcing or avalanching process of subcritical crack growth during brittle failure (Kilburn and Sammonds, 2005). However, the first hypothesis is unable to explain why linearity is present in undisturbed (i.e. bonded) samples, but not in those where interparticle bonds have been broken or are not present. This is manifested by evidence from failures moving on pre-existing shear planes or as a result of non-brittle mechanisms, which have been shown to undergo exponential increases in strain rate, giving an asymptotic trend in Λ-t space. The implication of this observation is that ductile or non-brittle failures will never develop into catastrophic failures, which holds a tight control over the manner of sediment generation. Tertiary movement represents a period in which strain weakening becomes progressively dominant, leading to complete rupture surface development and thus to final failure (Petley, 2004). The nature of the tertiary movement phase in landslides is primarily dependent on the basal deformation process, but is complicated by, for example, slope topography and discontinuities (Bachmann *et al.*, 2004).

Analysis of landslide movement patterns has shown that the hyperbolic increase in velocity is characteristic of strain development through brittle failure and the formation of a localized shear surface (Saito, 1969), resulting in the extensive bulk strains in the failing mass, which may be responsible for the increase in precursory phenomena at the slope surface due to high differential strains Suwa (1991). As a result of these insights, a model has been generated for the processes of failure in both brittle and ductile landslide systems, based upon the three-phase creep model and the stress ratio (the ratio between the peak shear strength for the normal stress state and the applied shear stress). The ultimate and peak rates of strain in a failing slope, and hence material delivery, are therefore a function of the dominant failure mechanism. Central to this suggestion are several key implications: first, that strain state and hence time may be as equally important as stress state in understanding landslide failure mechanisms, and hence sediment generation. Implicit therefore is the idea that slope failure is strain-dependent, and hence a scale-dependent critical strain, for example a shear zone length, may be a key determinant of the susceptibility to failure. This concept echoes observations of scaling properties in landslide limited relief, and may provide a physical basis for power-law behaviour in landslide inventories. Second, that during the final period of failure (tertiary creep), landslide movement adheres to a linear trend in Λ-t space, regardless of pore pressure variations or effective normal stress state. The controls on the form of this model in the geometrical complexity of real slopes remain poorly understood. The implication for sediment yield is that first-time brittle failures afford the opportunity for effectively instantaneous delivery of large volumes of sediment to the slope toe and beyond, whereas for ductile or reactivations this rate is moderated.

3.2.2 Magnitude, Frequency and Erosion

3.2.2.1 Landslide Magnitude

Deriving sediment yields from steep topography remains problematic, and as a result has received only limited, truly quantitative studies from direct observation (cf. Rapp, 1960; Caine and Swanson, 1989; Schrott *et al.*, 2002; Otto and Dikau, 2004; Beylich *et al.*, 2005; Slaymaker, 2008), resulting in a reliance on exploiting the scaling properties of landslides to model landslide contribution. Landslide flux volumes have been shown to exhibit power-law (Kelsey *et al.*, 1995) or fractal (Hovius, Stark and Allen, 1997) distributions. Failed material is deposited on or at the bottom of the slope, and is commonly not completely removed (eroded) from the scar. For this reason, Guzzetti *et al.* (2009) caution against the use of erosion or denudation rates, instead favouring landslide mobilization rates (φL) (e.g. Hovius, Stark and Allen, 1997; Martin *et al.*, 2002; Guthrie and Evans, 2004; Korup, 2005; Imaizumi and Sidle, 2007; Guzzetti *et al.*, 2009). Landslides commonly dominate erosion and strongly influence slope morphology in steep weak terrain. There remains no universal geomorphological transport law for landslides, greatly confounded by the applicability of universal laws between various failure types. This arises from the inherent difficulties of documenting landslide processes and in linking geomechanical models based upon idealized static conditions to the dynamic landscape. Instability may ultimately be a function of intrinsically stochastic forcing, notably driven by precipitation and earthquakes.

The geometry of a specific landslide is dictated by the variability and combined influence upon effective stress of the slope material, topography and the nature of environmental forcing and trigger. Depth of sliding can be a function of weathering, material strength, infiltration depth, the presence and configuration of discontinuities, and the interaction of environmental forcing or triggers. The plan form geometry of sliding is commonly associated with the spatial heterogeneity of these factors, in combination with lateral restrictors, such as root cohesion, or structural control and confinement that partition the landscape into (more-) stable and less-stable slopes. It is widely recognized that landslides result from the interaction between rock strength (cohesion c and the effective angle of internal friction Φ), slope angles (β), relief, pore pressure and seismic acceleration. In a simplified steady-state limit equilibrium slope stability model, as applied by Schmidt and Montgomery (1995), in the absence of pore pressure or dynamic (seismic) loading, the maximum slope height and hence local relief can be defined by:

$$H_c = \frac{4c}{\lambda} \frac{\sin\beta\cos\phi}{[1-\cos(\beta-\phi)]} \tag{3.1}$$

where, H_c is maximum hillslope height, and λ is the unit weight of the slope material (Burbank and Anderson, 2001). As such, under this closely coupled model, slope-parallel denudation occurs by planar failure, whereby peaks erode at a rate equal to that of channels, hence this approach can be employed to estimate denudation rate.

Lower bounds on failure size relate closely to tensile rock strength, material cohesion, density and orientation of discontinuities, and the removal of surface debris through means other than landsliding, such as mass wasting, deflation, slope wash, chemical slope denudation or micobiological weathering. These fine-scale processes are widely studied as controls on the incremental retreat of high-alpine rockwalls (cf. Krautblatter and Dikau, 2007). Rates of backwearing exhibit high local heterogeneity, implying that simple rates of retreat are not an adequate way of constraining sediment supply, and that the processes that trigger material release (termed *primary rockfalls*) are different to those which ultimately mobilize the released mass (*secondary rockfalls*) (Dorren, 2003), a concept that could be up-scaled to large-scale landslides more widely.

3.2.2.2 Landslide Magnitude and Frequency

Interest in landslide magnitude frequency distributions has arisen from a desire to constrain landslide susceptibility for hazard assessments, in parallel with a theoretical interest in landscape evolution studies using power-law behaviour (Benda and Dunne, 1997). The magnitude–frequency signature defines the contribution of the various landslide event sizes to the sediment cascade. Scaling behaviour allows short-term geomorphological observation to be extrapolated over timescales pertinent to landscape evolution (Wolman and Miller, 1960). Landslide magnitude–frequency relations have long been recognized to display power-law (fractal) scaling properties whereby the number of landslides is a negative power function of the landslide size (Fuyii, 1969; Sugai, Ohmori and Hirano, 1994; Hovius, Stark and Allen, 1997), which transcend a range of morphometric attributes of failures, from scar area (Hovius *et al.*, 2000), or area of disturbance including the plan extent of the deposit (Pelletier *et al.*, 1997; Guzzetti *et al.*, 2002), to displaced volume (Hungr, Evans and Hazzard, 1999; Dai and Lee, 2002). The pattern in the distribution is also sustained across a range of timescales, from temporal extensive inventories (Pelletier *et al.*, 1997) to single-trigger multifailure events (Harp and Jibson, 1996). If the exponential coefficient b is greater than 2, then the bulk of the cumulative landslide sediment volume is comprised of small failures; hence small-failure (rockfalls/mass wasting) mechanisms dominant sediment yield. Conversely if b is less than 2, large landslides, although few in number, dominate the cumulative volume generated (Brardinoni and Church, 2004). When customarily plotted on logarithmic scales such distributions display additional elements, notably the much debated *rollover* in frequency of smaller landslides (Malamud *et al.*, 2004) (Figure 3.3). The consistent presence of the rollover defines some interesting qualities that occur across a variety of inventories, under a range a triggering mechanisms and

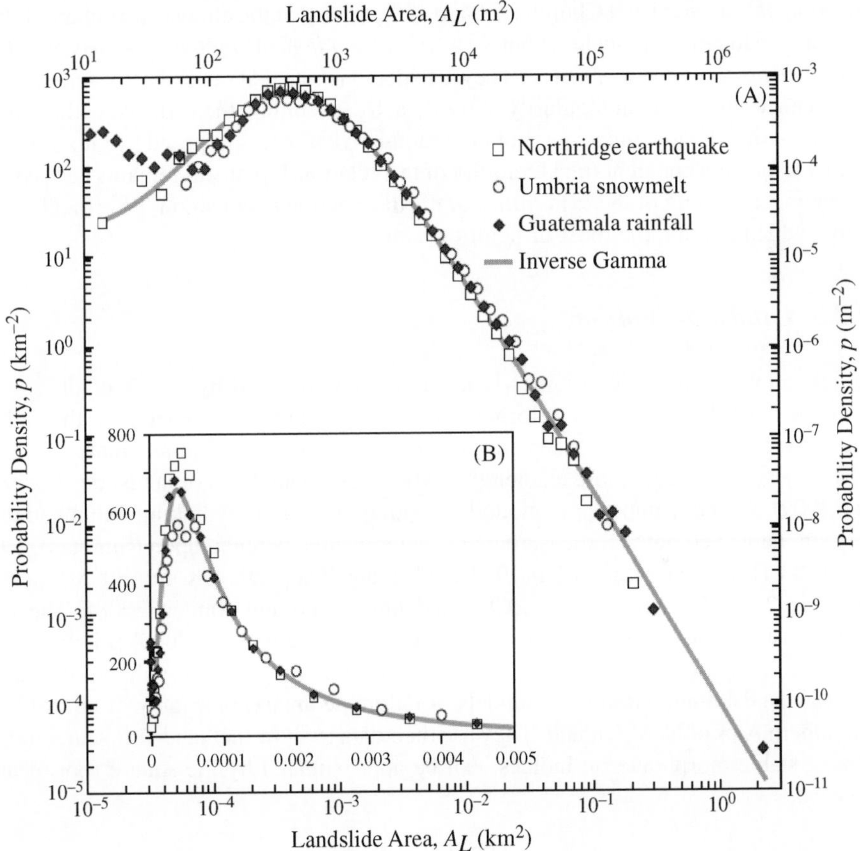

Figure 3.3 Dependence of landslide probability densities p on landslide area AL, for three landslide inventories: (1) 11 111 landslides triggered by the 17 January 1994 Northridge earthquake in California, USA (open squares) (Harp and Jibson, 1995, 1996); (2) 4233 landslides triggered by a snowmelt event in the Umbria region of Italy in January 1997 (open circles) (Cardinali *et al.*, 2000); (3) 9594 landslides triggered by heavy rainfall from Hurricane Mitch in Guatemala in late October and early November 1998 (closed diamonds) (Bucknam *et al.*, 2001). Probability densities are given on logarithmic axes (A) and linear axes (B). Also included is our proposed landslide probability distribution. This is the best fit to the three landslide inventories of the three-parameter inverse-gamma distribution
(after Malamud *et al.*, 2004).

environments. Debate surrounds the source of this rollover, with some citing an artefact of survey resolution (Hovius *et al.*, 2000) or scale of observation (Stark and Hovius, 2001). Others explain this physically; for example, Pelletier *et al.* (1997) attributed it to slope material strength, in particular a shift from the dependence of failure on friction for large landslides to cohesion-controlled failure for small landslides. The peak of the rollover implies a universal mean area landslide, and a predictable population of failures in any given multifailure event, independent of the size of the event (Malamud *et al.*, 2004).

This approach is particularly appealing in this context as it permits total volumes to be predicted where there is known undersampling of the full size spectrum, either due to infrequency, fragmented data or limited persistence. The display of this type of data is contentious; Brardinoni and Church (2004, p 120) argue that the almost customary plotting of the data on logarithmic scales is purely a '*trick for adjudicating the goodness of fit from an assumed power law distribution*'. They also demonstrate that the choice of sampling bin size influences the apparent regularity of the data, by subsuming the variance within binned totals, favouring the more recent use of non-cumulative plots as proposed by Guzzetti *et al.* (2002). Despite the apparent transferability of this relationship, it still remains necessary to have an understanding of the extremities of the distribution to constrain the critical central portion, which is often the most difficult to sample.

3.2.2.3 Landslide Volume

The derivation of landslide volumes is inherently complicated by a lack of differential topographic data, the need for information on the subsurface geometry of the failure, occlusion and shadowing in steep terrain and the residence of failed material in the landslide scar, each made more challenging with large populations of landslides (Guzzetti *et al.*, 2009). As such, empirically modelling scaling behaviour by linking volume to more easily attainable geometrical measurements, notably area, is relied upon (Simonett, 1967; Innes, 1983; Hovius, Stark and Allen, 1997; Guthrie and Evans, 2004; Korup, 2005; Imaizumi and Sidle, 2007; Guzzetti *et al.*, 2008; Imaizumi, Sidle and Kamei, 2008). The most comprehensive review of this subject is provided by Guzzetti *et al.* (2009) (Figure 3.4).

Hovius, Stark and Allen (1997) developed an estimate of the total volume of material eroded by landsliding, based upon models of failure geometry, in a study of landslides in the Southern Alps of New Zealand. They use their data to show that landslide scar width, as the most stable morphometric indices, can be approximated by the square root of area. Hence:

$$n_c(l \geq l_c) = \kappa l_c^{-2\beta} \tag{3.2}$$

where, n_c is the number of landslides, l is the scale length, and κ and β are constants. Field assessment derived a linear width-to-depth relationship for mean landslide mass thickness t:

$$t(l) = \varepsilon l \tag{3.3}$$

The volume discharge at any given length scale (l) is therefore defined as:

$$V(l) = n(l)A(l)t(l) = \varepsilon l^3 n(l) \tag{3.4}$$

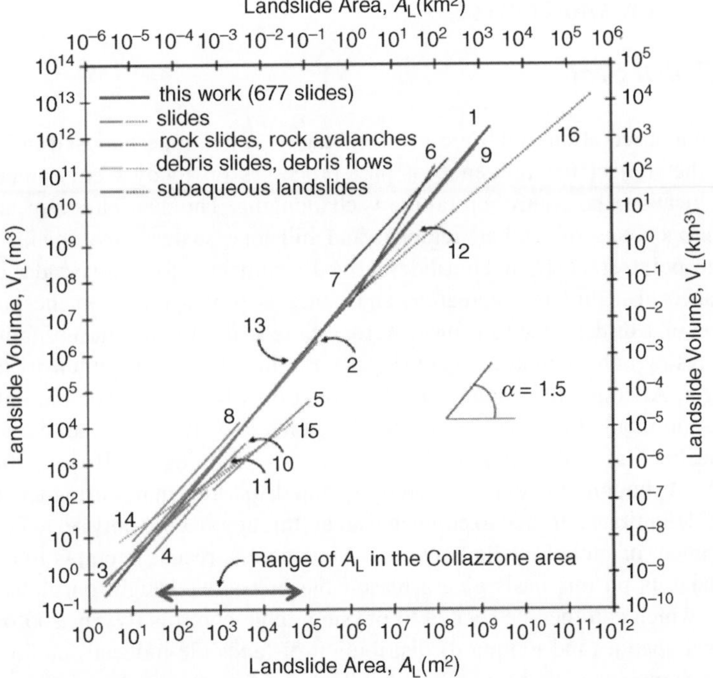

Figure 3.4 Empirical relationships proposed in the literature to link landslide area AL (x-axis) to landslide volume VL (y-axis). Colours show relationships for different types of landslides. Continuous lines are dependencies proposed by authors. Dotted lines are dependencies obtained from empirical data published in the literature or made available to the authors (after Guzzetti *et al.*, 2009).

where $n(l)$ is the number distribution of landslides of length (or width) l, obtained from the cumulative distribution of Equation 3.4, by $n(l) = \mathrm{d}n_c/\mathrm{d}l$. The volume delivered from the area under consideration A_r is therefore:

$$V = 2\beta\varepsilon\kappa \int_{L_0}^{L_l} l^{2-2\beta}\,\mathrm{d}l \tag{3.5}$$

where L_l is the maximum possible width of a landslide in the region and L_0 is the minimum. These models have been refined with a more elegant length–depth scaling relationship, and a precise definition of the length scaling limits over which landsliding can occur. Guzzetti *et al.* (2009) refine this using a global catalogue of 677 slides (as defined by Cruden and Varnes, 1996) in which they measure both area and volume to derive a scaling exponent of 1.450 across eight orders of magnitude of area and twelve orders of magnitude of length, and hence landslides almost described self-similar behaviour (where $\alpha = 1.5$) (Figure 3.4). The authors differentiate between the scaling properties of slides, rock slides and avalanches, debris slides and flows, and subaqeous slides. Guzzetti *et al.* (2009) use this to argue that the relationships identified are principally geometrical, and not significantly influenced by geomorphological or mechanical properties of the landslide material, or by landslide type, but this has yet to be verified on a large-event inventory that exhibits multiple lithologies and failure types.

3.2.3 Location and Control

3.2.3.1 Global Scale

Invariably, our observations today are an assemblage of relict, superimposed and superseded landforms that reflect the influence of past processes on today's environment (9), so describing location and control on failure is challenging. The distribution of landslides is highly heterogeneous on global, regional and hillslope scales, but remains critical to defining the point of entry of landslide-derived sediment into the cascade, defines the travel distance of sediments thereafter, and hence is proportionate to the potential for modification of sediment during transit. Across these scales the location of failure closely relates to physiographic conditioning of slopes for failure, but also the intensity distribution of the trigger. Attempts to characterize the global distribution are notable only by their absence, so commonly there is tendency to rely upon relatively crude secondary indicators, such as landslide-induced damage or fatalities (Petley, Dunning and Rosser, 2005; Nadim *et al.*, 2006). Although inherently biased by spatial disparities in reporting and nomenclature, and by definition restricted to populated areas, this approach is only set to improve with the development of global media networks. Nonetheless, recent attempts to collate long temporal datasets on this basis give a reasonable first-order insight on global landslide occurrence, which can then be related to physiographic controls (Oven, 2005).

The global spatial (and temporal) distribution of landslide-induced fatalities strongly reflects the dominance of the combined influence of relative relief and precipitation, delimiting the concentration of failures in the world's major mountain ranges, exacerbated by intense precipitation, and moderated by population density and vulnerability (Petley, Dunning and Rosser, 2005; Figure 3.5). Notably, there is concentration on the middle Himalayan arc, central China, southern and southeastern and central Asia, and the Andes. The physiographically controlled distribution is reaffirmed by the absence of events across low-relief continental interiors, yet still retains the clarity to depict regional-scale foci such

Figure 3.5 2008 recorded global landslide fatalities (International Landslide Centre, Durham University).

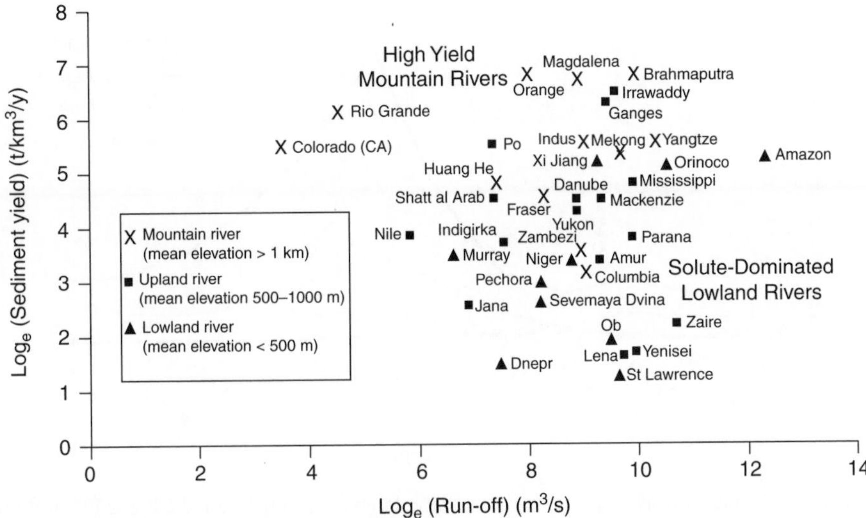

Figure 3.6 Global sediment yields, indicating the contribution of landslide dominated catchment of mountainous rivers
(data provided by P. Allan).

as the African Rift system, or the European Alps. Oven (2005) modelled the physiographic conditions at the site of each landslide-induced fatality, and extrapolated this across the globe. The analysis shows good association to global sediment yields derived from rivers (Milliman and Meade, 1983; Jha, Subramanian and Sitasawad, 1988; Milliman and Syvitski, 1992; Burbank and Anderson, 2001) (Figure 3.6).

Temporally, the longest-scale evidence for variations in landslide sediment yields are derived from palaeoenvironmental studies (Figure 3.7). For example, McCarroll, Shakesby and Matthews (1998, 2001) note accelerated rockfall rates in talus aggradation in response to climatic deterioration during the late Holocene. On the short-term (annual), the global database mirrors mesoscale climatic forcing, such as the southern Asian and eastern Asian monsoon and typhoon seasons, as indicated by persistent year on year seasonality (Petley, Dunning and Rosser, 2005) (Figure 3.7). Attempts to quantify the contribution of slope failures to total fatalities in large complex events such as Kashmir 2005 or Sichuan 2008, are inhibited by the complexity of the event. In Kashmir Petley *et al.* (2006) estimated a total of 79 000 (25%) were landslide induced, yet it is notable that an accelerated frequency of fatalities has been observed in the subsequent years.

3.2.3.2 Orogen Scale

At the orogen scale, the location of landsliding is closely linked to the control on mountain building and relief generation, notably precipitation and seismicity. The distribution of landsliding has been studied both vertically with altitude (Bull, 2009) and relief (Montgomery and Brandon, 2002; Korup *et al.*, 2007), and geographically with respect to structure (Guzzetti, Cardinali and Reichenbach, 1996) and forcing (Hovius, Stark and Allen, 1997; Petley, Petley and Allison, 2008). Montgomery and Brandon (2002) identified

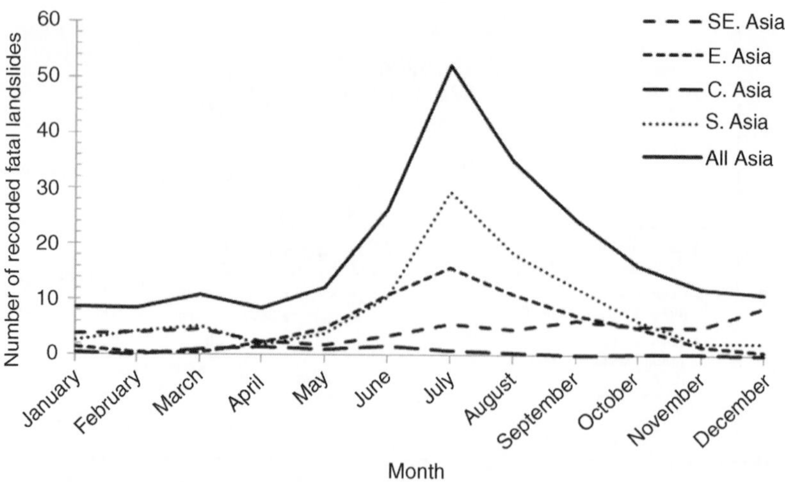

Figure 3.7 The monthly average occurrence of fatal landslides in the period 2002–2007. Note that the data does not include the Kashmir earthquake (International Landslide Centre, Durham University).

orogen-scale limiting relief at 1500 m, and altitude that differentiated those regions with the highest rates of bedrock landsliding driven erosion from those areas with sufficiently high material strength to maintain relief. They summarize this relationship in the model relating mean local relief to erosion rate, as shown in Equation 3.6:

$$E = E_o + \frac{K\bar{H}}{[1-(\bar{H}/H_c)^2]} \tag{3.6}$$

where E is the erosion rate, H is the mean local relief, E_o is the erosion rate due to chemical weathering, K, is a constant, and H_c is the limiting relief (after Korup *et al.*, 2007).

In a study of 300 of the largest known landslides spanning the Himalaya, the European Alps, the Tien Shan–Pamir, the Andes and New Zealand, Korup *et al.* (2007) observed that globally in excess of 50% of the largest landslides occurred in the steepest 15% of mountainous terrain (Figure 3.8). In the Himalaya more than 50% of landslide events occurred in the steepest 6%, and >70% of the total landslide volume was found in the steepest 5%. At this scale the vertical distribution of environmental forcing plays a key role, defining the dominant preparation and triggering processes of material release. For example, Bull (2009) identifies the potential role of the ELA in determining the dominance of rockwall retreat mechanisms on high-altitude free faces.

The geographical distribution of landsliding at the orogen scale has been closely linked to mesoscale climatic forcing, notably highly consistent patterns of orographic rainfall in the Himalayas (Anders *et al.*, 2006), the Alps (Frei and Schär, 1998) and the Olympic Mountains (Anders *et al.*, 2004), and has moved towards a theory of orographic precipitation (Smith, 2006). The links between meso- to local-scale climate and erosion processes (Reiners *et al.*, 2003) are believed to be highly sensitive to spatial and temporal distribution of precipitation (Lague, 2003). Enhanced stress relaxation has also been identified as a contributor to orogen-scale variability in rockfall activity rates (André, 1997). Research has

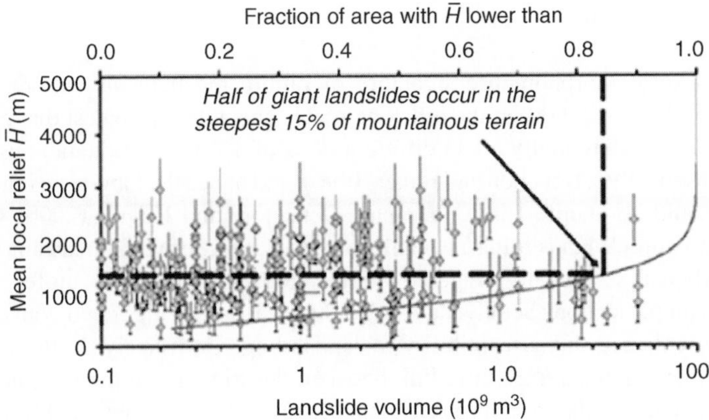

Figure 3.8 Relationship between mean local relief and landslide volume (grey points with 1σ bars). The grey curve is the sum of cumulative distribution of H in all mountain regions studied (after: Korup *et al.*, 2007).

also identified close spatial relationships between the proximity of landsliding to the spatial focusing of seismic triggers (Keefer, 1984). Keefer (1984, 1994) proposed an empirical power-law relation between the logarithm of the total volume of landslides generated by an earthquake and the earthquake moment magnitude. Malamud *et al.* (2004) uses the seismic intensity factor I_M of Kossobokov *et al.* (2000) to model the potential contribution of landslides in terms of volume rate due to global seismicity based upon the relationship between the distribution of earthquakes, as described by the Gutenberg–Richter parameter, and maximum earthquake moment magnitude (Gutenberg and Richter, 1954; Guzzetti *et al.*, 2002). This allows erosion rates to be obtained from the scaling properties of landslides and the moment magnitude of earthquakes in any given region. Malamud *et al.* (2004) use this approach to demonstrate the potential contribution of seismically induced landslides in active subduction zones to be within the range of 0.2–7 mm yr^{-1}, and contrast this with predictions for areas adjacent to strike-slip faults in the range 0.01–0.7 mm yr^{-1}.

In a study of landslides triggered by ruptures of three thrust faults, the Northridge Earthquake California, the Chi-Chi earthquake in Taiwan and two earthquakes in the Finisterre Mountains in Papua New Guinea, landslide densities were shown to be highest in those areas that experienced the strongest ground acceleration, and tended to decay with distance from the epicentre (Meunier, Hovius and Haines, 2007). In the cases of California and Taiwan, the density of landsliding was linearly correlated with the vertical and horizontal components of peak ground acceleration; hence landslide density on a regional scale is proposed to be proportional to seismic attenuation. In a study of the aftermath of the 1999 Chi-Chi earthquake Lin *et al.* (2008) showed that this relationship is retained in the subsequent rates and patterns of sediment redistribution. Wider observations, particularly in light of the well documented and more recent 2005 Kashmir and 2008 Sichuan earthquakes, suggest a significant control of geology, topography and rupture dynamics on landslide distribution (cf. Kirby, Whipple and Harkins, 2008; Owen *et al.*, 2008; Huang and Li, 2009; Yin, Wang and Sun, 2009).

3.2.3.3 Local Scale

The regional scale morphometric signature of failure mechanisms was recognized by Densmore and Hovius (2000) at the hillslope scale. The authors suggest that precipitation and earthquakes preferentially affect different parts of a slope in mountainous terrain (cf. Iverson and Reid, 1992; Bouchon and Baker, 1996), and hence the long-term impact of these processes should be manifest in topography. Densmore and Hovius (2000) suggest that storm-driven bedrock landsliding near hillslope toes will derive and sustain inner gorge structures, whereas seismically triggered failures produce more planar, uniform hillslopes. Topographic amplification (Sepúlveda *et al.*, 2005) on valley ridges and convex breaks of slopes act to enhance the susceptibility to landsliding. Conversely, in the case of rock avalanches, there is apparently little link between the trigger and the morphology of the deposit, particularly in the case of rock avalanches (Strom, 1998; Hewitt, 2002), supporting the use of scar morphology as a better indicator or artefact of the trigger. For example, rock avalanche scars appear to be much deeper than the more shallow source areas of those associated with aseismic triggers (McSaveney and Davies, 2004; Dunning, 2006), although at the finer scale a close association of rockfall size-dependence upon seismicity has been observed (Vidrih and Ribicic, 2004).

Inasmuch as the trigger reflects location, studies have identified close correlation between ameliorative geological conditions for landslide occurrence, and ultimate landslide mechanism and form. In a study of the *bedrock topographic signature*, Korup (2006a) uses plots of probability distributions of two slope metrics, slope angle (β) and slope gradient ($\tan\beta$) to identify median slope angles across a region of variable uplift, erosion, landsliding density, precipitation and glaciation, with the only unifying variable being rock type; the suggestion arises that local hillslope evolution adjusts to rock mass strength irrespective of the intensity of tectonic and climatic forcing (Figure 3.9). Korup (2006a) cautions that apparent correlations between the density and location of contemporary landsliding may not be indicative of the long-term erosion rates particularly in small catchments where the frequency of one geomorphological process in small ($<100\,\text{km}^2$) catchments greatly affects yield (Gomi, Sidle and Swanson, 2004). This is supported by observations of exponential growth of landslide density of earthquake-triggered landslides with mean slope angle in post-Niigata Earthquake, Japan (Yamagishi and Iwashashi, 2007), as opposed to a more linear increase with rainfall-induced failures.

Local-scale controls on location of landsliding are diverse and well understood, ranging from lithology (Guzzetti, Cardinali and Reichenbach, 1996), tectonic structures and lineaments (Eisbacher, 1971; Friedman, Kwon and Losert, 2003; Brideau, Yan and Stead, 2009), local bedrock outcropping (Tarolli, Borga and Fontana, 2008), fluvial incision promoting feedbacks ranging from channel uplift to knick-point migration (Bartarya and Sah, 1995; Bigi *et al.*, 2006), glacial debutressing (Oppikofer, Jaboyedoff and Keusen, 2008) and human modification. The finest level of detail in the spatial distribution and connectivity of sediment transfer is provided by a sediment budgets approach (cf. Johnson and Warburton, 2006) or hypsometric *toposequences* (Schrott *et al.*, 2002; Rasemann, 2004; Otto, Goetz and Schrott, 2008). A topographic succession of landforms is passed by a virtual gravity-controlled particle and used to identify key altitudinal levels of transport and deposition, that represent the structure of the sediment cascade in the wider catchment, and which can in turn be associated with sub-systems in the catchment. The distribution of depositional landforms differs at each level, dependent upon the process regime and the

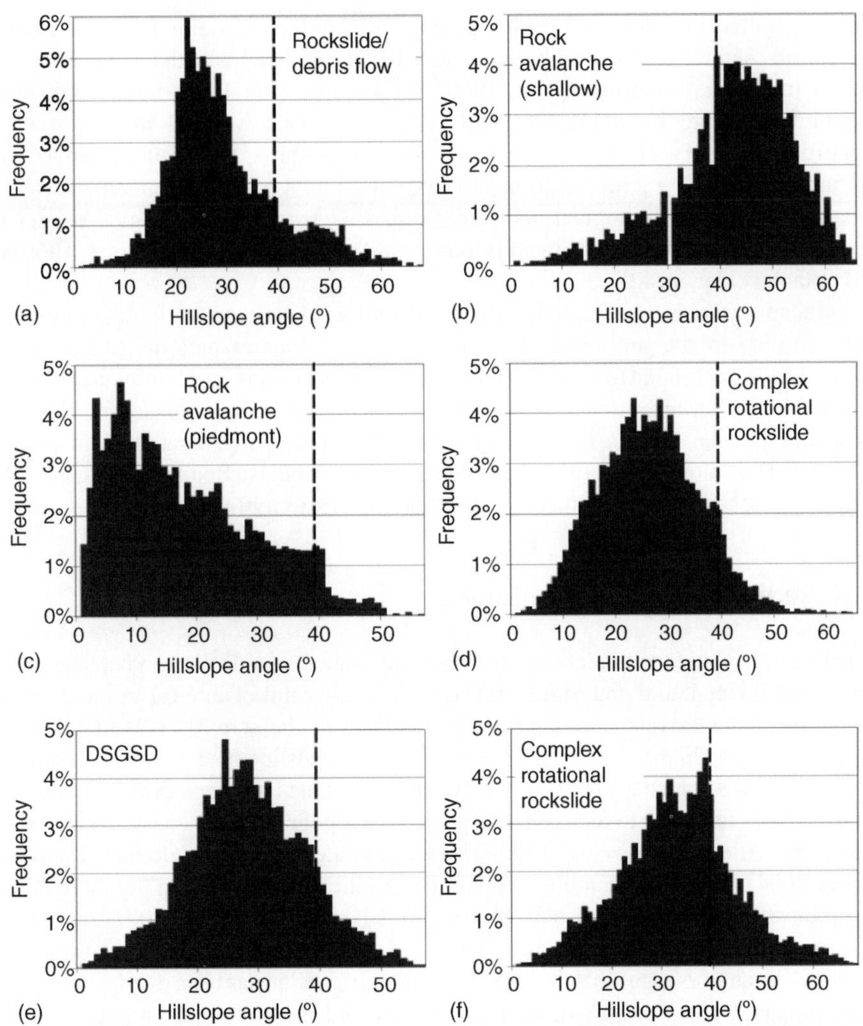

Figure 3.9 Area-normalized slope histograms derived from a 25-m DEM of large bedrock landslides. (a) Ruera, Copland R. (b) Mount Adams, Poerua R. (c) Selbourne Spur, Waiatoto R. (d) Misty Peak, Karangarua R. (e) Bealy Range, Haast R. (f) Hyperia, Waiatoto R. Dashed lines denote regional modal slope $\varphi_{mod} = 39°$ (all derived at 75-m length scale). Note that shallow failures (b) and those with low displacement of rock masses (f) have modes at φ_{mod} (after Korup 2006a).

landform dependency on altitude and relief. Using this approach, Otto, Goetz and Schrott (2008) suggest that the age of depositional features, by definition, increases with distance from ridges to valleys, where larger valleys have been deglaciated for longer than shorter valleys. The complexity of depositional landforms is suggested to increase downslope, an observation that is supported by the common dominance of single landforms in hanging valleys versus the compound nature of deposits of different process origin in large valley bottoms. Many processes are also significant contributors to the reworking of previously failed material, and derive greater bulk sediment yields from increasingly finer calibre sediments as clast volume to surface area ratios reduce. Demonstrated controls include rain storms (Rapp, 1960; Sandersen *et al.*, 1996; Sass, 1998), daily temperature cycling including ice melt (Gardner, 1980), vertical migration of the permafrost boundary (Haeberli, 1999), macrogelivation – the exploitation of rock structures by ice, and microgelivation – the exploitation of rock independent of structure (Tricart, 1956), microbiological action (André, 1997), root prying (Sass, Wetzel and Friedmann, 2006), animal movement (Gorers and Presen, 1998), creep (Rapp, 1960) and wind (Blackwelder, 1942).

Significant effort has been made in trying to relate short-term variations in environmental forcing to the timing of mass movements. Temporal patterns of failures are commonly closely related to the distribution of triggering events, predominantly storms or earthquakes, which are often assumed to occur stochastically in space and time (e.g. Benda and Dunne, 1997). Time series of shallow failures generated from physical modelling (Densmore *et al.*, 1997) and physics-based models (Benda and Dunne, 1997) are highly stochastic, as explained by an inherent sensitivity to local conditions.

3.2.4 Sediment Calibre and Facies

The nature of landslide sediments controls their subsequent reactivation and uptake into the fluvial system (Imaizumi and Sidle, 2007). Despite a wealth of interest in the grain-size distribution of rock avalanche deposits, particularly in the stability of valley-blocking failures (Dunning, 2006), the expertise and interest in landslide mechanisms and sediment in engineering geology has not been widely reflected in the landscape evolution literature, which has had a tendency to focus on bulk yield from topographically constrained landslide geometries (Allen and Hovius, 1998; Hovius and Leeder, 1998; Hovius, Lague and Dadson, 2004). This is made more surprising given the historic focus in fluvial hydrology upon grain size, sorting and transport. Molnar, Anderson and Anderson (2007) argue for the significance of the role of fracturing of rock in optimizing the efficacy of post-failure transport. The authors argue that the role of tectonics in erosion has been overly emphasized, when compared with the role of rock fracturing, and make a strong argument for the influence, and hence consideration of, disintegration of rock in erosion. Only a limited literature addresses the characteristics of landslide sediments or (micro-) facies (McSaveney, Davies and Hodgson, 2000). Of note is Bertran and Texier's (1999) detailed review of such deposits, categorized by failure type, velocity and rheological characteristics that led to emplacement, which could be treated as of equal importance in a consideration of sediment grade and mobilization. Here, this categorization is developed to consider landslide sediment condition in the sediment cascade.

3.2.4.1 Low Sediment Concentration, Interstitial Fluid = Air (e.g. Rockfall Deposits)

These deposits are characterized by a random fabric with a coarse clast-supported matrix that exhibits limited structure, with only a weak orientation tendency parallel to the slope (Bertran, Francou and Texier, 1995; Bertran *et al.*, 1997). Modifications to clast size result from impact fragmentation, controlled by impact surface material, orientation and the resulting coefficient of restitution. Deposits are characterized by high fluid mobility, the preferential removal of fines, and relative stability controlled by clast-to-clast contacts. The retention of these features in the landscape is commonly reduced by co-emplacement of other less stable mass movement deposits, burial and removal.

3.2.4.2 High Sediment Concentration, Interstitial Fluid = Air. (e.g. Grain Flows and Rock Avalanches)

Deposits arising from grain flows (dry cohesionless flows of debris on steep slopes (Whitehouse and McSaveney, 1983) commonly form levées, fine-grained channel deposits, and frontal lobes, typically exhibiting highly permeable stratified deposits with inverse grading due to dispersive pressure and dynamic sieving while in transport, clast imbrication and slope parallel orientation (Bertran *et al.*, 1997). These materials are therefore easily reworked by low levels of surface wash or flow.

Rock avalanches have unique sedimentological features. The interior of the deposit is commonly highly fragmented to a silty/sand grain size, with limited mixing that commonly preserves the *in situ* structure of the source. They exhibit *in situ* fractured clasts, a basal layer with an abundant cataclastic fine matrix often with distinct zones of shear, and have high fissural porosity (Bertran *et al.*, 1997; Dunning, 2006). The deposit surface commonly has a coarse-clastic carapace, leading to a common misunderstanding of the deposit stability (Dunning, 2004). Such deposits therefore can represent long-term persistent features in the landscape, altering river profiles and building sediment wedges (Korup, 2006), and can lead to a catastrophic release of highly mobile sediment into a fluvial system. Considerable debate surrounds the mechanisms responsible for the generation of this unique sedimentology and accounts for the extensive runout of rock avalanches. For example, Kent (1966) suggested that entrained air maintains a distance between the blocks, reducing friction; Shreve (1968) suggested that the landslide 'lifts off' and is lubricated allowing air to be trapped in the basal region; Goguel (1978) suggested that frictional heating generates steam from pore water which provides buoyancy to the failing mass; and Melosh (1979) suggested acoustic energy in the form of high-pressure vibrations is trapped in the basal region again reducing friction. Most recently, Davies, McSaveney and Deganutti (2007) examine the potential for point loading of debris promoting fragmentation, which then provides energy to the mass and reduces friction during movement.

3.2.4.3 High Sediment Concentration, Interstitial Fluid = Water or Water + Fine Particles

Liquefaction (e.g. Debris Flows) These deposits are characterized by lateral levées with crude stratification of coarse openwork lenses in a diamicton (channel lags, pavements due to post-depositional run-off activity) (Van Steijn *et al.*, 1995). Downslope, sequential deposition of lobes build fan sequences, and crude horizontal to slope-parallel bedding with frequent intercalation of palaeosols and overland flow deposits. Clast fabric shows a weak to moderate orientation to flow direction, which is often porous initially after deposition but poorly connected and subject to reworking. Lateral levées are mechanically strong, and as such their presence acts to channelize subsequent flows (Nieuwenhuijzen and Van Steijn, 1990).

No Liquefaction – Sliding Plane (e.g. Earth Slides and Flows) The material strength of earth-slide deposits determines the degree to which failure and movement fragments the deposit. When the disintegration of the failed mass is low, material is commonly tilted in the upslope direction. In more dynamic failures in rock, cataclastic interactions lead to fracturing and fragmentation. Soils that have undergone creep show initially a reduced size and then a complete closure of void spaces. In failed material with intergranular bonds broken, clay domains orientate increasingly with time and microcracks develop.

In each of these subdivisions, there is clear control of failure type and material, which in turn influences the character of the landslide deposit and constrains its subsequent mobilization. Field-parameterized landslide transport laws should take into account the link between bedrock, landslide mechanism and sediment character modification. Sediment transfer results in different sediment storage configurations on slopes and in valley bottoms, ranging from talus and debris cones, alluvial fans, and rockfall deposits. The roles of each of these storage types in the wider sediment cascade and the variability in space and time are poorly constrained (Dietrich and Dunne, 1978; Schrott *et al.*, 2002). This arises in part from variable residence times and episodic removal, in addition to the buffering capacity of a catchment to hold sediment. Hewitt (2009) presents this in respect to massive rock avalanches in what he terms the 'landscape interruption epicycle'. In such instances, the coseismic release of large amounts of sediment into the fluvial system lacking in transport capacity may result in protracted removal histories, obscuring the climatic history. Coupling between various sources and sinks is therefore of vital importance in understanding the function landsliding in sediment cascades (Caine and Swanson, 1989). Our understanding is limited by our ability to accurately quantify the volumes of sediment in catchments (Hoffmann and Schrott, 2002), and the infrequent nature of formative events which result in only fragmented preserved chronologies.[[]]

3.2.5 Time

Landslides can act as very non-uniform and non-linear supplies of sediment to a fluvial system (Kelsey, 1980), with apparently stochastic temporal distributions leading to the assumption of local fine-scaled controls on failure. Time is probably the least well constrained parameter in landslide sediment generation and efflux. To address this, some general characteristics of landsliding occurrence can be considered. First, despite the wide range of triggering factors in a diversity of environments, resulting inventory and morphology diagnostics remain consistent. Second, the trigger of many slope failures remains elusive, with lag times or no apparent trigger, yet much work is focused upon establishing critical rainfall thresholds, or seismic shaking characteristics to confine when any given event may occur. After a multifailure event, such as post-2008 Sichuan earthquake, many slopes fail, yet many remain intact, albeit intrinsically damaged. In these instances the subsequent sequencing of environmental conditions has profound geomorphological implications for sediment mobilization, as has been dramatically witnessed in post-Chi Chi Taiwan (cf. Lin *et al.*, 2008).

A critical consideration is the stress (and strain) history of slopes within the landscape, and recognition of the time-dependence of processes in action. It can be assumed that all slopes have the ability to fail, when exposed to sufficient levels of forcing, as a function of the accumulation of damage. It therefore may be realistic to assume that not all slopes in a landscape are at equal levels of damage, or maturity, despite equal geological age. Hence, under the conditions of a catastrophic trigger, some slopes will fail in *clearing events* (Densmore *et al.*, 1997), yet others may remain superficially intact. This framework is developed in the model depicted in Figure 3.10.

During period 'a' the rock mass undergoes a background level of surface material detachment as a product of weathering and physical detachment of surface material. Many slopes may never experience sufficient environmental forcing to deviate from this state. Environmental events or perturbations such as storms instigate an increased number of (relatively small) landslides and rockfalls. The result is a redistribution of stress within the rock mass, leading to progressive damage accumulation via microcrack development (Figure 3.10, stage 'b'). Damage incurred is considered both cumulative and largely irreversible, although it may reduce when dilated joints are recemented by infill. Periods of relative quiescence in rockfall activity are be mirrored by events of variable intensity, duration and distribution (Figure 3.10, stage 'c'), each of which act to push the rock mass towards a critical strain-related threshold in response, marked in Figure 3.10 by the dashed horizontal line. Allen (2008) defines activities during stages a–c as 'buffered' or 'reactive' dependent on the ratio of response time to repeat time of perturbation, and 'steady' or 'transient' dependent on the ratio of response time to time elapsed since the most recent change in boundary conditions. In some instances, the rock mass may reside in a condition very close to instability in the absence of an event to instigate the transition to instability (Figure 3.10, stage 'd'). During particularly acute environmental conditions, this period may be rapid. The rock mass may, however, exist in a state close to this threshold for a significant period without entering the final phase of failure, potentially explaining why many failures occur without an apparent trigger. After the threshold is crossed, here defined as the point after which the dominant controls on rock mass failure shift from external to internal, the rock mass enters a final tertiary phase of failure, characterized by a hyperbolic

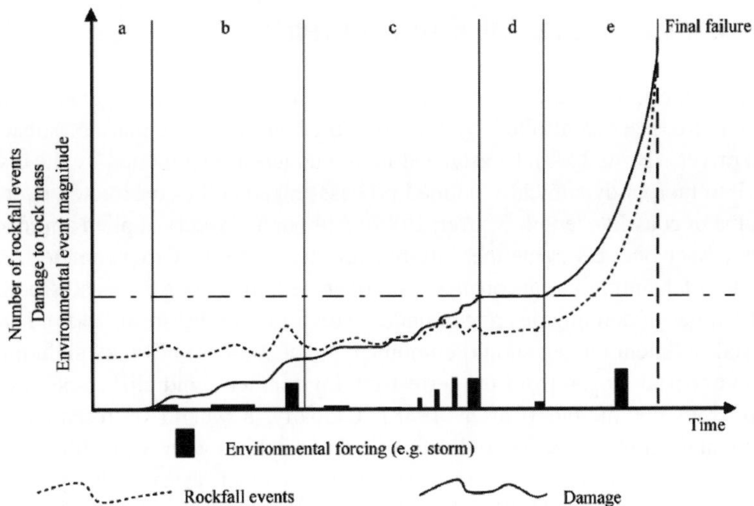

Figure 3.10 Conceptual model of progressive accumulation of damage in a failing slope (adapted from Rosser *et al.*, 2007).

increase in the accumulation of strain (Figure 3.10, stage 'e'). Rosser *et al.* (2007), amongst others, have observed the occurrence of increasingly large rockfalls in the months prior to slope failure, reflecting this pattern. Environmental events during this period have little or no discernable influence on the rate of strain accumulation; the failure would appear to adhere to a linear relationship with inverse velocity against time space. All sections of the landscape could be suggested to reside at some point on this curve as a function of the ratio of accumulated damage versus critical level required for failure at each location. The time between the point of transition from the secondary to tertiary phase of behaviour here appears to be scale dependent (Petley, Petley and Allison, 2008) but it remains difficult to identify the shift in behaviour, particularly in the early tail of the hyperbolic curve given the variability of the baseline rate of rockfall activity. Notwithstanding this, the magnitude and timing of precursors is scale dependent. The relationship between small and large events in brittle systems has been previously addressed, particularly with reference to earthquake triggering (Main, 2000; Helmstetter *et al.*, 2004; Malamud *et al.*, 2004).

This pattern of behaviour is attributed to time-dependent processes that result in a progressive accumulation of damage (Brideau, Yan and Stead, 2009). These processes are as diverse as they are numerous, but each acts at geological significant rates comparable to those of mass wasting, weathering and uplift. These include brittle movement driven by, for example, communition (Locat *et al.*, 2006), ductile movement due to stratigraphic sequencing (Evans, 1983), creep driven by subcritical crack growth (Kemeny, 2003), fatigue in response to cyclic loading (Potyondy, 2007), thermal cycling (Nagaraja Rao and Murthry, 2001), tectonic faulting (Bergbauer and Pollard, 2004), anthropogenic alteration from excavation or blasting (Hoek and Bray, 1981), physico-chemical modification (Borrelli, Greco and Gulla, 2007), and geomorphological changes that alter landscape-scale stress distributions, such as fluvial erosion (Kinakin and Stead, 2005; Bachmann, Bouissou and Chemenda, 2004, 2006).

3.3 The Rock-Slope Contribution to Sediment Cascades in Coastal Environments

Coasts represent one of the most complex environments, with sediment generation, storage and transfer processes controlled by the combined influence of marine, subaerial and terrestrial drivers (Bird, 2008). Coastal sediment budgets are facilitated by the division of the coast into morphodynamically defined process-linked cells, constructed over units of area, volume or coastline length (Carter, 1988). Although a widely applied approach in the coastal zone, such budgets are neither intuitive nor straightforward to parameterize, given a common lack of quantitative data on processes over medium to long timescales, confounded by the difficulties of defining discrete boundary conditions (Nordstrom, 2000). A regional-scale coastal sediment budget should combine distinct sources and sinks including rivers, glaciers and coastal slopes from the terrestrial environment, and cliffs, shelves, ice and organisms from the marine (Carter, 1988). Globally, a wealth of research has been undertaken on dynamic erosional and depositional coastlines with 'soft' lithology, including their interface with the near- and offshore environment, as extensively reviewed by Carter (1988), Nordstrom (2000) and Bird (2008), amongst others. Significantly, however, Naylor *et al.* in press observe that on average, but with the notable exceptions of Zenkovich (1967) and Trenhaile (1997), 'hard rock' coasts only attract 8.5% of coastal textbook length. Although hard rock cliffs make up *c.* 80% of the world's coast (Emery and Kuhn, 1982), form some of the most iconic coastal landforms, contribute extensively to the littoral sediment budget, and support and defend critical infrastructure, arguably limited research on the mechanisms of sediment generation and flux is available. Our understanding of hard rock coasts is mainly inhibited by a lack of direct process monitoring data; perpetuated in part by a focus on risk management (e.g. predicting coastline retreat rather than geomorphological process rates). Where annualized rates of retreat are comparable to the levels of measurement error, an understanding of episodic coastal sediment release is inevitably derived (Lim, Mills and Rosser, 2009). Sherman and Gares (2002, p. 2) argue that 'because these systems tend to change slowly, they are more difficult to study on human time scales. However, the wide-spread occurrence of these coastal features demands our attention'; a problem confounded by the challenging nature of these environments. It is widely anticipated that in future we will observe enhanced rates of cliff change, yet data to support this hypothesis remains sparse. Although increased coastal landsliding may locally accelerate retreat, a by-product of this may be to offset the effects of sea-level rise on down-drift beaches, as residence times for failed material at cliff toes are commonly low given high rates of mechanical and chemical breakdown (Bird, 2008). Equally, rocky coasts are predicted to be particularly susceptible to increases in the intensity and frequency of transformative storms (Hume and Schalk, 1967; Hansom, 2001), yet although models of future climate continue to be improved, there remains a lack of basic fundamental research on the mechanisms and relative contribution of climate- and marine-driven controls on rock-cliff erosion (May and Heeps, 1985).

Modelling approaches to coastal cliff dynamics are twofold. First, deterministic prediction models of cliff erosion in response to future sea-level rise are widely applied, but attract criticism for their lack of consideration of mechanisms driving change and the potential uncertainties in the predicted recession scenarios (Moore, 2000; Lim *et al.*, 2005). Second, the uncertainty associated with coastal prediction is commonly addressed probabilistically, allowing risk-based economic appraisals to be undertaken, as is the case in the UK Shoreline Management Plans (SMPs) (DEFRA, 2005). Such models range from statistical analysis of past cliff recession, to the more process-oriented approaches that focus upon numerical approximations of the erosion mechanism (Bray and Hooke, 1997; Kogure *et al.*, 2006). The best results have been obtained by combining the two (Walkden and Hall, 2005); intermediary forms of modelling, which use stochastic process models, have demonstrated a close synergy with observed historic changes (Hall *et al.*, 2002), yet the potential for discrepancies between historic and future process conditions remains problematic, rendering such approaches often only speculative. A key opportunity to address these challenges is presented by developments in data capture in the coastal environment, such as the application of terrestrial LiDAR to cliff monitoring (Rosser *et al.*, 2005). A large body of work on rock coasts arose from the high-precision measurement of downwearing and abrasion (Robinson, 1977; Stephenson and Naylor, in press). More recently wide-area monitoring approaches are providing data of high precision with coverage at the whole landform scale. Data from these approaches have questioned the role and relative contribution of wave-notch development to whole cliff face erosion, implying an equally if not more significant role for subaerial drivers in the often aggressive maritime cliff environment (Lim *et al.*, in press). Research of this nature has suggested that the magnitude and frequency of sediment from coastal cliffs mirrors that in non-marine environments (Dong and Guzzetti, 2005). Data from an extensive series of sites on the northeast UK coast demonstrate that, although failures in excess of 1000 m^3 were recorded, the major proportion of coastal cliff rockfall number and cumulative volume can be attributed to iterative failures below 0.1 m^3 (Lim, Mills and Rosser, 2009), which adhere to a linear trend in a non-cumulative distribution of log volume, extending down to changes as small as 1.25×10^{-4} m^3 (rockfall dimensions equivalent to $0.05\,m \times 0.05\,m \times 0.05\,m$). Lim, Mills and Rosser (2009) also note a significant relationship between the cliff lithology and the resulting rate of rockfall, giving rise to, over the short term, variable rates of erosion across the cliff face that must, by definition, converge over longer timescales to maintain profile parallel retreat (Brunsden and Jones, 1980). This observation mirrors the variable rates of cliff recession observed in the seminal work by Sunamura (1992), which the author attributes to variable lithological resistance to abrasion (1 mm yr^{-1} in granite; 1 mm to 1 cm yr^{-1} on limestone; 1 cm yr^{-1} on shale; 10 cm to 1 m yr^{-1} on chalk and Tertiary sedimentary rocks; and at least 10 m yr^{-1} on volcanic ash). It remains clear, however, that coastal rock slopes represent a relatively poorly understood contribution to the wider sediment cascade, both in terms of bulk rates of process and the governing mechanisms.

3.4 Summary and Conclusions

Landsliding provides and conditions the input of sediment volumes to the far-field fluvial system. Landslide movement operates on a scale comparable to weathering and uplift. What is increasingly apparent is the role with which the mechanism of failure conditions the sediment prior to leaving the hillslope. At present there is a seeming mismatch between the relatively intense study of rock avalanche sedimentology, yet little consideration of equivalent processes in the landscape evolution literature. Critically, a 1 km^3 rock fall as a rigid coherent mass behaves very differently in the sediment cascade than a 1 km^3 highly fragmented fine-grained rock avalanche deposit. Landslide mechanism is of paramount importance for defining where, when, how and what sediment is delivered into the sediment cascade. The distribution of landsliding closely reflects the geological and topographic controls on failure, superimposed by the distribution and intensity of precipitation, each of which shows considerable heterogeneity. Few quantitative studies of sediment budgets have been undertaken in active mountain environments at a scale relevant to topographic evolution. There is an increasing desire to combine mechanistic understanding of the controls on slopes failure and their characteristics with wider scale sediment dynamics and landscape evolution.

References

Allen, P.A. (2008) *Time Scales of Tectonic Landscapes and their Sediment Routing Systems*. Geological Society Publishing House, Bath, Special Publication 296, pp. 7–28.

Allen, P.A. and Hovius, N. (1998) Sediment supply from landslide-dominated catchments: implications for basin-margin fans. *Basin Research*, **10**, 19–35.

Anders, A.M., Roe, G.H., Durran, D.R. *et al.* (2004) Precipitation and the form of mountain ranges. *Bulletin of the American Meteorological Society*, **85**, 498–499.

Anders, A.M., Roe, G.H., Hallet, B. *et al.* (2006) Spatial patterns of precipitation and topography in the Himalaya, in *Tectonics, Climate and Landscape Evolution*, (eds S.D. Willett, N. Hovius, M.T. Brandon and D. Fisher), Geological Society of America, Boulder, CO, Special Paper 398.

André, M.-F. (1997) Holocene rockwall retreat in Svalbard: a triple-rate evolution. *Earth Surface Processes and Landforms*, **22**, 423–440.

Bachmann, D., Bouissou, S. and Chemenda, A. (2004) Influence of weathering and pre-existing large scale fractures on gravitational slope failure: insights from 3-D physical modelling. *Natural Hazards and Earth System Sciences*, **4**, 711–717.

Bachmann, D., Bouissou, S. and Chemenda, A. (2006) Influence of large scale topography on gravitational rock mass movements: new insights from physical modeling. *Geophysical Research Letters*, **33**, L21406.

Bartarya, S.K. and Sah, M.P. (1995) Landslide induced river bed uplift in the Tal Valley of Garhwal Himalaya, India. *Geomorphology*, **12**, 109–121.

Benda, L. and Dunne, T. (1997) Stochastic forcing of sediment supply to the channel network from landsliding and debris flow. *Water Resources Research*, **33**, 2849–2863.

Bergbauer, S. and Pollard, D.D. (2004) A new conceptual fold-fracture model including prefolding joints, based on the Emigrant Gap anticline, Wyoming. *Geological Society of America Bulletin*, **116**, 294–307.

Bertran, P. and Texier, J.-P. (1999) Facies and microfacies of slope deposits. *Catena*, **35**, 99–121.

Bertran, P., Francou, B. and Texier, J.P. (1995) Stratified slope deposits: the stone-banked sheets and lobes model, in *Steepland Geomorphology* (ed. O. Slaymaker), John Wiley and Sons, Chichester, pp. 147–169.

Bertran, P., Hétu, B., Texier, J.P. and Van Steijn, H. (1997) Fabric of subaerial slope deposits. *Sedimentology*, **44**, 1–16.

Beylich, A.A., Molau, U., Luthbom, K. and Gintz, D. (2005) Rates of chemical and mechanical fluvial denudation in an arctic oceanic periglacial environment, Latnjavagge drainage basin, northernmost Swedish Lapland. *Arctic, Antarctic and Alpine Research*, **37**, 75–87.

Bigi, A., Hasbargen, L.E., Montanari, A. and Paola, C. (2006) Knickpoints and hillslope failures: Interactions in a steady-state experimental landscape, in *Tectonics Climate and Landscape Evolution* (eds S.D. Willett, N. Hovius, M.T. Brandon and D.M. Fisher), Geological Society of America, Boulder, CO, Special Paper 398, p. 295. doi: 10.1130/2006.2398(18).

Bird, E. (2008) *Coastal Geomorphology*, 2nd edn, John Wiley and Sons. Chichester.

Bjerrum, L. (1967) Progressive failure in slopes of overconsolidated plastic clay and clay shales. *Journal of the Soil Mechanics and Foundations Division, Proceedings of the American Society of Civil Engineers*, **93** (SM5), 2–49.

Blackwelder, E. (1942) The process of mountain sculpturing by rolling debris. *Journal of Geomorphology*, **4**, 324–328.

Borrelli, L., Greco, R. and Gulla, G. (2007) Weathering grade of rock masses as a predisposing factor to slope instabilities: reconnaissance and control procedures. *Geomorphology*, **87** (3), 158–175.

Bouchon, M. and Baker, J. (1996) Seismic response of a hill: the example of Tarzana, California. *Bulletin of the Seismological Society of America*, **86**, 66–72.

Brardinoni, F. and Church, M. (2004) Representing the landslide magnitude–frequency relation, Capilano River Basin, British Columbia. *Earth Surface Processes and Landforms*, **29** (115–124), 2004.

Bray, M.J. and Hooke, J.M. (1997) Prediction of soft-cliff retreat with accelerating sea-level rise. *Journal of Coastal Research*, **13**, 453–467.

Brideau, M.-A., Yan, M. and Stead, D. (2009) The role of tectonic damage in brittle fracture in the development of large rockslope failures. *Geomorphology*, **103** (1), 30–49. doi: 10.1016/j.geomorph.2008.04.010.

Brunsden, D. and Jones, D.K.C. (1980) Relative time scales and formative events in coastal landslide systems. *Zeitschrift fur Geomorphologie*, **34**, 1–19.

Bucknam, R.C., Coe, J.A., Chavarría, M.M. *et al.* (2001) Landslides triggered by Hurricane Mitch in Guatemala. Inventory and Discussion. *U.S. Geological Survey Open File Report* 01-443. p. 39.

Bull, W.B. (2009) *Tectonically Active Landscapes*, Wiley-Blackwell, Oxford.

Burbank, D. (2002) Rates of erosion and their implications for exhumation. *Mineralogical Magazine*, **66**, 25–52.

Burbank, D.W. and Anderson, R. (2001) *Tectonic Geomorphology*, Blackwell Publishing, Oxford.

Burbank, .D.W., Leland, J., Fielding, E. *et al.* (1996) Bedrock incision, rock uplift, and threshold hillslopes in the northwestern Himalaya. *Nature*, **379**, 505–510.

Caine, N. and Swanson, F.J. (1989) Geomorphic coupling of hillslope and channel systems in two small mountain basins. *Zeitschrift für Geomorphologie Neue Folge*, **33** (2), 189–203.

Cardinali, M., Ardizzone, F., Galli, M. *et al.* (2000) Landslides triggered by rapid snow melting: the December 1996–January 1997 event in Central Italy, in *Mediterranean Storms, Proceedings Plinius Conference'99, Maratea, 14–16 October 1999* (eds P. Claps and F. Siccardi), CNR GNDCI publication number 2012, Editoriale Bios, Cosenza, pp. 439–448.

Carson, M.A. and Petley, D.J. (1970) The existence of threshold hillslopes in the denudation of the landscape. *Transactions of the Institute of British Geographers*, **49**, 71–95.

Carter, R.W.G. (1988) *Coastal Environments: An Introduction to the Physical, Ecological and Cultural Systems of Coastlines*, Academic Press, London.

Crosta, G.B., Imposimato, S. and Roddeman, D.G. (2003) Numerical modelling of large landslides stability and runout. *Natural Hazards and Earth System Sciences*, **3** (6), 523–538.

Cruden, D.M. and Varnes, D.J. (1996) Landslide types and processes, in *Landslides: Investigation and Mitigation* (eds A.K. Turner and R.L. Shuster), Transp Res Board, Spec Rep 247, pp. 36–75.

Dai, F.C. and Lee, C.F. (2002) Landslide characteristics and slope instability modeling using GIS, Lantau Island, Hong Kong. *Geomorphology*, **42**, 213–228.

Davies, T.R., McSaveney, M.J. and Deganutti, A.M. (2007) Dynamic rock fragmentation causes low rock-on-rock friction. 1st Canada–U.S. Rock Mechanics Symposium, 27–31 May 2007, Vancouver, Canada, p. 8.

DEFRA (2005) Making space for water: Taking forward a new Government strategy for flood and coastal erosion risk management in England: First Government response to the autumn 2004 Making space for water consultation exercise. Available online: (http://www.defra.gov.uk/environ/fcd/policy/strategy/firstresponse.pdf) accessed: 18th June 2009.

Densmore, A.L. and Hovius, N. (2000) Topographic fingerprints of bedrock landslides. *Geology*, **28**, 371–374.

Densmore, A.L. Anderson, R.S., McAdoo, B.G. and Ellis, M.A. (1997) Hillslope evolution by bedrock landslides. *Science*, **275** (5298), 369–372.

Dietrich, W.E. and Dunne, T. (1978) Sediment budget for a small catchment in mountainous terrain. *Zeitschrift für Geomorphologie Supplement Band*, **29**, 191–206.

Dikau, R., Brunsden, D., Schrott, L. and Ibsen, M. (eds) (1996) *Landslide Recognition: Identification, Movement and Causes*, John Wiley and Sons, Chichester, p. 251.

Dong, P. and Guzzetti, F. (2005) Frequency-size statistics of coastal soft-cliff erosion. *ASCE, Journal of Waterways, Port, Coastal and Ocean Engineering*, **131** (1), 37–42.

Dorren, L.K.A. (2003) A review of rockfall mechanics and modelling approaches. *Progress in Physical Geography*, **27** (1), 69–87.

Dunning, S.A. (2004) The grain-size distribution of rock-avalanche deposits. Abstract volume *NATO Advanced Research Workshop: Security of Natural and Artificial Rockslide Dams*, Bishkek, Kyrgyzstan, 7–13 June 2004 (eds S.G. Evans and A. Strom), pp. 38–43.

Dunning, S.A. (2006) The grain-size distribution of rock-avalanche deposits in valley confined settings. *The Italian Journal of Engineering Geology and Environment*, **1**, 117–121.

Eisbacher, G.H. (1971) Natural slope failure, northeastern Skeena Mountains. *Canadian Geotechnical Journal*, **8**, 384–390.

Emery, K.O. and Kuhn, G.G. (1982) Sea cliffs: their processes, profiles, and classification. *Geological Society of America Bulletin*, **93**, 644–654.

Erismann, T.H. and Abele, G. (2001) *Dynamics of Rockslides and Rockfalls*, Springer-Verlag, Berlin Heidelberg, ISBN 3-540-67198-6, p. 316.

Evans, S.G. (1983) *Landslides in layered volcanic successions with particular reference to the Tertiary Rocks of south central British Columbia.* PhD Thesis, University of Alberta, Edmonton, Alberta.

Frei, C. and Schär, C. (1998) A precipitation climatology of the Alps from high-resolution rain-gauge observations. *International Journal of Climatology*, **18**, 873–900.

Friedman, S.J., Kwon, G. and Losert, W. (2003) Granular memory and its effect on the triggering and distribution of rock avalanche events. *Journal of Geophysical Research*, **108** (B8), 2380. doi: 10.1029/2002JB002174.

Fuyii, Y. (1969) Frequency distribution of landslides caused by heavy rainfall. *Journal Seismological Society Japan*, **22**, 244–247.

Gardner, J. (1980) Frequency, magnitude and spatial distribution of mountain rockfalls and rockslides in the Highwood Pass area, Alberta, in *Thresholds in Geomorphology* (eds D.R. Coates and J.D. Vitek), Allen & Unwin, London, pp. 267–295.

Goguel, J. (1978) Scale-dependent rockslides mechanisms, with emphasis on the role of pore fluid vaporization, in *Rockslides and Avalanches. 1. Natural Phenomena* (ed. B. Voight), Elsevier, Amsterdam, pp. 693–705.

Gomi, T., Sidle, R.C. and Swanson, D.N. (2004) Hydrogeomorphological linkages of sediment transport in headwater streams, Maybeso Experimental Forest, southwest Alaska. *Hydrological Processes*, **18**, 667–683. doi: 10.1002/hyp.1366.

Gorers, G. and Presen, J. (1998) Field experiment on the transport of rock fragments by animal trampling on scree slopes. *Geomorphology*, **23**, 193–203.

Gutenberg, B. and Richter, C.F. (1954) *Seismicity of the Earth*, Princeton Univ. Press.

Guthrie, R.H. and Evans, S.G. (2004) Analysis of landslide frequency and characteristics in a natural system, coastal British Columbia. *Earth Surface Processes and Landforms*, **29**, 1321–1339.

Guzzetti, F., Cardinali, M. and Reichenbach, P. (1996) The influence of structural setting and lithology on landslide type and pattern. *Environmental & Engineering Geoscience*, **2**, 531–555.

Guzzetti, F., Malamud, B.D., Turcotte, D.L. and Reichenbach, P. (2002) Power-law correlations of landslide areas in central Italy. *Earth and Planetary Science Letters*, **195** (3–4), 169–183. doi: 10.1016/S0012-821X(01)00589-1.

Guzzetti, F., Ardizzone, F., Cardinali, M. *et al.* (2008) Distribution of landslides in the Upper Tiber River basin, Central Italy. *Geomorphology*, **96**, 105–122.

Guzzetti, F., Ardizzone, F., Cardinali, M. *et al.* (2009) Landslide volumes and landslide mobilization rates in Umbria, central Italy. *Earth and Planetary Science Letters*, **279**, 222–229. doi: 10.1016/j.epsl.2009.01.005.

Haeberli, W. (1999) Hangstabilitätsprobleme im Zusammenhang mit Gletscherschwund und Permafrostdegradation im Hochgebirge. *Relief, Boden, Paläoklima*, **14**, 11–30.

Hall, J.W., Meadowcroft, I.C., Lee, E.M. and van Gelder, P.H.A.J.M. (2002) Stochastic simulation of episodic soft coastal cliff recession. *Coastal Engineering*, **46** (3), 159–174.

Hansom, J.D. (2001) Coastal sensitivity to environmental change: a view from the beach. *Catena*, **42**, 291–305.

Harp, E.L. and Jibson, R.W. (1995) Inventory of landslides triggered by the 1994 Northridge, California earthquake: *U.S. Geological Survey Open-File Report 95-213*, 17p.

Harp, E.L. and Jibson, R.W. (1996) Landslides triggered by the 1994 Northridge, California earthquake. *Bulletin of the Seismological Society of America*, **86**, 319–332.

Heim, A. (1932) *Bergsturz und Menschenleben*, Fretz and Wasmuth Verlag, Zurich, p. 218.

Heimsath, A.M., Dietrich, W.E., Nishiizumi, K. and Finkel, R.C. (1997) The soil production function and landscape equilibrium. *Nature*, **388**, 358–361.

Helmstetter, A. and Sornette, D. (2004) Slider block friction model for landslides; application to Vaiont and La Clapiere landslides. *Journal of Geophysical Research*, **109** (B2), 210–225.

Helmstetter, A., Sornette, D., Grasso, J.-R. *et al.* (2004) Slider-block friction model for landslides: implication for prediction of mountain collapse. *Journal of Geophysical Research*, **109**, B02409. doi: 10.1029/2002JB002160, (http://www.arxiv.org/abs/cond-mat/0208413).

Hewitt, K. (2002) Styles of rock avalanche depositional complexes in very rugged terrain, Karakoram Himalaya, Pakistan, in *Catastrophic Landslides: Effects, Occurrence, and Mechanisms, Reviews in Engineering Geology* (eds S.G. Evans and J. DeGraff), Geological Society of America, Boulder, CO, pp. 345–378.

Hewitt, K. (2009) Rock avalanches that travel onto glaciersand related developments, Karakoram Himalaya, Inner Asia. *Geomorphology*, **103** (1), 66–79.

Hoek, E. and Bray, J.W. (1981) *Rock Slope Engineering*, 3rd edn, Institution of Mining and Metallurgy, London, p. 360.

Hoffmann, T. and Schrott, L. (2002) Modelling sediment thickness and rockwall retreat in an Alpine valley using 2D-seismic refraction (Reintal, Bavarian Alps). *Zeitschrift für Geomorphologie Supplement Band*, **127**, 153–173.

Hovius, N. and Leeder, M. (1998) Clastic sediment supply to basins. *Basin Research*, **10**, 1–5.

Hovius, N., Lague, D. and Dadson, S. (2004) Processes, rates and patterns of mountain-belt erosion, in *Mountain Geomorphology* (eds P.N. Owens and O. Slaymaker), Arnold.

Hovius, N., Stark, C.P. and Allen, P.A. (1997) Sediment flux from a mountain belt derived by landslide mapping. *Geology*, **25** (3), 231–234.

Hovius, N., Stark, C.P., Hao-Tsu, C. and Jium-Chuan, L. (2000) Supply and removal of sediment in a landslide-dominated mountain belt: central range, Taiwan. *Geology*, **108** (1), 73–89.

Howard, A.D. (1997) Badland morphology and evolution: interpretation using a simulation model. *Earth Surface Processes and Landforms*, **22**, 211–227.

Huang, R. and Li, W. (2009) Development and distribution of geohazards triggered by the 5.12 Wenchuan Earthquake in China Science in China Series E. *Technological Sciences*, **52**, 810–819.

Hume, J.D. and Schalk, M. (1967) Shoreline processes near Barrow. Alaska. A comparison of the normal and catastrophic. *Arctic*, **20**, 86–103.

Hungr, O., Evans, S.G. and Hazzard, J. (1999) Magnitude and frequency of rock falls and rock slides along the main transportation corridors of southwestern British Colombia. *Canadian Geotechnical Journal*, **36**, 224–238.

Hunter, G. and Fell, R. (2003) Travel distance angle for "rapid" landslides in constructed and natural soil slopes. *Canadian Geotechnical Journal*, **40**, 1123–1141.

Imaizumi, F. and Sidle, R.C. (2007) Linkage of sediment supply and transport processes in Miyagawa Dam catchment, Japan. *Journal of Geophysical Research Earth Surface*, **112**, F03012.

Imaizumi, F., Sidle, R.C. and Kamei, R. (2008) Effects of forest harvesting on the occurrence of landslides and debris flows in steep terrain of central Japan. *Earth Surface Processes and Landforms*, **33**, 827–840.

Innes, J.N. (1983) Lichenometric dating of debris-flow deposits in the Scottish Highlands. *Earth Surface Processes and Landforms*, **8**, 579–588.

Iverson, R.M. and Reid, M.E. (1992) Gravity-driven groundwater flow and slope failure potential, 1, elastic effective-stress model. *Water Resources Research*, **28**, 925–938.

Jha, P.K., Subramanian, V. and Sitasawad, R. (1988) Chemical and sediment mass transfer in the Yamuna River – a tributary of the Ganges system. *Journal of Hydrology*, **104**, 237–246.

Johnson, R.M. and Warburton, J. (2006) Variability in sediment supply, transfer and deposition in an upland torrent system: Iron Crag, northern England. *Earth Surface Processes and Landforms*, **31** (7), 844–861.

Keefer, D.K. (1984) Landslides caused by earthquake. *Geological Society of America Bulletin*, **95**, 406–421.

Keefer, D.K. (1994) The importance of earthquake-induced landslides to long-term slope erosion and slope-failure hazards in seismically active regions. *Geology*, **10**, 265–284.

Kelsey, H.M. (1980) A sediment budget and an analysis of geomorphological process in the Van Duzen River basin, north coastal California, 1941–1975. *Geological Society of America Bulletin*, **91**, 1119–1216.

Kelsey, H.M., Coghlan, M., Pitlick, J. and Best, D. (1995) Geomorphological analysis of streamside landslides in the Redwood Greek Basin Northwestern California. *U.S. Geological Survey Professional Paper* 1454-J, pp. J1–J12.

Kemeny, J. (2003) The time-dependent reduction of sliding cohesion due to rock bridges along discontinuities: a fracture mechanics approach. *Rock Mechanics and Rock Engineering*, **36**, 27–38.

Kent, P.E. (1966) The transport mechanism in catastrophic rock falls. *Journal of Geology*, **74**, 79–83.

Kilburn, C.R.J. and Sammonds, P.R. (2005) Maximum warning times for imminent volcanic eruptions. *Geophysical Research Letters*, **32**, L24313. doi: 10.1029/2005GL024184.

Kilburn, C. and Sorensen, S.-A. (1998) Runout lengths of sturzstroms: The control of initial conditions and of fragment dynamics. *Journal of Geophysical Research*, **103**, 17877–17884.

Kinakin, D. and Stead, D. (2005) Analysis of the distributions of stress in natural ridge forms: implications for the deformation mechanisms of rock slopes and the formation of sackung. *Geomorphology*, **65**, 85–100.

Kirby, E., Whipple, K. and Harkins, N. (2008) Topography reveals seismic hazard. *Nature Geoscience*, **1**, 485–487.

Kogure, T., Aoki, H., Maekado, A. *et al.* (2006) Effect of the development of notches and tension cracks on instability of limestone coastal cliffs in the Ryukyus, Japan. *Geomorphology*, **80**, 236–244.

Korup, O. (2005) Geomorphological imprint of landslides on alpine river systems, southwest New Zealand. *Earth Surface Processes and Landforms*, **30**, 783–300.

Korup, O. (2006a) Rock-slope failure and the river long profile. *Geology*, **34**, 45–48.

Korup, O. (2006b) Effects of large deep-seated landslides on hillslope morphology, western Southern Alps, New Zealand. *Journal of Geophysical Research*, **111**, F01018. doi: 10.1029/2004JF000242.

Korup, O., Clague, J.J., Hermanns, R.L. *et al.* (2007) Giant landslides, topography, and erosion. *Earth and Planetary Science Letters*, **261**, 578–589.

Kossobokov, V.G., Keilis-Borok, V.I., Turcotte, D.L. and Malamud, B.D. (2000) Implications of a statistical physics approach for earthquake hazard assessment and forecasting. *Pure and Applied Geophysics*, **157**, 2323–2349.

Krautblatter, M. and Dikau, R. (2007) Towards a uniform concept for the comparison and extrapolation of rockwall retreat and rockfall supply. *Geografiska Annaler, Series A: Physical Geography*, **89** (1), 21–40.

Lague, D. (2003) Constraints on the long-term colluvial erosion law by analyzing slope-area relationships at various tectonic uplift rates in the Siwaliks Hills (Nepal). *Journal of Geophysical Research*, **108**, 2129. doi: 10.1029/2002JB001893.

Leroueil, S. (2001) Natural slopes and cuts: movements and failure mechanisms. *Geotechnique*, **51** (3), 197–243.

Lim, M., Mills, J. and Rosser, N.J. (2009) Laser scanning surveying of linear features: considerations and applications, in *Laser Scanning for the Environmental Sciences*, 1st edn (eds G. Heritage, M. Charlton and A. Large), Blackwell Publishing, Oxford.

Lim, M., Petley, D.N., Rosser, N.J. *et al.* (2005) Combined digital photogrammetry and time-of-flight laser scanning for monitoring cliff evolution. *Photogrammetric Record*, **20**, 109–129.

Lim, M., Rosser, N.J., Allison, R.J. and Petley, D.N. (in press) Erosional processes in the hard rock coastal cliffs at Staithes, North Yorkshire. *Geomorphology*. doi: 10.1016/j.geomorph.2009.02.011.

Lin, G.-W., Chen, H., Hovius, N. *et al.* (2008) Effects of earthquake and cyclone sequencing on landsliding and fluvial sediment transfer in a mountain catchment. *Earth Surface Processes and Landforms*, **33** (9), 1354. doi: 10.1002/esp.1716.

Locat, P., Couture, R., Leroueil, S. *et al.* (2006) Fragmentation energy in rock avalanches. *Canadian Geotechnical Journal*, **43**, 830–851.

Main, I. (2000) A damage mechanics model for power-law creep and earthquake aftershock sequences. *Geophysical Journal International*, **142** (1), 151–161.

Malamud, B.D., Turcotte, D.L., Guzzetti, F. and Reichenbach, P. (2004) Landslides, earthquakes and erosion. *Earth and Planetary Science Letters*, **229** (1–2), 45–59.

Martin, Y., Rood, K., Schwab, J.W. and Church, M. (2002) Sediment transfer by shallow landsliding in the Queen Charlotte Islands, British Columbia. *Canadian Journal of Earth Sciences*, **39** (2), 189–205.

May, V.J. and Heeps, C. (1985) The nature and rates of change on a chalk coastline. *Zeitschrift für Geomorphologie Neue Folge, Supplement Band.*, **57**, 81–94.

McCarroll, D., Shakesby, R.A. and Matthews, J.A. (1998) Spatial and temporal pattern of Late Holocene rockfall activity on a Norwegian talus slope: a lichenometric and simulation-modelling approach. *Arctic and Alpine Research*, **30**, 51–60.

McCarroll, D., Shakesby, R.A. and Matthews, J.A. (2001) Enhanced rockfall activity during the Little Ice Age: further lichenometric evidence from a Norwegian talus. *Permafrost and Periglacial Processes*, **12**, 157–164.

McSaveney, M.J. and Davies, T.R. (2004) Does the big question have an answer? – potential landslide size. Abstract volume *NATO Advanced Research Workshop: Security of Natural and Artificial Rockslide Dams, Bishkek*, Kyrgyzstan, 7–13 June 2004 (eds S.G. Evans and A. Strom).

McSaveney, M.J., Davies, T.R. and Hodgson, K.A. (2000) A contrast in deposit style and process between large and small rock avalanches, in *Landslides in Research, Theory and Practice Proceedings 8th International Symposium on Landslides 26–30 June* (eds E. Bromhead, N. Dixon and M.L. Ibsen), Thomas Telford, Cardiff, pp. 1052–1058.

Melosh, H.J. (1979) Acoustic fluidization – a new geologic process? *Journal of Geophysical Research*, **84**, 7513–7520.

Meunier, P., Hovius, N. and Haines, A.J. (2007) Regional patterns of earthquake-triggered landslides and their relation to ground motion. *Geophysical Research Letters*, **34**, L20408.

Milliman, J.D. and Meade, R.H. (1983) World-wide delivery of river sediments to the oceans. *Journal of Geology*, **91**, 1–21.

Milliman, J.D. and Syvitski, J.P.M. (1992) Geomorphological/tectonic control of sediment discharge to the ocean: the importance of small mountainous rivers. *Journal of Geology*, **100**, 525–544.

Molnar, P., Anderson, R.S. and Anderson, S.P. (2007) Tectonics, fracturing of rock, and erosion. *Journal of Geophysical Research*, **112**, F03014. doi: 10.1029/2005FJ000433.

Montgomery, D.R. (2001) Slope distributions, thresholds hillslopes, and steady-state topography. *American Journal of Science*, **301**, 432–452.

Montgomery, D.R. and Brandon, M.T. (2002) Topographic controls on erosion rates in tectonically active mountain ranges. *Earth and Planetary Science Letters*, **201**, 481–489.

Moore, L.J. (2000) Shoreline mapping techniques. *Journal of Coastal Research*, **16**, 111–124.

Nadim, F., Kjekstad, O., Peduzzi, P. *et al.* (2006) Global landslide and avalanche hotspots. *Landslides*, **3**, 159–173. doi: org/10.1007/s10346-006-0036-1.

Nagaraja Rao, G.M. and Murthry, C.R.L. (2001) Dual role of microcracks: toughening and degradation. *Canadian Geotechnical Journal*, **38**, 427–440.

Naylor, L.A., Stephenson, W.J. and Trenhaile, A.S. (in press) Rock coast geomorphology: Recent advances and future research directions. *Geomorphology*. doi: 10.1016/j.geomorph. 2009.02.004.

Nieuwenhuijzen, M.E. and Van Steijn, H. (1990) Alpine debris flows and their sedimentary properties. A case study from the French Alps. *Permafrost and Periglacial Processes*, **1**, 111–128.

Nordstrom, K.F. (2000) *Beaches and Dunes of Developed Coasts*, Cambridge University Press, Cambridge.

Oppikofer, T., Jaboyedoff, M. and Keusen, H.-R. (2008) Collapse at the eastern Eiger flank in the Swiss Alps. *Nature Geoscience*, **1**, 531–535. doi: 10.1038/ngeo258.

Otto, J.C. and Dikau, R. (2004) Geomorphologic system analysis of a high mountain valley in the swiss alps. *Zeitschrift für Geomorphologie*, **48** (3), 323–341.

Otto, J.-C., Goetz, J. and Schrott, L. (2008) Sediment storage in Alpine sedimentary systems – quantification and scaling issues. *Sediment Dynamics in Changing Environments*, International Association of Hydrological Sciences, Wallingford, IAHS-AISH Publication 325, pp. 258–265.

Oven, K.J. (2005) *The analysis of the spatial patterns and controls governing the global occurrence of fatal landslides*. Unpublished MSc Thesis, University of Durham, UK.

Owen, L.A., Kamp, U., Khattak, G.A. *et al.* (2008) Landslides triggered by the 8 October 2005 Kashmir Earthquake. *Geomorphology*, **94** (1–2), 1–9.

Pelletier, J.D., Malamud, B.D., Blodgett, T. and Turcotte, D.L. (1997) Scale-invariance of soil moisture variability and its implications for the frequency–size distribution of landslides. *Engineering Geology*, **48**, 255–268.

Petley, D. (2004) The evolution of large slope failures: mechanisms of rupture propagation. *Natural Hazards and Earth System Sciences*, **4**, 147–152.

Petley, D.N., Bulmer, M.H.K. and Murphy, W. (2002) Patterns of movement in rotational and translational landslides. *Geology*, **30** (8), 719–722.

Petley, D.N., Dunning, S.A. and Rosser, N.J. (2005) The analysis of global landslide risk through the creation of a database of worldwide landslide fatalities, in *Landslide Risk Management* (eds O. Hungr, R. Fell, R. Couture and E. Eberhardt), A.T. Balkema, Amsterdam, pp. 367–374.

Petley, D.N., Petley, D.J. and Allison, R.J. (2008) Temporal prediction in landslides – understanding the Saito effect. *10th International Symposium on Landslides and Engineered Slopes, Xi'an, China.*

Petley, D.N., Dunning, S.A., Rosser, N.J. and Kausar, A.B. (2006) Incipient earthquakes in the Jhelum Valley, in Pakistan Following the 8th October 2005 Earthquake. Frontiers of Science, in *Disaster Mitigation of Debris Flows, Slope Failures and Landslides* (ed. H. Marui), Universal Academy Press, Tokyo, Frontiers of Science Series 47, pp. 47–56.

Potyondy, D.O. (2007) Simulating stress corrosion with a bounded-particle model for rock. *International Journal of Rock Mechanics and Mining Sciences*, **44**, 677–691.

Rapp, A. (1960) Recent development of mountain slopes in Kärkevagge and surroundings, Northern Scandinavia. *Annales De Geographie*, **42A**, 65–200.

Rasemann, S. (2004) Geomorphometrische Struktureines mesoskaligen alpinen Geosystems. *Bonner Geographische Abhandlungen*, Heft **111**. Retrieved 16 January 2007 from http://hss.ulb.unibonnde/diss_online/math_nat_fak/2003/rasemann_stefan/index.htm.

Reiners, P.W., Ehlers, T.A., Mitchell, S.G. and Montgomery, D.R. (2003) Coupled spatial variations in precipitation and long-term erosion rates across the Washington Cascades. *Nature*, **426**, 645–647.

Robinson, L.A. (1977) Marine erosive processes at the cliff foot. *Marine Geology*, **23**, 257–271.

Rosser, N.J., Lim, N., Petley, D.N. *et al.* (2007) Patterns of precursory rockfall prior to slope failure. *Journal of Geophysical Research*, **112**, F04014. doi: 10.1029/2006JF000642.

Rosser, N.J., Petley, D.N., Lim, M. *et al.* (2005) Terrestrial laser scanning for monitoring the process of hard rock coastal cliff erosion. *Quarterly Journal of Engineering Geology*, **38**, 363–375.

Safran, E.B., Bierman, P.R., Aalto, R. *et al.* (2005) Erosion rates driven by channel network incision in the Bolivian Andes. *Earth Surface Processes and Landforms*, **30**, 1007–1024.

Saito, M. (1969) Forecasting time of slope failure by tertiary creep. *Proceedings, 7th International Conference on Soil Mechanics and Foundation Engineering,* Mexico City, vol. 2, pp. 677–683.

Sandersen, F., Bakkeoi, S., Hestnes, E. and Lied, K. (1996) *The influence of meteorological factors on the initiation of debris flows, rockfalls, rockslides and rockmass stability, Proceedings of the VII International Symposium on Landslides, Trondheim*, Vol. **1**, Balkema, Rotterdam, pp. 97–114.

Sass, O. (1998) Die Steuerung von Steinschlagmenge durch Mikroklima, Gesteinsfeuchte und Gesteinseigenschaften im westlichen Karwendelgebirge. *Münchner Geographische Abhandlungen Reihe B*, **29**, 347–359.

Sass, O., Wetzel, K. and Friedmann, A. (2006) Landscape dynamics of sub-alpine forest fire slopes in the Northern Alps. *Zeitschrift für Geomorphologie. Neue Folge*, **142**, 207–227.

Sassa, K. (1988) Geotechnical model for the motion of landslides, in *Landslides. Proceedings of the Vth International Symposium on Landslides, Lausanne* (ed. C. Bonnard), Balkema, Rotterdam, pp. 37–56.

Schlunegger, F., Detzner, K. and Olsson, D. (2002) The evolution towards steady state erosion in a soil-mantled drainainge basin: Semi-quantitative data from a transient landscape in the Swiss Alps. *Geomorphology*, **43**, 55–76. doi: 10.1016/s0169-555X(01)00120-9.

Schmidt, K.M. and Montgomery, D.R. (1995) Limits to relief. *Science*, **270**, 617–620.

Schrott, L., Niederheide, A., Hankammer, M. *et al.* (2002) Sediment storage in a mountain catchment: geomorphological coupling and temporal variability (Reintal, Bavarian Alps, Germany). *Zeitschrift für Geomorphologie Supplement Band*, **127**, 175–196.

Sepülveda, S.A., Murphy, W., Jibson, R.W. and Petley, D.N. (2005) Topographic controls on co-seismic rock slides during the 1999 Chi-Chi earthquake, Taiwan. *Quarterly Journal of Engineering Geology and Hydrogeology*, **38**, 189–196.

Sherman, D.J. and Gares, P.A. (2002) The geomorphology of coastal environments. *Geomorphology*, **48**, 1–6.

Shreve, R.L. (1968) The Blackhawk landslide. *Geological Society of America, Special Paper*, **108**, 1–47.

Simonett, D.S. (1967) Landslide distribution and earthquakes in the Bewani and Torricelli Mountains, New Guinea, in *Landform Studies from Australia and New Guinea* (eds J.N. Jennings and J.A. Mabbutt), Cambridge University Press, Cambridge, pp. 64–84.

Slaymaker, O. (2008) Sediment budget and sediment flux studies under accelerating global change in cold environments. *Zeitschrift für Geomorphologie, Supplementbände*, **52**, 123–148.

Slaymaker, O. and Spencer, T. (1998) *Physical Geography and Global Environmental Change*, Addison Wesley Longman, Harlow, p. 292.

Smith, R.B. (2006) Progress on the theory of orographic precipitation, in *Tectonics, Climate and Landscape Evolution* (eds S.D. Willett, N. Hovius, M.T. Brandon and D. Fisher), Special Paper 398, Geological Society of America, Boulder, CO.

Sobel, E.R., Oskin, M., Burbank, D. and Mikolaichuk, A.V. (2006) Exhumation of basement-cored uplifts: Example of the Kyrgyz Range quantified with apatite fission track thermochronology. *Tectonics*, **25**. doi: 10.1029/2005TC001809.

Stark, C.P. and Hovius, N. (2001) The characterization of landslide size distributions. *Geophysical research Letters*, **28**, 1091–1094.

Stephenson, W.J. and Naylor, L.A. (in press) Rock coast geomorphology. *Geomorphology*. doi: 10.1016/j.geomorph.2009.02.013.

Stock, J.D. and Dietrich, W.E. (2006) Erosion of steepland valleys by debris flows. *GSA Bulletin*, **118** (9/10), 1125–1148.

Strom, A.L. (1998) Giant ancient rockslide and rock avalanche in the Tien Shan Mountains, Kyrgyzstan. *Landslide News*, **11**, 20–23.

Sugai, T., Ohmori, H. and Hirano, M. (1994) Rock control on the magnitude–frequency distribution of landslides. *Transactions of the Japanese Geomorphology Union*, **15**, 233–251.

Sunamura, T. (1992) *Geomorphology of Rocky Coasts*, John Wiley and Sons, Chichester.

Suwa, H. (1991) Visual observed failure of a rock slope in Japan. *Landslide News*, **5**, 8–9.

Tarolli, P., Borga, M. and Fontana, G.D. (2008) Analysing the influence of upslope bedrock outcrops on shallow landsliding. *Geomorphology*, **93** (304), 186–200.

Terzaghi, K. (1950) Mechanisms of Landslides. Geological Society of America, Boulder, CO, Berkeley Volume.

Tricart, J. (1956) Ètude expérimentale du problème de la gélivation. *Biuletyn Peryglacjalny*, **4**, 285–318.

Trenhaile, A.S. (1997) *The Geomorphology of Rock Coasts*, Clarendon Press, Oxford.

Walkden, M.J.A. and Hall, J.W. (2005) A predictive mesoscale model of the erosion and profile development of soft rock shores. *Coastal Engineering*, **52** (6), 535–563.

Van Asch, T.W.J., Malet, J.-P., van Beek, L.P.H. and Armitrano, D. (2007) Techniuqes, issues and advances in numerical modelling of landslide hazard. *Bulletin de la Societe Geologique de France*, **178** (2), 65–88. doi: 10.2113/gssgfbull.178.2.65.

Van Steijn, H., Bertran, P., Francou, B. *et al.* (1995) Review of models for genetical interpretation of stratified slope deposits. *Permafrost and Periglacial Processes*, **6**, 125–146.

Varnes, D.J. (1978) Slope movement. Types and processes, in *Landslides. Analysis and Control*, National Academy of Science, Washington, DC, Special Report 176, pp. 11–33.

Vidrih, R. and Ribicic, M. (2004) The earthquake on July 12, 2004 in Upper Soca territory (NW Slovenia) – preliminary geological and seismological characteristics. *Geologija*, **47** (2), 199–220.

Whipple, K., Kirby, E. and Brocklehurst, S. (1999) Geomorphological limits to climate-induced increases in topographic relief. *Nature*, **401**, 39–43. doi: 19.1038/43375.

Whitehouse, I.E. and McSaveney, M.J. (1983) Diachronous talus surfaces in the Southern Alps, New Zealand, and their implications to talus accumulation. *Arctic and Alpine Research*, **15** (1), 53–64.

Wolman, M.G. and Miller, J.P. (1960) Magnitude and frequency of forces in geomorphological processes. *Journal of Geology*, **68**, 54–74.

Yamagishi, H. and Iwashashi, J. (2007) Comparison between the two triggered landslides in Mid-Niigata, Japan by July 13 heavy rainfall and October 23 intensive earthquakes in 2004. *Landslides*, **4** (4), 1612–510.

Yin, Y., Wang, F. and Sun, P. (2009) Landslide hazards triggered by the 2008 Wenchuan earthquake, Sichuan, China. *Landslides*. doi: 10.1007/s10346-009-0148-5.

Zenkovich, V.P. (1967) *Processes of Coastal Development*, Oliver and Boyd, London.

4

Sediment Cascades in Active Landscapes

Tim R. H. Davies[1] and Oliver Korup[2]

[1]Department of Geological Sciences, University of Canterbury, Christchurch, New Zealand
[2]Swiss Federal Institute for Forest, Snow and Landscape Research, CH-7260 Davos, Switzerland

4.1 Introduction

Many of Earth's tectonically active landscapes are mountainous and aesthetically dramatic, and socially valued because of their natural and cultural significance. As a result, many formerly sparsely populated mountain landscapes are now more and more used for tourism and recreation. This growing human presence in areas of high geological process activity increases the range and intensity of potential threats to society, jeopardizing sustainable development in many mountain belts (Ives and Messerli, 1989). In this context, 'natural' hazards are the result of society occupying geomorphically active terrain, characterized by episodic and sometimes catastrophic movement of water and sediment. It follows that the viability of societies in active landscapes depends on their ability to recognize, understand, and adapt to these landscape, in order to avoid loss of lives, housing, and infrastructure.

The evolution of active landscapes is driven by competing rates of rock uplift and erosion. Topographic relief and the steepness of hillslopes partly result from this interplay, and at the same time set tight controls on processes of erosion and sediment transport. Bedrock incision by rivers responds to rock uplift, thus driving topographic relief development (Whipple, 2004; Finnegan *et al.*, 2008), while fluvial dynamics are affected by rates of sediment supply from upstream and from adjacent sources such as terraces or hillslopes (Korup, 2005). Here we show that these rates can be drastically impacted by formation and failure of the sediment reservoirs between which sediment cascades. Storages range in size from individual river bars to large valley trains, and their effects on river and landscape dynamics vary accordingly.

Sediment Cascades: An Integrated Approach Edited by Timothy Burt and Robert Allison

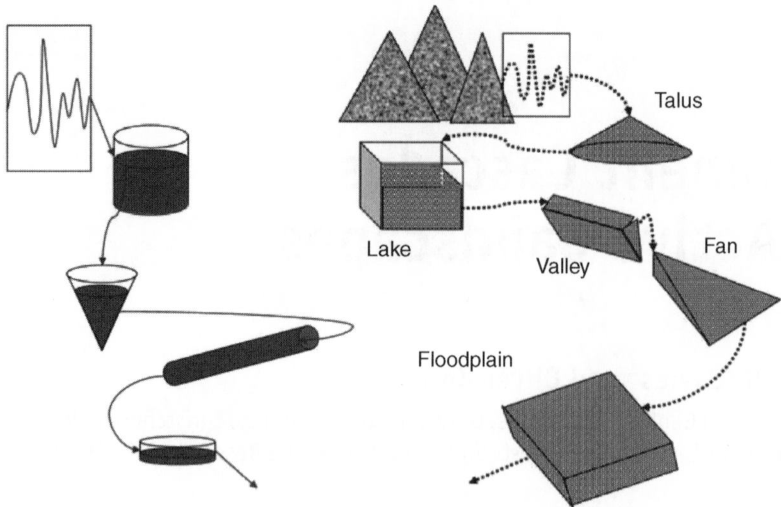

Figure 4.1 In cascades of water (left) and sediment (right), the transfer sequences between the storages depend on the sequences of inputs and on the characteristics of the storages. Water flows are continuous, whereas sediment transfers are very intermittent

In this chapter we explore the effects of sediment cascades on geomorphological process rates and landform development in tectonically active landscapes, with an emphasis on collisional mountain belts. These cascades result from the fact that sediment generation, transport, and deposition in active mountain ranges vary dramatically in time and space (Figure 4.1; Kirchner *et al.,* 2001). This also implies that the volumes, morphologies, and residence times of sediment storages are similarly variable, but data to test this quantitatively are sparse. This lack of reference or benchmark data is a major shortcoming, especially since rapid changes to sediment storages such as catastrophic erosion or deposition can have serious consequences for human society and investments located on them or in their vicinity.

We begin by outlining general concepts of landscape evolution, followed by a brief overview of sediment production, transport, and storage; then focus on the chain of storages through which sediment moves to eventually reach base level. We emphasize several characteristics of the sediment cascade (Figure 4.1), and their implications for process rates and landform dynamics in mountain rivers, based on the fundamental concept of sediment continuity (mass conservation). Finally, we discuss the implications of sediment cascades and their behaviour for human land use in active landscapes.

4.2 Dynamics of Active Landscapes

Active landscapes form by the processes of rock uplift and erosion, transport, and deposition of sediment. Any landscape which remains active for $> 10^6$ years is likely to achieve a dynamic equilibrium between rates of rock uplift U and erosion E over these timescales:

$$-\frac{\partial z}{\partial t} = U + E = 0, \tag{4.1}$$

where z is the height of the land surface with reference to a fixed datum, and t is time. Uplift in non-volcanic settings results from tectonic plate motion and isostasy, and appears to remain somewhat steady at $>10^6$ year timescales (e.g. Valensise and Ward, 1991). This implies that if E matches U in the long term, topography remains invariant in a statistical sense over that timescale. If this were not the case, the difference between U and E would cause changes in relief at a scale not observed on Earth's surface. For example, if U exceeded E in the Southern Alps of New Zealand consistently by as little as 1 mm yr^{-1}, then in its 10^7 years of existence (Tippett and Kamp, 1995) the range would have achieved an elevation of 10 km above sea level. In fact, its maximum elevation is *c.* 4 km, and there is no evidence for significantly greater mountain-peak elevations in the past. Hence any long-term difference between U and E must be <1 mm yr^{-1} in this mountain belt. Elevation and topographic relief are therefore limited, and the assumption that this particular landscape reflects a large-scale, long-term dynamic equilibrium between U and E seems reasonable. Obviously, though, in order for topography to be created in the first place, U must have exceeded E for some geologically significant period of time.

Equation 4.1 implies constancy of sediment flux, that is the rate at which eroded sediment is transported across a landscape. This depends on the power available to the transporting agents, which is in turn determined by local slope gradients and the flux rate of the transporting media such as water, ice, and wind. In a dynamic equilibrium, therefore, active landscapes must tend towards a set of gradients that enable transport processes to carry sediment to base level at the rate at which it is being created by uplift and weathering (Lu, Moran and Sivapalan, 2005).

An important corollary is that any change to the geometry of landforms also changes the power of transporting agents, and topography may control erosion rates (Figure 4.2). In the case of rivers, for example, this causes changes in sediment flux q_s. In its simplest one-dimensional form this can be stated as

$$-\frac{\partial z}{\partial t} = \frac{1}{(1-P)}\frac{\partial q_s}{\partial x}, \tag{4.2}$$

where P is the sediment porosity of the material transported, and x is the coordinate in the transport direction. Equation 4.2 derives from the physical principle of conservation of mass, and is referred to as the *sediment continuity* equation. It states that, in the absence of net uplift, changes to land surface (or in this case channel) elevation z by either erosion or deposition require changes in sediment flux q_s, which in Equation 4.2 is measured per unit channel width [$L^2 T^{-1}$]. In other words, eroded material must be exported from, and/or accumulated material must be imported to, the point of landscape in question. Integrating Equation 4.2 over a particular spatial and temporal scale of interest (e.g. a catchment) yields

$$V_e = \frac{\rho_o}{\rho_r} V_o + \frac{\rho_s}{\rho_r} V_s, \tag{4.3}$$

where V_e, V_o, and V_s are the volumes of sediment eroded, exported beyond a defined point in space (e.g. the drainage-basin outlet), and stored for the timescale of interest, respectively; ρ_r, ρ_o, and ρ_s are bulk densities of bedrock, exported sediment, and stored sediment, respectively. The ratio V_o/V_e is termed the *sediment delivery ratio* (*SDR*), and used as a quantitative indicator of spatially averaged sediment storage for a specific region over

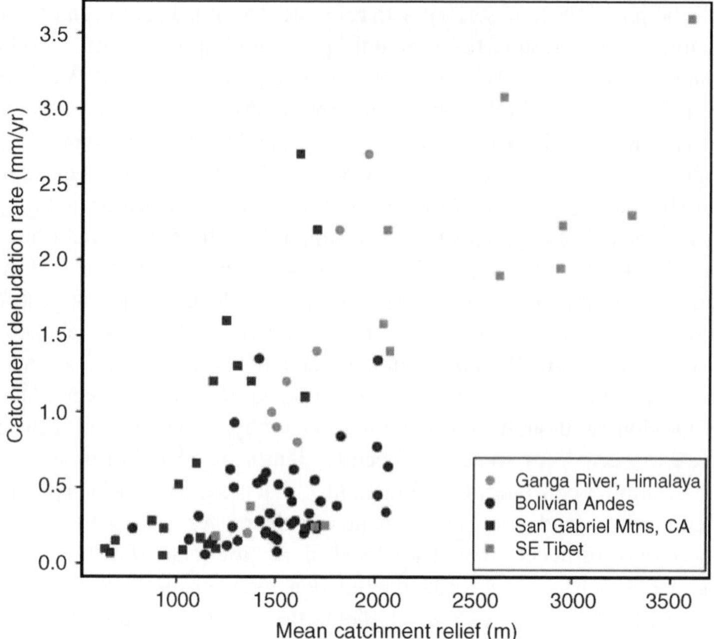

Figure 4.2 Estimates of mean 10^3-year catchment denudation rates derived from detrital cosmogenic nuclide samples of river sands increase with mean catchment relief, expressed here as the average maximum elevation difference in a 5-km radius for a given catchment using the SRTM30 digital elevation model
Data compiled from Vance *et al.* (2003), Safran *et al.* (2005), Binnie *et al.* (2007), and Finnegan *et al.* (2008).

a defined period of time (Lu, Moran and Sivapalan, 2005). Although the SDR is a common measure in sediment budget studies, it is debated because of its poorly defined frame of reference (Parsons *et al.*, 2006). A few alternatives have been suggested to quantify the *efficiency* of individual sediment storage units. The trapping efficiency T_i of sediment in lakes, for example, can be approximated by:

$$T_i = 1 - \mathrm{e}^{-x\omega_i/hv}, \tag{4.4}$$

where ω_i is the settling velocity of a particle of size i, h is the local lake depth, and v is the flow velocity in the x direction. Griffiths and McSaveney (1986) used this approach to estimate the sediment trapping efficiency and aggradation history of an alluvial fan bordering the Southern Alps of New Zealand.

In particular, we note the implication of mass conservation, that

$$q_{\mathrm{so}} = q_{\mathrm{si}} + \mathrm{d}V_{\mathrm{s}}/\mathrm{d}t \tag{4.5}$$

where q_{si} and q_{so} are sediment flux rates upstream and downstream of a storage of volume V_{s} respectively: the rate of sediment supply to a downstream river reach is affected by the rate of change of volume of upstream storage (Lu, Moran and Sivapalan, 2005). Examples of this effect are the river bed degradation that commonly takes place downstream of natural dams, and the aggradation following failure of a landslide dam (see Figure 4.13C). Moreover, rapid reworking and entrainment of stored floodplain or terrace sediments may significantly modulate the sediment yield of small catchments, temporarily outweighing the effects of high hillslope erosion rates (see Figure 4.10). These effects attenuate with distance downstream, as the intervening channel storage volume either replenishes the sediment supply deficit or accommodates the supply excess.

4.3 Generation of Rock Debris

In active mountain ranges crustal rocks are uplifted many kilometres above base level. In order to be eroded and transported back to base level, these rocks must first become fractured and disaggregated. This is achieved mainly through physical, chemical, and biological weathering, as well as by the mechanical fracturing inherent in tectonic stresses during uplift and advection (Molnar, Anderson and Anderson, 2007). Intact rock exposed to the atmosphere becomes fractured by, amongst others, earthquake shaking, temperature variations, freeze–thaw cycles, growth of plant roots, and chemical activity. These (near-) surface processes mainly depend on climate, vegetation, and rock type, and lead to the development of a regolith (e.g. Taylor and Eggleton, 2001).

Further means to fragment rock include stress corrosion (Molnar, Anderson and Anderson, 2007), which is the gradual growth of microcracks in intact rock at stress levels much lower than the failure strength. The growth rate of microcracks is determined by the net stress in the rock. When crack networks coalesce they form separate rock fragments that are more readily erodible than intact masses of bedrock. Stress corrosion causes all steep rock walls eventually to become fragmented, and fail under gravity, generating spontaneous large deep-seated landslides, and thus large volumes of sediment. However, this may take *c.* 10^6 years in the case of very competent rocks (Augustinus, 1995). Strong seismic shaking can also generate very high short-term local stresses deep inside mountain edifices through internal reflection of seismic waves and edifice resonance to particular wave frequencies. Field data indicate that along mountain ridges, ground acceleration can be an order of magnitude higher than in adjacent flat terrain (Buech, 2008). This is corroborated by the observation that large earthquake-triggered landslides frequently lower drainage divides (Korup, 2006b), and leave bowl-shaped detachment areas on upper hillslopes (Jibson *et al.*, 2006) similar in shape and location to cirques (Turnbull and Davies, 2006).

Rates of sediment production are normally measured for individual catchments, and may vary over at least three orders of magnitude at the mountain-belt scale (e.g. Aalto, Dunne and Guyot, 2006). There is growing evidence that rare, very large landslides contribute significantly to these rates (Korup *et al.*, 2007; Hewitt, Clague and Orwin, 2008). By contrast, fluid-driven erosion agents (wind, rivers, and glaciers) appear to create relatively little new rock debris. Nevertheless rivers and glaciers are very effective *transporters* of available sediment. Although advancing glaciers readily entrain valley-floor sediments, their role in eroding bedrock has mainly been quantified indirectly through mass balances in proglacial lakes or gauged suspended sediment yields (Evans and Church, 2000). Hence researchers often refer to 'sediment evacuation' rather than 'glacial erosion' (Hallet, Hunter and Bogen, 1996). The long-term efficacy of glacial erosion is nevertheless demonstrated by deep troughs and fiords, which are generally (and somewhat paradoxically) well preserved in very strong rocks. Yet the maximum rate of long-term glacial bedrock erosion in these troughs, as inferred from cosmogenic exposure dating, appears to be of the order of 10^0 mm yr^{-1} (Fabel and Harbor, 1999), which is decisively lower than maximum uplift rates in active mountain belts such as the Himalayas, Taiwan, or the Southern Alps of New Zealand over comparable timescales. Whether glaciers and glaciated landscapes erode more rapidly than non-glaciated ones is still controversial (Hallet, Hunter and Bogen, 1996; Brozovic, Burbank and Meigs, 1997); for example, limited evidence points to surging glaciers abrading bedrock at $>10^1$ mm yr^{-1} in the short term (Humphrey and Raymond, 1994).

4.4 Sediment Transport

Sediment transport occurs through gravity and fluid flow, involving water, ice, and/or wind. These agents have to overcome frictional resistance before sediment transport can occur. Dry, granular unconsolidated debris will begin to move under gravity at slope angles $\geq 35°$; however, some large landslides start moving at significantly lower angles, and may run out for extraordinary distances (McSaveney and Davies, 2007). The presence of water can substantially reduce the frictional resistance to gravitational debris transport via fluctuating pore-water pressures (Iverson and Denlinger, 2001). Fluid motion under gravity occurs much more readily than that of solids, because water, ice, and air have very low to nil resistance to applied shear, and can flow under low pressure or topographical gradients. Fluid erosion of sediment is typically modelled as a non-linear function of the excess shear stress, which encompasses the effects of local slope gradient and fluid flux (Whipple, 2004; Davies and McSaveney, 2008), albeit at different characteristic transport distances and effectiveness. Rates of sediment transport vary by at least five orders of magnitude in steep mountainous terrain (Figure 4.3), depend among others factors on study area and sampling interval, and are notoriously difficult to measure in the field (Kirchner *et al.*, 2001).

Glacial erosion is often modelled as a function of ice velocity (Hallet, 1996), but clearly also depends on parameters such as sediment availability (Huerta and Winberry, 2007), bedrock topography and basal ice temperature. Mountain glaciers are limited in extent to scales of 10^2 km, so the debris they carry normally remains close to an orogen, especially where large ice-scoured lake basins bounded by terminal moraines act as substantial natural sediment traps (Figure 4.4). In contrast, wind and particularly water can carry debris over $>10^3$ km. The contribution of wind to the overall sediment flux in active mountain belts is expected to be negligible relative to those of gravity and water.

Debris flows involve both gravity and fluid transport, and consist of water-saturated debris that moves rapidly *en masse* on low slopes ($> 3°$; Marchi, Pasuto and Tecca, 1993); volumes can be up to 10^7 m^3 (Plafker and Ericksen, 1978; Haeberli *et al.*, 2004). Catastrophic outburst flows following sudden failure of natural dams often involve debris-flow phases, and produce some of the largest sediment pulses documented in active mountain belts (Korup and Tweed, 2007). Landslides, glaciers, and debris flows can transport blocks of up to 10^2 m in diameter, whereas even the steepest river flows rarely move rocks of dimension greater than their flow depth, typically 10^0 m. Beyond the mountain range, rivers are limited to transporting material up to gravel size; wind transport is limited to grain sizes $< 10^{-3}$ m.

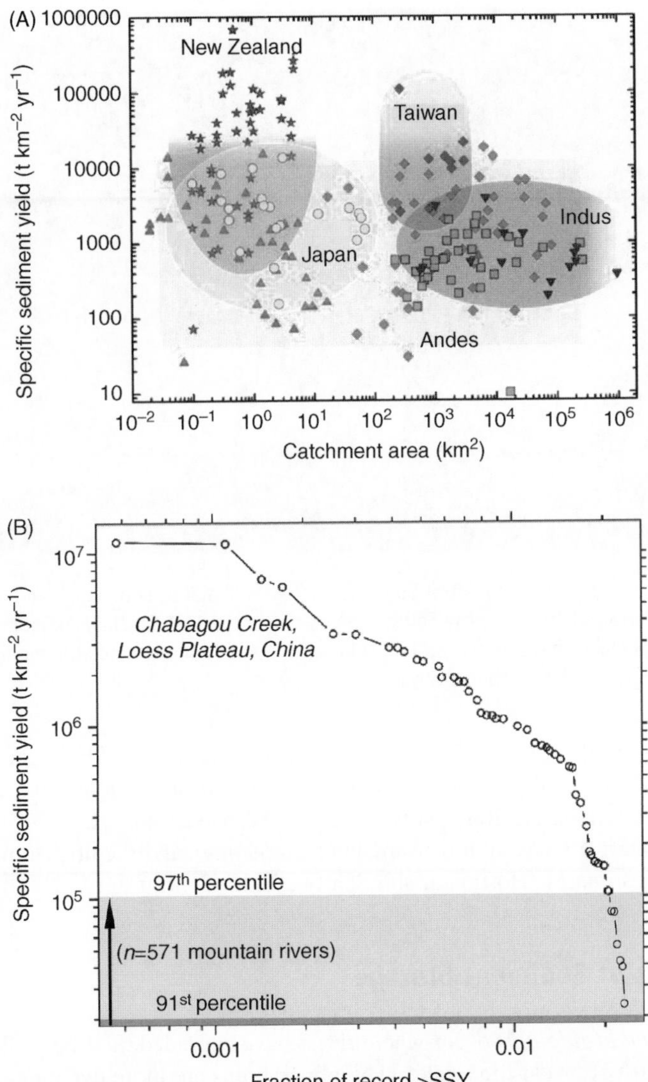

Figure 4.3 (A) Specific sediment yield from selected mountain rivers of the world as a function of catchment area. (B) Six-year record (1960–1969) of suspended specific sediment yield (SSY) from Chabagou Creek, Loess Plateau, China (data from Fang *et al.*, 2008), as a function of the fraction of time that a given value of SSY is exceeded. Grey shaded areas are 91st and 97th percentiles of SSY from a sample of $n = 571$ mountain rivers throughout the world. Excessive sediment yields typically peak during stormflows, which make up $< 2\%$ of the total flow record

4.5 Deposition and Storage of Sediment

The location, size, form and composition of sediment storages influence the dynamics of individual mountain rivers. Even at the mountain-belt scale, the formation of large moraine-dammed lakes or alluvial fans along a mountain front may alter base levels, and thus

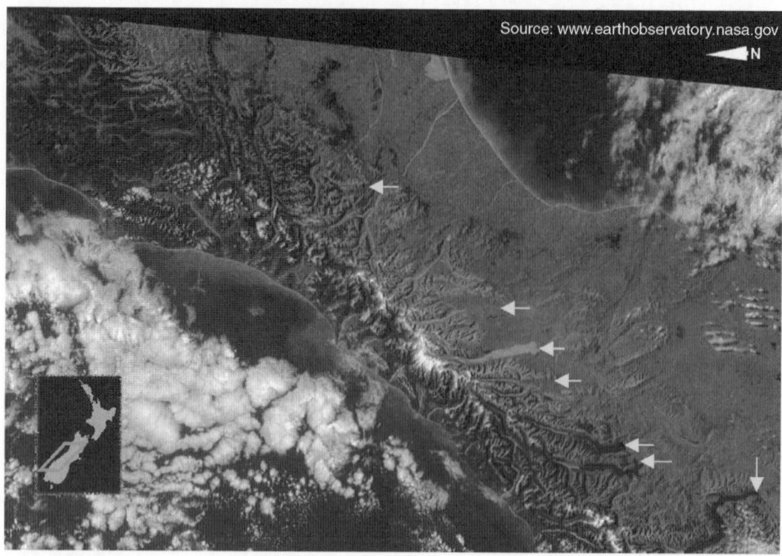

Figure 4.4 Satellite image of central South Island, New Zealand, showing several large moraine-dammed lakes (arrowed) fringing the Southern Alps. Since the Last Glacial Maximum (c. ~20,000 yr BP) many of these lakes have partly infilled. Many active and formerly glaciated mountain belts host such major sediment sinks at their fringes

modulate rates of fluvial erosion (e.g. Carretier and Lucazeau, 2005). Surprisingly few studies have systematically quantified sediment storage in large catchments, let alone entire mountain belts (Hinderer, 2001; Wasson, 2003; Wang, Li and He, 2007). Mapping of sediment storage is often straightforward, but methodological difficulties limit the accuracy of volumetric estimates (Hoffmann and Schrott, 2002; Schrott *et al.*, 2003).

4.5.1 Types of Sediment Storage

Floodplains and braidplains occur where rivers have aggraded their beds; floodplains are inundated at 10^1–10^2-year timescales, while braidplains are more dynamic. Data for such fluvial storage in tectonically active mountain belts are sparse (Reinfelds and Nanson, 1993). For example, in the New Zealand Southern Alps the total area of active channel bed increases systematically with catchment area, while it seems to *decrease* with increasing rates of uplift and precipitation (Figure 4.5). Reduction of sediment supply or base-level fall causes floodplain dissection and *terrace* formation (Figure 4.6). Importantly, even steep low-order catchments can become infilled with large amounts of hillslope-derived debris, and some of the highest rates of aggradation (up to $0.9\,\mathrm{m\,yr^{-1}}$ between 1957 and 2002) have been reported from debris-flow-prone ravines in Jinsha River, China (Wang, Li and He, 2007).

Fans develop where river-channel gradients and lateral confinement (and hence sediment transport capacity) decrease dramatically. Debris flows also form fans, and their great destruction potential means that identifying debris-flow fans is extremely important (Davies, 1997; Davies and McSaveney, 2008). Tributary alluvial fans can significantly

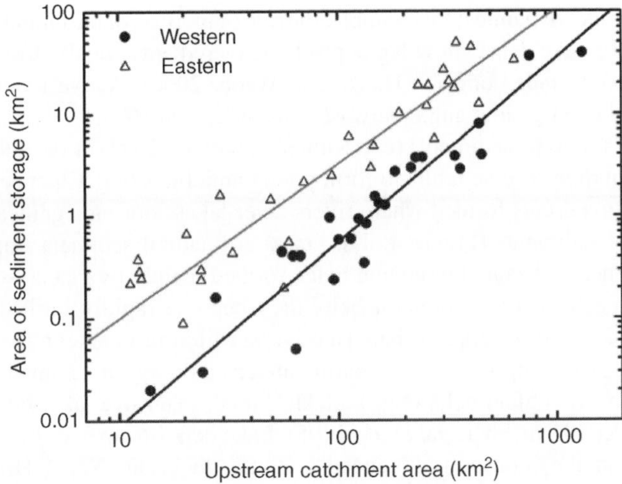

Figure 4.5 Relationship between active channel area of gravel-bed rivers and catchment area for selected drainage basins in the Southern Alps west (filled circles) and east (open triangles) of the divide. Solid lines are best fit from non-linear geometric mean regressions. Rates of uplift and precipitation are an order of magnitude higher in the western catchments. East-draining catchments tend to have larger valley trains, and hence higher potential for reworking and storing sediment, for a given catchment area

Figure 4.6 Types of sediment storage in tectonically active mountain belts. (A). Fluvial terraces, Broken River, Southern Alps, New Zealand. (B). Rock-fall scree, Alamedin River, Kyrgyz Tien Shan. (C). Late Quaternary lake sediments, Kyrgyz Tien Shan. (D). River delta, Lake Wanaka, Southern Alps, New Zealand. (E). Terraces formed in backwater sediments, Landquart River, Swiss Alps. (F). Reworked rock-avalanche debris, Crow River, Southern Alps, New Zealand. (G). Infilled low-order channels, At-Bashy River, Kyrgyz Tien Shan

modulate trunk-river dynamics; for example, the rate of incision of the Grand Canyon, USA, is moderated by coarse debris-flow lag deposits delivered episodically from side canyons forming long-lived grade controls (Hanks and Webb, 2006). Active mountain belts are typically fringed by large low-angle outwash fans, which on 10^5–10^7-year timescales may be stored as molasse-type sediments (e.g. Kuhlemann *et al.*, 2002) or recycled and uplifted along range-bounding reverse faults to form young anticlines (e.g. Champel *et al.*, 2002).

Deltas are alluvial fans formed where rivers enter lakes, storing significant amounts of incoming coarse sediments (Figure 4.6D). *Lakes* are natural sediment traps particularly abundant in formerly glaciated mountain belts, formed mainly by glacial scour or natural dams; in tectonically active mountain belts they become rapidly infilled (Figure 4.7). Einsele and Hinderer (1997) approximated the average lifetime of lakes by using the ratio of their surface area to the upstream contributing area as a proxy for sediment yield. In arid, tectonically active mountain belts (e.g. Ladakh, India), extensive lake sediments may be preserved for over 10^4 yr (Phartiyal *et al.*, 2005). Lake deposits up to c. 1 million years old may be found in the monsoon-influenced Kathmandu basin, Nepal Himalaya (Sakai *et al.*, 2006).

Landslide deposits accumulate on hillslopes, terraces, and floodplains, directly in the drainage network, in lakes and in fjords. The deposit locations are crucial in determining the sediment delivery ratio by landsliding in a given area of interest, and several classifications have been proposed in this regard (Korup, 2005). Few studies have detailed the relative (hillslope or catchment) locations of landslide deposits, and it is often simply assumed that the bulk of landslide sediment directly enters the fluvial drainage network. Extremely large ($>10^9$ m^3) landslide deposits can infill valley floors and cause drainage systems to reorganize (e.g. Korup, 2004; Korup and Tweed, 2007; Hewitt, Clague and Orwin, 2008). At smaller scale, *scree* (*talus*) forms through gradual accumulation of repeated small rockfalls.

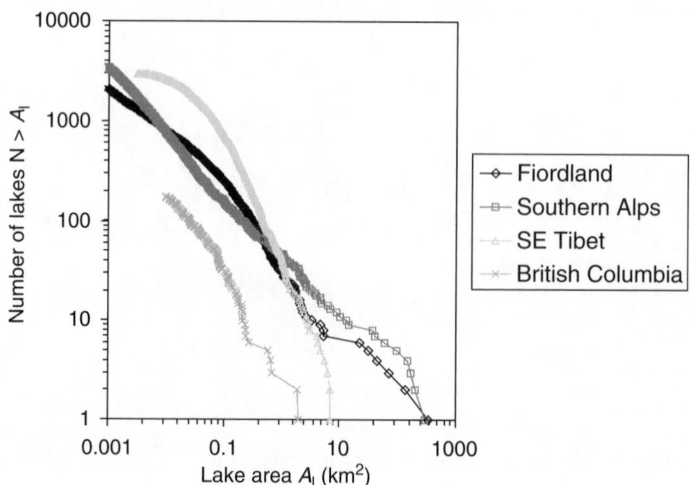

Figure 4.7 Cumulative size-frequency distribution of lake area in tectonically active and deglaciated mountain belts, i.e. Fiordland and the Southern Alps of New Zealand, Yarlung Tsangpo catchment near Namche Barwa, southeast Tibet, and British Columbia (moraine-dammed lakes only, McKillop and Clague, 2007).

Scree cones inclined at 30–40° typically mantle the toes of densely jointed rock slopes, and if undercut by a river can be copious sources of sediment (Figure 4.6B and F).

Moraines are glacial deposits; the most prominent are terminal moraines formed at glacier snouts, which in many mountain belts impound lakes, trapping (glacio-) fluvial sediment, which may be released catastrophically (Clague and Evans, 2000; Figure 4.4). Lateral and medial moraines are generally smaller and less significant to sediment translation along the glacier. However, during glacial cycles, large quantities of moraine material may reach adjacent lowlands. This can have significant large-scale and long-term effects on overall orogen dynamics, as discussed below. Although moraines are typically used as palaeoclimatic indicators (Benn and Evans, 1998), some moraines result from landslide-driven glacier advances (Hewitt, 2008; Tovar, Shulmeister and Davies, 2008).

Formation of sediment storages may cause further deposition of sediment through backwater effects, such as blockage of a river by a landslide deposit, moraine or tributary fan. Sediment will accumulate upstream of the blockage as a pro- and aggrading flood – or braidplain deposit. Importantly, such sediment storage can *propagate upstream* as well as *downstream*, through gradual backwater aggradation as well as catastrophic deposition of outburst flow deposits, respectively (Korup and Tweed, 2007). Depending on the valley gradient and geometry, these *consequent* storages can be very much larger in volume than the causative storage.

4.5.2 Effect of Sediment Storage on Incision and Relief

In tectonically active landscapes, bedrock rivers respond to uplift by incision, which increases hillslope relief, modulates edifice stresses, and increases the propensity for landsliding (Korup *et al.*, 2007). Sediment storage in the form of major valley fills can substantially inhibit this bedrock incision (Korup, 2006), but it also allows the river to attack hillslope portions that were previously out of reach.

4.5.3 Effect on Crustal Stresses

Denudation and export of sediment from mountain ranges relieves crustal stress and affects uplift rates (e.g. Beaumont, Fullsack and Hamilton, 1992; Koons, 1995; Zeitler *et al.*, 2001; Pysklywec, 2006). Conversely, build-up of sediment storage volumes at geological timescales ($> 10^5$ yr) in mountain river systems retains in the orogen substantial masses of sediment that would otherwise have been exported to foreland basins. Given sufficient storage volume and appropriate crustal strength, this may affect the crustal deformation locally, especially over longer timescales chiefly perturbed by glacial–interglacial cycles (see Section 4.6).

4.5.4 Storage Longevity

The longevity of sediment storage in mountainous terrain is highly variable and largely unpredictable. Specifically, the longevity of landslide dams, though much studied, remains obscure (Nash *et al.*, 2008). The most significant characteristic of sediment storage is its ability to *fail catastrophically*, suddenly releasing large quantities of water and sediment into the river system. Lowland storages are less prone to sudden failure, but the most dramatic way lowland rivers can be perturbed is by impulsive release of sediment from mountain storage upstream. From Equation 4.5, if $q_i < q_o$ for sufficient time, $V_s \rightarrow 0$ and the storage is destroyed. This can be achieved at a wide range of rates, and it is useful to distinguish between two end-members:

- *Gradual erosion* of storage takes place on $> 10^0$-year timescales and occurs by repeated fluvial or glacial entrainment or undermining – for example, headward knickpoint migration along a steep facet of the storage.

- *Catastrophic erosion or 'failure'* involves the rapid ($< 10^{-1}$ yr) destruction and emptying of sediment storages, and may result from fluvial, glacial, or mass-movement erosion, or seismic ground shaking.

An emblematic mechanism is the catastrophic failure of natural dams, and the release of large volumes of water from their associated lakes. Quantifying the resistance to failure of such dams is a key concern for hazard assessments, and also determines their geomorphological effects on the fluvial system. Storages comprising mainly fine sediment are more likely to be eroded than those containing substantial proportions of coarse material, due to differences in strength and permeability, all other influences held constant. Care is needed in inferring subsurface material from surface appearance, however; many large rockslide dams comprise large boulders of intact rock, typically set in a matrix of intensely fragmented, finer particles (McSaveney and Davies, 2007), and covered by a coarse carapace of less-fragmented rock (Dunning, 2004) that at first glance appears able to armour them.

The position of sediment storage in a catchment determines the availability of erosional processes to remove it. Examples of large and persistent storages include deposits of large rock-slope failures, many of which have been identified in the Indus River, Karakoram (Hewitt, 1999, 2009; Hewitt, Clague and Orwin, 2008). Notwithstanding the extremely high rates of uplift and river incision into bedrock, most are Late Quaternary features that have caused the accumulation of sediment wedges, containing up to 10^{10} m^3 of debris, in deeply dissected valleys. Landslide-dammed lakes are natural sediment traps, so age information further allows inferences about the long-term denudation rates of catchments upstream of these barriers (Korup and Tweed, 2007). Intriguingly, many of these landslide dams have evidently survived until their accommodation space was full of sediment, despite inferred denudation rates of 5–10 mm yr^{-1} (Figures 4.8 and 4.9).

4.6 Effects of Sediment Storage and its Failure on Downstream Sediment Flux

4.6.1 Processes

The formation and failure of sediment storages have important impacts on sediment transfer rates and hence on the landforms of the mountain valleys they occupy (Figure 4.10).

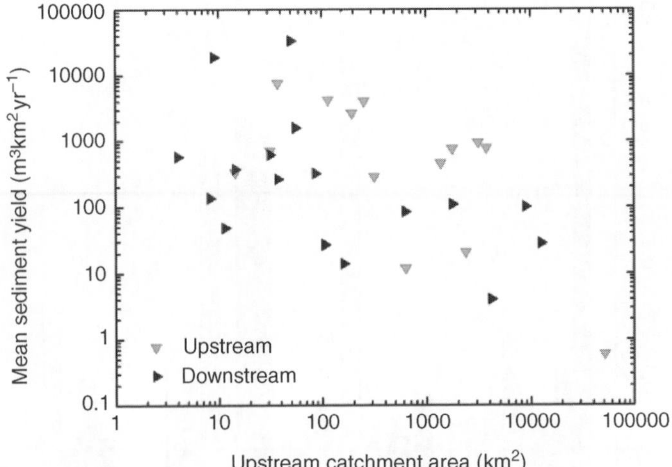

Figure 4.8 Sediment yields into natural reservoirs ("upstream"), and from failed landslide dams ("downstream") averaged over the time since initial formation of selected river blocking landslides, as a function of upstream catchment area. Upstream yields reveal high denudation rates and little storage in smaller catchments. Downstream pulses of excessive sediment are mainly prominent in upper catchment positions
Data from Korup and Tweed (2007).

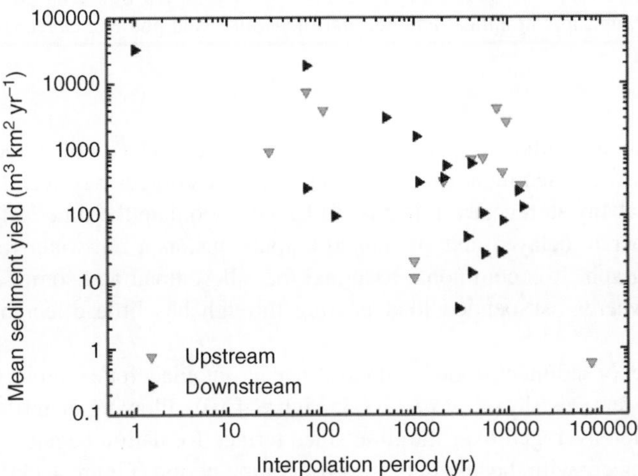

Figure 4.9 Sediment yields into natural reservoirs ("upstream"), and from failed landslide dams ("downstream") averaged over the time since initial formation of selected river blocking landslides ("interpolation period"). Massive sediment pulses from dam failures can be expected on 10^1 year timescales, although the millennial-scale contribution can still be significant. High upstream yields over such timescales indicate persistent high denudation rates
Data from Korup and Tweed (2007).

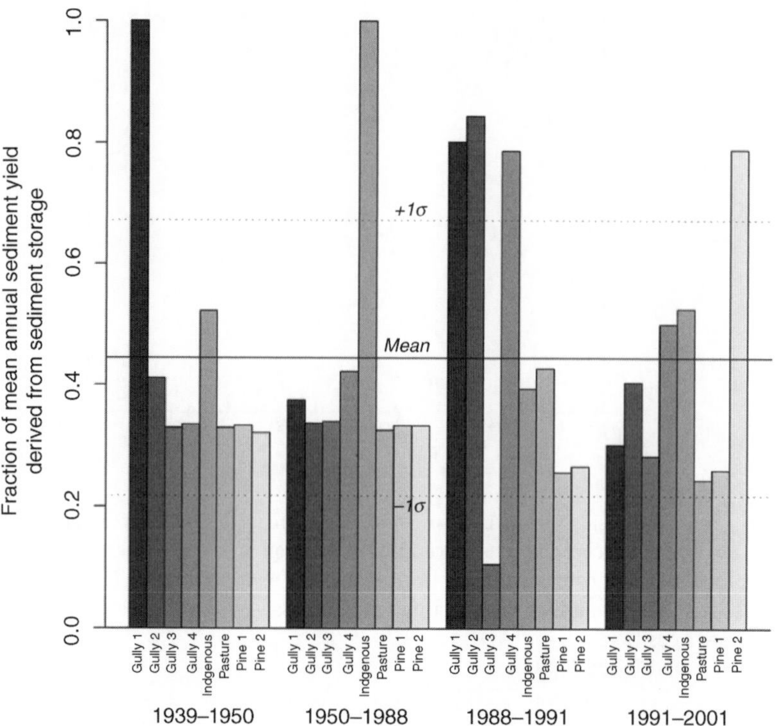

Figure 4.10 Fraction of mean annual sediment yield derived from valley-floor sediment storage for $n = 8$ small landslide-affected headwater catchments, Weraamaia, North Island, New Zealand (data from Kasai *et al.*, 2005). The fraction of annual sediment yield fed by fluvial erosion of sediment storage is highly episodic at annual and decadal timescales, and poorly characterized by the mean value

A sediment storage will trap most of the incoming sediment until its accommodation space is filled. While sediment routing through natural storages has been little studied, it seems clear that any storage, even if it is 'full', will modulate the time series of sediment inputs and generate delayed and attenuated outputs, just as a full water reservoir affects a flood hydrograph. It is commonly assumed that all bedload and some suspended load are trapped, whereas suspended load passing through has little effect on downstream morphology.

Average rates of sediment storage reduction can be obtained, for example, from dated and incised fluvial terraces (Kasai *et al.*, 2005; Figure 4.10), allowing quantification of rates of fluvial erosion averaged over the time since terrace formation began. Typically, these rates scale inversely with the length of the observation period (Figure 4.11), because of the inability to trace in the geological record smaller events over increasing longer periods of time, as they are being overprinted by larger events. This *erosional censoring* inhibits a complete reconstruction of the former magnitudes and frequencies of processes. The non-linear power-law trend that is typically observed in such records (Figure 4.11) highlights the limited utility of *mean* process rates, as the magnitude of process rate clearly scales with interpolation period.

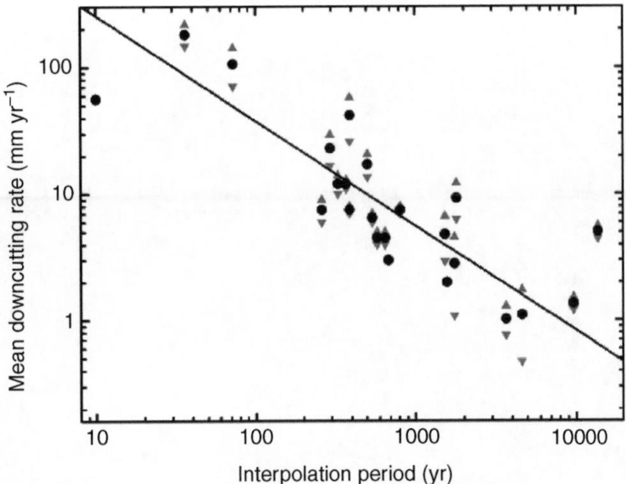

Figure 4.11 Rates of fluvial downcutting into fluvial fill terraces averaged over the time span since erosion began. Data are from Korup (2004) and other sources for terraces in the western Southern Alps, New Zealand. Filled circles are mean rates, and grey triangles are minimum and maximum rate estimates. Note systematic decrease of fluvial incision rates as a function of interpolation period

Many storage failures cause sudden and substantial sediment inputs to a system (e.g. Meigs *et al.*, 2006), and some of the highest estimated sediment yields (*c.* $10^5\,m^3\,km^{-2}\,yr^{-1}$; Korup and Tweed, 2007) arise from such catastrophic releases of sediment. With negligible consequent storage, the input is likely to be short-term and intense, especially in headwater catchments (Figures 4.8 and 4.9). With substantial consequent storage, the effect is likely to be less intense and longer-term, but the total sediment volume will be greater so the downstream geomorphological effects might also be greater. Davies and Korup (2007) showed that sudden sediment inputs to alluvial fan heads fringing the western Southern Alps, New Zealand, helped maintain disequilibrium conditions throughout the Holocene due to the episodic formation and failure of upstream sediment storages (Figures 4.12 and 4.13).

Our overview leads to a picture of mountain landscape evolution being modulated by sediment storages, between which sediment cascades, in a manner determined by the sediment production rate and the characteristics of intermediate storages. Storages are limited in longevity, and are liable to fail and disappear, depending on the recurrence of events that favour widespread erosion such as earthquakes, major storms, glaciations. These transient dynamics significantly complicate the long-term assumption of mountain belt evolution where U equals E in the long term (Equation 4.1).

4.6.2 Significance

The importance of sediment storage may be gauged from its regional occurrence. For example, the rapidly denuding (E *c.* $10\,mm\,yr^{-1}$) western Southern Alps, New Zealand, feature surprisingly few continuous, deeply incised bedrock rivers, although rates of U *c.* $10\,mm\,yr^{-1}$ along the active range-bounding Alpine Fault force steep river gradients.

Figure 4.12 The Tatare fan, foreland of the western Southern Alps, New Zealand, built up behind the Waiho Loop before overtopping it at c. 1 ka. The Tatare fan has always been higher than the Waiho River (until the past 20 yr), creating a natural abutment, and protected most of the Waiho Loop from erosion by the Waiho River
(photograph courtesy of George Denton).

Yet most river valley profiles are stepped, representing a cascade of alternating gorges and sediment storages, mainly controlled by rapid infilling of glacially scoured bedrock basins or naturally dammed lakes (Figure 4.14). First-order estimates of sediment volumes stored in 20 Holocene moraines accumulate to $>1.4 \times 10^8\,m^3$; nearly twice as much ($>2.5 \times 10^8\,m^3$) is stored behind just three major moraine dams. Eroding these volumes at current sediment yield rates estimated for these catchments would take nearly 100 years, simplistically assuming no other sediment sources.

4.6.3 Effects on Sediment Regimes

We distinguish between three distinct downstream sediment delivery regimes that occur sequentially as storages are created and destroyed:

- The *unmodified regime* stores sediment only in river channels and floodplains, and its behaviour in response to sediment inputs can be analysed or modelled on that basis.

- The *pre-failure regime* is due to emplacement of a temporary sediment sink, which traps significant fractions of incoming material until it is full, thereafter delaying and attenuating downstream delivery from a specific sediment input. The consequent storage surface will act as an aggrading alluvial fan or braidplain. Analysis requires understanding of how both prograding and fully developed alluvial fans and braidplains respond to specific sediment inputs.

Figure 4.13 (A) Poerua landslide dam, New Zealand, two months after failure (1999). Note fine material comprising the dam core. Depth of eroded section is 50 m. (B and C). Sediment released from dam is being reworked downstream, overwhelming and aggrading on highly productive farmland. The two photographs were taken in opposite directions; arrows show the same farmhouse

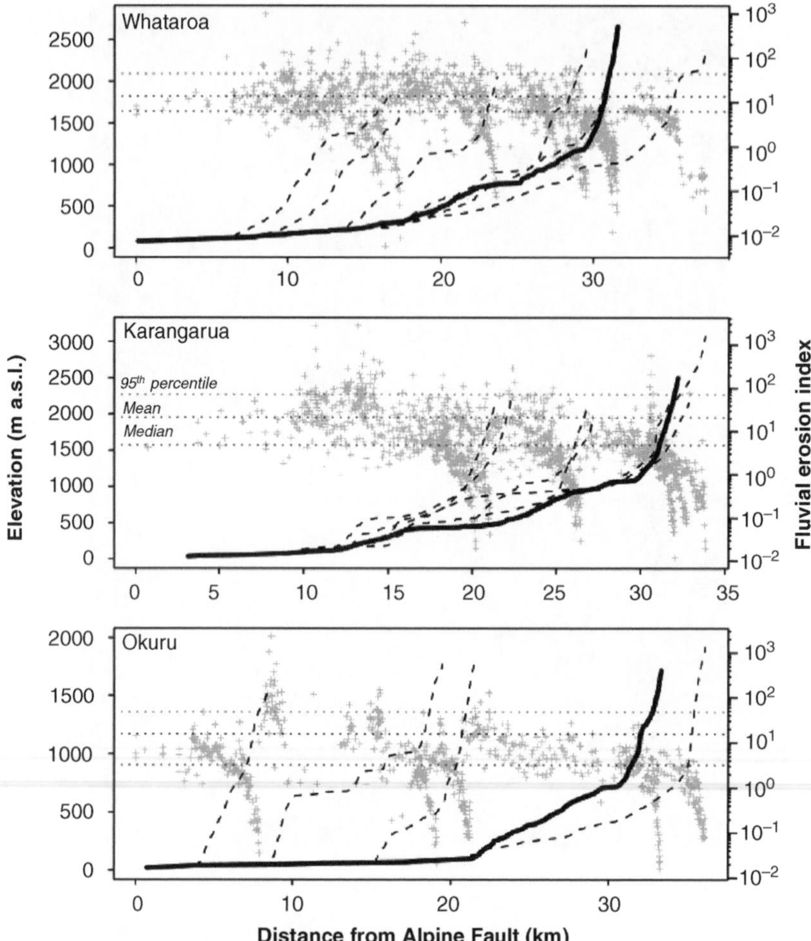

Figure 4.14 Stepped river longitudinal profiles, western Southern Alps, New Zealand (names refer to trunk streams shown in thick black; dashed lines are tributaries); flat reaches correspond to sediment storage such as floodplains and terraces, whereas steeper reaches are bedrock gorges. Fluvial erosion index (grey [$m^3 s^{-1}$]; together with median, mean, and 95th percentile) is based on the product of local channel gradient and discharge estimated from regional flood frequency

- The *post-failure regime* involves release of some or all of the stored sediment downstream, and the time-distribution of this release depends on when failure occurs.

The effects of storage failure time on downstream sediment delivery are sketched in Figure 4.15 which compares sediment delivery rates to the unmodified rate (dotted line). Only persistent storage that becomes and remains filled with sediment can force an upstream migration of base-level increase through backwater aggradation. Subsequent base-level fall following failure of the storage will also be transmitted upstream by means of knick-point recession. It is important to realize that these are, geologically speaking, transient effects. Ultimately, most of the stored sediment will be reworked down through the system.

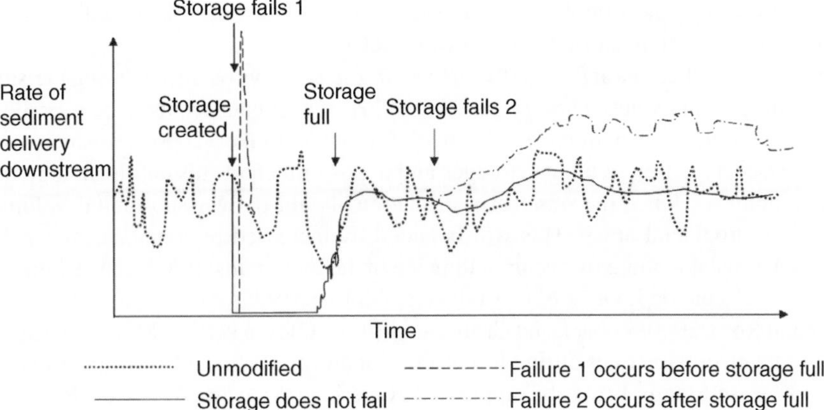

Figure 4.15 Downstream sediment delivery sequence as a function of when the storage fails. If failure occurs soon after emplacement (1), there will be a sudden and short-lived sediment pulse fed by material from the storage (dashed line). If failure occurs after the consequent storage is full (2), the sediment release will be greater still in volume, of much longer duration and lower in intensity (chain dotted line). Ultimately, the storage may be largely eroded and the unmodified sediment delivery regime re-established. If the storage does not fail, sediment delivery will only resume once it is full. After that the average delivery rate will attain the unmodified rate, but the large surface area of the storage will smooth the sediment delivery rate (full line) compared to the unmodified time series

4.6.4 Timescales of Perturbation

Our overview highlights that the effects of sediment storage dynamics on the overall sediment flux depend strongly on the timescale of interest (Figures 4.9–4.11 and 4.15). At *short (10^0 to 10^1 yr) timescales* we expect storages to damp out fluctuating sediment input within the cascade, whereas episodic major sediment inputs may cause substantial disturbances (Figure 4.13).

At *intermediate (10^2 to 10^3 yr) timescales* disturbances by high-magnitude/low-frequency events become increasingly important. In particular, many tectonically active mountain belts experience major ($M > 8$) earthquakes at intervals of the order of 10^2–10^3 years (Goldfinger *et al.*, 2008). These typically increase sediment production through landsliding, catastrophic formation of valley fills, blocking river channels and failure of existing sediment storages, and also force river-channel response through local base-level change. Earthquake-induced sediment storages may be ephemeral, and a significant portion may fail soon after formation, resulting in dramatic increases in sediment delivery to downstream river reaches, floodplains and foreland fans, which become and remain aggraded for years, if not centuries, after the earthquake (e.g. Keefer, 1999; Dadson *et al.*, 2004). Once such major sediment pulses have passed through the system, rivers will become incised into the newly aggraded (foreland) surfaces. This scenario is recorded in conspicuous metre-scale aggradation of much of the *c.* 100 km^2 Whataroa fan in Westland, New Zealand, and attributed to the 1620 Alpine Fault earthquake (Berryman *et al.*, 2001). The very young ages of nearly all foreland floodplain and fan soils in Westland support this notion (Davies and Korup, 2007). Coastal dune sequences in South Westland with seaward-decreasing ages associated with reconstructed dates of

Alpine Fault earthquakes imply that sediment delivery to the coast increased dramatically following major earthquakes (Wells and Goff, 2007).

Repeated perturbations at *long (10^4 to 10^5 yr) timescales* have occurred during Pleistocene glacial–interglacial cycles. The generation and flux of sediment seems to be related to the onset and decline of glaciations (e.g. Barnard, Owen and Finkel, 2006). Advancing valley glaciers evacuate valley sediment storages and deposit much of this debris in large latero-frontal moraines. Hence the onset of a glaciation should deliver substantial volumes of sediment to proglacial areas. This will be aided by increased regional seismicity due to differential crustal loading by accumulating ice in the mountains, proglacial sediment, and any effects of concomitant sea-level fall (e.g. Mörner, 1995; Arvidsson, 1996; Stewart, Sauber and Rose, 2000; Zoback and Grollimund, 2001; Olesen *et al.*, 2004). However, there are few data on erosion rates during full glacial conditions. While a landscape is under full glaciation, seismic activity may be suppressed by ice-load (Stewart, Sauber and Rose, 2000). The response of mountain peaks to seismic shaking will be reduced if partly capped by ice. The occurrence of bedrock landslides is further limited by reduced availability of exposed rock slopes. Removal of weathered material should therefore be slower during glacials, and the rate of uplift may exceed it, resulting in increasing summit elevations with potential for positive feedback to ice accumulation (Rother and Shulmeister, 2006). Deglaciation re-exposes valley flanks to subaerial erosion, leaving elevated peaks particularly susceptible to large-scale landsliding, including that due to increased seismic activity. In Scandinavia, for example, the final stages of the Last Glacial Maximum (LGM) were accompanied by unusually high rates of fault activity (Turpeinen *et al.*, 2008). Valley glacier retreat is commonly associated with the formation of proglacial lakes that trap meltwater sediments and hillslope debris (Figure 4.7). In the European Alps, Hinderer (2001) found that Lateglacial denudation outpaced Holocene denudation by a factor of about two. Post-glacial sediment yield is typically out of equilibrium with long-term erosion rates in formerly glaciated mountain belts (Church and Slaymaker, 1989). There is a growing notion that the most pronounced changes to sediment flux are linked to oscillations of glaciers (e.g. Barnard, Owen and Finkel, 2006). Historic deglaciation in Icy Bay, Alaska, showed 10^0–10^1-year sediment fluxes equivalent to E *c.* 30 mm yr^{-1} that clearly outpace long-term rates (Koppes and Hallet, 2006; Meigs *et al.*, 2006). Figure 4.16 summarizes a possible sediment delivery sequence to the foreland that might be expected from a mountain belt as a result of episodic seismic activity and a glacial–deglacial cycle.

4.7 Implications for Sustainable Development

People living in active mountain belts generally occupy terrain that suits agriculture, transport and building; this is often gently sloping land at low elevation (Figure 4.6E). Such terrain forms mainly by sedimentation, and thus comprises storage elements of the sediment cascade. Our discussion indicates that sediment storages associated with active mountain belts are vulnerable to catastrophic perturbations of the sediment cascade. To be sustainable in the broadest sense, human activities on these storages need to be adapted to cope with the intrinsic dynamics of the land on which they take place. Timescales of human experience are generally much shorter than the recurrence interval of catastrophic events. Temporary geomorphological quiescence appears benign, and can

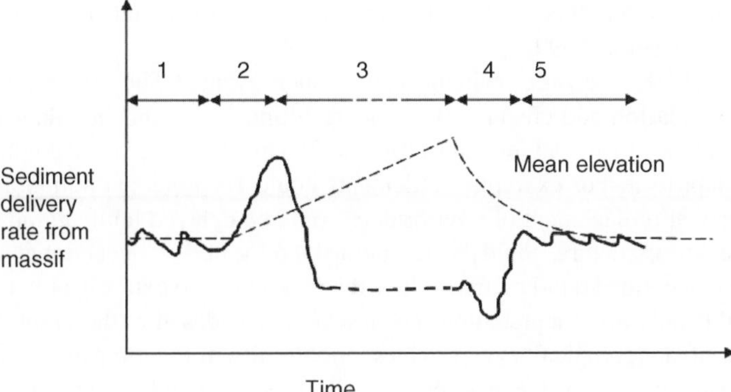

Figure 4.16 Sediment delivery rate from massif to foreland. 1 = preglacial, with perturbations due to major seismic events; 2 = onset of glaciation, with valley storages evacuated, and debris generation reducing as ice encasement of landscape increases; 3 = full glaciation, with debris production suppressed by ice encasement, and uplift exceeding denudation; 4 = deglaciation, with debris production very high, but most goes to refill valley storages; 5 = postglacial, with perturbations due to major seismic events

foster the belief that mountain valleys may be developed in a sustainable way in the long term.

Numerous mountain communities dwell on sediment storages that have changed catastrophically. Famous examples include the cities of Yungay (Peru; sited on landslide material) and Armero (Columbia; sited on lahar deposits) that were historically obliterated by just the types of events that formed the storages (Plafker and Ericksen, 1978; Voight, 1990). Countless mountain villages in both developed and less developed countries occupy small, steep alluvial fans that occasionally experience devastating debris flows.

More distant, but still vulnerable to the dynamics of sediment cascades, are settlements on fans and floodplains of mountain forelands. Many of them are intensely developed, and may appear unthreatened by geomorphological processes in the mountains. The Gangetic plains in India, which have formed from sediments delivered from the tectonically highly active Himalayas, feature classic examples. The catastrophic 2008 avulsion of the Kosi River on its alluvial fan is typical of the foreland processes expected to result from sediment delivery from active mountain belts. Such rivers naturally move to and fro across their fans, and the Kosi is recorded to have translated 140 km westwards in the past 250 years (Sinha, 1996). The 2008 avulsion shifted the artificially confined river about 100 km back to the east, highlighting the natural dynamics of rivers emerging from mountainous terrain.

A traditional approach to dealing with perceived hazards is to modify the geomorphological processes 'responsible'. One of the most widespread strategies is to modify river geometry so that it is compatible with human use of the floodplain. Davies and McSaveney (2006) showed that the bedload transport capacity of a self-formed river is a local maximum, and that river modifications such as width reduction by stopbanking or constraint of natural avulsion, reduce bedload capacity and cause river-bed aggradation. Thus even during periods of relatively steady sediment supply, human interference can exacerbate river problems (Kao and Liu, 2002). Catastrophic sediment pulses triggered by earthquakes pose even greater threats, and this has been illustrated by numerous landslide-triggering events worldwide (e.g. Keefer, 1999).

How can a society become resilient against such events? The effects might develop gradually; aggradation and channel avulsions resulting from landslide-dam failures may peak years after the event (Davies and Korup, 2007) – in which case local communities may be able to adapt to it. For example, settlements could be protected temporarily by ring-bunds; permanent replacement of river bridges could be delayed until aggradation waned; and valuable infrastructure could be temporarily relocated. Conventional river-control measures cannot control rapid metre-scale bed aggradation, so extensive (re-)development will be possible only after degradation sets in several decades after the event. What occurs in the blink of a geological eye provides opportunity for a prepared, adaptable and determined society to survive and prosper, if it is prepared to accept decadal-scale disruption. Alternatively, simply abandoning the area affected for the disturbance period may be preferable on economic, if not on social, grounds.

In summary, the dynamics of sediment cascades present severe problems to sustainable development in and close to mountain ranges. The sharp contrast between the ability of sediment storages to attenuate the flux of debris on the one hand, and to be the source of destructive sediment pulses on the other, may not always be evident. In this context, adaptability seems more relevant to long-term occupation of land than sustainability, because the latter has implications of maintaining the *status quo* – which is somewhat unrealistic in tectonically active landscapes, whose natural processes are far too powerful to be reliably controlled. The vital requirement for successful and planned adaptability is fore-knowledge of the situations to which communities will be required to adapt, and recognition of the role of sediment storages in modulating sediment delivery through cascades is crucial in this respect.

4.8 Summary and Conclusion

1. Sediment transfer through tectonically active mountain belts is modulated by sediment storages with volumes up to $>10^{10}\,m^3$ and residence times ranging between 10^0 and 10^6 years. A key characteristic of sediment storages in active mountains is their susceptibility to catastrophic formation and failure through a range of geomorphological processes.

2. Tectonic pre-fracturing, seismic shaking, stress corrosion, and weathering are capable of generating large-volume failures of mountain edifices. Landsliding appears to dominate sediment production and hence processes of geomorphological hillslope-channel coupling. Fluvial and glacial erosion may produce lesser amounts of rock debris, but generally transfer sediment across much larger distances.

3. Sediment transport by gravity is unlimited in capacity but cannot act over very long distances. Glaciers can transport very large volumes and sizes of rocks, but, with the exception of ice sheets covering low-gradient terrain, not for very large distances beyond the mountain belts they originate from. Water can transport smaller-sized sediment for distances set by regional base level.

4. The formation and failure of sediment storages affects both upstream and downstream river-system dynamics at a range of temporal and spatial scales, and has the potential to cause catastrophic perturbations of sediment flux.

5. Sustainable use of fluvially formed landscapes within or close to tectonically active mountains requires knowledge of the system dynamics resulting from the potentially catastrophic failure of sediment cascades. This calls for a certain degree of preparation and adaptation, rather than just reliance on engineering countermeasures.

References

Aalto, R., Dunne, T. and Guyot, J.L. (2006) Geomorphic controls on Andean denudation rates. *Journal of Geology*, **114**, 85–99.

Arvidsson, R. (1996) Fennoscandian earthquakes: whole crustal rupturing related to glacial rebound. *Science*, **274**, 744–745.

Augustinus, P.C. (1995) Glacial valley cross-profile development: the influence of in situ rock stress and rock mass strength, with examples from the Southern Alps, New Zealand. *Geomorphology*, **14**, 87–97.

Barnard, P.L., Owen, L.A. and Finkel, R.C. (2006) Quaternary fans and terraces in the Khumbu Himal south of Mount Everest: their characteristics, age and formation. *Journal of the Geological Society of London*, **163**, 383–399.

Beaumont, C., Fullsack, P. and Hamilton, J. (1992) Erosional control of active compressional orogens, in *Thrust Tectonics* (ed. K.P. McClay), Chapman and Hall, New York, pp. 1–18.

Benn, D.I. and Evans, D.J.A. (1998) *Glaciers and Glaciation*, Arnold, London.

Berryman, K., Alloway, B., Almond, P. *et al.* (2001) Alpine fault rupture and landscape evolution in Westland, New Zealand. Proceedings 5th International Conference on Geomorphology, Tokyo 2001.

Binnie, S.A., Phillips, W.M., Summerfield, M.A. and Fifield, L.K. (2007) Tectonic uplift, threshold hillslopes, and denudation rates in a developing mountain range. *Geology*, **35**, 743–746.

Brozovic, N., Burbank, D.W. and Meigs, A.J. (1997) Climatic limits to landscape development in the Northwestern Himalaya. *Science*, **276**, 571–574.

Buech, F. (2008) Seismic response of Little Red Hill – towards an understanding of topographic effects on ground motion and rock slope failure. PhD thesis, University of Canterbury, New Zealand, p. 148.

Carretier, S. and Lucazeau, F. (2005) How does alluvial sedimentation at range fronts modify the erosional dynamics of mountain catchments? *Basin Research*, **17**, 361–381.

Champel, B., van der Beek, P., Mugnier, J.L. and Leturmy, P. (2002) Growth and lateral propagation of fault-related folds in the Siwaliks of western Nepal: Rates, mechanisms, and geomorphic signature. *Journal of Geophysical Research*, **107** (B6), 2111. doi: 10.1029/2001JB000578.

Church, M. and Slaymaker, O. (1989) Disequilibrium of Holocene sediment yield in glaciated British Columbia. *Nature*, **337**, 452–454.

Clague, J.J. and Evans, S.G. (2000) A review of catastrophic drainage of moraine-dammed lakes in British Columbia. *Quaternary Science Reviews*, **19**, 1763–1783.

Dadson, S.J., Hovius, N., Chen, H.W. *et al.* (2004) Earthquake-triggered increase in sediment yield from an active mountain belt. *Geology*, **32**, 733–736.

Davies, T.R.H. (1997) Using hydroscience and hydrotechnical engineering to reduce debris flow hazards. Proceedings 1st International Conference on Debris Flow Hazards Mitigation, American Society of Civil Engineers, San Francisco, pp. 787–810.

Davies, T.R.H. and Korup, O. (2007) Persistent alluvial fanhead trenching resulting from large, infrequent sediment inputs. *Earth Surface Processes and Landforms*, **32**, 725–742.

Davies, T.R.H. and McSaveney, M.J. (2006) Geomorphic constraints on the management of bedload-dominated rivers. *Journal of Hydrology (NZ)*, **45**, 63–82.

Davies, T.R.H. and McSaveney, M.J. (2008) Principles of sustainable development on fans. *Journal of Hydrology (NZ)*, **47**, 43–65.

Dunning, S.A. (2004) Rock avalanches in high mountains – a sedimentological approach. PhD Thesis, University of Luton, UK.

Einsele, G. and Hinderer, M. (1997) Terrestrial sediment yield and the lifetimes of reservoirs, lakes, and larger basins. *Geologische Rundschau*, **86**, 288–310.

Evans, M. and Church, M. (2000) A method for error analysis of sediment yields derived from estimates of lacustrine sediment accumulation. *Earth Surface Processes and Landforms*, **25**, 1257–1267.

Fabel, D. and Harbor, J. (1999) The use of in-situ produced cosmogenic radionuclides in glaciology and glacial geomorphology. *Annals of Glaciology*, **28**, 103–110.

Fang, H.Y., Cai, Q.C., Chen, H. and Li, Q.Y. (2008) Temporal changes in suspended sediment transport in a gullied loess basin: the lower Chabagou Creek on the Loess Plateau in China. *Earth Surface Processes and Landforms*, **33**, 1977–1992.

Finnegan, N.J., Hallet, B., Montgomery, D.R. *et al.* (2008) Coupling of rock uplift and river incision in the Namche Barwa-Gyala Peri massif, Tibet. *Geological Society of America Bulletin*, **120**, 142–155.

Goldfinger, C., Grijalva, K., Bürgmann, R. *et al.* (2008) Late Holocene rupture of the northern San Andreas fault and possible stress linkage to the Cascadia Subduction Zone. *Bulletin of the Seismological Society of America*, **98**, 861–889.

Griffiths, G.A. and McSaveney, M.J. (1986) Sedimentation and river containment on Waitangitaona alluvial fan–South Westland, New Zealand. *Zeitschrift für Geomorphologie*, **30**, 215–230.

Haeberli, W., Huggel, C., Kääb, A. *et al.* (2004) The Kolka–Karmadon rock/ice slide of 20 September 2002: An extraordinary event of historical dimensions in North Ossetia, Russian Caucasus. *Journal of Glaciology*, **50**, 533–546.

Hallet, B. (1996) Glacial quarrying, a simple theoretical model. *Annals of Glaciology*, **22**, 1–8.

Hallet, B., Hunter, L. and Bogen, J. (1996) Rates of erosion and sediment evacuation by glaciers: a review of field data and their implications. *Global and Planetary Change*, **12**, 213–235.

Hanks, T.C. and Webb, R.H. (2006) Effects of tributary debris on the longitudinal profile of the Colorado River in Gran Canyon. *Journal of Geophysical Research*, **111**, F02020. doi: 10.1029/2004JF000257

Hewitt, K. (1999) Quaternary moraines versus catastrophic rock avalanches in the Karakorum Himalaya, northern Pakistan. *Quaternary Research*, **51**, 220–237.

Hewitt, K. (2008) Rock avalanches that travel onto glaciers and related developments, Karakoram Himalaya, Inner Asia. *Geomorphology*, **103** (1), 66–79.

Hewitt, K. (2009) Catastrophic rock slope failures and late Quaternary developments in the Nanga Parbat–Haramosh Massif, upper Indus basin, northern Pakistan. *Quaternary Science Reviews*, **28**, 1055–1069. doi: 10.1016/j.quascirev.2008.12.

Hewitt, K., Clague, J.J. and Orwin, J.F. (2008) Legacies of catastrophic rock slope failures in mountain landscapes. *Earth Science Reviews*, **87**, 1–38.

Hinderer, M. (2001) Late Quaternary denudation of the Alps, valley and lake fillings and modern river loads. *Geodinamica Acta*, **14**, 231–263.

Hoffmann, T. and Schrott, L. (2002) Modelling sediment thickness and rockfall retreat in an Alpine valley using 2D-seismic refraction (Reintal, Bavarian Alps). *Zeitschrift für Geomorphologie N.F.*, **127**, 153–173.

Huerta, A.D. and Winberry, J.P. (2007) Slow erosion by a fast glacier. *Eos (Transactions of the American Geophysical Union*), 88 (Fall Meeting Suppl.), Abstract C52A-07.

Humphrey, N.F. and Raymond, C.F. (1994) Hydrology, erosion and sediment production in a surging glacier – Variegated Glacier, Alaska, 1982–83. *Journal of Glaciology*, **40**, 539–552.

Iverson, R.M. and Denlinger, R.P. (2001) Flow of variably fluidized granular masses across three-dimensional terrain I. Coulomb mixture theory. *Journal of Geophysical Research B*, **106**, 537–552.

Ives, J.D. and Messerli, B. (1989) *The Himalayan Dilemma*, Routledge, London, p. 295.

Jibson, R.W., Harp, E.L., Schulz, W. and Keefer, D.K. (2006) Large rock avalanches triggered by the M 7.9 Denali Fault, Alaska, earthquake of 3 November 2002. *Engineering Geology*, **83**, 144–160.

Kao, S.J. and Liu, K.K. (2002) Exacerbation of erosion induced by human perturbation in a typical Oceania watershed: Insight from 45 years of hydrological records from the Lanyang-Hsi River, northeastern Taiwan. *Global Biogeochemical Cycles*, **16**, 1016. doi: 10.1029/2000GB001334

Kasai, M., Brierley, G.J., Page, M.J. *et al.* (2005) Impacts of land use change on patterns of sediment flux in Weraamaia catchment, New Zealand. *Catena*, **64**, 27–60.

Keefer, D.K. (1999) Earthquake-induced landslides and their effects on alluvial fans. *Journal of Sedimentary Research*, **69**, 84–104.

Kirchner, J.W., Finkel, R.C., Riebe, C.S. *et al.* (2001) Mountain erosion over 10 yr, 10 k.y., and 10 m.y. time scales. *Geology*, **29**, 591–594.

Koons, P.O. (1995) Modelling the topographic evolution of collisional belts. *Annual Review of Earth and Planetary Sciences*, **23**, 375–408.

Koppes, M. and Hallet, B. (2006) Erosion rates during deglaciation in Icy Bay, Alaska. *Journal of Geophysical Research*, **111**, F02023. doi: 10.1029/2005JF000349

Korup, O. (2004) Landslide-induced river channel avulsions in mountain catchments of southwest New Zealand. *Geomorphology*, **63**, 57–80.

Korup, O. (2005) Geomorphic imprint of landslides on alpine river systems, southwest New Zealand. *Earth Surface Processes and Landforms*, **30**, 783–300.

Korup, O. (2006a) Rock-slope failure and the river long profile. *Geology*, **34**, 45–48.

Korup, O. (2006b) Effects of large deep-seated landslides on hillslope morphology, western Southern Alps, New Zealand. *Journal of Geophysical Research*, **111**, F01018. doi: 10.1029/2004JF000242

Korup, O. and Tweed, F. (2007) Ice, moraine, and landslide dams in mountainous terrain. *Quaternary Science Reviews*, **26**, 3406–3422.

Korup, O., Clague, J.J., Hermanns, R.L., Hewitt, K., Strom, A.L. and Weidinger, J.T. (2007) Giant landslides, topography, and erosion. *Earth and Planetary Science Letters*, **261**, 578–589.

Kuhlemann, J., Frisch, W., Székely, B. *et al.* (2002) Post-collisional sediment budget history of the Alps: tectonic versus climatic control. *International Journal of Earth Science*, **91**, 818–837.

Lu, H., Moran, C.J. and Sivapalan, M. (2005) A theoretical exploration of catchment-scale sediment delivery. *Water Resources Research*, **41**, W09415. doi: 10.1029/2005WR004018

Marchi, L., Pasuto, A. and Tecca, P.R. (1993) Flow processes on alluvial fans in the Eastern Italian Alps. *Zeitschrift für Geomorphologie*, **37**, 447–458.

McKillop, R.J. and Clague, J.J. (2007) Statistical, remote sensing-based approach for estimating the probability of catastrophic drainage from moraine-dammed lakes in southwestern British Columbia. *Global and Planetary Change*, **56**, 153–171.

McSaveney, M.J. and Davies, T.R. (2007) Rockslides and their motion, in *Progress in Landslide Science* (eds K. Sassa, H. Fukuoka, F. Wang and G. Wang), Springer, pp. 113–134.

Meigs, A., Krugh, W.C., Davis, K. and Bank, G. (2006) Ultra-rapid landscape response and sediment yield following glacier retreat, Icy Bay, southern Alaska. *Geomorphology*, **78**, 207–221.

Molnar, P., Anderson, R.S. and Anderson, S.P. (2007) Tectonics, fracturing of rock, and erosion. *Journal of Geophysical Research*, **112**, F03014. doi: 10.1029/2005JF000433

Mörner, N.-A. (1995) Paleoseismicity – the Swedish case. *Quaternary International*, **25**, 75–79.

Nash, T., Bell, D.H., Davies, T.R. and Nathan, S. (2008) Analysis of the formation and failure of Ram Creek landslide dam, South Island, New Zealand. *New Zealand Journal of Geology & Geophysics*, **51**, 187–193.

Olesen, O., Blikra, L.H., Braathen, A. *et al.* (2004) Neotectonic deformation in Norway and its implications: a review. *Norwegian Journal of Geology*, **84**, 3–34.

Parsons, A.J., Wainwright, J., Brazier, R.E. and Powell, D.M. (2006) Is sediment delivery a fallacy? *Earth Surface Processes and Landforms*, **31**, 1325–1328.

Phartiyal, B., Sharma, A., Upadhyay, R. *et al.* (2005) Quaternary geology, tectonics and distribution of palaeo- and present fluvio/glacio lacustrine deposits in Ladakh, NW Indian Himalaya–a study based on field observations. *Geomorphology*, **65**, 241–256.

Plafker, G. and Ericksen, G.E. (1978) Nevados Huascaran avalanches, Peru, in *Rockslides and Avalanches, 1 Natural Phenomena* (ed. B. Voight), Elsevier, pp. 277–314.

Pysklywec, R.N. (2006) Surface erosion control on the evolution of the deep lithosphere. *Geology*, **34**, 225–228.

Reinfelds, I. and Nanson, G. (1993) Formation of braided river floodplains. *Sedimentology*, **40**, 1113–1127.

Rother, H. and Shulmeister, J. (2006) Synoptic climate change as a driver of late Quaternary glaciations in the mid-latitudes of the Southern Hemisphere. *Climate of the Past*, **2**, 12–19.

Safran, E.B., Bierman, P.R., Aalto, R. *et al.* (2005) Erosion rates driven by channel network incision in the Bolivian Andes. *Earth Surface Processes and Landforms*, **30**, 1007–1024.

Sakai, H., Sakai, H., Yahagi, W., Fujii, R., Hayashi, T. and Upreti, B.N. (2006) Pleistocene rapid uplift of the Himalayan frontal ranges recorded in the Kathmandu and Siwalik basins. *Palaeogeography Palaeoclimatology and Palaeoecology*, **241**, 16–27.

Schrott, L., Hufschmidt, G., Hankammer, M. *et al.* (2003) Spatial distribution of sediment storage types and quantification of valley fill deposits in an alpine basin, Reintal, Bavarian Alps, Germany. *Geomorphology*, **55**, 45–63.

Sinha, R. (1996) Channel avulsion and floodplain structure in the Gandak-Kosi interfan, north Bihar plains, India. *Zeitschrift für Geomorphologie N.F.*, **103**, 249–268.

Stewart, I.S., Sauber, J. and Rose, J. (2000) Glacio-seismotectonics: ice sheets, crustal deformation and seismicity. *Quaternary Science Reviews*, **19**, 367–1389.

Taylor, G. and Eggleton, R.A. (2001) *Regolith Geology and Geomorphology: Nature and Process*, John Wiley and Sons, p. 375.

Tippett, J.M. and Kamp, P.J.J. (1995) Geomorphic evolution of the Southern Alps, New Zealand. *Earth Surface Processes and Landforms*, **20** (2), 177–192.

Tovar, D.S., Shulmeister, J. and Davies, T.R. (2008) Evidence for a landslide origin of New Zealand's Waiho Loop, Moraine. *Nature Geosciences*, **1**, 524–526.

Turnbull, J.M. and Davies, T.R.H. (2006) A mass movement origin for cirques. *Zeitschrift für Geomorphologie*, **31**, 129–1148.

Turpeinen, H., Hampel, A., Karow, T. and Maniatis, G. (2008) Effect of ice sheet growth and melting on the slip evolution of thrust faults. *Earth and Planetary Science Letters*, **269**, 230–241.

Valensise, G. and Ward, S.N. (1991) Long-term uplift of the Santa Cruz coastline in response to repeated earthquakes along the San Andreas Fault. *Bulletin – Seismological Society of America*, **81** (5), 1694–1704.

Vance, D., Bickle, M., Ivy-Ochs, S. and Kubik, P.W. (2003) Erosion and exhumation in the Himalaya from cosmogenic isotope inventories of river sediments. *Earth and Planetary Science Letters*, **206**, 273–288.

Voight, B. (1990) The 1985 Nevado del Ruiz volcano catastrophe: anatomy and retrospection. *Journal of Volcanology and Geothermal Research*, **42**, 51–188.

Wang, Z.Y., Li, Y. and He, Y. (2007) Sediment budget of the Yangtze River. *Water Resources Research*, **43**, W04401. doi: 10.1029/2006WR005012

Wasson, R.J. (2003) A sediment budget for the Ganga-Brahmaputra catchment. *Current Science*, **84**, 1041–1047.

Wells, A. and Goff, J.R. (2007) Coastal dunes in Westland, New Zealand, provide a record of paleoseismic activity on the Alpine fault. *Geology*, **35**, 731–734.

Whipple, K.X. (2004) Bedrock rivers and the geomorphology of active orogens. *Annual Review of Earth and Planetary Science*, **32**, 151–185.

Zeitler, P.K., Meltzer, A.S. and Koons, P.O. (2001) Erosion, Himalayan geodynamics and the geomorphology of metamorphism. *GSA Today*, **11**, 4–9.

Zoback, M.D. and Grollimund, B. (2001) Impact of deglaciation on present-day intraplate seismicity in eastern North America and Western Europe. *Comptes Rendus de l'Academie des Sciences, Serie II. Sciences de la Terre et des Planetes*, **333**, 23–33.

5

Pacific Rim Steeplands

Basil Gomez,[1] Michael J. Page[2] and Noel A. Trustrum[2]

[1]Geomorphology Laboratory, Indiana State University, Terre Haute, IN 47809, USA
[2]GNS Science, PO Box 30-368, Lower Hutt 6315, New Zealand

5.1 Introduction

There is no universally accepted definition of 'steepland', which may be characterized as terrain in which the hillslopes have a gradient >12% (*c.* 7°) and are mantled by shallow, immature soils that support natural vegetation, forest or pasture. The depth and maturity of the soil profile are constrained by the frequency with which mass movements, such as landslides triggered by heavy rainstorms or seismic activity, occur and the rate of recovery on scar surfaces (Wright and Mella, 1963; Shimokawa, 1984; Trustrum and DeRose, 1988; DeRose, Trustrum and Blaschke, 1993; Dietrich *et al.*, 1995; D'Odorico and Fagherazzi, 2003; Wakatsuki, Tanaka and Matsukura, 2005). Steeplands are evident in all parts of the world, but the many small mountainous river basins draining into the Pacific Ocean account for *c.* 13% of Earth's total land area and have very high sediment loads (Milliman and Syvitski, 1992); not only by virtue of their position above active plate boundaries and their exposure to tropical and mid-latitude storms (Hicks, Gomez and Trustrum, 2000; Wheatcroft and Sommerfield, 2005), but also because in many cases their integrity has been compromised by the activities of humans (Nakamura, Swanson and Wandzell, 2000).

At timescales (of decades to millennia) relevant to coupling within fluvial systems, geodynamics create the antecedent conditions which active tectonics and climate act on, and human activities modify (e.g. Lock *et al.*, 2006). Much of the *c.* 135 000 km of coastline around the Pacific Rim lies above active plate boundaries, and active tectonics in the chaotic mountainous terrains of the circum-Pacific orogenic belts is associated with earthquakes that occur along the boundary between the subducting oceanic lithosphere and overriding plate (e.g. Reyners, 1998; Nicol and Beavan, 2003). Rainfall is the primary climate driver of mass movements (Crozier, 1997), and the effects of unsustainable human activities first began to appreciably accelerate erosion processes and impact fluvial systems in Pacific Rim steeplands 150–200 years ago (Mount, 1995; Nunn, 2007).

Sediment Cascades: An Integrated Approach Edited by Timothy Burt and Robert Allison

Figure 5.1 Examples of accelerated erosion in the Waipaoa and Waiapu river basins, New Zealand. (a) Storm-driven shallow landsliding on pastoral hillslopes, Ngatapa – Mangatoetoe Stream in foreground. (b) Gully erosion in the Mangatu Forest – Tarndale gully complex. (c) Gully erosion and streambed aggradation along the Taupuaeroa River (black shading). See accompanying imagery and map for locations (note the wide-well defined channel of the Waiapu River and its tributaries, which have all aggraded in the historic period)

The focus of this chapter is on two widespread accelerated erosion processes; shallow landslides (initiated at the contact between soil and bedrock) and gully erosion (Figure 5.1), that are an outcome of land-use change and have the potential to impact the sediment dispersal system over timescales of decades to centuries (Kelsey, Lamberson and Madej, 1987; May and Gresswell, 2003; Imaizumi, Sidle and Kamei, 2008). First, we consider the conditions that promote rock, regolith (the layer of weathered material that mantles the bedrock) and soil movement, and gully development on steepland hillslopes. Then, we examine the fluvial response to accelerated erosion as it is manifest in the observational and encrypted in the depositional records. Finally, we contemplate the implications that changing patterns of sediment production and dispersal in steepland basins have for the Anthropocene. Comprehensive field data on erosion and sediment transport have only been obtained from a small number of steepland river basins, and for this reason the examples we rely on are drawn primarily from those counties (Australia, Japan, New Zealand, Taiwan and the USA) with sufficiently long historical records and where intensive studies have been undertaken.

5.2 Potential for Mass Movement

Slopes that develop on fractured rock masses are not controlled by the mass strength of the underlying rock (cf. Selby, 1982). Instead, they behave as a densely packed mass of cohesionless material and cannot remain vertical after they are oversteepened. Fluvial incision of uplifting rock masses destabilizes hillslopes (Densmore *et al.*, 1997), and the continued incision of a river channel into the land surface has a tendency to produce straight hillslopes and V-shaped valleys. The stability of infinitely long slopes is governed by the relationship between the downslope component of the shear force, $\gamma_r Y\, l \cos\alpha \sin\alpha$, acting on a potential failure (joint or bedding) plane and shear force along that plane, $\gamma_r l\, Y \cos^2\alpha \tan\varphi$ (Figure 5.2a), where: l is the downslope length; Y is the depth to the failure plane; γ_r is the specific weight of the rock; α is the slope angle; and φ is the angle of internal friction, which depends on the packing density and typically varies between 30° and 60°. The limiting condition is $\tan\alpha = \tan\varphi$, and as downcutting proceeds mass movements maintain the hillslopes at an angle ($\alpha = \varphi$) that is contingent upon the rate of downcutting relative to the rate of weathering and debris removal (Burbank *et al.*, 1996). If Y is also the

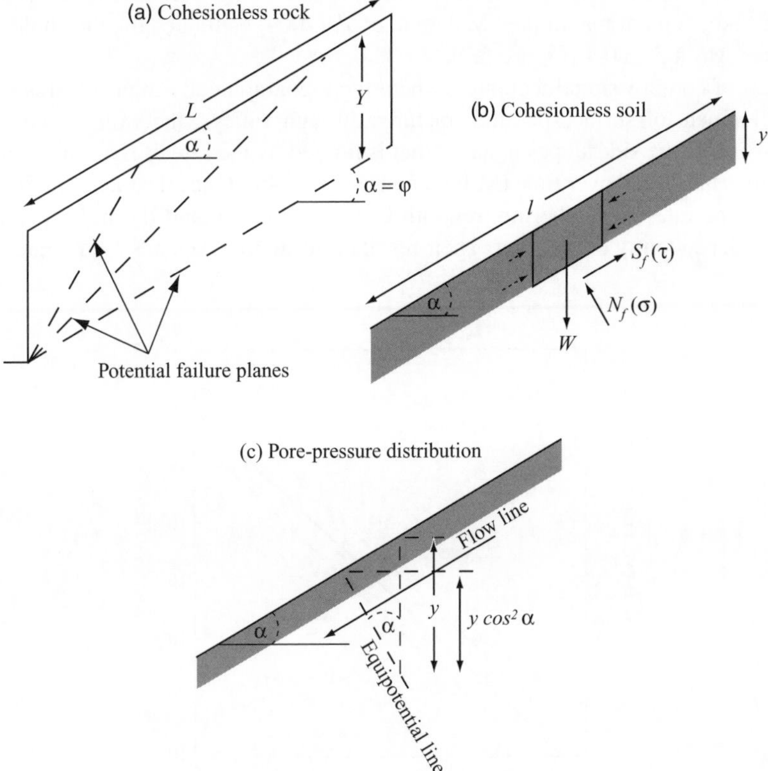

Figure 5.2 Definition diagrams for the stability analysis of an (a) infinite slope of cohesionless rock; (b) infinite slope of cohesionless soil; and (c) the pore-water pressure distribution with slope-parallel groundwater flow. The descriptions follow those presented in standard textbooks on the mechanics of earth materials, such as Carson (1971) and Middleton and Wilcock (1994)

depth of downcutting, $LY \cos \alpha$ represents the approximate volume of material that will be removed. Thus the potential for large volumes of rock to be mobilized increases as Y increases and the angle of the failure plane approaches the angle of friction (Cruden, 1976), and in mountain belts where relief is accentuated by active tectonics, landscape development is controlled by deep-seated landslides (Anderson, 1994; Burbank *et al.*, 1996; Hovius *et al.*, 1998; Roering, Kirchner and Dietrich, 2005; Hovius and Stark, 2006; Korup *et al.*, 2007).

Hillslopes fail during earthquakes and rainstorms which are of sufficient magnitude to affect the stability of the substrate and trigger landslides (Martin *et al.*, 2002). Earthquake-triggered landslides occur because seismic shaking reduces the strength and/or cohesion of rock and soil. Landslides tend to be distributed across the entire slope profile (Densmore and Hovius, 2000), but in areas where topography amplifies seismic energy the steepest upslope segments are predisposed to failure (Chang, Chiang and Hsu, 2007; Yamagishi and Iwahashi, 2007; Lin *et al.*, 2008). Earthquake moment magnitude, M, determines the area affected by landslides, which increases from 0 at $M \approx 4.0$ to 500 000 km^2 at $M = 9.2$ (Keefer, 1984). Based on the small number of earthquakes for which comprehensive landside inventories are available (cf. Harp and Jibson, 1995), by relating the magnitude of the landslide event (which is a function of seismological and geomorphological conditions) to earthquake moment magnitude, Malamud *et al.* (2004) estimated the threshold moment magnitude was 4.3 ± 0.4 (Figure 5.3).

In areas of compressional tectonics, where the topographically induced stresses favour the development of surface-parallel fractures through valleys, the valleys may grow by downwasting of the side slopes at a rate that is limited by the rate of river incision, so that their form is preserved over time (Miller and Dunne, 1996). Once the rate of incision slows relative to the rate of weathering, regolith and soil develop, and the hillslopes begin to develop independently of the river. Their resultant form now depends on the rate at which

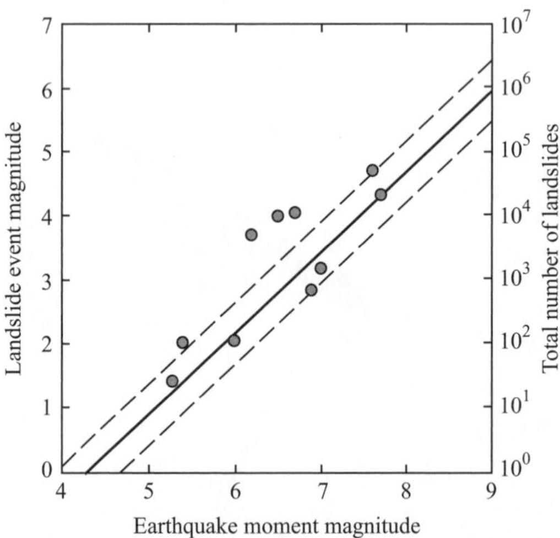

Figure 5.3 Landslide event magnitude, m_L, and total number of landslides as a function of earthquake moment magnitude, M (after Malamud *et al.*, 2004). Solid line is the least square best fit, $m_L = 1.27M - 5.45(\pm 0.46)$, and dashed lines are the ± 0.52 error bounds

transport-limited processes operate. The condition for failure, τ_c, may be defined with reference to the Mohr–Coulomb criterion

$$\tau_c = C + \sigma \tan \varphi \tag{5.1}$$

where: τ and σ are the shear and normal stresses acting on a plane orientated at an angle, α, to the horizontal; and C is the cohesion (i.e. the shear strength under zero normal stress). The shear stresses are maximized at $2\alpha = 90°$, but $\alpha_c = [90° - \varphi]/2$ is the critical angle for failure.

For the case of an infinitely long hillslope, mantled by homogeneous soil and inclined at an angle, α, to the horizontal, and neglecting the pressures acting on the up- and downslope sides (Figure 5.2b), the forces acting on a segment with a downslope length, l, and depth, y, are due to its weight, $W = \gamma_s y\, l \cos \alpha$; and the normal force, $N_f = W \cos \alpha = [\gamma_s y\, (l \cos \alpha)] \cos \alpha$, and shear force, $T_f = W \sin \alpha = \gamma_s y\, l \cos \alpha \sin \alpha$ (i.e. the normal and shear stresses), acting on the surface along which failure potentially will occur, where; γ_s is the specific weight of the soil. For a cohesionless soil $C = 0$, and the force resisting failure, $S_f = N_f \tan \varphi = \gamma_s y\, l \cos^2 \alpha \tan \varphi$ (i.e. the shear strength). If $C \neq 0$, the limiting condition (when $T_f = S_f$) is $\gamma_s y \cos \alpha \sin \alpha = C + \gamma_s y \cos^2 \alpha \tan \varphi$ or

$$\tan \alpha = \left[\frac{C}{\gamma_s\, y \cos^2 \alpha}\right] + \tan \varphi \tag{5.2}$$

That is, a slope may become unstable if α or γ_s increase, or C or φ decrease.

Groundwater pressures also determine how soil and regolith respond to stress (Figure 5.2c), and Terzaghi (1943) showed that the total normal stress should be apportioned into the effective normal stress, σ_e, and the pore pressure, p, where: $\sigma_e = [\sigma - p]$. Thus the Mohr–Coulomb criterion may be rewritten as

$$\tau_c = C + \sigma_e \tan \varphi \tag{5.3}$$

The effect is to reduce τ and decrease σ_e as p increases. Redefining the force resisting failure in terms of σ_e, $S_f = l\,[C + (\gamma_s y \cos^2 \alpha - p) \tan \varphi]$, so that

$$\tan \alpha = \left[\frac{C}{\gamma_s y \cos^2 \alpha}\right] + \left[1 - \frac{p}{\gamma_s\, y \cos^2 \alpha}\right] \tan \varphi \tag{5.4}$$

When $p = 0$ and $C = 0$, the limiting condition is again $\tan \alpha = \tan \varphi$.

Near the bottom of a hillslope, where for example, if the water table is at the ground surface and the flow lines parallel the hillslope gradient, $p = \gamma\, y \cos^2 \alpha$, where: γ is the specific weight of water; and $y \cos^2 \alpha$ is the head of water. If $C = 0$, substitution for p yields

$$\tan \alpha = \left[1 - \frac{\gamma}{\gamma_s}\right] \tan \varphi \tag{5.5}$$

For typical values of γ and γ_s $[1 - \gamma/\gamma_s] \approx 0.5$ and so the potential effect of groundwater flow is to reduce the critical angle for slopes in saturated cohesionless materials by *c.* 50%

although, because p does not affect C, the actual reduction in most earth materials is much smaller.

Iverson and Major (1986) examined the conditions under which groundwater seepage forces cause instability in a soil mass. They observed that the susceptibility of a cohesionless hillslope to failure is determined by the direction and magnitude of the hydraulic gradient relative to the angle of the hillslope and the density and frictional resistance of the soil. Three dimensionless parameters govern the Coulomb failure potential of saturated, cohesionless, infinite homogeneous hillslopes: $z = i\gamma/(\gamma_s - \gamma)$, the ratio of seepage force magnitude to the submerged unit weight of the soil, where i is the magnitude of the hydraulic gradient or seepage vector; the angle $\alpha - \varphi$; and the angle $\lambda + \varphi$, where λ is the angle of the seepage vector measured with respect to an outward-directed surface-normal vector (Figure 5.4a). Slope stability is minimized by seepage in the direction $\lambda = 90° - \varphi$. On steep hillslopes where $\alpha \approx \varphi$, horizontal seepage promotes instability and in some conditions vertical seepage also has the potential to transform a Coulomb failure into a flowing mass (Iverson and Major, 1986). Iverson (2000) subsequently used a simplified analytical solution of the Richards equation (which describes the vertical movement of water in unsaturated porous media) to model infiltration and link slope failure to the groundwater pressure head responses to rainfall over varying periods of time. Before rainfall commences, efficient drainage on steep hillslopes mantled by thin soils accentuates the factor of safety (the ratio between the resisting and driving forces), which declines rapidly as the pressure head increases during intense rainstorms (Figure 5.4b). The slope fails abruptly and the landslide accelerates downslope if the material is porous enough to permit pore-water pressures to rise rapidly during failure, attain levels that balance the normal stresses and liquefy the soil (Iverson *et al.*, 2000).

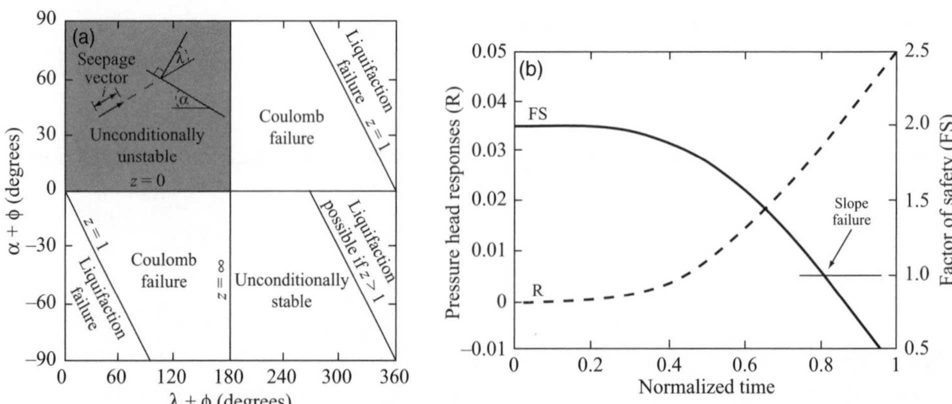

Figure 5.4 (a) Definition diagram and distribution of domains in which preferred modes of slope failure occur; based on the normalized relationship between the limiting slope angle, α, the seepage direction, λ, and the angle of internal friction, φ for specified values of the normalized seepage magnitude, z (after Iverson and Major, 1986). (b) Relationship between the pressure-head responses to rainfall, the dimensionless factor of safety, and normalized time for conditions when rainfall of normalized duration triggers shallow landslides on steep slopes (after Iverson, 2000). During intense rainfall slope failure occurs abruptly as the pressure-head increase promotes a rapid decline in the factor of safety

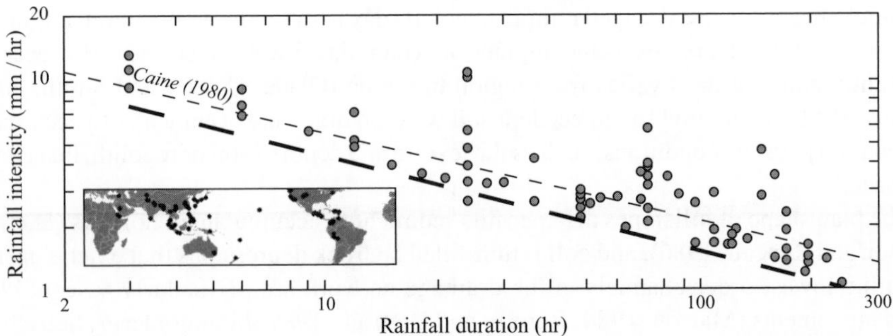

Figure 5.5 Rainfall intensity–duration threshold ($I = 12.45\ D^{-0.42}$; where: I is rainfall intensity and D is rainfall duration) for landslides derived from Tropical Rainfall Measuring Mission (TRMM)-based Multi-satellite Precipitation Analysis (TMPA) over the latitude band 50° N–S (after Hong, Adler and Huffman, 2006). Inset shows the locations where landslides occurred, and Caine's (1980) global threshold is $I = 14.82\ D^{-0.39}$

In areas where records of past rainstorm-triggered landslides exist, the empirical analysis of storm characteristics provides an alternative to using a mechanistic approach to determine the potential for hillslope failure, the assumption being that there is a link between the occurrence of landslides and identifiable transient factors, such as rainfall intensity, duration or total (Tsai, 2008). The relationship between different rainfall parameters is used to define the threshold values that trigger mass movements (Figure 5.5). Studies undertaken in areas where detailed hydrogeomorphic data are available suggest spatial variations in rainfall and terrain susceptibility to landsliding make the conditions that promote failures in specific locales difficult to isolate (Casadei, Dietrich and Miller, 2003), but the approach has been used to develop models that describe the regional sensitivity to mass movements triggered by rainfall (Finlay, Fell and Maguire, 1997; Godt, Baum and Chleborad, 2006). Experience also suggests the conditions below which hillslopes are stable, and above which mass movements may be initiated, differ substantially from those that inevitably generate mass movements. This is because most portions of hillslopes drain rapidly under the influence of gravity and antecedent rainfall is required to bring regolith and soil to field capacity or increase pore-water pressures along the basal slip surface (Campbell, 1975; Reid, 1994). Consequently, knowledge of the accumulated rainfall and rainstorm characteristics may be required to describe the conditions that initiate storm-driven failures (Church and Miles, 1987; Wieczorek, 1987).

Crozier (1999) developed an empirical model to account for the contributions made by antecedent soil water and event water, represented by the soil water status and the daily rainfall, respectively. Negative values of the soil water status index conditions below field capacity, and positive values the gravitational (ground)water that preferentially accumulates in certain locations on hillslopes, such as hollows (Crozier, Vaughn and Tippett, 1990), where the contours are concave and flow lines converge. It is necessary to know the storage capacity (porosity and depth) of the regolith and soil, the rate of evapotranspiration, and the rate of drainage in excess of precipitation to ascertain the soil-water status. To characterize the conditions under which rainfall-triggered landslides occur as hillslopes in Wellington, New Zealand, get wetter, Crozier (1999) calculated the antecedent soil-water status by accumulating the excess decayed rainfall over a 10-day period. The daily rainfall required to trigger landslides decreases as the amount of accumulated water increases. However, the relevant number of days varies from region to region (Glade, Crozier and Smith, 2000), because the time required for antecedent soil water to drain and event water to accumulate depends on physical conditions, such as the texture and depth of soil or regolith, that change in space.

The plan shape of hillslopes also permits sediment to accumulate in hollows (Dietrich, Wilson and Reneau, 1986), and colluvium-filled bedrock depressions that extend upslope from the lowest order channels in the drainage network are particularly susceptible to mass movements (Marron, 1984; Reneau and Dietrich, 1987; Montgomery, Dietrich and Heffner, 2002). GIS-driven models provide a spatially distributed perspective on the topographic attributes that are important to the movement and distribution of water on hillslopes. The topographic wetness index ($TWI = \ln(a/b \tan\theta)$, where: a is the local upslope contributing area; b is the unit contour length; and θ is the local slope), for example, is a measure of the degree to which water can accumulate at a site (sites where the index is higher, due to a large specific catchment area (a/b) or low slope, are more likely to be saturated with water than dry) that can be derived from digital elevation models and is often used to quantify the control topography exerts on hydrological processes (Beven and Kirkby, 1979; O'Loughlin, 1986). To determine the critical steady-state rainfall required to trigger slope instability at any point in a landscape, Montgomery *et al.* (1998) assumed the flow infiltrated to an impermeable layer and followed topographically determined flowpaths, and the saturated conductivity did not vary with depth. In circumstances where there is no overland flow, the local wetness, $w = h/z$ (where: h is the thickness of the saturated soil above the impermeable layer; and z is the total thickness of the soil), and the relative saturation of the soil profile can be expressed as $h/z = R\, a/(b\, T' \sin\theta)$, where: R is the steady-state rainfall; and T is the depth-integrated soil transmissivity. Considering again the forces acting on an infinitely long hillslope, the condition for failure is $\rho_s g\, z \cos\theta \sin\theta = C' + [\rho_s - (h/z)\rho]\, g\, z \cos^2\theta \tan\varphi$, where ρ is the bulk density of water; g is the acceleration due to gravity; ρ_s is the bulk density of the soil and C' is the effective cohesion of the soil, including root strength. The critical steady-state rainfall, R_c, required to initiate landslides, specified in terms of the relative saturation of the soil profile and condition for failure, is

$$R_{\mathrm{crit}} = \frac{T \sin\theta}{(a/b)} \left[\frac{C'}{\rho g z \cos^2\theta \tan\varphi} + \frac{\rho_s}{\rho} \left(1 - \frac{\tan\theta}{\tan\varphi} \right) \right] \tag{5.6}$$

Montgomery, Sullivan and Greenberg (1998) demonstrated there was a strong correlation between landslide frequency and the calculated critical steady-state rainfall in 14 Pacific Northwest catchments, and Montgomery and Dietrich (1994) model predictions for cohesionless soils were also consistent with the spatial patterns of observed landslide scars, as were predictions made with an enhanced version of their model that accounted for the effect storm intensity and duration have on slope stability (Rosso, Rulli and Vannucchi, 2006). The implication is that, in some regions, topographically driven convergence of near-surface runoff is a primary control on shallow landsliding.

As well as causing Coulomb failure or liquefaction, water flowing through and emerging from regolith and soil can initiate motion of individual particles by generating a critical drag force (Howard and McLane, 1988). An additional requirement is that the flow must be able to transport sediment away from the seepage face. Thus, for a permanent channel to develop continuity has to be maintained between the localized erosion processes that produce sediment at the seepage face and the fluvial transport processes that remove it. Consider, for example, the self-driven process of colluvial infilling and hollow evacuation that Hack and Goodlett (1960), Marron (1984), Dietrich and Dunne (1978) and Reneau *et al.* (1989, 1990) observed, which occurs over timescales of thousands of years. The position of the channel head in a hollow merely fluctuates about the extremes imposed by protracted episodes of filling and sudden evacuation due to landsliding initiated under conditions that are set by the terrain and the prevailing climatic regime (Sidle, 1987; Crozier, Vaughn and Tippett, 1990; Yamada, 1999).

Dietrich, Wilson and Reneau (1986) derived a functional expression for the unchannelled source basin length, L, of colluvium-mantled bedrock depressions with shallow groundwater flow systems in which the depth of saturation does not exceed the soil depth and $[\rho_s - \rho]\rho_s^{-1} \tan\varphi < \tan\theta \leq \varphi$

$$L = \frac{\rho_s K z}{c \rho R'} \sin^2\theta \left[\frac{1}{\tan\theta} - \frac{1}{\tan\varphi}\right] \tag{5.7}$$

where: K is the saturated hydraulic conductivity; c is the ratio of valley width to hollow width and R' is the constant rainfall rate that governs the rate of recharge. For a given rainfall rate, hydraulic conductivity and angle of internal friction, the length of unchanneled source basins declines as hollow gradient increases. Soil strength, as determined by the angle of internal friction, exerts a strong influence on this relation and source basin length declines rapidly as θ approaches φ. For a given hollow gradient and angle of internal friction, source-basin length increases as the hydraulic conductivity increases relative to the rainfall rate accumulated over a period of 10^1–10^2 days. Source-basin length declines with increasing rainfall over this characteristic period, and so a long-term change in the precipitation regime is required to effect a permanent change in the location of the channel head (Reneau and Dietrich, 1987; Reneau *et al.*, 1990; Oguchi, 1996).

Steepland gullies are permanent, relatively deep and rapidly eroding incisions on hillslopes that are often coupled to stream channels (Figure 5.1b), although they can also exist in isolation (Wells and Andriamihaja, 1993). Gullies may be generated by overland as well as subsurface flow and their location is commonly explained in terms of threshold behaviour set by topography, even though there is not always a continuous transition from ungullied to gullied hillslopes (Mosley, 1972; Swanson, Kondolf and Boison, 1989; Prosser and Abernethy, 1996; Parkner *et al.*, 2007). A similar threshold exists for vegetation, the type, density and condition of which all affect the susceptibility of hillslopes to gullying (Prosser and Slade, 1994). For this reason, gully initiation is often associated with forest clearance and the expansion and intensification of agriculture in the historic period (Eyles, 1977; Prosser, Chappell and Gillespie, 1994; Prosser, 1996; Prosser and Winchester, 1996; Marden *et al.*, 2005), although gully development can also occur within areas of undisturbed indigenous forest during high-magnitude rainfall events (Parkner *et al.*, 2007).

Some gullies may be controlled by or inherited from pipe networks which form loci for channel extension (Jones, 1990). Surface collapse can transform a pipe into a channel (Gibbs, 1945; Swanson, Kondolf and Boison, 1989), and the discharge from pipes can also trigger the mass movements that are required to initiate and sustain a surface channel (Pierson, 1983; McDonnell, 1990). However, not all pipes have the potential to develop into gullies, because their potential for growth and enlargement is limited by the ability of pipeflow to entrain and evacuate sediment (Crouch, 1976; Hagerty, 1991).

Thornes (1984) assumed that the rate at which gullies migrate upslope is proportional to the discharge of overland flow over the headcut boundary, and portrayed gullies as shockwaves migrating through hillslopes. The rate of headcut migration is proportional to the drainage area upslope, so the rate of gully retreat decreases over time (Seginer, 1966), because the source-basin area and discharge decline as the gully erodes headward. Smaller disturbances also propagate upslope at a faster rate than larger disturbances (Begin, 1983). At some point gullies stabilize. The limit may be set by a threshold condition but negative feedback mechanisms, such as a reduction in the rate at which debris is removed from the gully boundary, and competition for groundwater or surface flow also regulate gully growth (Faulkner, 1974). If the amount of overland flow is insignificant, the magnitude and direction of seepage are governed by the geometry of the groundwater flow net, surface topography, hydraulic conductivity and recharge rate (Iverson and Major, 1986). Consequently, gully evolution varies with the magnitude and frequency of the rainfall events which determine when the mass movements that evacuate sediment from the seepage face occur (Betts, Trustrum and De Rose, 2003).

5.3 Response to Mass Wasting

There is a fundamental difference in the catchment response to mass movements triggered by earthquakes and rainstorms. Coseismic landslides occur independently of the rainfall events that deliver sediment to channels, and debris may remain on hillslopes for years or decades before it is remobilized by subsequent storms (Dadson *et al.*, 2004; Lin *et al.*, 2008). Storm-triggered landslides occur concurrently with heavy rainfall and high river discharges when coupling between hillslopes and channels is enhanced, and there is greater potential for landslide debris to be mobilized into debris flows and enter tributary drainages (Figure 5.6). Landslide intensity increases with both earthquake and storm magnitude and in steepland drainage basins (cf. Figure 5.3), where spatial variations in the erosion process that provide the dominant surface control and event effectiveness are minimized, the supply of sediment to stream channels increases dramatically during large-magnitude, low-frequency rainstorms that are able to mobilize large volumes of sediment on hillslopes (Pain and Hoskin, 1970; Brown and Ritter, 1971; Page *et al.*, 1999; Milliman and Kao, 2005).

Earthquake-triggered landslides can inject large volumes of debris directly into high-order channels, which either blocks valleys or is dispersed downstream if the failures liquefy and turn into debris flows (Adams, 1981; Costa and Schuster, 1988; Schuster *et al.*, 1996). Depending on the size of the debris and transport capacity of the river, the co-seismic deposit may be quickly effaced or reworked over a protracted period of time (Pain and Bowler, 1973; Pearce and Watson, 1986). Rates of mass wasting and sediment production on hillslopes in seismically active regions may be $>200\ m^3\ km^2\ yr^{-1}$ (Keefer, 1994), and landsliding can sustain long-term denudation rates that are as high as $9\ mm\ yr^{-1}$ (Hovius, Stark and Allen, 1997; Dadson *et al.*, 2003; Fuller *et al.*, 2003). However, it remains that most sediment which has a co-seismic origin is delivered to stream channels during post-seismic storms that exploit the increased probability for slope failure in epicentral basins and remobilize unconsolidated landslide debris in colluvial and alluvial stores (Lin *et al.*, 2008).

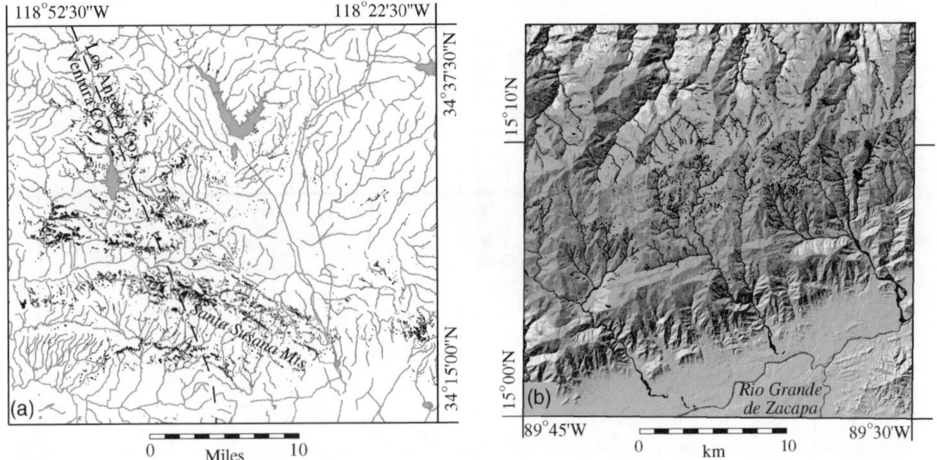

Figure 5.6 Concentrations of landslides (black shading) triggered by: (a) the 17 January 1994 Northridge, California ($M = 6.7$), earthquake (after Harp and Jibson, 1995); and (b) heavy rainfall (200–600 mm between 25 October and 6 November) in Guatemala from Hurricane Mitch (after Bucknam *et al.*, 2001), when many landslides mobilized into debris flows that entered tributary drainages

Earthquake-triggered landslides have the potential to impact sediment fluxes for many decades (Koi *et al.*, 2008). If landslide-dammed lakes form, both event and long-term sediment discharges will be reduced, but the general tendency is for suspended sediment fluxes to increase dramatically after earthquakes and relax to aseismic conditions as the amount of landslide debris delivered directly to stream channels declines and colluvium in storage on the lower portions of hillslopes close to stream channels is depleted (Lin *et al.*, 2008). After the $M = 7.6$ 1 Chi-Chi earthquake in 1999 the average annual suspended sediment discharge of the Choshui River, Taiwan, increased by a factor of 2.6 from 54 to 143 Mt yr^{-1} (Dadson *et al.*, 2004), and there were larger post-seismic increases in higher order tributaries (Lin *et al.*, 2008). Suspended sediment concentrations were higher than normal in floods of all magnitudes, but anomalously high concentrations occurred during tropical cyclones when the increase in transport capacity matches the increase in sediment availability (Figure 5.7a). At such times suspended sediment concentrations at the mouths of Taiwanese rivers can attain hyperpycnal concentrations ($\geq$40 000 mg L^{-1}). It is estimated that typhoon-generated hyperpycnal plumes account for 30–42% of the island's cumulative suspended sediment discharge to the coastal ocean (Dadson *et al.*, 2005). Individual rivers may discharge $\geq$ 80% of their suspended sediment load at hyperpycnal concentrations (Figure 5.7b and c), and in the Choshui River, seven short-lived exceptionally high discharge events generated 48.5% of the 970 Mt of suspended sediment discharged between 1980 and 2001 in <30 days (Milliman and Kao, 2005).

It should be noted that the accuracy with which suspended sediment discharge can be estimated depends on the extent to which measured concentrations reflect the amount of sediment the river transports over the entire range of water discharge. If continuous or closely spaced observations of water discharge and suspended sediment concentration are available, the calculation may be reasonably straightforward, but such data are not available for most rivers and long-term statistics, such as the mean annual suspended sediment

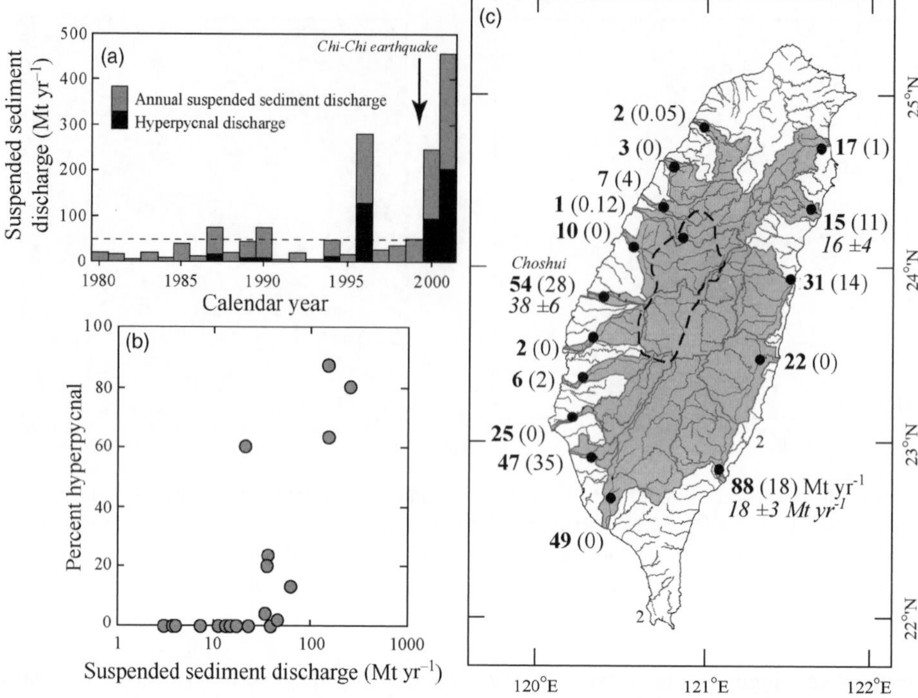

Figure 5.7 Mean annual suspended sediment discharge from Taiwanese rivers. (a) Annual suspended sediment discharge of the Choshui River at Ziuchian Bridge, 1980–2001 (after Milliman and Kao, 2005). Black shading indicates the estimated amount of suspended sediment discharged at hyperpycnal concentrations (in 1996 super typhoon Herb generated an estimated 130 Mt of suspended sediment, 96% of which was discharged at hyperpycnal concentrations), and the horizontal dashed line denotes the mean annual suspended sediment discharge (42 Mt yr^{-1}). (b) Percentage of the annual suspended sediment discharge of the Choshui River at Ziuchian Bridge, 1980–2001, that is discharged at hyperpycnal concentrations (after Milliman and Kao, 2005). (c) Mean annual suspended sediment discharge (bold type) and the estimated amount of suspended sediment discharged at hyperpycnal concentrations (in parentheses) from the 16 primary rivers in Taiwan (after Dadson *et al.*, 2003, 2005). Figures in italics are revised estimates of the long-term mean suspended sediment discharge calculated using the Kao, Lee and Milliman (2005) optimal rating curve method (see text for discussion). Shading denotes their drainage areas, solid dots the location of hydrometric stations closest to the coast, and the dashed line delimits the area where landslides initiated by the Chi-Chi earthquake disturbed >2% of the land surface (after Kao, Lee and Milliman, 2005 and Lin *et al.*, 2008).

discharge, usually are derived from event-based information about the kinematic aspects of the suspended load (Hicks and Gomez, 2003). A common methodology for determining the long-term average suspended sediment discharge is to construct a sediment rating to represent the suspended sediment concentration as a continuous function of water discharge (Ferguson, 1987; Crawford, 1991; Asselman, 2000; Syvitski *et al.*, 2000), by plotting concurrent measurements of suspended sediment concentration against water discharge in log–log space (Figure 5.8). The conditional mean relation can then be combined with the water discharge record for the same period to determine the mean annual suspended

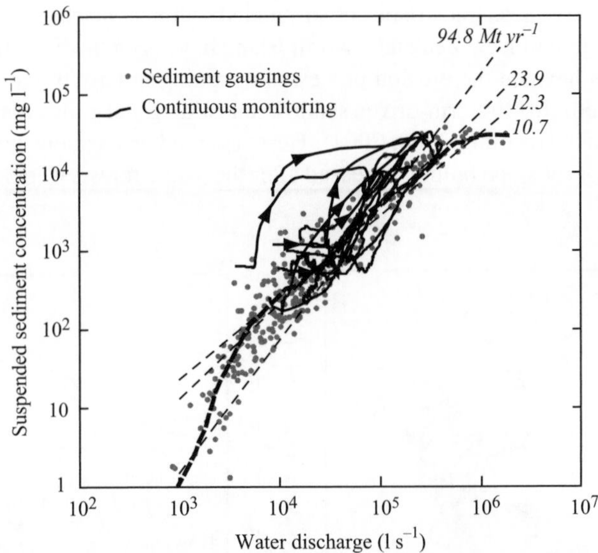

Figure 5.8 Relationship between suspended sediment concentration and water discharge in the Waipaoa River at Kanakanaia, New Zealand, obtained from 301 gaugings made between 1962 and 1996 (solid dots), and continuous monitoring of six runoff events (solid lines) in November 1997, July and December 1998, and January, June and December 1999 showing clockwise hysteresis (after Hicks, Gomez and Trustrum, 2004). The thin dashed lines are linear rating relationships based on different lengths of record that take the simple power form $C = aQ^b$ (where: C is suspended sediment concentration; Q is water discharge; and a and b are empirical coefficients), and the thick dashed line is a rating model that uses all the gauging data derived using the Locally Weighted Scatterplot Smoothing (LOWESS) curve-fitting procedure following correction for log–log bias (after Hicks, Gomez and Trustrum, 2000). Figures in italics are the estimated mean annual suspended sediment discharge derived from the different rating models

sediment discharge. Assuming there is no inherent bias in the data collection and sampling is representative of the entire range of discharges, stages and seasons (cf. Kao *et al.*, 2005), scatter in the data that is independent of discharge (homoscedastic) arises because the sediment supply is non-stationary (Figure 5.8); as is the case during the recovery period following exceptionally large events that cause extreme erosion (Trustrum *et al.*, 1999; Dadson *et al.*, 2005).

Bias corrections and curve-fitting techniques can be employed to improve the accuracy of the rating relationship (Hicks, Gomez and Trustrum, 2000), but rivers in Taiwan are so dynamic that the rating relationship may remain stationary for only a short period of time. To account for this variability Kao, Lee and Milliman (2005) used seasonal rating curves applied to hourly water discharges to calculate the suspended sediment discharge during typhoons and other episodic events, and their long-term estimates are smaller than those calculated using Dadson *et al.* (2003) monthly weighted average method (Figure 5.7c). Long-term estimates of the mean annual suspended sediment discharge of other steepland rivers that have been derived using different rating models are equally variable (Figure 5.8), and serve to highlight the sensitivity of rating relationships to extreme events in the flow record (Hicks, Gomez and Trustrum, 2000).

The spatially varying characteristics of suspended sediment discharge in rivers draining the East Cape region of New Zealand's North Island have been linked to the influence that threshold effects have on the erosion processes operating in terrain that exhibits varying degrees of susceptibility to storm-driven shallow landsliding and gully erosion (Figure 5.9a; Hicks, Gomez and Trustrum, 2000, 2004). Here, as in other locations around the Pacific Rim, the potential for slope failure increased after the native forests were logged or cleared

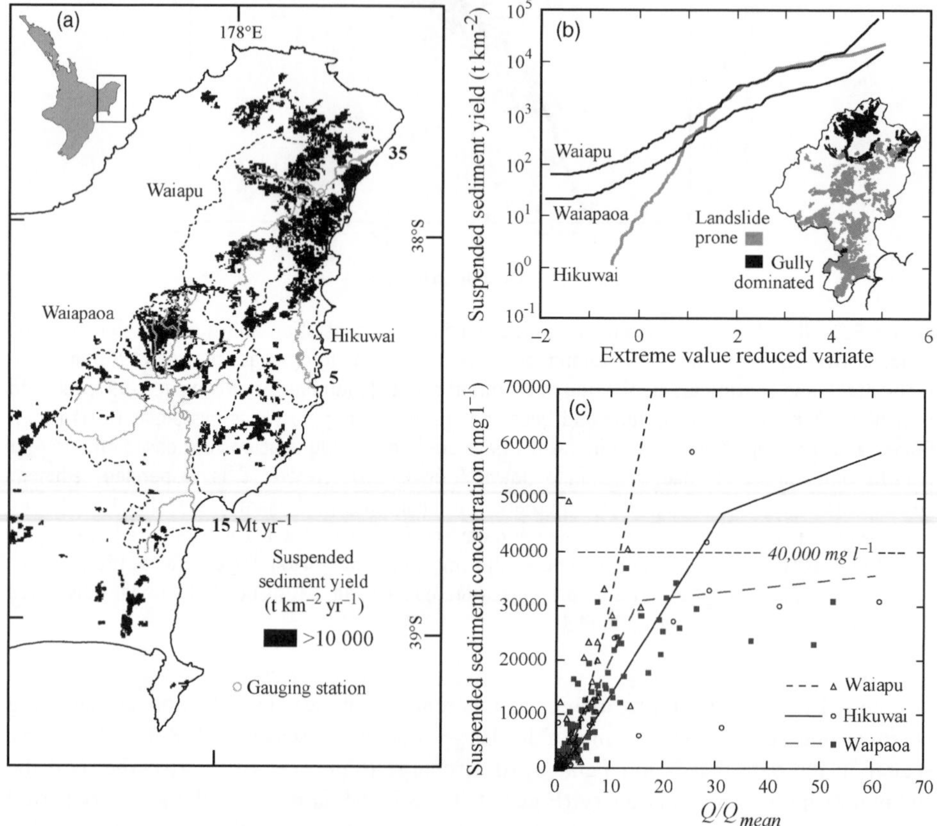

Figure 5.9 (a) Estimates of the specific suspended yield to rivers in the East Cape region, North Island, New Zealand (after Hicks and Shankar, 2003), and location of the Waiapoa, Hikuwai and Waiapu drainage basins and mainstem gauging stations. Bold figures indicate the estimated mean annual suspended sediment discharge to the coastal ocean from each river basin. (b) Magnitude-frequency relationships for event suspended sediment yields at the mainstem gauging stations on the Waiapoa, Hikuwai and Waiapu rivers (after Hicks, Gomez and Trustrum, 2004). Inset shows the distribution of landslide prone and gully dominated terrain types in the Waipaoa River basin (after Jessen *et al.*, 1999). (c) Liner relations between suspended sediment concentration and water discharge, Q, normalized by mean discharge, Q_{mean}, in the Waipaoa, Hikuwai and Waiapu rivers (after Hicks, Gomez and Trustrum, 2004). The lines are step-functions approximating the rating curves which were derived using the bias-corrected LOWESS technique developed in Hicks, Gomez and Trustrum (2000), and 40 000 mg L^{-1} is the threshold concentration for hyperpycnal discharges at the river mouths

and the land converted to pasture or used for growing crops, because the cohesion of soil and regolith declines as tree roots decay and the succeeding vegetation intercepts and transpires less water (Bosch and Hewlett, 1982; O'Loughlin and Ziemer, 1982; Beschta *et al.*, 2000; Johnson, Swanston and McGee, 2000; Montgomery *et al.*, 2000; Schmidt *et al.*, 2001; Roering *et al.*, 2003). The stability of hillslopes is dependent on the interrelationship between lithology, structure and topography (Gage and Black, 1979), and the specific suspended sediment yield to rivers from terrain that is particularly susceptible to shallow landsliding or gully erosion can exceed 10 000 t $km^2 yr^{-1}$ (Figure 5.9a). Through sediment supply and erosion threshold effects these erosion processes generate characteristic signatures in rating relationships between suspended sediment concentration and water discharge and in the magnitude–frequency characteristics of event sediment yields (Trustrum *et al.*, 1999; Hicks, Gomez and Trustrum, 2000). Most gullies are connected directly to river channels (Figure 5.1b). Sediment storage in feeder channels and on alluvial fans moderates sediment delivery (Marutani *et al.*, 1999; Kasai *et al.*, 2001), but they generate sediment during events of all scales. Landslides can be triggered if the regional rainfall threshold of *c*. 200 mm in < 72 h is exceeded, and the intensity of landsliding increases with increasing storm magnitude (Page *et al.*, 1999; Reid and Page, 2002).

Rating-based magnitude–frequency relationships show how the signature of these different erosion processes is imprinted on the suspended sediment yield from river basins with similar deforestation histories (Figure 5.9b). The offset of individual relations reflects overall sediment availability in the basin, which varies with lithology, structure and rainfall, and the gradient of a relation reflects the relative sediment availability as a function of return period (expressed in terms of the extreme-value reduced-variate, $yT = -\ln(-\ln[1 - 1/T'])$, where: T' ($= 12T$) is the return period on a monthly basis). The Waiapu River basin experiences the highest annual rainfall and contains the highest percentage of terrain (21.7%) that is susceptible to gully erosion. Consistent with the abundant sediment supply from the intensely gullied terrain during all scales of runoff events, the magnitude–frequency relation for the Waiapu River is flatter and plots higher than those for the other rivers. Terrain in the Hikuawi River basin is predisposed to shallow landsliding, and the relation for Hikuwai River has the steepest gradient. This is consistent with the relatively low yield from landslide-prone terrain during frequent, subannual events, but the plot flattens as the suspended sediment yield becomes comparable to that of the Waiapu River during events with return periods greater than *c*. 1 yr. The relation for the Waipaoa River occupies an intermediate position. Compared to the Waiapu River basin, annual rainfall is lower and there are less than half the number of active gullies, but sediment is also generated by landsliding and the suspended sediment yield approaches that of the Hikuwai River during high-magnitude low-frequency storms which greatly exceed the threshold for landsliding and can induce non-stationarity in sediment relationships for up to three years (Hicks, Gomez and Trustrum, 2000). The highest measured suspended sediment concentrations in the Hikuwai and Waiapu Rivers are comparable to those Taiwanese Rivers (Figure 5.9c), and it is estimated that *c*. 60% of the long-term suspended sediment yield of these two rivers could be discharge at hyperpycnal concentrations (Hicks, Gomez and Trustrum, 2004).

The potential that steepland rivers have for discharging suspended sediment at hyperpycnal concentrations is enhanced by high relief and the small transmission losses that occur along the length of the drainage network because their floodplains are narrow and discontinuous (Milliman and Syvitski, 1992; Mulder and Syvitski, 1995). Rivers in semi-arid Southern California, USA, also have been reported to discharge suspended sediment at hyperpycnal concentrations during major storms (Warrick and Milliman, 2003), and the streamflow and sediment flux respond to the alternating decadal-scale changes in climate that occur in association with the Pacific–North American climate pattern (Inman and Jenkins, 1999). In this environment the rate of sediment production also increases after Santa Ana fires; in ensuing years winter storms deposit large volumes of sediment in channels and during major storms sediment that has accumulated at the base of hillslopes and on the banks and bed of ephemeral channels is scoured by surface runoff and debris flows (Scott and Williams, 1978; Swanson, 1981; Keller, Valentine and Gibbs, 1997). Wildfires similarly have been observed to accelerate hillslope erosion in the Transverse and Peninsular ranges of southern California and the Oregon Coast Range (Lavé and Burbank, 2004; Roering and Gerber, 2005; Warrick and Rubin, 2007; Hunsinger *et al.*, 2008). The significance of hyperpycnal discharges is that they provide a mechanism for rapidly dispersing sediment over a wide area and beyond the locus of deposition on the middle shelf (Wheatcroft *et al.*, 1997; Mulder *et al.*, 2003; Warrick and Fong, 2004). For this reason, or because the flood deposits are reworked, the stratigraphic evidence for their occurrence on river-fed shelves is elusive (Sommerfield and Wheatcroft, 2007; Gomez, Carter and Trustrum, 2007; Kettner *et al.*, 2009).

Major storms affect the geometry of stream channels and in southern California, where landslides centre on the locus of maximum rainfall intensity, the impact of mass wasting is most pronounced where colluvial inputs locally overwhelm the transport capacity of the fluvial system at delivery points (Nolan and Marron, 1985). There is a much more widespread response to storm-driven landsliding in northern California, where mass wasting tends to be concentrated along stream channels which impinge on colluvial hillslopes that are prone to failure when undercut by high flows, and the close coupling between hillslopes and channels generates very large amounts of sediment that overload streams (Nolan and Marron, 1995). The Eel River has the highest recorded average annual suspended yield of any river in the USA with a drainage area $> 8000\,km^2$ (Meade, Yuzyk and Day, 1990). It has a mean annual suspended sediment discharge of *c.* 18 Mt and during two major storms in December 1964 and January 1965 160 Mt of suspended sediment were discharged in a 30-day period (Brown and Ritter, 1971; Wheatcroft and Sommerfield, 2005).

Elevated sediment loadings in northern Californian rivers are in part a product of timber harvesting, which exacerbates natural hillslope instability and makes terrain more susceptible to erosion (Figure 5.10a and b), and in the Redwood Creek drainage sediment yields from tributaries that have been extensively harvested are 10 times greater than those from basins where no trees had been removed (Nolan and Janda, 1981). The major floods of 1964–5, 1972 and 1975, which had a recurrence interval >10 yr, caused extensive mass wasting and severe aggradation (Madej, 1987). Aggradation increases the influence that relatively low flows have on channel morphology and enhances their ability to transport sediment, because the accompanying decrease in the median particle size of the bed material reduces the discharge required to initiate gravel transport (Lisle, 1982). The increased capacity for bedload transport at low and moderate flows reduces the depth of pools, and the resulting streamwise reduction in morphological and hydraulic contrasts facilitates the removal of flood deposits from the channel as it seeks to re-establish the pre-flood hydraulic geometry (Figure 5.10c). In Redwood Creek, 60–100% of the sediment stored in low-order channels is removed within 10 years of a major flood but depending on the deposit's residual

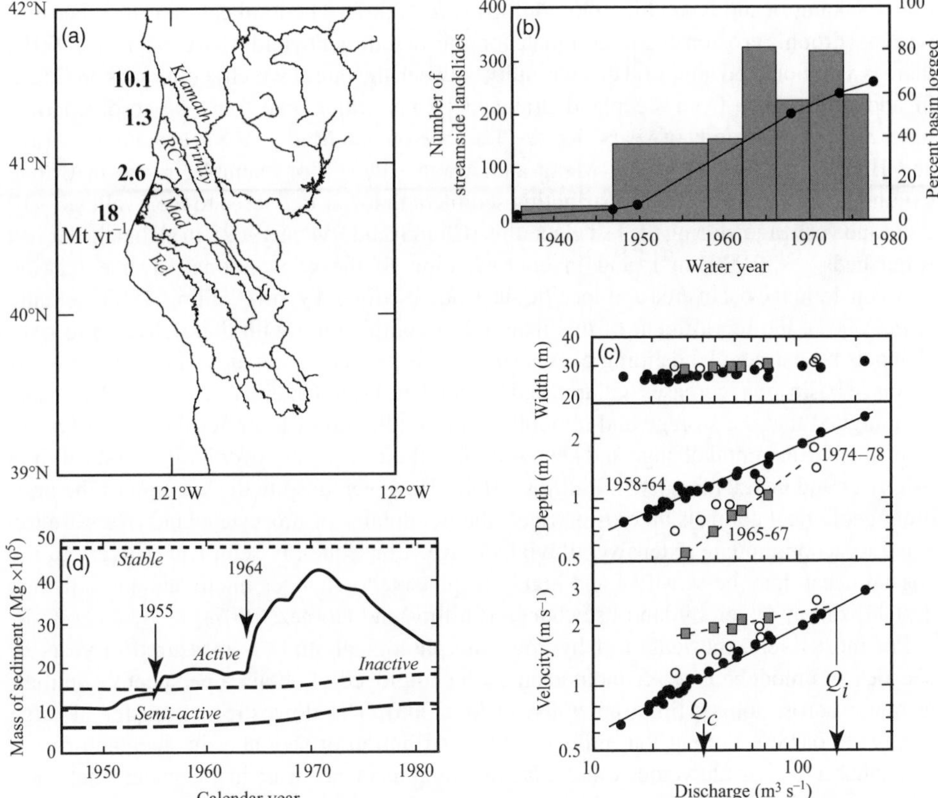

Figure 5.10 (a) Mean annual suspended sediment discharge from rivers in northern California (after Wheatcroft and Sommerfield, 2005). RC = Redwood Creek and the Trinity River contributes 6.8 Mt of sediment to the Klamath River. (b) Land use (solid dots) and streamside landsliding in the Redwood Creek drainage basin, 1936–1981 (after Nolan and Marron, 1995), where accelerated mass wasting was triggered by major storms in 1953, 1955, 1964, 1972 and 1975. (c) Hydraulic geometry relations for the North Fork Trinity River prior to the December 1964 flood (solid dots), at the peak of aggradation (open circles), and during the recovery period (shaded squares), where: Q_c is the discharge required to initiate gravel transport and Q_i is the equivalent discharge prior to the flood, corresponding to the convergence of the hydraulic geometries prior to and at the peak of aggradation, which has a recurrence interval of 1.5 yr (after Lisle, 1982). (d) Changes in the mass of sediment stored in different sediment reservoirs within the middle reach of Redwood Creek, between 1947 and 1982, modelled as a Markov process (after Kelsey, Lamberson and Madej, 1987). Major storms (1955 and 1964) mobilize the stable reservoir and initiate changes in the inactive and semi-active reservoirs, and the average amount of time required to flush particles from these three reservoirs is *c.* 750 yr. Sediment in the stable reservoir is isolated from the river by tectonic uplift

mobility (Figure 5.10d), which is set by the particle size and elevation (relative to the active channel) of the fill, sediment in downstream reaches may remain in storage for >100 years and its residence time thus exceeds the recurrence interval of the floods which deposited it (Madej, 1987). This is also the case for the fine-grained alluvium deposited on the floodplains of steepland rivers, which tends to stay in storage for much longer periods of time (Nakamura and Kikuchi, 1996; Phillips *et al.*, 2007). The implication is that, because rare catastrophic erosion events dominate long-term sediment yields (Kirchner *et al.*, 2001), there is a pronounced temporal discontinuity between the rate at which sediment is produced in and transported from steepland drainage basins which may only be resolved over timescales of thousands of years (Kelsey, Lamberson and Madej, 1987; Nakamura, Araya and Higashi, 1987; Nakamura, Maita and Araya, 1995). For example, rivers in British Columbia, Canada, are still redistributing sediment that was delivered to their valleys > 10 thousand years ago, during the last glaciation (Church and Ryder, 1972). In all but the largest watersheds (>30 000 km^2), and in contravention to the conventional inverse relation between drainage basin area and specific suspended sediment yield (Walling, 1983; deVente *et al.*, 2007), the recruitment of this material in combination with that delivered to river channels by massive landsliding causes specific suspended sediment yields to increase as drainage basin area increases (Church and Slaymaker, 1989; Church *et al.*, 1999). It has also been argued that the storage and remobilization of alluvium can buffer the fluvial system against environmental change and keep sediment fluxes constant over long periods of time (Métivier and Gaudemar, 1999; Phillips, 2003). However, despite the strength of the first-order geological controls on erosion rates, the floodplains of most steepland rivers are too small to accommodate extensive alluvial storage. Consequently sediment delivery to the coastal ocean may be sensitive and highly responsive to changes in climate and hillslope destabilization driven by land-use change (Phillips and Gomez, 2007).

For most rivers, the length of hydrogeomorphological study is measured in years or decades and models or proxy methods must be employed to obtain a perspective on their response across longer timeframes (Syvitski, 2003). The downstream transfer of large inputs of coarse sediment that influence the distribution of sediment in steepland rivers, which has a low virtual velocity and relatively long residence time in the channel network, can be modelled using the statistical properties of the sediment supply and sediment storage on valley floors or the equations governing water flow and sediment continuity (Benda and Dunne, 1997a, 1997b; Lisle *et al.*, 2001; Cui and Parker, 2005). Episodic mass wasting delivers pulses of sediment to stream channels and sediment movement in these models is characterized as a bed-material wave, or temporary zone of accumulation that loses mass to abrasion and suspension as it is routed through the drainage network and is dispersed at some characteristic particle velocity. A sediment wave in the Navarro River, California, that was generated by a single streamside landslide, dispersed within a few years (Sutherland *et al.*, 2002), but it is estimated that the aggradation wave in Redwood Creek that formed in response to the storm-driven basin-wide increase in sediment after commercial timber harvesting commenced may take several decades to disperse (Madej and Ozaki, 1996). In areas where land-use change initiates a disturbance that is disseminated over the entire landscape, the new sources of sediment cause aggradation throughout the river system (Figure 5.1c), and the alluvium may remain in storage until the sediment supply is reduced and there is a basin-wide shift to a degradational state (cf. Liébault and Piégay, 2001).

If the amounts of sediment generated by different erosion processes, stored in and moved between different reservoirs can be quantified, it may be possible to develop a sediment budget for a drainage basin that can be used to predict the sediment yield at its outlet (Dietrich *et al.*, 1982). Not enough sediment budgets have been constructed to allow the accuracy and reproducibility of the general approach to be evaluated, but the accuracy with which the sediment yield can be estimated in large part depends on the extent to which the processes that regulate sediment delivery and convey sediment to the basin outlet can be depicted. This necessarily requires an appreciation of the relative roles that infrequent, large-magnitude events and (the cumulative effect of) frequent events of small and moderate magnitude play in the production and dispersal of sediment (Pain and Hoskin, 1970). Process frequencies profoundly influence long-term sediment yields and the effectiveness of events varies between and within basins. A disproportionately large amount of the suspended sediment load of northern California rivers is transported during major floods (Nolan *et al.*, 1987). In other rivers large floods have a diminished role and, for example, in the Waipaoa River 86% of the suspended sediment load is transported by events with a < 10 yr return period (Trustrum *et al.*, 1999). Major storms generate and mobilize large amounts of sediment (Figure 5.1a), but much of this material goes directly into storage and is remobilized by small events which have a high enough frequency of occurrence to cumulatively outweigh large, less frequent events as transporters of sediment. The threshold of landscape sensitivity to erosion also shifts as successive rainfall events strip regolith from hillslopes. Without sufficient time for recovery, terrain resistance progressively increases, so that the effectiveness of a flood of given magnitude depends on the timing of previous floods (Beven, 1981; Crozier and Preston, 1998). In spite of these uncertainties, sediment budgets have proved to be a useful tool for characterizing end-of-basin sediment yields and the relative contributions made by different erosion processes (Kelsey, 1980; Roberts and Church, 1986; Page, Trustrum and Dymond, 1994). Sediment budgets also have been used to identify problematic land management patterns and practices and GIS-driven sediment budgets, that permit patterns of erosion and sediment transport to be resolved at different spatial scales (Prosser *et al.*, 2001; Kinsey-Henderson, Post and Prosser, 2005; McKergow *et al.*, 2005), can be used to identify critical areas of erosion potential in catchments where only limited measurements previously have been made (Figure 5.11).

A much longer time window on the events that influence end-of-basin sediment yields is provided by the record of deposition preserved in upland lakes and on river-fed continental shelves (Page, Trustrum and DeRose, 1994; Lamy *et al.*, 2001; Gomez *et al.*, 2004a; Rein *et al.*, 2005; Gomez, Carter and Trustrum, 2007; Sommerfield and Wheatcroft, 2007; Conroy *et al.*, 2008). Over geological timescales climate patterns have been shown to exert a first-order control on the morphology of the Andes (Montgomery, Balco and Willett, 2001), and in the Holocene erosion processes and sediment delivery to the coastal ocean have been influenced by El Niño–Southern Oscillation (ENSO)-driven variations in the precipitation regime. The systematic long-term growth in the strength of ENSO during the mid-Holocene has been linked to radiative forcing (Clement, Seager and Cane, 2000; Moy *et al.*, 2002), and sediment archives in Ecuador, Peru and Chile all suggest the characteristics of sediment delivered to lakes and the coastal ocean changed in the middle and late Holocene as the frequency of large ENSO events capable of initiating mass movements increased (Figure 5.12). Shallow landsliding triggered by high-intensity storms is a fundamental erosion process throughout the East Cape region, North Island, New Zealand, where intercorrelated terrestrial and marine sediment archives also record the mid-Holocene increase in ENSO-related storm activity (Gomez *et al.*, 2004a). Changes in sediment properties and fluxes (Figure 5.12e), driven by variations in the frequency of geomorphologically effective storm activity, record periods of landscape instability, intensified hillslope erosion and changing sediment source dynamics (Gomez, Carter and Trustrum, 2007; Phillips and Gomez, 2007). However, the removal of the native forests had a more

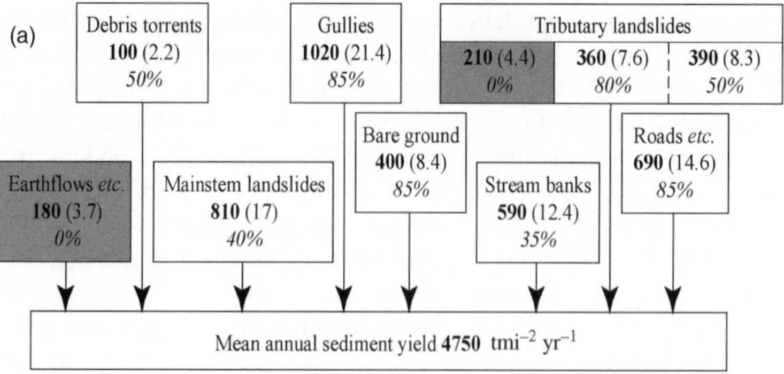

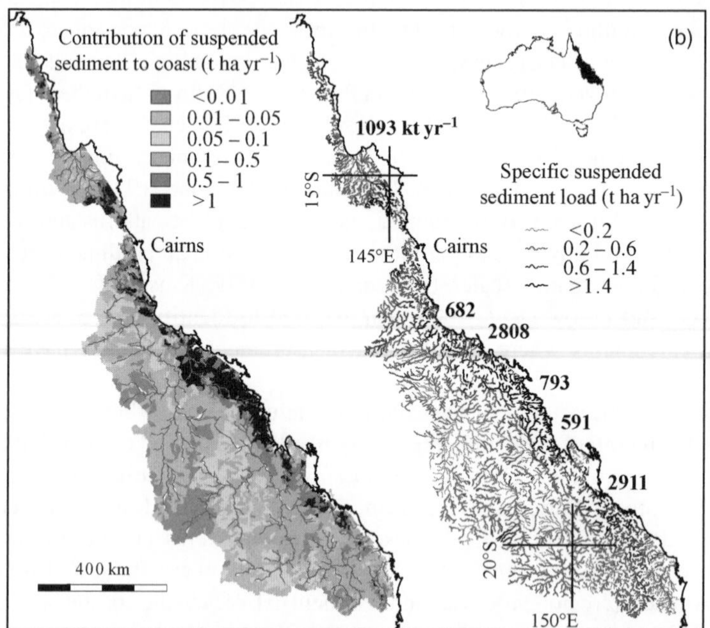

Figure 5.11 (a) Sediment budget (1954–1997) for the 285 mile^2 Redwood Creek drainage basin, California (after EPA, 1998). Bold figures are the sediment yields for each source; figures in parentheses are per cent of the estimated mean annual sediment yield; and figures in italics are per cent that is estimated as controllable (which amount to 60% of the total sediment yield) and are the result of commercial timber harvesting. Shading denotes natural sources. Tributary landslides caused by human activities are apportioned into road-related and harvest-related sources. There is no storage term because it is assumed that the long-term input equals the long-term output (EPA, 1998), and bedload is estimated to account for 10–30% of the total sediment load (Lisle and Madej, 1992). (b) Source and amount of suspended sediment discharged by rivers draining the 423 000 km^2 area adjacent to the Great Barrier Reef, Australia, estimated from a spatially distributed sediment budget (after McKergow *et al.*, 2005). Figures in bold type denote the amount of suspended sediment exported by the Normanby, Herbert, Burdekin, O'Connell, Plane and Fitzroy rivers, that collectively generate 59% of the 15.1 Mt of suspended sediment exported to the coast (70% of which is derived from 20% of the total catchment area)

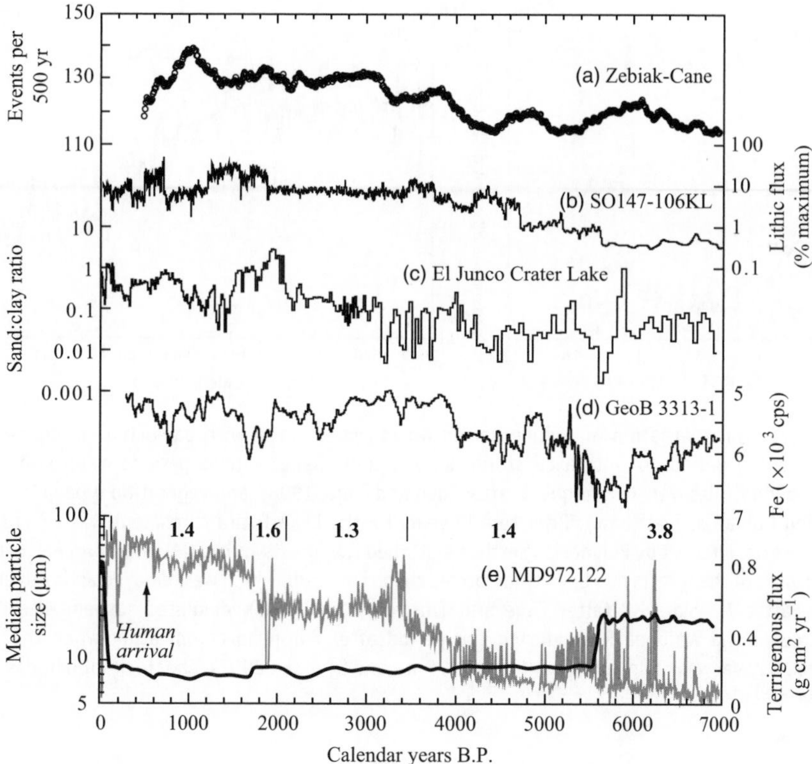

Figure 5.12 Changing sediment characteristics (proxies for El Niño-driven flood events) in lacustrine and marine sediment cores that are interpreted to be driven by long-term growth in the strength of El Niño–Southern Oscillation during the mid-Holocene represented as: (a) The ensemble mean of six simulations with the Zebiak–Cane model (cf. Clement, Seager and Cane, 2000) forced by orbital variations (Amy Clement, personal communication). Variations in: (b) the flux of fine-grained lithics from catchments along the coast of Peru (after Rein *et al.*, 2005); (c) the size distribution of material eroded from the slopes around El Junco Carter Lake, Galápagos Islands, Ecuador (after Conroy *et al.*, 2008); (d) iron (Fe) concentration delivered by rivers to the Chilean continental slope (NB: the inverted *y*-axis scale; after Lamy *et al.*, 2001); and (e) the median particle size of sediment delivered to Poverty Shelf, New Zealand (after Gomez *et al.*, 2004a). Bold numbers indicate the sedimentation rate, which progressively increased from 2.2 to 5.5 mm yr^{-1} after European colonization (Gomez, Carter and Trustrum, 2007). The solid black line indicates the terrigenous sediment flux (after Phillips and Gomez, 2007), which declined after the major phase of downcutting ended *c.* 5580 calendar yr BP. The calibre of the sediment also changed at this time, when landsliding supplanted fluvial incision as the dominant mode of sediment production (Gomez *et al.*, 2004a)

profound influence on sediment source dynamics and drainage basin suspended sediment fluxes (Figure 5.13), and the legacy of deforestation has left a distinctive signal in the region's terrestrial and marine sediment archives. Concordant with the regional history of timber harvesting and changing land use (Sommerfield and Wheatcroft, 2007), the size and amount of suspended sediment delivered to the northern California continental shelf also changed in the historic period.

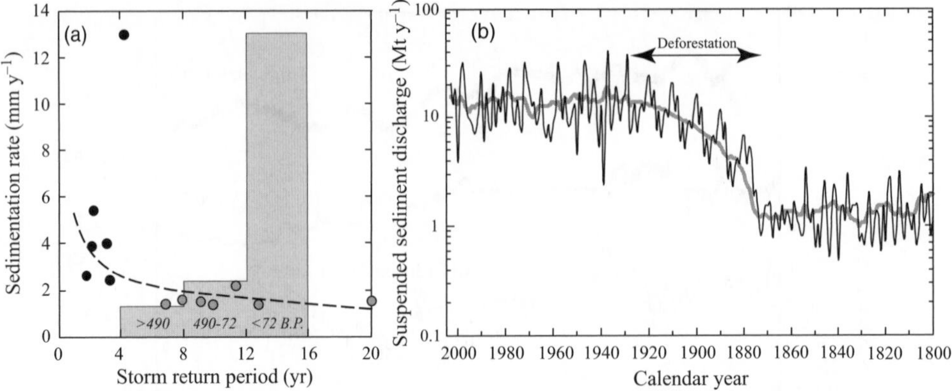

Figure 5.13 (a) Variation of sedimentation rate in Lake Tutira with paleoclimate regime (solid circles denote periods of increased storm activity and shaded circles periods when storms that triggered landslides were less frequent; after Eden and Page, 1998), and vegetation type in the period following human arrival (bars). Prior to 490 years BP the Lake Tutira catchment was covered with native forests. Fires set by Polynesian settlers disturbed the forests and by the time European colonists arrived most of the forests had been transformed to native scrub, which was removed and converted to pasture after 72 years BP (after Page and Trustrum, 1997). (b) Simulated suspended sediment discharge of the Waipaoa River during the period after European colonization when the native vegetation cover was removed (after Kettner, Gomez and Syvitski, 2007). The thick solid line is a 25 yr running mean

5.4 Implications for the Anthropocene

Globally, it is estimated that humans have increased sediment transport by rivers through soil erosion by 2.3 ± 0.6 billion metric tons per year (Syvitski *et al.*, 2005), and that the accumulation of alluvium in higher-order tributary channels and on floodplains is the most important geomorphological process currently shaping Earth's landscape (Wilkinson and McElroy, 2007). The extent to which patterns of sediment production and dispersal in steepland river systems have changed during the Anthropocene depends in large part on the influence a single environmental factor, land cover, has on the sediment delivery process. In Taiwan, where there is little evidence that any single factor controls sediment load and where erodible lithologies, earthquakes and typhoon-generated floods combine to generate large quantities of sediment, agriculture and urbanization have elevated sediment yields in eastern watersheds, whereas the sediment loads of western rivers appear to be unaffected by human activities (Kao and Liu, 2002; Kao and Milliman, 2008; Liu *et al.*, 2008). Landslides and bank erosion generate $> 80\%$ of the sediment loads of rivers in northern California and the natural instability of riparian hillslopes has been adversely impacted by timber harvesting (Brown and Ritter, 1971; Kelsey, 1980; Nolan and Janda, 1995), so that sediment production and delivery to the coastal ocean appear to have increased during the historic period (Sommerfield and Wheatcroft, 2007). Soils under exotic grassland in southern California also lack root reinforcement and are more susceptible to failure than hillslopes covered by native sage, but larger shallow landslides which convert to debris flows on sage-covered hillslopes deliver more sediment to valley floors (Gabet and Dunne, 2002). In New Zealand, a comparison of measured soil depths in first-order steepland basins showed there was a tenfold increase in the erosion rate after deforestation (DeRose, Trustrum and Blaschke, 1993), lake sediment records document the acceleration of hillslope erosion after European colonists removed the native forest and converted the land to pasture (Figure 5.13a), and simulations suggest that the suspended sediment discharge of the Waipaoa River increased from 2.3 ± 4.5 to 14.9 ± 8.7 Mt yr^{-1} in the late nineteenth and early twentieth centuries (Figure 5.13b; Kettner, Gomez and Syvitski, 2007). Despite strong first-order geological controls on erosion, deforestation had such a profound impact on the suspended sediment load of the Waipaoa River because it lowered the threshold of landscape sensitivity to erosion and destabilized the drainage basin mass balance system, and so impacted sediment production and dispersal across the entire spectrum of events regulating sediment delivery to and transport in stream channels (Gomez, Carter and Trustrum, 2007; Phillips and Gomez, 2007).

The processes that drive sediment fluxes are spatially and temporally variable over a large range of scales, and it has been suggested that disturbance events and topography can interact to create patches of intense erosional activity (Miller, Luce and Benda, 2003; cf. Figure 5.6). Geomorphological effectiveness and the off-site consequences of disturbance events are also contingent upon the scale and location of impediments to sediment conveyance interposed between hillslopes, channels and the basin outlet (Fryirs *et al.*, 2007; Reid *et al.*, 2007). Although the recurrence interval of bankfull discharge is several times that of the effective discharge (Nolan *et al.*, 1987; Gomez, Carter and Trustrum, 2007), elevated suspended sediment concentrations in steepland rivers promote high rates of overbank deposition and the rate of sediment accumulation on the Waipaoa River floodplain increased by a factor of three in the historic period (Gomez *et al.*, 1998; Gomez, Carter and Trustrum, 2007). Much of this sediment has the potential to be preserved for a significant period of time (millennia) in the alluvial record, but it remains that floodplain storage is equivalent to only *c.* 2% of the Waipaoa River's annual suspended sediment load. Thus, as in other small, tectonically active, steepland river systems, alluvial buffering has little impact on the contemporary suspended sediment flux at the basin outlet (Phillips, Marden and Gomez, 2007). The impact of dispersive bed-material waves created by episodic influxes of coarse sediment to headwater channels diminishes rapidly with time and distance downstream (Lisle *et al.*, 2001), but bedload has a low characteristic velocity and if the effects of accelerated hillslope erosion are broadly disseminated across the landscape, higher order stream channels may take decades or centuries to revert to their pre-disturbance condition (Benda and Dunne, 1997; Kelsey, Lamberson and Madej, 1987; cf. Figure 5.1c).

Reforestation has proved to be an effective strategy for improving slope stability in steepland river basins (Phillips and Marden, 2004). In New Zealand, the East Coast Forestry Project was set up to reduce erosion through sustainable land management, the objective being to reforest 600 km^2 of severely eroding and potentially erodible land (MAF, 2005). There was a 62% decrease in sediment production from gullies along Te Weraroa Stream, a headwater tributary to the Waipaoa River, after exotic forest plantings commenced in 1962 and exotic reforestation of 140 km^2 of land in the headwaters reduced the amount of gully-derived sediment by 40% (from 3.2 to 1.2 Mt yr^{-1}), which amounts to *c.* 8% of the Waipaoa River's annual suspended sediment load (Gomez *et al.*, 2003a; Marden *et al.*, 2005). That said, *c.* 75% of the active gullies in the Waipaoa River basin have yet to be reforested and *c.* 26% (2170 km^2) of land in the East Cape region as a whole is susceptible to mass movements and regolith stripping, which not only decrease land productivity, but also has a significant impact on the national carbon budget because processes of sediment production and delivery determine the origin, amount and age of riverine particulate organic carbon (POC) loads (Blaschke, Trustrum and Hicks, 2000; Leithold, Blair and Perkey, 2006; Scott *et al.*, 2006).

Steepland rivers contribute a disproportionately large amount of riverine sediment and, by association, carbon to the coastal ocean (Lyons *et al.*, 2002; Komada, Druffel and Trumbore, 2004; Goldsmith *et al.*, 2008; Hilton, Galy and Hovius, 2008). Very large POC fluxes are a characteristic feature of those areas where rapid uplift exposes soft or highly fractured rocks to tropical cyclones (Hilton *et al.*, 2008b), and New Zealand's rivers yield $10 \pm 3\,\text{t C km}^{-2}\,\text{yr}^{-1}$ of POC to the coastal ocean (Scott *et al.*, 2006). This is 10 times the global average, and the total riverine organic carbon yield is equivalent to 45% of New Zealand's total fossil fuel CO_2 emissions (globally, this ratio is less than 10%). Rivers in the East Cape region carry a mixture of modern-plant- and ancient-rock-derived organic carbon (Gomez *et al.*, 2004b; Leithold, Blair and Perkey, 2006). At low suspended sediment concentrations (and low discharges), modern organic carbon delivered to channels by sheetwash dominates, but as the concentration increases, more and more carbon is contributed directly from erosion of the weathering mantle and the percentage POC stabilizes at a low value that approaches the carbon content of the bedrock (Figure 5.14a). Bedrock is a source of fossil POC which has little impact on CO_2 drawdown and long-term C cycling and it remains that, even though sediments that accumulate on river-fed continental margins retain their terrigenous carbon (Leithold and Blair, 2001), the high sedimentation rates also provide optimal conditions for the burial of non-fossil POC in the coastal ocean (Hilton *et al.*, 2008b). Consequently, steepland rivers function as the delivery mechanism to a carbon sink.

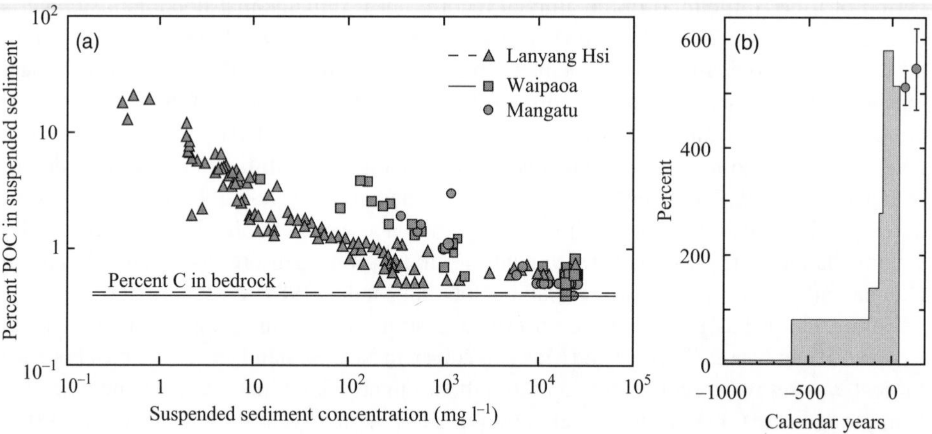

Figure 5.14 (a) Relationship between particulate organic carbon (POC) loading and suspended sediment concentration in rivers in New Zealand and Taiwan (after Gomez *et al.*, 2003b; and Kao and Liu, 1996). (b) Percentage change (relative to the time when the landscape was subject only to perturbations by natural events) in the suspended sediment load of the Waipaoa River immediately before and during the Anthropocene. The marked increase in the period after −135 calendar years (AD 1820) is associated with the arrival of European colonists, who cleared the native vegetation (cf. Figures 5.12e and 5.13b). Zero calendar years equate with AD 1955, solid dots indicate the mean and the error bars plus/minus one standard deviation of the potential values for the 2030s and 2080s (after Kettner, Gomez and Syvitski, 2008).

We conclude this chapter by making two observations about the sediment flux from steepland catchments and its consequences for management decisions. First, it is predicted that human impact and the legacy of historic land-use change likely will continue to exert the dominant influence on the sediment loads of steepland rivers in the twenty-first entury (Figure 5.14b). Second, a large proportion of this sediment is derived from erosional 'hot spots' that can be stabilized by erosion control treatments or reforestation. Targeted restoration programmes cannot completely eliminate erosion (Madej, 2001; Reid and Page, 2002; Marden *et al.*, 2005), but their adoption has the potential to significantly reduce the amount of storm-generated sediment and obviates the need to resort to indiscriminate and potentially more controversial erosion-control measures.

References

Adams, J. (1981) Earthquake-dammed lakes in New Zealand. *Geology*, **9**, 215–219.

Anderson, R.S. (1994) Evolution of the Santa Cruz Mountains, California, through tectonic growth and geomorphic decay. *Journal of Geophysical Research*, **99**, 20161–20179.

Asselman, N.E.M. (2000) Fitting and interpretation of sediment rating curves. *Journal of Hydrology*, **234**, 228–248.

Begin, Z.B. (1983) Application of 'diffusion' degradation to some aspects of drainage net development, in *Badland Geomorphology and Piping* (eds R.B. Bryan and A. Yair), GeoBooks, Norwich, pp. 169–181.

Benda, L. and Dunne, T. (1997a) Stochastic forcing of sediment supply to channel networks from landsliding and debris flow. *Water Resources Research*, **33**, 2849–2863.

Benda, L. and Dunne, T. (1997b) Stochastic forcing of sediment routing and storage in channel networks. *Water Resources Research*, **33**, 2865–2880.

Betts, H.D., Trustrum, N.A. and De Rose, R.C. (2003) Geomorphic changes in a complex gully system measured from sequential digital elevation models, and implications for management. *Earth Surface Processes and Landforms*, **28**, 1043–1058.

Beven, K. (1981) The effect of ordering on the geomorphic effectiveness of hydrologic events. *International Association of Hydrological Sciences Publication*, **132**, 510–526.

Beven, K. and Kirkby, M.J. (1979) A physically-based variable contributing area model of basin hydrology. *Hydrological Sciences Bulletin*, **24**, 43–69.

Beschta, R.L., Pyles, M.R., Skaugset, A.E. and Surfleet, C.G. (2000) Peakflow responses to forest practices in the western Cascades of Oregon, USA. *Journal of Hydrology*, **233**, 102–120.

Blaschke, P.M., Trustrum, N.A. and Hicks, D.L. (2000) Impacts of mass movement erosion on land productivity: a review. *Progress in Physical geography*, **24**, 21–52.

Bosch, J.M. and Hewlett, J.D. (1982) A review of catchment experiments to determine the effect of vegetation changes on water yield and evapotranspiration. *Journal of Hydrology*, **55**, 3–23.

Brown, W.M. and Ritter, J.R. (1971) Sediment transport and turbidity in the Eel River basin, California. *U.S. Geological Survey Water-Supply Paper*, **1986**, 70.

Burbank, D.W., Leland, J., Fielding, E. *et al.* (1996) Bedrock incision, rock uplift and threshold hillsopes in the northwestern Himalayas. *Nature*, **379**, 505–510.

Bucknam, R.C., Coe, J.A., Chavarría, M.M. *et al.* (2001) Landslides triggered by Hurricane Mitch in Guatemala – inventory and discussion. *U.S. Geological Survey Open-File Report*, 01-443, 38.

Caine, N. (1980) The rainfall intensity-duration control of shallow landslides and debris flows. *Geografiska Annaler*, **62**, 23–27.

Campbell, R.H. (1975) Soil slips, debris flows and rainstorms in the Santa Monica mountains, southern California. *U.S. Geological Survey Professional Paper*, **851**, 51.

Carson, M.A. (1971) *The Mechanics of Erosion*, Pion, London, p. 174.

Casadei, M., Dietrich, W.E. and Miller, N.L. (2003) Testing a model for predicting the timing and location of shallow landslide initiation in soil-mantled landscapes. *Earth Surface Processes and Landforms*, **28**, 925–950.

Chang, K., Chiang, S. and Hsu, M. (2007) Modeling typhoon- and earthquake-induced landslides in a mountainous watershed using logistic regression. *Geomorphology*, **89**, 335–347.

Church, M.A. and Miles, M.J. (1987) Meteorological antecedents to debris flow in southwestern British Columbia; some case studies. *Geological Society of America Reviews in Engineering Geology*, **7**, 63–79.

Church, M.A. and Ryder, J.M. (1972) Paraglacial Sedimentation: A Consideration of Fluvial Processes Conditioned by Glaciation. *Geological Society of America Bulletin*, **83**, 3059–3072.

Church, M.A. and Slaymaker, O. (1989) Disequilibrium of Holocene sediment yield in glaciated British Columbia. *Nature*, **337**, 452–454.

Church, M.A., Ham, D., Hassan, M. and Slaymaker, O. (1999) Fluvial clastic sediment yield in Canada: scaled analysis. *Canadian Journal of Earth Sciences*, **36**, 1267–1280.

Clement, A.C., Seager, R. and Cane, M.A. (2000) Suppression of El Niño during the mid-Holocene by changes in Earth's orbit. *Paleoceanography*, **15**, 731–737.

Conroy, J.L., Overpeck, J.T., Cole, J.E. *et al.* (2008) Holocene changes in eastern tropical Pacific climate inferred from a Galápagos lake sediment record. *Quaternary Science Reviews*, **27**, 1166–1180.

Costa, J.E. and Schuster, R.L. (1988) The formation and failure of natural dams. *Geological Society of America Bulletin*, **100**, 1054–1068.

Crawford, C.R. (1991) Estimation of suspended-sediment rating curves and mean suspended-sediment loads. *Journal of Hydrology*, **129**, 331–348.

Crouch, R.J. (1976) Tunnel erosion - a review. *Journal of Soil Conservation Service, New South Wales*, **32**, 98–111.

Crozier, M.J. (1997) The climate-landslide couple: a Southern Hemisphere perspective, in *Rapid Mass Movement as a Source of Climatic Evidence for the Holocene* (eds J.A. Matthews, D. Brunsden, B. Frenzel*et al.*), Gustav Fischer Verlag, Stuttgart, pp. 333–354.

Crozier, M.J. (1999) Prediction of rainfall-triggered landslides: A test of the antecedent water status model. *Earth Surface Processes and Landforms*, **24**, 825–833.

Crozier, M.J. and Preston, N.J. (1998) Modelling changes in terrain resistance as a component of landform evolution in unstable hill country. *Lecture Notes in Earth Sciences*, **78**, 267–284.

Crozier, M.J., Vaughn, E.E. and Tippett, J.M. (1990) The relative instability of colluvium-filled bedrock depressions. *Earth Surface Processes and Landforms*, **15**, 326–339.

Cruden, D.M. (1976) Major rock slides in the Rockies. *Canadian Geotechnical Journal*, **13**, 8–20.

Cui, Y. and Parker, G. (2005) Numerical model of sediment pulses and sediment supply disturbances in mountain rivers. *Journal of Hydraulic Engineering*, **131**, 646–656.

Dadson, S.J., Hovius, N., Chen, H. *et al.* (2003) Links between erosion, runoff variability and seismicity in the Taiwan orogen. *Nature*, **426**, 648–651.

Dadson, S.J., Hovius, N., Chen, H. *et al.* (2004) Earthquake-driven increase in sediment delivery from an active mountain belt. *Geology*, **32**, 733–736.

Dadson, S.J., Hovius, N., Pegg, S. *et al.* (2005) Hyperpycnal river flows from an active mountain belt. *Journal of Geophysical Research*, **110**, F04016. doi: 10.1029/2004JF000244.

Densmore, A.L. and Hovius, N. (2000) Topographic fingerprint of bedrock landslides. *Geology*, **28**, 371–374.

Densmore, A.L., Anderson, R.S., McAdoo, B.G. and Ellis, M.A. (1997) Hillslope evolution by bedrock landslides. *Science*, **275**, 369–372.

DeRose, R.C., Trustrum, N.A. and Blaschke, P.M. (1993) Post-deforestation soil loss from steepland hillslopes in Taranaki, New Zealand. *Earth Surface Processes and Landforms*, **18**, 131–144.

DeVente, J., Poesen, J., Arabkhedri, M. and Verstraeten, G. (2007) The sediment delivery problem revisited. *Progress in Physical Geography*, **31**, 155–178.

Dietrich, W.E. and Dunne, T. (1978) Sediment budget for a small catchment in mountainous terrain. *Zeitschrift für Geomorphologie Supplement Band*, **29**, 191–206.

Dietrich, W.E., Wilson, C.J. and Reneau, S.L. (1986) Hollows, colluvium, and landslides in soil-mantled landscapes, in *Hillslope Processes* (ed. A.D. Abrahams), Allen and Unwin, Boston, pp. 361–388.

Dietrich, W.E., Dunne, T., Humphrey, N.F. and Reid, L.M. (1982) Construction of sediment budgets for drainage basins, Pacific Northwest Forest and Range Experiment Station. U.S. Department of Agriculture, Forest Service, General Technical Report PNW-141, pp. 5–23.

Dietrich, W.E., Reiss, R., Hsu, M.L. and Montgomery, D.R. (1995) A process-based model for colluvial soil depth and shallow landslides using digital elevation data. *Hydrological Processes*, **9**, 383–400.

D'Odorico, P. and Fagherazzi, S. (2003) A probabilistic model of rainfall-triggered shallow landslides in hollows: a long-term analysis. *Water Resources. Research*, **39**, 1262. doi: 10.1029/2002WR001595.

Eden, D.N. and Page, M.J. (1998) Palaeoclimatic implications of a storm erosion record from late Holocene lake sediments, North Island, New Zealand. *Palaeogeography, Palaeoclimatology and Palaeoecology*, **139**, 37–58.

EPA (1998) Total maximum daily load for sediment Redwood Creek, California. U.S. Environmental Protection Agency Region 9, San Francisco, California, p. 73.

Eyles, R.J. (1977) Changes in drainage networks since 1820, Southern Tablelands, NSW. *Australian Geographer*, **13**, 377–387.

Faulkner, H. (1974) An allometric growth model for competitive gullies. *Zeitschrift für Geomorphologie Supplement Band*, **21**, 76–87.

Ferguson, R.I. (1987) Accuracy and precision of methods for estimating river loads. *Earth Surface Processes and Landforms*, **12**, 95–104.

Finlay, P.J., Fell, R. and Maguire, P.K. (1997) The relationship between the probability of landslide occurrence and rainfall. *Canadian Geotechnical Journal*, **34**, 811–824.

Fryirs, K.A., Brierely, G.J., Preston, N.J. and Kasi, M. (2007) Buffers, barriers and blankets: the (dis) connectivity of catchment scale sediment cascades. *Catena*, **70**, 49–67.

Fuller, C.W., Willett, S.D., Hovius, N. and Slingerland, R. (2003) Erosion rates for Taiwan mountain basins: new determinations from suspended sediment records and a stochastic model of their temporal variation. *Journal of Geology*, **111**, 71–87.

Gabet, E.J. and Dunne, T. (2002) Landslides on coastal sage-scrub and grassland hillslopes in a severe El Niño winter: the effects of vegetation conversion on sediment delivery. *Geological Society of America Bulletin*, **114**, 983–990.

Gage, M. and Black, R.D. (1979) Slope stability and geological investigations at Mangatu State Forest. Technical Paper Number 66, Forest Research Institute, New Zealand Forest Service, p. 47.

Gibbs, H.S. (1945) Tunnel-gully erosion on the Wither Hills, Marlborough, New Zealand. *New Zealand Journal of Science and Technology*, **27**, 135–146.

Glade, T., Crozier, M.J. and Smith, P. (2000) Applying probability determination to refine landslide-triggering rainfall thresholds using an empirical Antecedent Daily Rainfall Model. *Pure and Applied Geophysics*, **157**, 1059–1079.

Godt, J.W., Baum, R.L. and Chleborad, A.F. (2006) Rainfall characteristics for shallow landsliding in Seattle, Washington, USA. *Earth Surface Processes and Landforms*, **31**, 97–110.

Goldsmith, S.T., Carey, A.E., Lyons, W.B. *et al.* (2008) Extreme storm events, landscape denudation and carbon sequestration: Typhoon Mindulle, Choshui River, Taiwan. *Geology*, **36**, 483–486.

Gomez, B., Carter, L. and Trustrum, N.A. (2007) A 2400 yr record of natural events and anthropogenic impacts in inter-correlated terrestrial and marine sediment cores: Waipaoa sedimentary system, New Zealand. *Geological Society of America Bulletin*, **119**, 1415–1432.

Gomez, B., Eden, D.N., Peacock, D.H. and Pinkney, E.J. (1998) Floodplain construction by rapid vertical accretion: Waipaoa River, New Zealand. *Earth Surface Processes and Landforms*, **23**, 405–413.

Gomez, B., Banbury, K., Marden, M. *et al.* (2003a) Gully Erosion and Sediment Production: Te Weraroa Stream, New Zealand. *Water Resources Research*, **39**, 1187. doi: 10.1029/2002WR001342.

Gomez, B., Trustrum, N.A., Hicks, D.M. *et al.* (2003b) Production, storage and output of particulate organic carbon: Waipaoa River Basin, New Zealand. *Water Resources Research*, **39**, doi: 10.1029/2002WR001619.

Gomez, B., Carter, L., Trustrum, N.A. *et al.* (2004a) El Niño–Southern Oscillation signal associated with middle-Holocene climate change in intercorrelated terrestrial and marine sediment cores, North Island, New Zealand. *Geology*, **32**, 653–656.

Gomez, B., Brackley, H.L., Hicks, D.M. *et al.* (2004b) Organic carbon in floodplain alluvium: signature of historic variations in erosion processes associated with deforestation, Waipaoa River basin, New Zealand. *Journal of Geophysical Research*, **109**, F04011. doi: 10.1029/2004JF000154.

Hack, J.T. and Goodlett, J.C. (1960) Geomorphology and forest ecology of a mountain region in the central Appalachians. *U.S. Geological Survey Professional Paper*, **347**, 66.

Hagerty, D.J. (1991) Piping/sapping erosion. I: Basic considerations. *Journal of Hydraulic Engineering*, **117**, 991–1008.

Harp, E.L. and Jibson, R.L. (1995) Inventory of landslides triggered by the 1994 Northridge, California earthquake. *U.S. Geological Survey Open File Report*, **95-213**.

Hicks, D.M. and Gomez, B. (2003) Sediment Transport, in *Tools in Fluvial Geomorphology* (eds G.M. Kondolf and H. Piégay), John Wiley and Sons, pp. 425–461.

Hicks, D.M. and Shankar, U. (2003) Sediment from New Zealand rivers, NIWA Miscellaneous Chart 236. National Institute for Water and Atmospheric Research, Wellington.

Hicks, D.M., Gomez, B. and Trustrum, N.A. (2000) Erosion thresholds and suspended sediment yields: Waipaoa River Basin, New Zealand. *Water Resources Research*, **36**, 1129–1142.

Hicks, D.M., Gomez, B. and Trustrum, N.A. (2004) Event suspended sediment characteristics and the generation of hyperpycnal plumes at river mouths: East Coast Continental Margin, North Island, New Zealand. *Journal of Geology*, **112**, 471–485.

Hilton, R.G., Galy, A. and Hovius, N. (2008a) Riverine particulate organic carbon from an active mountain belt: importance of landslides. *Global Biogeochemical Cycles*, **22**, GB1017. doi: 10.1029/2006GB002905.

Hilton, R.G., Galy, A., Hovius, N. *et al.* (2008b) Typhoon–cyclone-driven erosion of the terrestrial biosphere from mountains. *Nature Geoscience*, **1**, 759–762.

Hong, Y., Adler, R.F. and Huffman, G.J. (2006) Evaluation of the potential of NASA multi-satellite precipitation analysis in global landslide hazard assessment. *Geophysical Research Letters*, **33**, L22402. doi: 10.1029/2006GL028010.

Hovius, N. and Stark, C.P. (2006) Landslide driven erosion and topographic evolution of active mountain belts, in *Landslides from Massive Rock Slope Failure* (eds S.G. Evans, G.S. Mugnozza, A. Strom and R.L. Hermanns), Springer, The Netherlands, pp. 573–590.

Hovius, N., Stark, C.P. and Allen, P.A. (1997) Sediment flux from a mountain belt derived by landslide mapping. *Geology*, **25**, 231–234.

Hovius, N., Stark, C.P., Tutton, M.A. and Abbott, L.D. (1998) Landslide-driven drainage network evolution in a pre-steady-state mountain belt: Finisterre Mountains, Papua New Guinea. *Geology*, **26**, 1071–1074.

Howard, A.D. and McLane, C.F. (1988) Erosion of cohesionless sediment by groundwater seepage. *Water Resources Research*, **24**, 1659–1674.

Hunsinger, G.B., Mitra, S., Warrick, J.A. and Alexander, C.R. (2008) Oceanic loading of wildfire-derived organic compounds from a small mountainous river. *Journal of Geophysical Research*, **113**, G02007. doi: 10.1029/2007JG000476.

Imaizumi, F., Sidle, R.C. and Kamei, R. (2008) Effects of forest harvesting on the occurrence of landslides and debris flows in steep terrain of central Japan. *Earth Surface Processes and Landforms*, **33**, 827–840.

Inman, D.L. and Jenkins, S.A. (1999) Climate change and the episodicity of sediment flux of small California rivers. *Journal of Geology*, **107**, 251–270.

Iverson, R.M. (2000) Landslide triggering by rain infiltration. *Water Resources Research*, **36**, 1897–1910.

Iverson, R.M. and Major, J.J. (1986) Groundwater seepage vectors and the potential for hillslope failure and debris flow mobilization. *Water Resources Research*, **22**, 1543–1548.

Iverson, R.M., Reid, M.E., Iverson, N.R. *et al.* (2000) Acute Sensitivity of Landslide Raters to Initial Soil Porosity. *Science*, **290**, 513–516.

Jessen, M.R., Crippen, T.F., Page, M.J. *et al.* (1999) Land use capability classification of the Gisborne (East Coast region. *Landcare Research Science Series*, **21**, 213.

Johnson, A.C., Swanston, D.N. and McGee, K.E. (2000) Landslide initiation, runout, and deposition within clearcuts and old-growth forests of Alaska. *Journal of the American Water Resources Association*, **36**, 17–30.

Jones, J.A.A. (1990) Piping effects in humid lands. *Geological Society of America Special Paper*, **252**, 111–138.

Kao, S.-J. and Liu, K.-K. (1996) Particulate organic carbon export from a subtropical mountainous river (Lanyang Hsi) in Taiwan. *Limnology and Oceanography*, **41**, 1749–1757.

Kao, S.-J. and Liu, K.-K. (2002) Exacerbation of erosion induced by human perturbation in a typical Oceania watershed: Insight from 45 years of hydrological records from the Lanyang-Hsi River, northeastern Taiwan. *Global Biogeocehmical Cycles*, **16**, doi: 10.1029/2000GB001334.

Kao, S.-J. and Milliman, J.D. (2008) Water and sediment discharge from small mountainous rivers, Taiwan: the roles of lithology, episodic events, and human activities. *Journal of Geology*, **116**, 431–448.

Kao, S.-J., Lee, T.-Y. and Milliman, J.D. (2005) Calculating highly fluctuated sediment fluxes from mountainous rivers in Taiwan. *Terrestrial Atmospheric and Ocean Sciences*, **16**, 653–675.

Kao, S.-J., Chan, S.-C., Kuo, C.-H. and Lee, K.-K. (2005) Transport dominated sediment loading in Taiwanese rivers: a case study from the Ma-an Stream. *Journal of Geology*, **113**, 217–225.

Kasai, M., Marutani, T., Reid, L.M. and Trustrum, N.A. (2001) Estimation of temporally averaged sediment delivery ratio using aggradational terraces in headwater catchments of the Waipaoa River, North Island. *New Zealand Earth Surface Processes and Landforms*, **26**, 1–16.

Keefer, D.K. (1984) Landslides caused by earthquakes. *Geological Society of America Bulletin*, **95**, 406–421.

Keefer, D.K. (1994) The importance of earthquake-induced landslides to longterm slope erosion and slope-failure hazards in seismically active regions. *Geomorphology*, **10**, 265–284.

Keller, E.A., Valentine, D.W. and Gibbs, D.R. (1997) Hydrological response of small watersheds following the Southern California Painted Cave Fire of June 1990. *Hydrological Processes*, **11**, 401–414.

Kelsey, H.M. (1980) A sediment budget and an analysis of geomorphic process in the Van Duzen River basin, north coastal California, 1941–1975: summary. *Geological Society of America Bulletin*, **91**, 190–195.

Kelsey, H., Lamberson, M.R. and Madej, M.A. (1987) Stochastic model for the long-term transport of stored sediment in a river channel. *Water Resources Research*, **23**, 1738–1750.

Kettner, A.J., Gomez, B. and Syvitski, J.P.M. (2007) Modeling suspended sediment discharge from the Waipaoa River system, New Zealand: the last 3000 years. *Water Resources Research*, **43**, W07411. doi: 10.1029/2006WR005570.

Kettner, A.J., Gomez, B. and Syvitski, J.P.M. (2008) Human catalysts or climate change: which will have a greater impact on the sediment load of the Waipaoa River in the 21st century? *International Association of Hydrological Sciences Publication*, **325**, 425–431.

Kettner, A.J., Gomez, B., Hutton, E.W.H. and Syvitski, J.P.M. (2009) Late Holocene dispersal and accumulation of terrigenous sediment on Poverty Shelf, New Zealand. *Basin Research*, **21**, 253–267.

Kinsey-Henderson, A.E., Post, D.A. and Prosser, I.P. (2005) Modelling sources of sediment at sub-catchment scale: an example from the Burdekin catchment, North Queensland, Australia. *Mathematics and Computers in Simulation*, **69**, 90–102.

Kirchner, J.W., Finkel, R.C., Riebe, C.S. *et al.* (2001) Mountain erosion over 10 yr, 10 k.y. and 10 m.y. time scales. *Geology*, **29**, 591–594.

Komada, T., Druffel, E.R.M. and Trumbore, S.E. (2004) Oceanic export of relict carbon by small mountainous rivers. *Geophysical Research Letters*, **31**, L07504. doi: 10.1029/2004GL019512.

Korup, O., Clague, J.J., Hermanns, R.L. *et al.* (2007) Giant landslides, topography and erosion. *Earth and Planetary Science Letters*, **261**, 578–589.

Koi, T., Hotta, N., Ishigaki, I. *et al.* (2008) Prolonged impact of earthquake-induced landslides on sediment yield in a mountain watershed: The Tanzawa region, Japan. *Geomorphology*, **101**, 692–702.

Lamy, F., Hebbeln, D., Röhl, U. and Wefer, G. (2001) Holocene rainfall variability in southern Chile: a marine record of latitudinal shifts of the Southern Westerlies. *Earth and Planetary Science Letters*, **185**, 369–382.

Lavé, J. and Burbank, D.W. (2004) Denudation processes and rates in the Transverse Ranges, southern California: erosional response of a transitional landscape to external and anthropogenic forcing. *Journal of Geophysical Research*, **109**, F01006. doi: 10.1029/2003JF000023.

Leithold, E.L. and Blair, N.E. (2001) Watershed control on the carbon loading of marine sedimentary particles. *Geochimica et Cosmochimica Acta*, **65**, 2231–2240.

Leithold, E.L., Blair, N.E. and Perkey, D.W. (2006) Geomorphic controls on the age of particulate organic carbon from small mountainous and upland rivers. *Global Biogeochemical Cycles*, **20**, doi: 10.1029/2005GB002677.

Liébault, F. and Piégay, H. (2001) Assessment of channel changes due to long-term bedload supply decrease. *Geomorphology*, **36**, 167–186.

Lin, G.-W., Chen, H., Hovius, N. *et al.* (2008) Effects of earthquake and cyclone sequencing on landsliding and fluvial sediment transfer in a mountain catchment. *Earth Surface Processes and Landforms*, **33**, 1354–1373.

Lisle, T.E. (1982) Effects of aggradation and degradation on riffle-pool morphology in natural gravel channels, northwestern California. *Water Resources Research*, **18**, 1643–1651.

Lisle, T.E. and Madej, M.A. (1992) Spatial variation in armouring in a channel with high sediment supply, in *Dynamics of Gravel-bed Rivers* (eds P. Billi, R.D. Hey, C.R. Thorne and P. Tacconi), John Wiley and Sons, Chichester, pp. 277–293.

Lisle, T.E., Cui, Y., Parker, G. *et al.* (2001) The dominance of dispersion in the evolution of bed material waves in gravel-bed rivers. *Earth Surface Processes and Landforms*, **26**, 1409–1420.

Liu, J.P., Liu, C.S., Xu, K.H. *et al.* (2008) Flux and fate of small mountainous rivers derived sediments into the Taiwan Strait. *Marine Geology*, **256**, 65–76.

Lock, J., Kelsey, H., Furlong, K. and Woolace, A. (2006) Late Neogene and Quaternary landscape evolution of the northern California Coast Ranges: evidence for Mendocino triple junction tectonics. *Geological Society of America Bulletin*, **118**, 1232–1246.

Lyons, W.B., Nezat, C.A., Carey, A.E. and Hicks, D.M. (2002) Organic carbon fluxes to the ocean from high-standing islands. *Geology*, **30**, 443–446.

Madej, M.A. (1987) Residence times of channel-stored sediment in Redwood Creek, northwestern California. *International Association of Hydrological Sciences Publication*, **165**, 429–438.

Madej, M.A. (2001) Erosion and sediment delivery following the removal of forest roads. *Earth Surface Processes and Landforms*, **26**, 175–190.

Madej, M.A. and Ozaki, V. (1996) Channel response to sediment wave propogation and movement, Redwood Creek, California, USA. *Earth Surface Processes and Landforms*, **21**, 911–927.

MAF (2005) The 2005 Review of the East Coast Forestry Project. Discussion Paper 37, Ministry of Agriculture and Forestry, Wellington, p. 47.

Malamud, B.D., Turcotte, D.L., Guzsetti, F. and Reichenbach, P. (2004) Landslides, earthquakes and erosion. *Earth and Planetary Science Letters*, **229**, 45–59.

Marden, M., Arnold, G., Gomez, B. and Rowan, D. (2005) Pre- and post-reforestation gully development in Mangatu Forest, East Coast, North Island, New Zealand. *Rivers Research and Applications*, **21**, 757–771.

Marron, D.M. (1984) Colluvium in bedrock hollows on steep slopes, Redwood Creek drainage basin, northwestern California. *Catena Supplement*, **6**, 59–68.

Martin, Y., Rood, K., Schwab, J.W. and Church, M.A. (2002) Sediment transfer by shallow landsliding in the Queen Charlotte Islands, British Columbia. *Canadian Journal of Earth Sciences*, **39**, 189–205.

Marutani, T., Kasai, M., Reid, L.M. and Trustrum, N.A. (1999) Influence of storm-related sediment storage on the sediment delivery from tributary catchments in the upper Waipaoa River, New Zealand. *Earth Surface Processes and Landforms*, **24**, 881–896.

May, C.L. and Gresswell, R.E. (2003) Processes and rates of sediment and wood accumulation in headwater streams of the Oregon Coast Range, USA. *Earth Surface Processes and Landforms*, **28**, 409–424.

McDonnell, J.J. (1990) The influence of macropores on debris flow initiation. *Quarterly Journal of Engineering Geology*, **23**, 325–331.

McKergow, L.A., Prosser, I.A., Hughes, A.O. and Brodie, J. (2005) Sources of sediment to the GreatBarrier Reef World heritage Area. *Marine Pollution Bulletin*, **51**, 200–211.

Meade, R.H., Yuzyk, T.R. and Day, T.J. (1990) Movement and storage of sediment in rivers of the United States and Canada, in *Surface Water Hydrology – the Geology of North America*, Vol. **0–1** (eds M.G. Wolman and H.C. Riggs) Geological Society of America, Boulder, Colorado, pp. 255–280.

Métivier, F. and Gaudemar, Y. (1999) Stability of output fuxes of large rivers in south and east Asia during the last 2 million years: implications on foodplain processes. *Basin Research*, **11**, 293–303.

Middleton, G.V. and Wilcock, P.R. (1994) *Mechanics in the Earth and Environmental Sciences*, Cambridge University Press, Cambridge, p. 459.

Miller, D.J. and Dunne, T. (1996) Topographic perturbations of regional stresses and consequent bedrock fracturing. *Journal of Geophysical Research*, **101**, 25,523–25,536.

Miller, D.J., Luce, C. and Benda, L. (2003) Time, space, and episodicity of physical disturbance in streams. *Forest Ecology and Management*, **178**, 121–140.

Milliman, J.D. and Kao, S.-J. (2005) Hyperpycnal discharge of fluvial sediment to the ocean: impact of super-typhoon Herb (1996) on Taiwanese rivers. *Journal of Geology*, **113**, 503–516.

Milliman, J.D. and Syvitski, J.P. (1992) Geomorphic/tectonic control of sediment discharge to the ocean: the importance of small mountainous rivers. *Journal of Geology*, **100**, 525–544.

Montgomery, D.R. and Dietrich, W.E. (1994) A physically based model for the topographic control on shallow landsliding. *Water Resources Research*, **30**, 1153–1171.

Montgomery, D.R., Balco, G. and Willett, S.D. (2001) Climate, tectonics and the morphology of the Andes. *Geology*, **29**, 579–582.

Montgomery, D.R., Dietrich, W.E. and Heffner, J.T. (2002) Piezometric response in shallow bedrock at CB1: implications for runoff generation and landsliding. *Water Resources Research*, **38**, 1274. doi: 10.1029/2002WR001429.

Montgomery, D.R., Sullivan, K. and Greenberg, H.M. (1998) Regional test of a model for shallow landsliding. *Hydrological Processes*, **12**, 943–955.

Montgomery, D.R., Schmidt, K.M., Greenberg, H.M. and Dietrich, H.M. (2000) Forest clearing and regional landsliding. *Geology*, **28**, 311–314.

Mosley, M.P. (1972) Evolution of a discontinuous gully system. *Annals of the Association of American Geographers*, **62**, 655–663.

Mount, J.F. (1995) *California Rivers and Streams*, University California Press, Berkeley, p. 359.

Moy, C.M., Seltzer, G.O., Seltzer, D.T. and Anderson, D.M. (2002) Variability of El Nino/Southern Oscillation activity at millennial time scales during the Holocene epoch. *Nature*, **420**, 162–165.

Mulder, T. and Syvitski, J.P.M. (1995) Turbidity currents generated at river mouths during exceptional discharges to the world oceans. *Journal of Geology*, **103**, 285–299.

Mulder, T., Syvitski, J.P.M., Migeon, S. *et al.* (2003) Marine hyperpycnal flows: initiation, behavior and related deposits: a review. *Marine and Petroleum Geology*, **20**, 861–882.

Nakamura, F. and Kikuchi, S.-I. (1996) Some methodological developments in the analysis of sediment transport processes using age distribution of floodplain deposits. *Geomorphology*, **16**, 139–148.

Nakamura, F., Araya, T. and Higashi, S. (1987) Influence of river channel morphology and sediment production on residence time and transport distance. *International Association of Hydrological Sciences Publication*, **165**, 355–364.

Nakamura, F., Maita, H. and Araya, T. (1995) Sediment routing analysis based on chronological changes in hillslope and riverbed morphologies. *Earth Surface Processes and Landforms*, **20**, 333–346.

Nakamura, F., Swanson, F.J. and Wondzell, S.M. (2000) Disturbance regimes of stream and riparian systems - a disturbance-cascade perspective. *Hydrological Processes*, **14**, 2849–2860.

Nicol, A. and Beavan, J. (2003) Shortening of an overriding plate and its implications for slip on a subduction thrust, central Hikurangi Margin, New Zealand. *Tectonics*, **22**, 1070. doi: 10.1029/2003TC001521.

Nolan, K.M. and Janda, R.J. (1981) Use of short-term water and suspended sediment discharge observations to assess impacts of logging on stream-sediment discharge in the Redwood Creek basin, northwestern California. *International Association of Hydrological Sciences Publication*, **132**, 415–437.

Nolan, K.M. and Janda, R.J. (1995) Impacts of logging on stream-sediment discharge in the Redwood Creek Basin, northwestern California. *U.S. Geological Survey Professional Paper*, **1454L**, p. 8.

Nolan, K.M. and Marron, D.C. (1985) Contrast in stream-channel response to major storms in two mountainous areas of California. *Geology*, **13**, 135–138.

Nolan, K.M. and Marron, D.C. (1995) History, causes and significance of changes in the channel geometry of Redwood Creek, northwestern California, 1936–1982. *U.S. Geological Survey Professional Paper*, **1454N**, 22.

Nolan, K.M., Lisle, T.E. and Kelsey, H.M. (1987) Bankfull discharge and sediment transport in northwestern California. *International Association of Hydrological Sciences Publication*, **165**, 439–449.

Nunn, P.D. (2007) *Climate Environment and Society in the Pacific during the Last Millennium*, Elsevier, p. 316.

Oguchi, T. (1996) factors affecting the magnitude of post-glacial hillslope incision in Japanese mountains. *Catena*, **26**, 171–186.

O'Loughlin, C. and Ziemer, R.R. (1982) The importance of root strength and deterioration rates upon edaphic stability, in steepland forests, in *Carbon uptake and Allocation in Subalpine Ecosystems as a Key to Management* (ed. R.H. Waring), Oregon State University, Corvallis, Oregon, pp. 70–78.

O'Loughlin, E.M. (1986) Prediction of surface saturation in natural catchments by topographic analysis. *Water Resources Research*, **22**, 794–804.

Page, M.J. and Trustrum, N.A. (1997) A late Holocene lake sediment record of the erosion response to land use change in a steepland catchment, New Zealand. *Zeitschrift für Geomorphologie*, **41**, 369–392.

Page, M.J., Trustrum, N.A. and DeRose, R.C. (1994) A high resolution record of storm- induced erosion from lake sediments, New Zealand. *Journal of Paleolimnology*, **11**, 333–348.

Page, M.J., Trustrum, N.A. and Dymond, J.R. (1994) Sediment budget to assess the geomorphic effect of a cyclonic storm, New Zealand. *Geomorphology*, **9**, 169–188.

Page, M.J., Reid, L.M. and Lynn, I.H. (1999) Sediment production from Cyclone Bola landslides, Waipaoa catchment. *Journal of Hydrology New Zealand*, **38**, 289–308.

Pain, C.F. and Bowler, J.M. (1973) Denudation following the November 1970 earthquake at Madang, Papua New Guinea. *Zeitschrift für Geomorphologie, Supplement Band*, **18**, 91–104.

Pain, C.F. and Hoskin, P.L. (1970) The movement of sediment in a channel in relation to magnitude and frequency concepts – a New Zealand example. *Earth Science Journal*, **4**, 17–23.

Parkner, T., Page, M.J., Marden, M. and Marutani, T. (2007) Gully systems under undisturbed indigenous forest, East Coast Region, New Zealand. *Geomorphology*, **84**, 241–253.

Pearce, A.J. and Watson, A.J. (1986) Effects of earthquake-induced landslides on sediment budget and transport over a 50-yr period. *Geology*, **14**, 52–55.

Phillips, C.J. and Marden, M. (2004) Reforestation schemes to manage regional landslide risk, in *Landslide Hazard and Risk* (eds T. Glade, M. Anderson and M.J. Crozier), John Wiley and Sons, Chichester, pp. 517–547.

Phillips, J.D. (2003) Alluvial storage and the long term stability of sediment yields. *Basin Research*, **15**, 153–163.

Phillips, J.D. and Gomez, B. (2007) Controls on sediment export from the Waipaoa River basin, New Zealand. *Basin Research*, **19**, 241–252.

Phillips, J.D., Marden, M. and Gomez, B. (2007) Residence time of alluvium in an aggrading fluvial system. *Earth Surface Processes and Landforms*, **32**, 307–316.

Pierson, T.C. (1983) Soil pipes and slope stability. *Quarterly Journal of Engineering Geology*, **16**, 1–11.

Prosser, I.P. (1996) Thresholds of channel initiation in historical and Holocene times, southeastern Australia, in *Hillslope Processes*, vol. **2** (eds M.G. Anderson and S.M. Brooks), John Wiley and Sons, Chichester, pp. 687–708.

Prosser, I.P. and Abernethy, B. (1996) Predicting the topographic limits to a gully network using a digital terrain model and process thresholds. *Water Resources Research*, **32**, 2289–2298.

Prosser, I.P. and Slade, C.J. (1994) Gully formation and the role of valley-floor vegetation, southeastern Australia. *Geology*, **22**, 1127–1130.

Prosser, I.P. and Winchester, S.J. (1996) History and processes of gully initiation and development in eastern Australia. *Zeitscrift für Geomorphologie Supplement Band*, **105**, 91–109.

Prosser, I.P., Chappell, J.M.A. and Gillespie, R. (1994) Holocene valley aggradation and gully erosion in headwater catchments, southeastern highland of Australia. *Earth Surface Processes and Landforms*, **19**, 465–480.

Prosser, I., Rustomji, P., Young, W. *et al.* (2001) Constructing river basin sediment budgets for the national land and water resources audit. CSIRO Land and Water Technical Report 15/01, p. 36, http://www.clw.csiro.au/publications/technical2001/tr15-01.pdf.

Reid, L.M. and Page, M.J. (2002) Magnitude and frequency of landsliding in a large New Zealand catchment. *Geomorphology*, **49**, 71–88.

Reid, M.E. (1994) A pore-pressure diffusion model for estimating landslide-inducing rainfall. *Journal of Geology*, **102**, 709–717.

Reid, S.C., Lane, S.N., Montgomery, D.R. and Brookes, C.J. (2007) Does hydrological connectivity improve modeling of coarse sediment delivery in upland environment? *Geomorphology*, **90**, 263–282.

Rein, B., Lückge, A., Reinhardt, L. *et al.* (2005) El Niño variability off Peru during the last 20m000 years. *Paleoceanography*, **20**, PA4003. doi: 10.1029/2004PA001099.

Reneau, S.L. and Dietrich, W.E. (1987) Size and location of colluvial landslides in a steep forested landscape. *International Association for Hydrological Sciences*, **165**, 39–48.

Reneau, S.L., Dietrich, W.E., Rubin, M. *et al.* (1989) Analysis of hillslope erosion rates using dated colluvial deposits. *Journal of Geology*, **97**, 45–63.

Reneau, S.L., Dietrich, W.E., Donahue, D.J. *et al.* (1990) Late Quaternary history of colluvial deposition and erosion in hollows, central California Coast ranges. *Geological Society of America Bulletin*, **102**, 969–982.

Reyners, M. (1998) Plate coupling and the hazard of large subduction thrust earthquakes at the Hikurangi subduction zone, New Zealand. *New Zealand Journal of Geology and Geophysics*, **41**, 343–354.

Roberts, R.G. and Church, M.A. (1986) The sediment budget in severely disturbed watersheds, Queen Charlotte Ranges, British Columbia. *Canadian Journal of Forest Research*, **16**, 1092–1106.

Roering, J.J. and Gerber, M. (2005) Fire and the evolution of steep, soil-mantled landscapes. *Geology*, **33**, 249–352.

Roering, J.J., Kirchner, J.W. and Dietrich, W.E. (2005) Characterizing structural and lithologic controls on deep-seated landsliding: Implications for topographic relief and landscape evolution in the Oregon Coast Range, USA. *Geological Society of America Bulletin*, **117**, 654–668.

Roering, J.J., Schmidt, K.M., Stock, J.D. *et al.* (2003) Shallow landsliding, root reinforcement, and the spatial distribution of trees in the Oregon Coast Range. *Canadian Geotechnical Journal*, **40**, 237–253.

Rosso, R., Rulli, M.C. and Vannucchi, G. (2006) A physically based model for the hydrologic control on shallow landsliding. *Water Resources Research*, **42**, W06410. doi: 10.1029/2005WR004369.

Schmidt, K.M., Roering, J.J., Stock, J.D. *et al.* (2001) The variability of root cohesion as an influence on shallow landslide susceptibility in the Oregon Coast Range. *Canadian Geotechnical Journal*, **38**, 995–1024.

Schuster, R.L., Nieto, A.S., O'Rourke, T.D. *et al.* (1996) Mass wasting triggered by the 5 March 1987 Ecuador earthquakes. *Engineering Geology*, **42**, 1–23.

Scott, D.T., Baisden, W.T., Davis-Colley, R.J. *et al.* (2006) Localized erosion affects national carbon budget. *Geophysical Research Letters*, **33**, L01402. doi: 10.129/2005GL024664.

Scott, K.M. and Williams, R.P. (1978) Erosion and sediment yields in the Transverse Ranges, Southern California. U.S. Geological Survey Professional Paper 1030, p. 37.

Seginer, I. (1966) Gully development and sediment yield. *Journal of Hydrology*, **4**, 236–253.

Selby, M.J. (1982) Controls on the stability and inclinations of hillslopes on hard rock. *Earth Surface Processes and Landforms*, **7**, 449–467.

Shimokawa, E. (1984) A natural recovery process of vegetation on landslide scars and landslide periodicity in forested drainage basins, in *Proceedings of the Symposium on Effects of Forest Land use on Erosion and Slope Stability* (eds C.L. O'Loughlin and A.J. Pearce), East-West Environmental Policy Institute, Honolulu, Hawaii, pp. 99–107.

Sidle, R.C. (1987) A dynamic model of slope stability in zero-order basins. *International Association for Hydrological Sciences Publication*, **165**, 101–110.

Sommerfield, C.K. and Wheatcroft, R.A. (2007) Late Holocene sediment accumulation on the northern California shelf: oceanic, fl uvial, and anthropogenic influences. *Geological Society of America Bulletin*, **119**, 1120–1134.

Sutherland, D.G., Ball, M.H., Hilton, S.J. and Lisle, T.E. (2002) Evolution of a landslide-induced sediment wave in the Navarro River, California. *Geological Society of America Bulletin*, **114**, 1036–1048.

Swanson, F.J. (1981) Fire and geomorphic processes. USDA Forest Service General Technical Report WO-26, pp. 401–420.

Swanson, M.L., Kondolf, G.M. and Boison, P.J. (1989) An example of rapid gully initiation and extension by subsurface erosion; coastal San Mateo county, California. *Geomorphology*, **2**, 393–403.

Syvitski, J.P.M. (2003) Supply and flux of sediment along hydrological pathways: research for the 21st century. *Global and Planetary Change*, **39**, 1–11.

Syvitski, J.P.M., Morehead, M.D., Bahr, D.B. and Mulder, T. (2000) Estimating fluvial sediment transport: the rating parameters. *Water Resources Research*, **36**, 2747–2760.

Syvitski, J.P.M., Vörösmarty, C.J., Kettner, A.J. and Green, P. (2005) Impact of humans on the flux of terrestrial sediment to the global coastal ocean. *Science*, **308**, 376–380.

Terzaghi, K. (1943) *Theoretical Soil Mechanics*, John Wiley and Sons, New York, p. 510.

Thornes, J.B. (1984) Gully growth and bifurcation. Proceedings of Conference XV, International Erosion Control Association, Denver, Colorado, pp. 131–140.

Trustrum, N.A. and DeRose, R.C. (1988) Soil depth-age relationship of landslides on deforested hillslopes, Taranaki, New Zealand. *Geomorphology*, **1**, 143–160.

Trustrum, N.A., Gomez, B., Reid, L.M. *et al.* (1999) Sediment production, storage and output: the relative role of large magnitude events in steepland catchment. *Zeitscrift für Geomorphologie, Supplement Band*, **115**, 71–86.

Tsai, T.-L. (2008) The influence of rainstorm pattern on shallow landslide. *Environmental Geology*, **53**, 1563–1569.

Wakatsuki, T., Tanaka, Y. and Matsukura, Y. (2005) Soil slips on weathering-limited slopes underlain by coarse-grained granite or fine-grained gneiss near Seoul, Republic of Korea. *Catena*, **60**, 181–203.

Walling, D.E. (1983) The sediment delivery problem. *Journal of Hydrology*, **65**, 209–237.

Warrick, J.A. and Milliman, J.D. (2003) Hyperpycnal sediment discharge from semiarid southern California rivers: Implications for coastal sediment budgets. *Geology*, **31**, 781–784.

Warrick, J.A. and Rubin, D.M. (2007) Suspended-sediment rating curve response to urbanization and wildfire, Santa Ana River, California. *Journal of Geophysical Research*, **112**, F02018. doi: 10.1029/2006JF000662.

Warrick, J.J. and Fong, D.A. (2004) Dispersal scaling from the world's rivers. *Geophysical Research Letters*, **31**, L04301. doi: 10.1029/2003GL019114.

Wells, N.A. and Andriamihaja, B. (1993) The initiation and growth of gullies in Madagascar: are humans to blame? *Geomorphology*, **8**, 1–46.

Wheatcroft, R.A. and Sommerfield, C.K. (2005) River sediment flux and shelf sediment accumulation rates on the Pacific Northwest margin. *Continental Shelf Research*, **25**, 311–332.

Wheatcroft, R.A., Sommerfield, C.K., Drake, D.E. *et al.* (1997) Rapid and widespread dispersal of flood sediment on the northern California continental margin. *Geology*, **25**, 163–166.

Wieczorek, G.F. (1987) Effect of Rainfall Intensity and Duration on Debris Flows in Central Santa Cruz Mountains, California. *Geological Society of America Reviews in Engineering Geology*, **7**, 93–104.

Wilkinson, B.H. and McElroy, B.J. (2007) The impact of humans on continental erosion and sedimentation. *Geological Society of America Bulletin*, **119**, 140–156.

Wright, C. and Mella, A. (1963) Modifications to the soil pattern of south-central Chile resulting from seismic and associated phenomena during the period May to August 1960. *Bulletin of the Seismological Society of America*, **53**, 1367–1402.

Yamada, S. (1999) The role of soil creep and slope failure in the landscape evolution of a head water basin: field measurements in a zero order basin of northern Japan. *Geomorphology*, **28**, 329–344.

Yamagishi, H. and Iwahashi, J. (2007) Comparison between the two triggered landslides in Mid-Niigata, Japan by July 13 heavy rainfall and October 23 intensive earthquakes in 2004. *Landslides*, **4**, 389–397.

6

Local Buffers to the Sediment Cascade: Debris Cones and Alluvial Fans

Adrian M. Harvey
Department of Geography, University of Liverpool, PO Box 147, Liverpool L69 3BX

6.1 Introduction

The sediment pathway from source to sink is not smooth and has been described as a 'jerky conveyor belt' (Ferguson, 1981). Indeed, major interruptions to the sediment cascade occur at a variety of temporal and spatial scales. They range from small, short-term and ephemeral zones of deposition (e.g. individual, within-channel gravel bars: see Chapter 9) related to individual floods, to larger more persistent features in the landscape. These latter features form the subject matter of this chapter.

Zones of sediment storage, debris cones and alluvial fans, often occur at changes or breaks in the boundary conditions of the sediment pathway, especially where two zones meet, such as the boundary between the hillslope and the channel system or within the channel system at tributary junctions or where there are topographic changes such as changes in valley-floor gradient.

Not only are such zones characterized by distinctive landforms (debris cones and alluvial fans), but they may have important effects on the dynamics of the system. The effectiveness with which sediment, especially coarse sediment, is trapped within these zones may result in within-system variations in the sediment flux, and exert a considerable influence on downstream channel processes and morphology. These 'coupling or buffering' relationships (Harvey, 2001, 2002a; see also Figure 6.1) tend to respond to variations in flood power and sediment supply from upstream, essentially expressing the critical stream power relationships (Bull, 1979) between actual power (flood-dependent) and critical power, defined as the power required to transport the sediment supplied either from upstream or from the hillslopes to the channel (sediment-dependant). Alluvial fans tend to be sensitive,

Sediment Cascades: An Integrated Approach Edited by Timothy Burt and Robert Allison

Figure 6.1 Gully-mouth debris cones: Greencombe Gills, Bowderdale, Howgill Fells, northwest England. Note the coupled cone on the left, feeding sediment into the stream and the non-coupled cone to the right, providing no sediment to the stream system. The stream channel downstream of the coupled debris cone is a sedimentation zone within which the otherwise single-thread channel switches to a braided regime
(photograph by A.M. Harvey).

with conditions hovering about the threshold of critical stream power. Under conditions of excess power, sediment tends to be removed, increasing the coupling through the system. Under conditions of excess sediment supply, deposition occurs, reducing the coupling.

For these reasons, such zones not only exert a major influence on the prevailing sediment flux, but in their stored sediments and expressed in their morphology, alluvial fans and debris cones may preserve a record of past variations in critical power relationships. This record may be a sensitive reflection of past environmental change in the upstream sediment source areas (Harvey, 1997a), more so perhaps than downstream fluvial sites, that tend to respond primarily to variations in flood power (Macklin and Lewin, 2003), modified by local hydraulic conditions, but whose record may be complicated by past variations in connectivity of the system.

As such, alluvial fans and debris cones have proven to be valuable sites in palaeoenvironmental reconstructions, especially in upland and mountain areas.

6.2 Occurrence and Scale of Debris Cones and Alluvial Fans

Downstream from upland or mountain areas, with high rates of sediment generation, there may be a loss of sediment transport capacity or stream power that results in deposition. Characteristic situations include slope-foot locations, mountain fronts, and gradient reductions where tributaries join a main stream. Such zones are often associated with decreased confinement (i.e. an increase in accommodation space), and result in deposition in the form of a debris cone or an alluvial fan. Excluding talus cones, which are fed by rockfalls or dirty snow avalanches, these features occur at a great range of spatial (and temporal) scales.

6.2.1 Small-Scale, Ephemeral Gully-Mouth Debris-Flow Lobes (Scale: Up to c. 10 m)

Sediment generated within gully systems by storms or other sediment-producing events (e.g. snowmelt) forms debris flows that pass down the gully channels. In situations close to a basal stream they may pass directly into the stream or may be deposited at the base of the gully as a debris-flow lobe or a small debris cone (Figure 6.2a), only to be partially or totally removed by a later stream flood and incorporated into the stream sediment.

In many cases there is likely to be a difference in the frequency or seasonality of sediment producing and removing events, leading to a situation of cyclic sediment build-up and periodic removal (flushing). In small catchments in western Colorado Faulkner (1987) identified a distinct seasonality, characterized by summer storms delivering sediment from gullied hillslopes to the channel network and late winter snowmelt floods flushing the system. In contrast, in the Howgill Fells, northwest England, Harvey (1977, 1987a) identified a dominantly winter sediment build-up and a sediment flushing event, once every 2–5 years, usually, but not always, as the result of a late summer storm. Similarly,

Figure 6.2 Scales of debris cones/alluvial fans (i). (a) Gully-mouth debris-flow lobes, Carlingill, Howgill Fells, northwest England. (b) Holocene debris cone, Langdale valley, Howgill Fells, northwest England. Note debris-flow lobes and levees. Sheep give approximate scale. (c) Tributary-junction canyon fan, Three Gorges section of the Yangtze river, China. Small boat, length *c.* 2.5 m for scale (photographs by A.M. Harvey).

Figure 6.2 *(Continued)*

Wells and Gutierrez (1982) identified winter sediment production and summer flushing on badlands in northern New Mexico. These seasonal variations reflect climatic contrasts, particularly the relative effectiveness of frost action, snowmelt and storm rainfall in sediment generation and removal. Such features as those described above characteristically relate to timescales of decades or less (Harvey, 2002a).

6.2.2 Tributary-Junction Debris Cones (Scales 10s to 100s m)

At the foot of large gully systems or steep tributaries where sediment transport is dominantly by debris flows, larger, persistent debris cones may be formed (Figure 6.2b). These are too large to be removed by individual stream floods, so preserve a stack of successive debris-flow deposits. They may be subject to distal erosion by 'toe trimming' (Leeder and Mack, 2001), or to trenching by fluvial incision along the axis of the cone. The depositional sequence may preserve a record of hillslope debris-flow activity, from sedimentation during periods of activity but surface stability and soil formation during periods of hillslope inactivity, characteristically over timescales of centuries during the late Holocene (Harvey, 2002a; also see later, Section 6.6).

6.2.3 Tributary-Junction Fluvial Fans: (Scales: 10s to 100s m)

These are similar to the category described above, but occur where larger tributary streams, that carry most of their sediment by fluvial traction, rather than as debris flows, join larger rivers. Such fans are often partially confined within the valley of the larger river (Figure 6.2c), but occur in protected situations, where floods on the larger river do not remove all the sediment supplied by the tributary stream (Gomez-Villar, Alvarez-Martinez and Garcia-Ruiz, 2006). In canyon settings, their style reflects the interaction between the flood power on the main river (Schmidt, 1990; Schmidt and Graf, 1990; Schmidt and Rubin, 1995) and the sediment supplied by the tributary (Wang *et al.*, 2008). Such features appear to relate to timescales of decades to centuries, over the late Holocene.

6.2.4 Classic Alluvial Fans: (Scales: 100s m to Several km)

These are much larger features, persistent in the landscape over millennia, characteristically dating from the late Quaternary. They occur in several characteristic topographic situations, at mountain fronts (Figure 6.3a) or on the margins of sedimentary basins and at tributary junctions (Figure 6.3b) at the margins of major valleys.

They range from relatively small debris-flow dominated fans to composite fans comprising debris-flow, sheetflood and channelized fluvial deposits, to fans dominated by channelized fluvial deposits. Generally, debris-flow fans are fed by smaller, steeper catchments and fluvial fans by larger less steep catchments (Kostaschuk, Macdonald and Putnam, 1986), though climate and source-rock lithology may also influence the range of sediments present.

Classic alluvial fans are common in arid environments (Bull, 1977; Blair and McPherson, 1994; Harvey, 1997a), and have been extensively described there, but can occur in any climatic environment, particularly where there is juxtaposition of a high sediment yielding mountain environment and a zone of reduced power and available accommodation space.

One variant of the classic alluvial fan is the fan delta, were the fan toes out into standing water (Figure 6.3c), either the sea or a lake. Such features differ from simple alluvial fans in that they may have a substantial subaqueous portion, and along the shoreline there is interaction between fan and shoreline processes and sediments.

Classic alluvial fans tend to be more complex and longer-term features than smaller fans, usually dating back into the late Pleistocene. Locational controls are the long-term geomorphological evolution, for example controlled by tectonics or glaciation, creating the juxtaposition of sediment source area and accommodation space. However, the sediment sequences and morphology tend to reflect variations in flood power, and especially sediment supply following major environmental change, which may include tectonic activity, but in most regions in the context of late Quaternary timescales are primarily controlled by Quaternary climatic fluctuations. For this reason the sedimentary and morphological

Figure 6.3 Scales of debris cones/alluvial fans (ii). (a) 'Classic' mountain-front alluvial fans, Cold Springs fans, Nevada. Telephone/electric poles for scale. (b) Tributary-junction fan, Wadi Al-Bih, UAE/Oman border. Note: debris cones (at right) tributary to the tributary-junction fan itself. Flow in Wadi Al-Bih is from left to right. (c) Fan delta, Peyto Lake, Alberta, Canada. Note main fluvially-dominant fan delta at the head of the lake (left) and small debris-cone fan delta (right) (photographs by A.M. Harvey). (d) Okavanga terminal fan, Botswana (reproduced from Google Earth). The fan is approximately 100 km wide

Figure 6.3 *(Continued)*

sequences preserved in 'classic' alluvial fans often preserve evidence for Quaternary environmental change.

6.2.5 Fluvial Megafans: (Scales 10s km)

At a significantly larger scale, at the margins of large sedimentary basins, where major piedmont rivers leave uplifting mountain ranges, huge megafans may be deposited. The Kosi megafan at the southern Himalayan margin is perhaps the best known (Gohain and Parkash, 1990), but others have been identified at the margins of The Andes (G. Weissmann, personal communication), within Alpine Europe (Gabris and Nagy, 2005) and in the Zagros Mountains, Iran (Arzani, 2005). Processes on these fans are dominated by the fluvial processes of large rivers, avulsion (channel switching) perhaps being the most important in controlling fan evolution (Gabris and Nagy, 2005). In some cases, especially where rivers enter a desert lowland, flood waters may infiltrate or evaporate, resulting in the deposition of all the river sediment load in a terminal fan (Stanistreet and McCarthy, 1993), such as that formed by the Okavanga River in northern Botswana (Figure 6.3d). These zones may be important zones where aeolian and fluvial processes interact (Al Farraj and Harvey, 2004), and may preserve a record of past aeolian/fluvial interactions that provides an important key to palaeoclimatic reconstructions of desert margins (Nanson, Chen and Price, 1995). Although recording Quaternary environmental change, megafans may date back to the late Tertiary and reflect major tectonic controls.

At all these spatial and temporal scales alluvial fans (and debris cones and debris-flow lobes) represent interruptions to the sediment cascade where sediment, particularly coarse sediment, is stored.

6.3 Processes on Fans

Fan morphology depends on the nature of the processes transporting sediment to the fan and on the mechanism(s) of deposition. Sediment transport to the fan may be by a range of debris-flow processes (Blair and McPherson, 1994) or by tractional processes in flowing water, that range from unconfined sheetflows to channelized fluvial processes within steep, single-thread, wandering, or braided channels. On larger, low-angle fluvial fans the channels may be braided, meandering or anastomosing (Gabris and Nagy, 2005).

Deposition occurs on a fan when transporting power falls below the minimum transportational threshold. In the debris-flow case this affects the runout distance of the flow on to the fan, and depends on the internal friction within the flow (dependent on clast size and the fluidity of the water–sediment mix; Whipple and Dunne, 1992), the mass or the thickness of the flow (that reduces as confinement or thickness decrease), the roughness of the surface over which the flow is moving, and the gradient. For traction flows the threshold depends on clast size and unit stream power, itself a function of gradient and depth. Depth decreases as confinement decreases, so that, for similar clast sizes and flood discharges, the threshold gradient for deposition by sheetflows is higher than that for confined channelized flows. The threshold gradient for debris-flow deposition is higher still (Whipple *et al.*, 1998).

Because the depositional gradient and morphology are dependent on the style of deposition, clast size, and flood power, the threshold of critical stream power (Bull, 1979) is fundamental. Alluvial fans are sensitive environments, hovering around this threshold, even within individual storm events (Wells and Harvey, 1987). If there is a change in water/sediment ratio across the debris-flow–tractional-flow threshold the transport and depositional processes may switch between debris flows and tractional processes. Over a longer period, if there is an environmental change, causing changes in flood regime or sediment volume or clast size, there may be a switch between depositional styles or between a depositional and an erosional regime. For this reason, alluvial fans can preserve a sensitive record of past environmental change within their sedimentary and morphological sequences.

There are implications also for fan morphology, because smaller, steeper catchments tend towards debris flows (Kostaschuk, Macdonald and Putnam, 1986), and for larger catchments, flood power tends to increase and clast size to decrease with increasing catchment size, depositional fan gradients tend to be steeper for smaller catchments and less for larger catchments, leading to the general inverse relationship between drainage area and fan gradient (Harvey, 1997a). This is one of the two basic morphometric relationships for alluvial fans, the other relating fan area to drainage area (Harvey, 1997a; Calvache, Viseras and Fernandez, 1997). These relationships are usually expressed as follows:

$$G = aA^{-b}$$

$$F = pA^{q}$$

where G is fan gradient, F is fan area (km^2) and A is drainage area (km^2). There are far too many morphometric studies in the literature to review them all here, but many studies of individual groups of fans show similarity in exponent b (normal range -0.15 to -0.35; Harvey, 1997a) but values of constant a show greater variation, reflecting the mix of sedimentation processes. Similarly there is only a narrow range for exponent q (generally between 0.7 and 1.1; Harvey, 1997a) but there is a wide range for constant p reflecting regional differences in sediment supply rates, fan age and fan history).

Morphometric analyses of alluvial fans, based on regression relationships, including analyses of residuals from the two primary regressions (above) (Silva *et al.*, 1992), have been used to identify the influence of the various factors affecting fan morphology, such as tectonic setting (e.g. Silva *et al.*, 1992; Calvache, Viseras and Fernandez, 1997), and base-level conditions (e.g. Harvey *et al.*, 1999a; Harvey, 2005).

The final group of processes that are important for fan geomorphology are those relating to post-depositional surface modification. Understanding these processes may be important in interpreting the past geodynamics of the fan and adjacent geomorphological systems. Older fan surfaces may have been subject to modification by erosion (Bull, 1991). In arid areas, dune sand may bury fan surfaces, and such interactions may have palaeoclimatic implications. More important perhaps, are modifications to the fan surface itself, by weathering processes, soil formation and biological activity. Each of these sets of processes may allow the relative dating and correlation of fan segments and aid the reconstruction of the erosional and depositional history of the fan and therefore of past sediment flux conditions. For example, in humid temperate upland and mountain environments the rate and degree of podzolization may be a sensitive age-indicator of surface age (Harvey, Alexander and James, 1984). In these areas lichenometry may also aid the dating of depositional surfaces. In semi-arid regions soil colour, iron oxide geochemistry or mineral magnetics (Harvey *et al.*, 1999a, 2003), and particularly the stage of pedogenic carbonate development (Machette, 1985) may aid the relative dating of fan surfaces. In arid areas these properties may also be useful, together with the degree of development of rock (desert) varnish (Dorn and Oberlander, 1982) and the degree of desert pavement development (Wells, McFadden and Dohrenwend, 1987; Al Farraj and Harvey, 2000; Harvey and Wells, 2003).

The correlation of fan surfaces and the relative dating of fan deposition can be tied into more precise absolute dating by the application of conventional sediment dating techniques, in the case of humid regions by conventional radiocarbon dating of buried soil horizons within or below the fan sediments (Harvey *et al.*, 1981; Harvey, 1996). Radiocarbon dating now has greater potential with the more recent development of AMS radiocarbon dating, which allows the dating of much smaller organic samples (Chiverrell, Harvey and Foster, 2007; Chiverrell *et al.*, 2008). In dry regions, the use of uranium-series dating of pedogenic carbonates (Candy *et al.*, 2003; Candy, Black and Sellwood, 2004) from soils capping the fan surfaces, and more particularly luminescence dating of the sediments themselves (e.g. Pope and Wilkinson, 2005; Robinson *et al.*, 2005) offers great scope for matching phases of sediment flux on fans and fan deposition with tectonic, climatic or human-induced environmental change. Another new technique with potential applications to the dating of sedimentary surfaces (Harbor, 1999), but not yet in widespread use in alluvial fan geomorphology, is cosmogenic nuclide dating.

6.4 Fan Style

There are several styles of alluvial fan morphology that reflect the dominant process regime. That regime responds to two fundamental sets of controls: proximal controls, those related to water and sediment supply from the drainage basin, and distal controls, essentially the base-level conditions affecting the fan toe.

Figure 6.4a illustrates common variations in fan style (after Harvey, 2002b, 2003). A fan may be a *relict* feature, under the present regime receiving no sediment, and even during floods, flood power is insufficient to cause erosion. A fan may be under an *aggradational* regime, receiving sediment from the fan apex downfan, and undergoing little or no erosion. Such a fan clearly acts as a buffer within the sediment cascade. A common fan style is that of the *telescopic* fan, where the proximal surfaces are incised by a fanhead trench, within which the channel has a lower gradient than the fan surface. The channel meets the fan surface at a mid-fan intersection point (Bowman, 1978; Wasson, 1979), downfan from which aggradation takes place. The distal end of this depositional lobe may extend the limit of the fan, in which case the fan is said to be *prograding*. Alternatively a fan may be under a *dissectional* regime, where, other than within the fanhead trench, incision may occur downfan from the intersection point (Harvey, 1987b), or at the fan toe in relation to base-level change (see below). Finally, local zones of incision may link together to form a fan *dissected throughout*, causing the fan surfaces to be abandoned, and establishing coupling through the system.

6.4.1 Proximal Controls

Apart from incision at the fan toe, triggered by base-level conditions, fan style reflects critical power relationships governed by the proximal controls of flood power and sediment

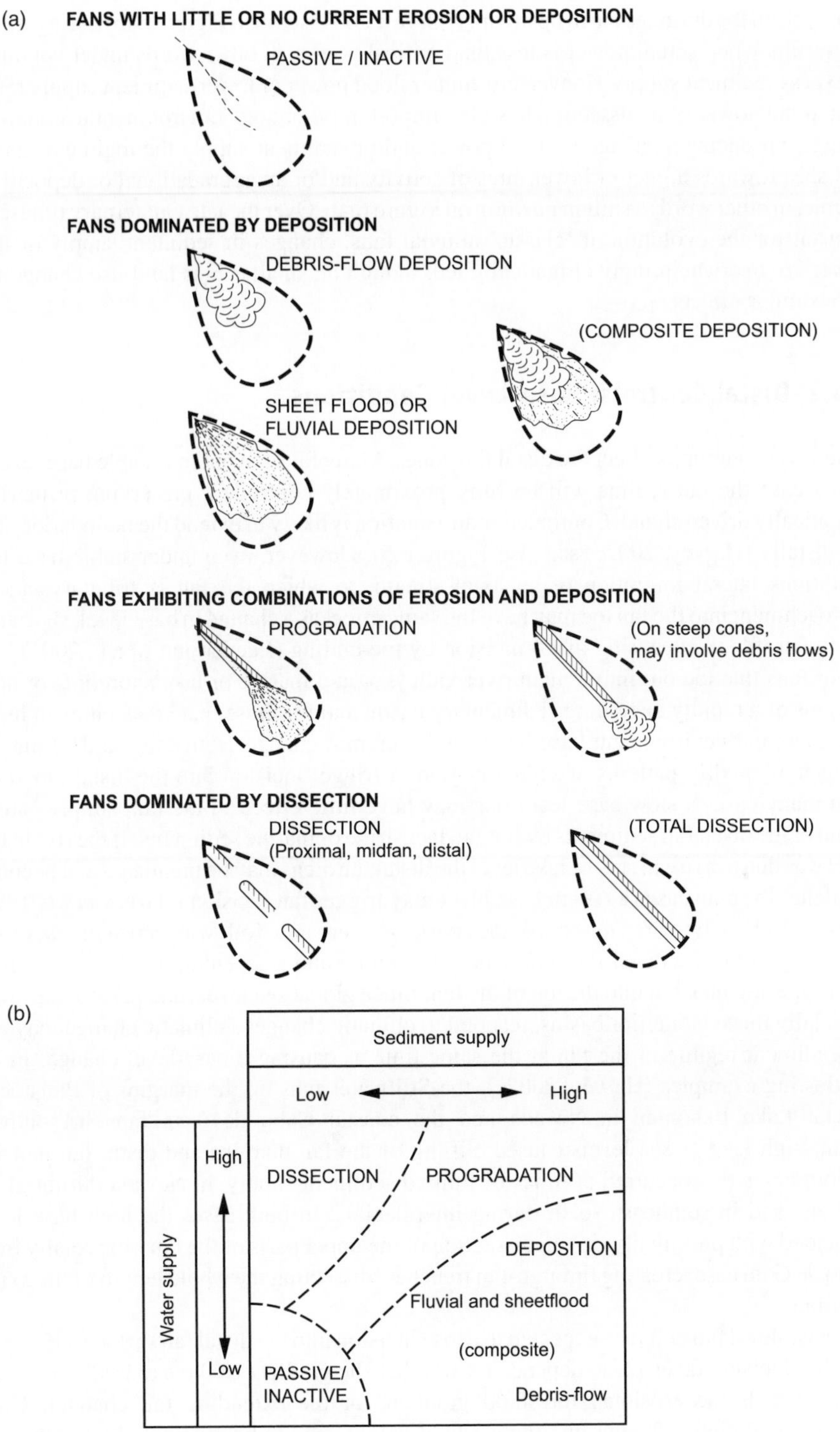

Figure 6.4 (a) Fan style; (b) influence of flood power and sediment supply on fan style (redrawn from Harvey, 2002b, 2003).

supply from the drainage basin upstream (Harvey, 2003; see also Figure 6.4b). An aggrading fan results when actual power is less than critical power, in other words under conditions of excess sediment supply. Conversely, higher flood power or lower sediment supply results in a trend towards a dissectional style. Importantly, should environmental conditions change, producing a change in flood power and/or sediment supply the regime of the fan will shift towards higher or lower rates of activity and/or an aggradational or depositional regime, in other words a shift in position on Figure 6.4b. Over the late Quaternary timescales relevant for the evolution of 'classic' alluvial fans, changes in sediment supply or flood power are overwhelmingly climatically led, though on smaller fans land-use change may have similar effects.

6.4.2 Distal Controls: Base-Level Conditions

Base-level conditions affect the distal fan zones. Many fans toe out to a stable base level, in which case the fan regime will be fully proximately controlled, preserving primarily a climatically driven signal. Continued sedimentation is likely to extend the boundaries of the fan distally (Harvey, 2002c; see also Figure 6.5). However, even under stable base-level conditions lateral migration of an axial stream, to which the fan is tributary, and its encroachment into the fan toe may have the same effect as a change in base level, shortening the fan profile, and causing distal incision by toe-cutting (Leeder and Mack, 2001).

For fans that toe out into a main river valley, or at a marine or lake shoreline, or at the margins of a rapidly aggrading sedimentary basin, and for those that cross faults in highly active tectonic environments base-level conditions may change. Both rising and falling base levels may modify patterns of sedimentation or trigger incision into the distal fan zones.

In many cases a slow base-level rise may have little effect on the fan, simply causing burial of the distal fan sediments by fluvial, lacustrine or marine sediments. If the rise in base level was due to a rise in sea- or lake level the distal fan areas may be inundated and become a fan delta. In some cases a *rise* in base level may trigger fan incision (Harvey *et al.*, 1999a; Harvey, 2002c). If the fan toes out at the coast, wave erosion, following a rise in sea- or lake level may erode the distal fan sediments, shortening and steepening the distal fan profile and triggering incision into the toe of the fan. Since global sea levels and palaeolake levels, especially those in interior basins, respond to climatic change, a climatic change may alter the sediment regime of the fan at the same time as causing a base-level change. In two contrasting examples (Harvey, 2002c), the Stillwater fans on the margins of Pleistocene pluvial Lake Lahontan in Nevada and the coastal Cabo de Gata fans in southeast Spain, high lake or sea levels caused cliffing of the fan margins and distal fan incision. In both cases this occurred at times of limited sediment supply, in Nevada during global glacials, and in southeast Spain during interglacials. In both cases the high base levels coincided with proximally-driven dissection of the upper parts of the fans, especially in the Cabo de Gata case, creating through-fan trenches, dissecting the whole length of the axis of each fan.

A base-level fall could be expected to cause incision into the distal fan surfaces. However, whether incision takes place depends on whether the gradient of the exposed sea- or lake floor exceeds the erosional threshold gradient for the extending fan channel. Under conditions of high sediment flux the gradient would need to be relatively high, otherwise a sediment lobe would simply prograde on to the exposed sea- or lake floor (Bowman, 1988).

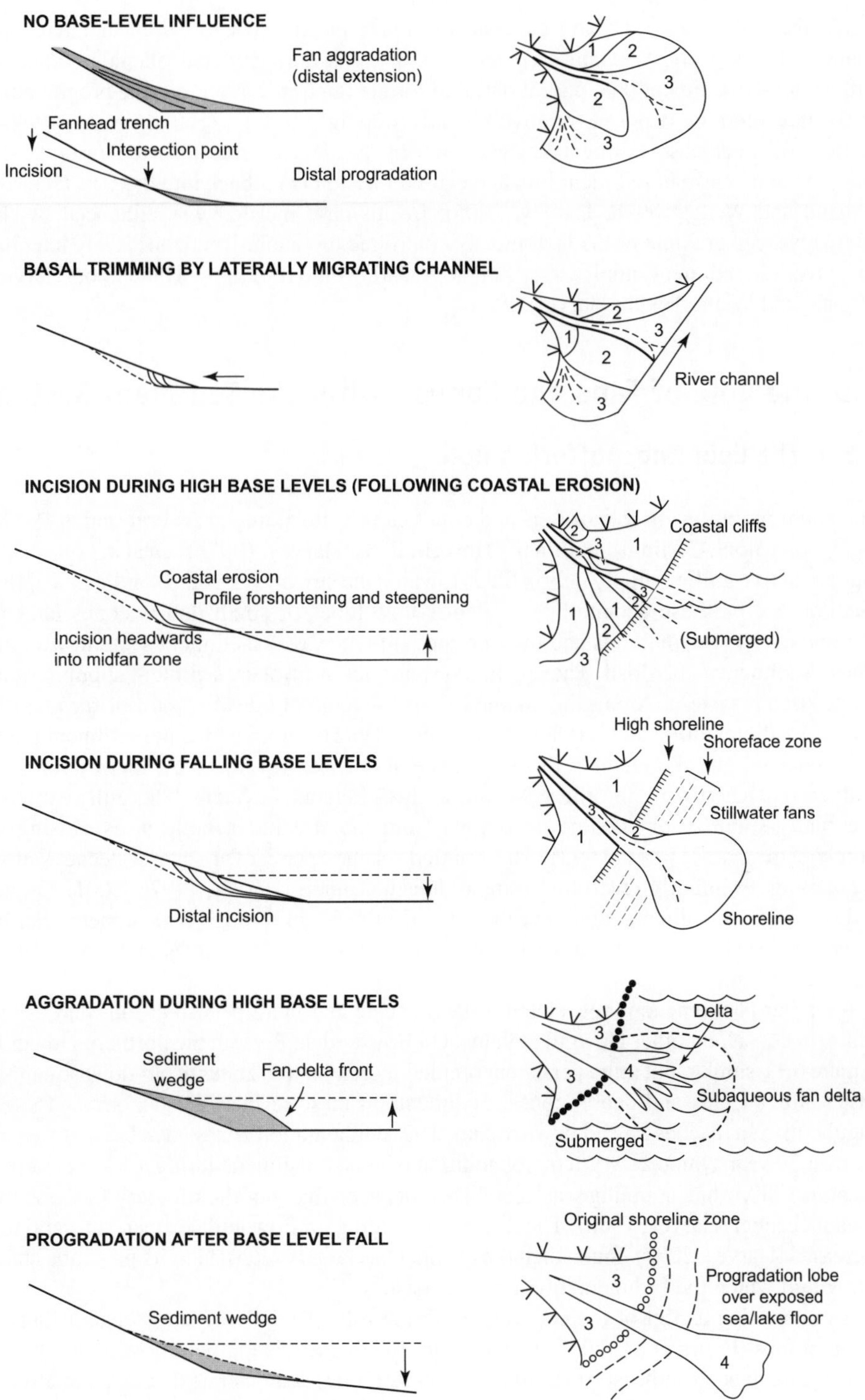

Figure 6.5 Influence of base-level conditions on fan profiles and plan views (modified from Harvey, 2002c, 2003).

Again the Stillwater and Cabo de Gata examples given above provide an interesting contrast. In the Cabo de Gata case falling sea levels during global glacials coincided with increased sediment flux, partial burial of former fan trenches and simple progradation of the fans onto the exposed (relatively gently sloping) sea floor (Harvey *et al.*, 1999a). In the Stillwater case falling lake levels during the late glacial occurred prior to the interglacial increase in sediment flux and caused incision to cutback into the fans (Harvey, Wigand and Wells, 1999b; Harvey, 2005). In this case incision was enhanced by the relatively steep gradient of the tectonically controlled mountain-front zone. Only later did an increase in sediment supply cause fan progradation onto the exposed lake floor (Harvey, Wigand and Wells, 1999b; Harvey, 2005).

6.5 The Role of Fans and Cones within the Sediment System

6.5.1 The Coupling/Buffering Role

The coupling/buffering role of fans and cones can be illustrated at several scales. At the local scale, along Carlingill within the Howgill Fells (Harvey, 1997b; see also Figure 6.6) modern active gullies supply debris flows towards the stream system. As indicated above (Section 6.2.1), these accumulate as debris-flow lobes or small debris cones later to be removed by stream floods and incorporated into the stream sediment system, with the coarse sediment as bedload. These gully systems act as primary sediment supply points to the stream system. Along the channel at or immediately downstream of these gully sites are sedimentation zones (Church and Jones, 1982), within which the sediment tends to be coarser, and the channel develops excessive width (Figure 6.6), locally forming braided reaches within an otherwise single-thread channel system. The gully systems eventually stabilize or become decoupled from the stream, in both cases ceasing to supply sediment (Harvey, 1992). The braided reaches persist for several decades after decoupling, eventually reverting to single-thread channels (Harvey, 1997b, 2001). On the valley floor past sedimentation zones can be identified and through lichenometry can be related to past phases of gully and debris cone activity (Harvey, Alexander and James, 1984; Harvey, 1997b).

A similar phenomenon, but related to larger scale and more persistent fans and debris cones, occurs along other Howgill streams. On Bowderdale Beck in the northern Howgills (Figure 6.1), similar, but more persistent braided sedimentation zones occur downstream of larger alluvial fans and debris cones. Sediments in these reaches are coarser and more angular than in reaches up and downstream. The sedimentation zones have been persistent for over 50 years, but received a major addition of coarse sediment during a 100-year flood event in 1982, when coupling was established between many of the tributary fans and the stream channel (Harvey, 1986). The degree of braiding on Bowderdale Beck substantially increased (Harvey, 2007). Since then the channel has largely returned to its pre-flood state, but within which the sedimentation zones persist.

An interesting aspect of the coupling/buffering role of relatively small-scale tributary junction fans (Fryirs *et al.*, 2007) is that their effectiveness is temporally variable. In the Hunter catchment, southeast Australia, Fryirs *et al.* (2007) demonstrate that under relatively frequent and low flood conditions the fans act as buffers, and only couple the system under rare, extreme flood conditions.

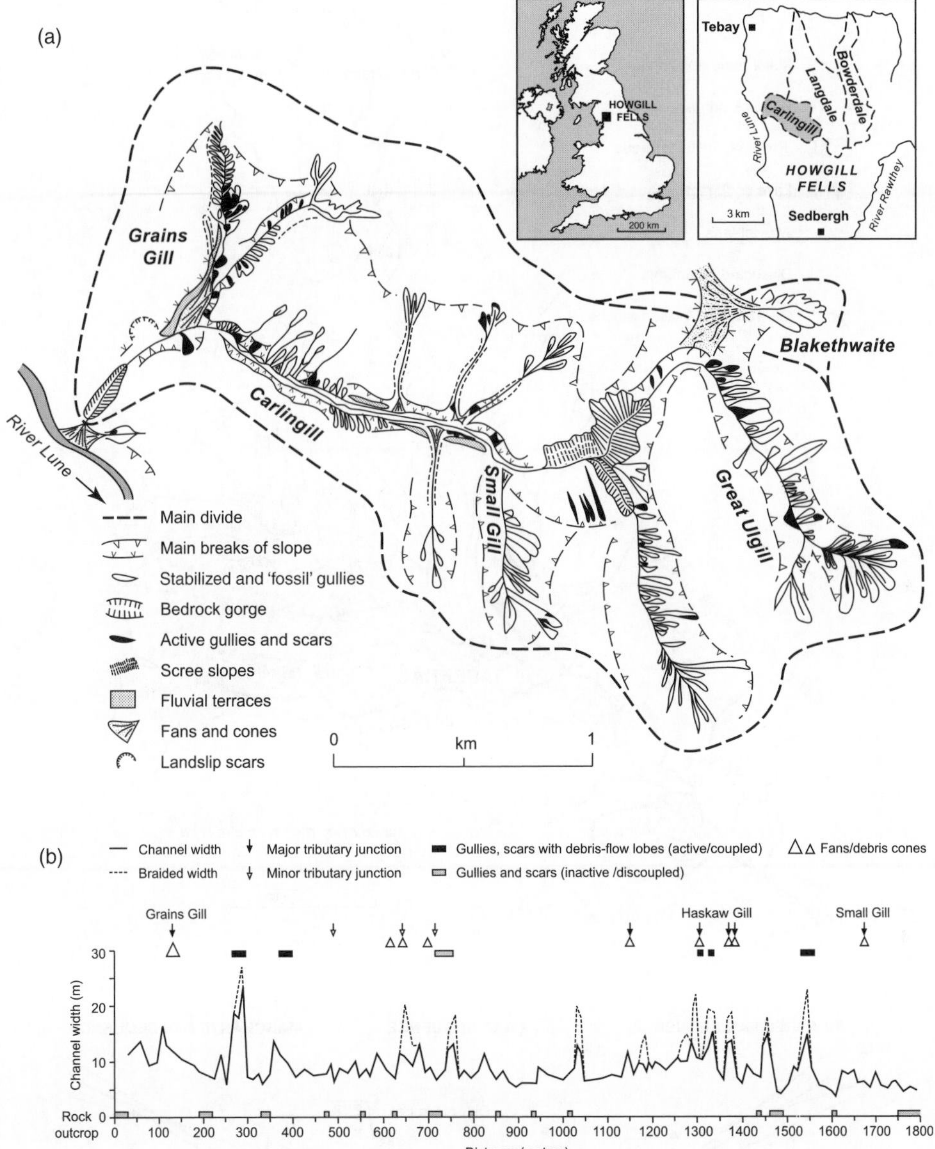

Figure 6.6 Carlingill, Howgill Fells, northwest England. (a) Map showing locations of tributary gully systems (insets show location within Great Britain). (b) Variation in channel widths downstream in relation to active and former sediment supply points

The coupling/buffering role of larger alluvial fan systems can be illustrated by the Quaternary Tabernas fans in southeast Spain (Figure 6.7). Three sets of mountain-front alluvial fans occur in the upper part of the Tabernas basin, each in contrasted local tectonic settings. The large Filabres fans are backfilled into valleys of the Sierra de los Filabres on a non-faulted tectonically relatively stable mountain front. Across the basin the small

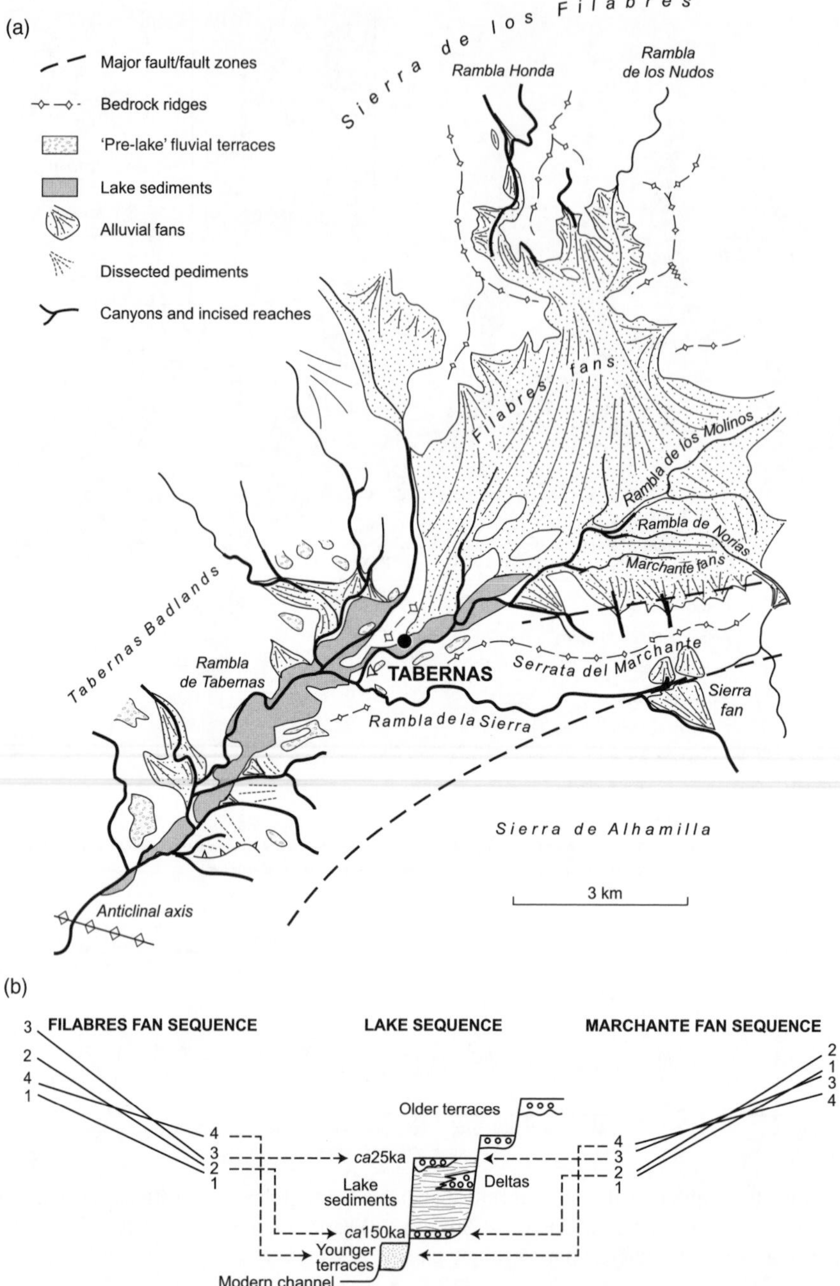

Figure 6.7 Tabernas basin, southeast Spain (modified from Harvey, 2002a). (a) Map illustrating the different coupling relationships on the three sets of fans, related to the headward progression of a tectonically induced incision wave. (b) Stratigraphical relationships (dates from Alexander *et al.*, 2008): 1, 2, 3 and 4 represent successive fan surfaces on the Filabres (left) and Marchante (right) fans
(modified from Harvey *et al.*, 2003).

Marchante fans occur along a mountain-front thrust fault to the north of the rising growth fold of the Serrata del Marchante. Further south, along the active major thrust fault zone bounding the uplifting Sierra de Alhamilla is the large Sierra fan. The whole system is drained by the Rambla de Tabernas and its tributaries, which to the west of Tabernas are deeply incised into the Tabernas badlands terrain (Harvey *et al.*, 2003).

During the late Pleistocene, from about 150 ka, this drainage was tectonically ponded by uplift along a growth fold to the west of the basin, forming a lacustrine and palustrine environment that backfilled into the upper part of the basin (Harvey *et al.*, 2003). Within the Filabres fans, fan sediments older than the lake are capped by palaeosols and buried by younger fan sediments. Sediments of a similar age form the extant but dissected fan surfaces in proximal positions on the Marchante and Sierra fans. At the time these sediments were deposited all the fans were undergoing aggradation and the fans acted as buffers within the fluvial system. This period was followed by fan dissection, at least in the proximal zones, forming fanhead trenches. It is not clear whether there was connectivity with the fluvial system downstream. During the existence of the lake system there was again no connectivity through the fans, although in some cases there were fan deltas at margins of the lake. There was another phase of fan aggradation, during the later part of that period. During the latest Pleistocene (Harvey *et al.*, 2003; Alexander *et al.*, 2008), the Rambla de Tabernas cut through the barrier at the distal end of the lake system, rapidly incising through the soft lacustrine and palustrine sediments into the Miocene marls below. The incision worked rapidly headwards during the Holocene. At the present time the incision has cut back through the Sierra fan, establishing connectivity from the Sierra de Alhamilla drainage into the Tabernas system. During the Holocene both the Marchante and Filabres fans have undergone fanhead trenching, but there is no continuity between the fanhead trenches and the incising Rambla de Tabernas system. The incision wave on the Rambla de Tabernas system has almost reached the toes of the Marchante fans and, at least in geomorphological timescales, coupling through these fans could be said to be imminent. The incision wave is still a long way from the Filabres fans, and at least for the present, there is no coupling from the headwaters of the Ramblas Honda and Los Nudos within the Sierra de los Filabres to the drainage of the Rambla de Tabernas downstream.

These coupling differences are expressed in the modern sediment properties within the streams tributary to the Rambla de Tabernas. The bed sediments of the Rambla Sierra comprise relatively coarse gravels and cobbles, derived in part from the Sierra de Alhamilla. Sediments from the Serrata del Marchante accumulate as lobate cobble bars on the distal surfaces of the Marchante fans. The sediments in the Rambla de los Molinos in this area are dominated by very fine gravel and sand, derived from reworked older distal Filabres fan sediments and the palustrine/lacustrine sediments. Modern sedimentation on the Filabres fans occurs during rare storm events as lobate gravel bars and sand sheets on the distal fan surfaces. Even during a *c.* 100-year storm event there was little evidence for much sediment crossing the distal Filabres fan surfaces into the Rambla de los Molinos (Harvey, 1984).

6.5.2 Sediment Preservation: Fans as Sensitive Indicators of Environmental Change

The Tabernas example (above) illustrates how fan sediments and morphology can help to identify past variations in sediment flux and connectivity. In that example only two dates help constrain the sequence. In other cases relative and absolute dating can provide greater precision in identifying past variations in sediment flux.

Several examples from upland northern Britain demonstrate that major changes in hillslope sediment supply to stream systems have occurred during the Holocene. In the Sottish highlands a major early Holocene paraglacial (*sensu* Ryder, 1971) phase of alluvial fan aggradation has been identified, followed by a late Holocene sediment pulse, reworking the alluvial fans and causing alluvial fan progradation (Brazier, Whittington and Ballantyne, 1988; Brazier and Ballantyne, 1989). This may have been human-induced, coincident with land-use changes in the Scottish Highlands, but was more likely associated with the climatic deterioration of the Little Ice Age in the seventeenth and eighteenth centuries AD (Brazier, Whittington and Ballantyne, 1988; Brazier and Ballantyne, 1989).

In northwest England and southwest Scotland, beyond the limits of the end-Pleistocene Loch Lomond stadial readvance, the end-Pleistocene was characterized by periglacial solifluction on the hillslopes, followed by early Holocene stabilization under an increasing woodland cover. It was only after the human-induced removal of this woodland cover in the later Holocene that the hillslopes were sufficiently sensitive for gully erosion to develop. The major periods of late Holocene gully erosion in this area, identified by radiocarbon dating of buried soils from beneath alluvial fan and debris cone deposits, correspond with major periods of human impact, to some extent with Iron Age/Romano-British activity and particularly with the tenth century AD Norse settlement, and later medieval expansion of grazing in the uplands (Chiverrell, Harvey and Foster, 2007; Chiverrell *et al.*, 2008; see also Figure 6.8). There is a later phase of hillslope erosion and minor accumulation of sediment on alluvial fans, poorly dated, but which appears to relate to the past few hundred years, coincident with the Little Ice Age, and therefore may have been triggered by climatic deterioration rather than by human-induced pressure on grazing land.

On larger fans, the 'classic' alluvial fans of the literature, where more relative and absolute dating evidence is present than in the Tabernas example (above), fan sediment and morphological sequences can be more precisely related to environmental changes, and the effects on the sediment flux inferred. The Basin and Range region of the American West has been the cradle of alluvial fan research, particularly by Hooke (1967), and Denny (1965) especially in the Death Valley region, and more recently by Bull (1977), Blair (1999) and Enzels, Wells and Lancaster (2003). Alluvial fan morphological and sedimentary sequences can be related to the results of numerous other studies dealing with late Quaternary environmental change. This is particularly true for fans bordering enclosed sedimentary basins within which large pluvial lakes formed in the late Pleistocene, for which much dated palaeoenvironmental information is available (see e.g. Grayson, 1993; Enzel, Wells and Lancaster, 2003). In many cases relative dating and correlation of fan surfaces has been achieved through analyses of surface characteristics, desert pavement and soil development (e.g. McFadden, Ritter and Wells, 1989), then absolute dating achieved by consideration of the spatial and stratigraphic relationships with dated lake shoreline features.

Studies of fans on the margins of palaeo Lake Mojave in southeast California (Wells, McFadden and Dohrenwend, 1987; Harvey and Wells, 1994, 2003) and palaeo Lake Lahontan in northeast Nevada (Harvey, Wigand and Wells, 1999b; Ritter, Miller and Husek-Wulforst, 2000; Harvey, 2002c, 2005) reveal major changes in sediment flux during the late Pleistocene and the Holocene, but also regional contrasts in the sequences.

On the Zzyzx fans on the margins of palaeo Lake Mojave and on the Stillwater fans on the margin of palaeo Lake Lahontan the proximal fan surfaces pre-date the late Pleistocene

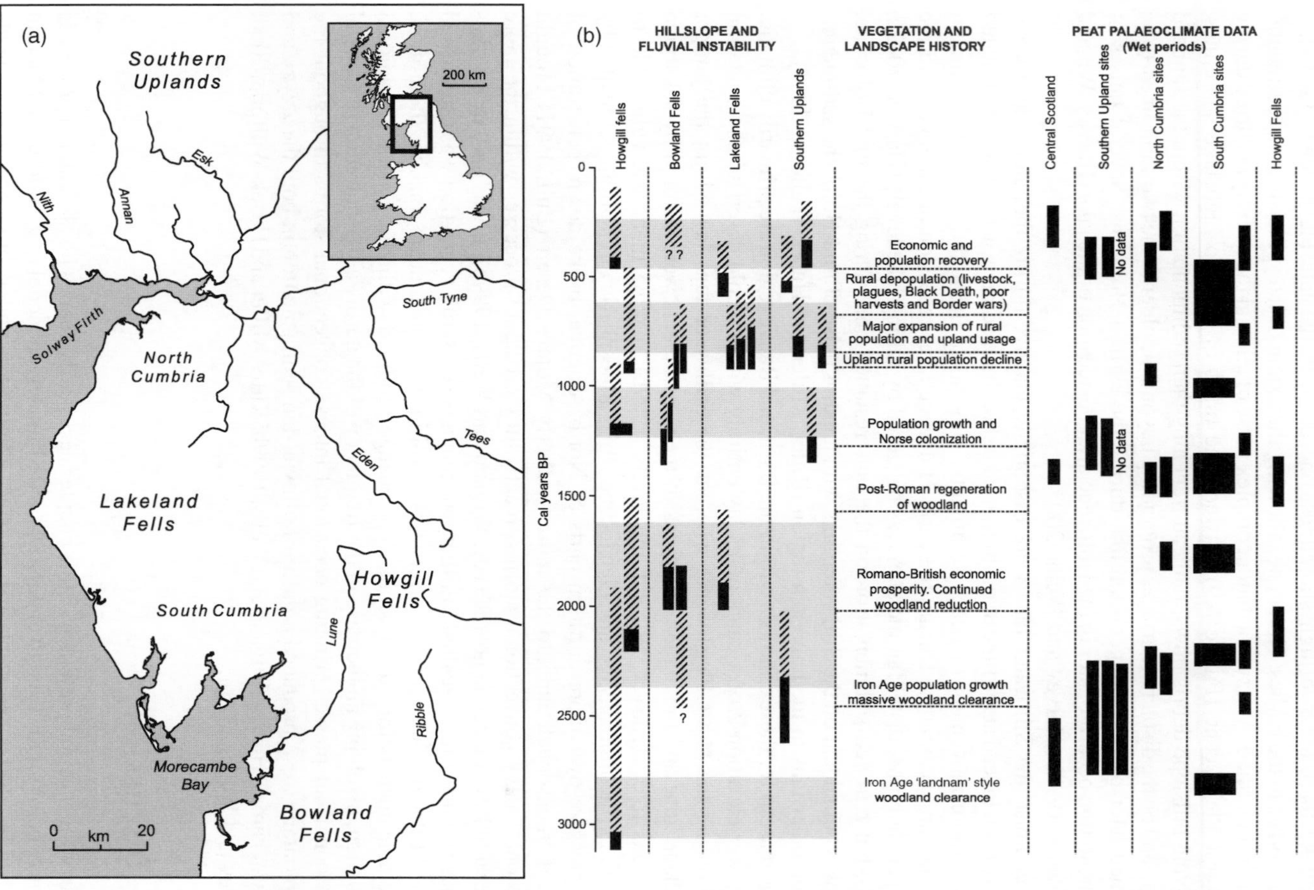

Figure 6.8 Summary of radiocarbon data from Holocene gully-mouth debris cones and alluvial fans in upland areas of northwest England and southwest Scotland, together with main phases of land-use change identified and main wet phases identified in the peat bog record (modified after Chiverrell, Harvey and Foster, 2007).

pluvial lakes. They show mature soil and desert pavement development, and in each case are cut by the older of two late Pleistocene lake shorelines. Around the margins of each palaeolake there are two high shorelines, one dated to the late Pleistocene (12–13 ka) and the other to the earliest Holocene, allowing the extent of latest Pleistocene sedimentation to be identified. At Zzyzx a major complex pulse of alluvial fan activity occurred during the latest Pleistocene (Figure 6.9) between the times of the two high lake shorelines, involving hillslope debris flows, fanhead trenching then sedimentation within the fanhead trench, and on the distal fan surfaces prograding beyond the older of the two shorelines, to be trimmed later by the younger shoreline during the earliest Holocene. Several pulses of sedimentation occurred later in the Holocene after desiccation of the lake (Harvey, Wigand and Wells, 1999b; Harvey and Wells, 2003).

In contrast, the Stillwater fans were essentially inactive during the latest Pleistocene (Figure 6.9). Indeed at the time of the higher lake shoreline sediment activity was dominated by longshore movement of beach sediments, rather than the addition of new sediment from the fan catchments. During the early Holocene after recession of the lake from the younger shoreline, distal fan incision was triggered by falling lake levels (Harvey, 2002c, 2005), but renewed sedimentation on the fans resumed only during the mid-Holocene. This was coincident on the Stillwater fans with a major pulse of mid-Holocene sedimentation on other fans in the Lahontan region (Ritter, Miller and Husek-Wulforst, 2000).

The reasons for these contrasts within the Basin and Range region are primarily climatic. The lower elevation Zzyzx catchments lay entirely within a cold desert scrub vegetation zone during the latest Pleistocene, whereas the Stillwater fans lay within the juniper woodland zone and the higher elevations of their catchments were covered by subalpine coniferous forest (Harvey, Wigand and Wells, 1999b), making them less vulnerable to erosion. Furthermore, as the late Pleistocene zonal atmospheric circulation, dominated by high pressure over more northern parts of North America, broke down, penetration of tropical 'monsoonal' air into the more southerly Mojave Desert (Bull, 1991) brought occasional heavy convectional storms, reactivating erosional processes within the more vulnerable Zzyzx catchments (Harvey, Wigand and Wells, 1999b). It was only in the mid-Holocene, during the so-called 'altithermal' that a desert scrub vegetation was established on the Stillwater fans and the lower slopes within their catchments, making them more vulnerable than before to erosion. This change, coupled with the general meridional circulation, involving an increase in the frequency of the northward penetration of tropical 'monsoonal' air masses bringing occasional intense convectional storms throughout the intermontane west, produced a pulse of alluvial-fan sedimentation in both the Zzyzx and Stillwater areas (Harvey, Wigand and Wells, 1999b; Ritter, Miller and Husek-Wulforst, 2000; Harvey and Wells, 2003).

6.6 Discussion

Much of the material presented in this chapter necessarily deals with alluvial fan processes and morphology. However, to understand and ultimately to model (e.g. Coulthard, Macklin and Kirkby, 2002) sediment flux through alluvial-fan and debris-cone environments requires consideration of the controls involved, their spatial–temporal relationships and their relations with an absolute chronology. Broadly, two sets of controls can be identified: those that are essentially passive within the timescale involved, and those that are dynamic

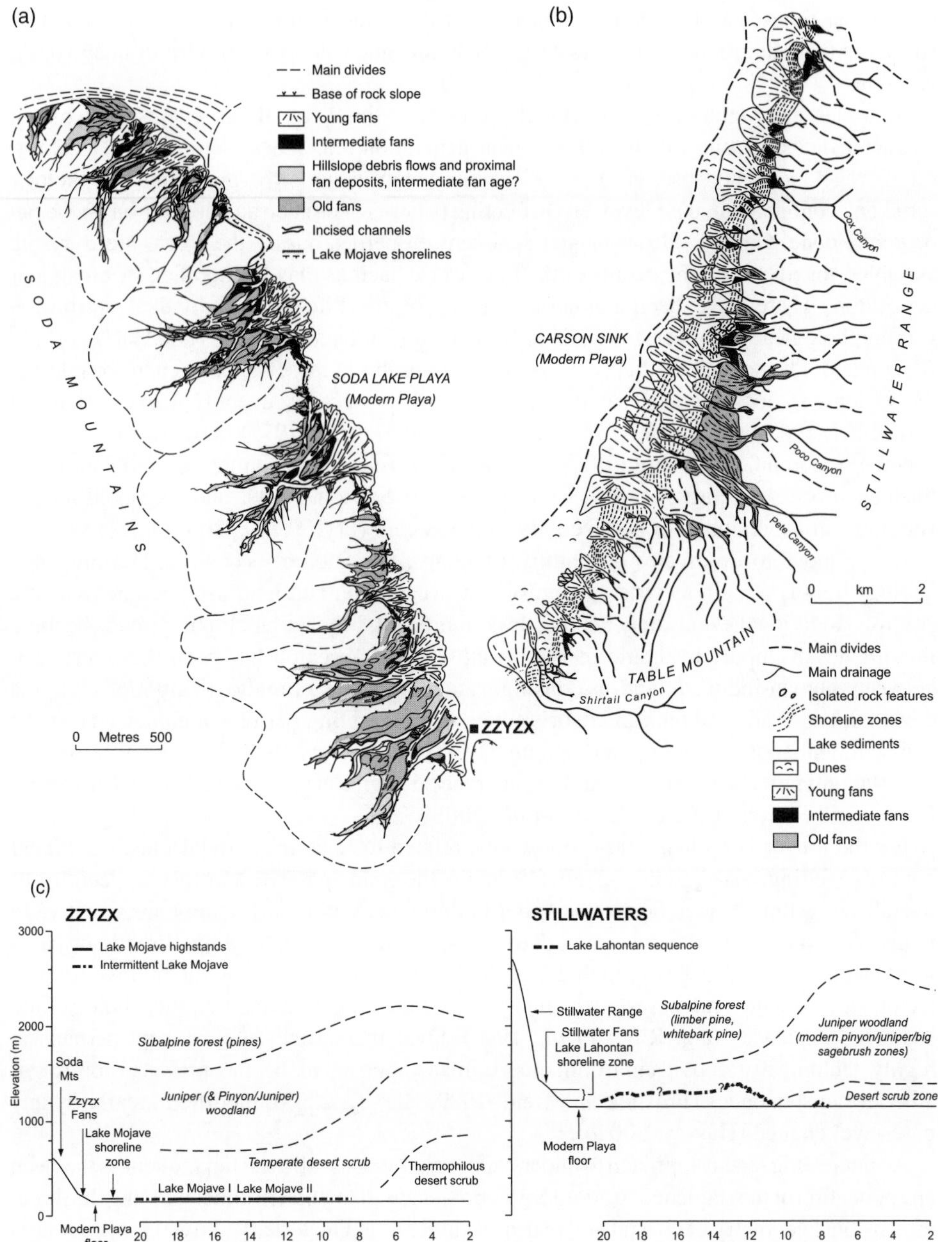

Figure 6.9 Late Pleistocene to Holocene alluvial fan activity on (a) the Zzyzx fans, Mojave Desert, California and (b) the Stillwater fans, Nevada, together with (c) reconstructed vegetation zones and palaeolake sequences
(modified from Harvey *et al.*, 1999b).

within that timescale. The status of these controls varies with the timescale in a manner similar to that identified many years ago by Schumm and Litchty (1965) for drainage basin variables.

At timescales related to individual processes, those of flood events, catchment characteristics affecting runoff and sediment generation (catchment setting, size, geology, relief, vegetation, land use, etc.) are all passive controls as is the gross fan morphology (gradient, confinement, base level, etc.) affecting patterns of erosion and deposition on the fan or debris cone. Only flood power and sediment properties during the storm are dynamic variables, together with the details of the fan surface itself as may be modified by erosion or deposition. There have been numerous studies of the effects of individual storms on geomorphological processes on debris cones and alluvial fans (e.g. Wells and Harvey, 1987; Blair and McPherson, 1998) that demonstrate the interplay of these factors. When considering the response to flood sequences over periods of perhaps decades, event frequency (Wolman and Miller, 1960) and recovery time (Brunsden and Thornes, 1979) between storm events become important. So too are differential thresholds related to sediment accumulation and flushing processes (Harvey, 1977, 1987a; see also Section 6.2.1), and threshold-related variations in the coupling or connectivity of the system (Fryirs *et al.*, 2007).

As we move to timescales of centuries perhaps, the catchment controls (setting, size, geology, relief) remain as passive controls, but vegetation and land use become dynamic controls. Individual events may become less significant, but secular climatic trends become important. An important theme in British alluvial fan research has been the interaction between human-induced land-use change and Holocene climatic fluctuations on the development of alluvial fans and debris cones as representing periods of enhanced erosion and sediment supply from upland catchments (Harvey *et al.*, 1981; Brazier, Whittington and Ballantyne, 1988; Brazier and Ballantyne, 1989; Harvey, 1996, 2001; Chiverrell, Harvey and Foster, 2007; Chiverrell *et al.*, 2008).

For the longer late Quaternary timescales, related to 'classic' alluvial fans, catchment controls (setting, size, geology, relief) remain the main passive controls, especially of overall fan geometry (see Section 6.3) but fan processes and morphology are sensitive to temporal variations in the dynamic controls, especially those caused by climatic or base-level change. There has been much debate in the research literature on the relative roles of tectonism, climate and base-level change in controlling alluvial fan morphology and sediment sequences (e.g. Ritter *et al.*, 1995). Over these timescales, except perhaps in highly tectonically active environments, climate appears to be the primary control of alluvial fan sequences (Frostick and Reid, 1989; Harvey, 2003), modified locally by any base-level change (Harvey, 2002c).

An interesting recent approach to understanding large-scale alluvial fan sequences has been an application of the sequence stratigraphy approach, fundamental in understanding the three-dimensional geometry of coastal and marine sediments, to Quaternary alluvial fan sequences in the central valley of California (Weissmann, Bennet and Lansdale, 2005). The sequence boundaries, rather than caused by sea-level change as in the conventional model, are represented by stable fan surfaces capped by buried soils, through which the fan channels were incised. At times of surface stability and soil formation (interglacials), sediment supply was limited. The intervening sediment sequences, when the fan surfaces became buried, represent major sediment pulses driven by glaciation within the catchments in the Sierra Nevada. Thus, climate is seen to be driving the system, with major switches between deposition and surface stability/fan-channel incision occurring in response to major climatic changes.

It is only over longer time periods extending back through the Pleistocene and perhaps into the late Neogene that tectonics may be seen as a major active control over alluvial fan evolution, on to which the Quaternary climatic signal is imposed. Even the previously passive drainage basin controls may exert a dynamic influence on fan evolution. Relief is not constant: tectonic uplift may increase relief and hence sediment yields, but relief may be progressively reduced in response to long-term erosion. Over these timescales there may be an early period dominated by debris flows on to mountain-front–basin-margin alluvial fans. As sediment yields decrease, the fans may become fluvially dominant, eventually trenching their channels and becoming part of an integrated river network (e.g. Garcia *et al.*, 2004; Silva *et al.*, 2008; Stokes, 2008). At these timescales even drainage area may not be constant. Major drainage reorganization by capture may have an impact on alluvial fan dynamics (Mather, Harvey and Stokes, 2000).

All the work reviewed so far in this chapter has been primarily qualitative. In order to assess the role of debris cones and alluvial fans as interruptions in the sediment cascade it has been important to identify the temporal and spatial scales over which the processes operate, and especially how the controls vary both temporally and spatially. Applying this knowledge to modelling the sediment cascade and testing sediment budget models would require the estimation of the volumes of sediment stored within debris cones and alluvial fans, together with sophisticated dating of sediment sequences. There have been few attempts to consider sediment budgets in relation to alluvial fans (see Oguchi and Ohmori, 1994; Oguchi, 1997). For individual flood events some attempts have been made to estimate volumes of deposition (Wells and Harvey, 1987), but over longer timescales and larger spatial scales subsurface information would also be needed. Perhaps there would be scope using the approach of Weissmann, Bennett and Lansdale (2005).

In the past, alluvial fan research has tended to follow different lines, with geologists/sedimentologists working on the past rock record, emphasizing tectonic controls. Geomorphologists have tended to work either on Quaternary fans or on details of erosion and deposition, in both cases emphasizing climatic controls (Harvey, Mather and Stokes, 2005). Provided that sufficient attention is devoted to the temporal and spatial aspects, some of the more recent approaches reviewed in this chapter could allow a quantitative treatment of debris cones and alluvial fans as part of a continuum within the sediment cascade, applicable to consideration of individual flood events, the generation of major sediment pulses by Quaternary climatic change and the long-term sediment budgets of tectonically active mountain regions.

Acknowledgements

I thank the cartographics section of the Department of Geography, University of Liverpool, particularly Sandra Mather, for producing the illustrations.

References

Alexander, R.W., Calvo-Cases, A., Arnau-Rosalen, E. *et al.* (2008) Erosion and stabilisation sequences in relation to base level changes in the El Cautivo badlands, SE Spain. *Geomorphology*, **100**, 83–90.

Al Farraj, A. and Harvey, A.M. (2000) Desert pavement characteristics on wadi terrace and alluvial fan surfaces: Wadi Al-Bih UAE and Oman. *Geomorphology*, **35**, 279–297.

Al Farraj, A. and Harvey, A.M. (2004) Late quaternary interactions between Aeolian and fluvial processes: a case study in the northern UAE. *Journal of Arid Environments*, **56**, 235–248.

Arzani, N. (2005) The fluvial megafan of the Abarfoh Basin (central Iran): an example of flash-flood sedimentation in arid lands, in *Alluvial Fans: Geomorphology, Sedimentology, Dynamics* (eds A.M. Harvey, A.E. Mather and M. Stokes), Geological Society, London, Special Publication 251, pp. 41–59.

Blair, T.C. (1999) Sedimentology of the debris-flow-dominated Warm Spring Canyon alluvial fan, Death Valley, California. *Sedimentology*, **46**, 941–965.

Blair, T.C. and McPherson, J.G. (1994) Alluvial fan processes and forms, in *Geomorphology of Desert Environments* (eds A.D. Abrahams and A.J. Parsons), Chapman & Hall, London, pp. 354–402.

Blair, T.C. and McPherson, J.G. (1998) Recent debris-flow processes and resultant forms and facies of the dolomite alluvial fan, Owens Valley, California. *Journal of Sedimentary Research*, **68**, 800–818.

Bowman, D. (1978) Determination of intersection points within a telescopic alluvial fan complex. *Earth Surface Processes*, **3**, 265–276.

Bowman, D. (1988) The declining but non-rejuvenating base lavel – the Lisan Lake, the Dead Sea, Israel. *Earth Surface Processes and Landforms*, **13**, 239–249.

Brazier, V. and Ballantyne, C.K. (1989) Late Holocene debris cone evolution in Glen Feshie, western Cairngorm Mountains, Scotland. *Transactions, Royal Society of Edinburgh, Earth Science*, **80**, 17–24.

Brazier, V., Whittington, G. and Ballantyne, C.K. (1988) Holocene debris cone evolution in Glen Etive, Western Grampian Highlands, Scotland. *Earth Surface Processes and Landforms*, **13**, 525–531.

Brunsden, D. and Thornes, J.B. (1979) Landscape sensitivity and change. *Institute of British Geographers Transactions, New Series*, **4**, 463–484.

Bull, W.B. (1977) The alluvial fan environment. *Progress in Physical Geography*, **1**, 222–270.

Bull, W.B. (1979) The threshold of critical power in streams. *Geological Society of America Bulletin*, **90**, 453–464.

Bull, W.B. (1991) *Geomorphic Responses to Climatic Change*, Oxford University Press, Oxford.

Calvache, M.L., Viseras, C. and Fernandez, J. (1997) Controls on fan development – evidence from fan morphometry and sedimentology: Sierra Nevada, Spain. *Geomorphology*, **21**, 69–84.

Candy, I., Black, S. and Sellwood, B.W. (2004) Quantifying time series of pedogenic calcrete formation using U-series disequilibria. *Sedimentary Geology*, **170**, 177–187.

Candy, I., Black, S., Sellwood, B.W. and Roan, J.S. (2003) Calcrete profile development in Quaternary alluvial sequences, southeast Spain: implications for using calcretes as a basis for landform chronologies. *Earth Surface Processes and Landforms*, **28**, 169–185.

Chiverrell, R.C., Harvey, A.M. and Foster, G.C. (2007) Hillslope gullying in the Solway Firth-Morecambe Bay region, Britain: responses to human impact and/or climatic deterioration? *Geomorphology*, **84**, 317–343.

Chiverrell, R.C., Harvey, A.M., Hunter (nee Miller), S.Y. *et al.* (2008) Late Holocene environmental change in the Howgill Fells. Northwest England. *Geomorphology*, **100**, 41–69.

Church, M.J. and Jones, D. (1982) Channel bars in gravel-bed rivers, in *Gravel-Bed Rivers* (eds R.D. Hey, J.C. Bathurst and C.R. Thorne), John Wiley and Sons, Chichester, pp. 238–291.

Coulthard, T.J., Macklin, M.G. and Kirkby, M.J. (2002) simulating upland rivercatchment and alluvial fan evolution. *Earth Surface Processes and Landforms*, **27**, 269–288.

Denny, C.S. (1965) Alluvial fans in the Death Valley region, California and Nevada. *US Geological Survey Professional Paper*, **466**, 59.

Dorn, R.I. and Oberlander, T.M. (1982) Rock varnish. *Progress in Physical Geography*, **6**, 317–367.

Enzel, Y., Wells, S.G. and Lancaster, N. (eds) (2003) *Palaeoenviroments and Palaeohydrology of the Mojave and Southern Great Basin Deserts*, Special Paper 368, Geological Society of America, Boulder, CO.

Faulkner, H. (1987) Spatial and temporal variation of sediment processes in the alpine semi-arid basin of Alkali C, Colorado, USA. *Geomorphology*, **9**, 203–222.

Ferguson, R.I. (1981) Channel forms and channel changes, in *British Rivers* (ed. J. Lewin), Allen and Unwin, London, pp. 90–125.

Frostick, L.E. and Reid, I. (1989) Climate versus tectonic controls of fan sequences: lessons from the Dead Sea, Israel. *Journal of the Geological Society, London*, **146**, 527–538.

Fryirs, K.A., Brierley, G.J., Preston, N.J. and Spencer, J. (2007) Catchment scale (dis)connectivity in sediment flux in the upper Hunter catchment, New South Wales, australia. *Geomorphology*, **84**, 297–316.

Gabris, G. and Nagy, B. (2005) Climate and tectonically controlled river style changes on the Sajo-Hernad alluvial fan (Hungary), in *Alluvial Fans: Geomorphology, Sedimentology, Dynamics* (eds A.M. Harvey, A.E. Mather and M. Stokes), Geological Society, London, Special Publication 251, pp. 61–67.

Garcia, A.F., Zhu, Z., Ku, T.L. *et al.* (2004) An incision wave in the geologic record, Alpujarran Corridor, southern Spain (Almeria). *Geomorphology*, **60**, 37–72.

Gohain, K. and Parkash, B. (1990) Morphology of the Kosi megafan, in *Alluvial Fans: A Field Approach* (eds A.H. Rachocki and M. Church), John Wiley and Sons, Chichester, pp. 151–178.

Gomez-Villar, A., Alvarez-Martinez, J. and Garcia-Ruiz, J.M. (2006) Factors influencing the presence or absence of tributary-junction fans in the Iberian Range, Spain. *Geomorphology*, **81**, 252–264.

Grayson, D.K. (1993) *The Desert's Past: A Natural History of the Great Basin*, Smithsonian Institution, Washington DC, p. 356.

Harbor, J. (1999) Cosmogenic isotopes in Geomorphology. *Geomorphology (Special Issue)*, **27**, 1–172.

Harvey, A.M. (1977) Event frequency in sediment production and channel change, in *River Channel Changes* (ed. K.J. Gregory), John Wiley and Sons, Chichester, pp. 301–315.

Harvey, A.M. (1984) Geomorphological response to an extreme flood: a case study from southeast Spain. *Earth Surface Processes and Landforms*, **9**, 267–279.

Harvey, A.M. (1986) Geomorphic effects of a 100-year storm in the Howgill Fells, northwest England. *Zeitschrift fur Geomorphologie*, **30**, 71–91.

Harvey, A.M. (1987a) Seasonality of processes on eroding gullies, a twelve year record of erosion rates, in *Processus et Mesure de l'Erosion* (eds A. Goddard and A. Rapp), CNRS, Paris, pp. 439–454.

Harvey, A.M. (1987b) Alluvial fan dissection: relationships between morphology and sedimentation, in *Desert Sediments, Ancient and Modern* (eds L. Frostick and I. Reid), Geological Society of London, Special Publication 35, Blackwell, Oxford, pp. 87–103.

Harvey, A.M. (1992) Process interactions, temporal scales and the development of gully systems: Howgill Fells, northwest England. *Geomorphology*, **5**, 323–344.

Harvey, A.M. (1996) Holocene hillslope gully systems in the Howgill Fells, Cumbria, in *Advances in Hillslope Processes*, vol. **2** (eds S.M. Brooks and M.G. Anderson), John Wiley and Sons, Chichester, pp. 731–752.

Harvey, A.M. (1997a) The role of alluvial fans in arid zone fluvial systems, in *Arid Zone Geomorphology: Process, Form and Change in Drylands* (ed. D.S.G. Thomas), John Wiley and Sons, Chichester, pp. 231–259.

Harvey, A.M. (1997b) Coupling between hillslope gully systems and stream channels in the Howgill Fells, northwest England: temporal implications. *Geomorphologie; Relief, Processes, Environment*, **1**, 3–20.

Harvey, A.M. (2001) Coupling between hillslopes and channels in upland fluvial systems: implications for landscape sensitivity, illustrated from the Howgill Fells, northwest England. *Catena*, **42**, 225–250.

Harvey, A.M. (2002a) Effective timescales of coupling within fluvial systems. *Geomorphology*, **44**, 175–201.

Harvey, A.M. (2002b) Factors influencing the geomorphology of alluvial fans: a review, in *Apertaciones a la Geomorfologia de Espana en el Inicio de Tercer Mileno* (eds A. Perez-Gonzalez, J. Vegas and M.J. Machado), Instituto Geologico y Minero de Espana, Madrid, pp. 59–75.

Harvey, A.M. (2002c) The role of base-level change on the dissection of alluvial fans: case studies from southeast Spain and Nevada. *Geomorphology*, **45**, 67–87.

Harvey, A.M. (2003) The response of dry region alluvial fans to Quaternary climatic change, in *Desertification in the Third Millennium* (eds A.S. Alsharhan, W.W. Wood, A.S. Goudie, A. Fowler and E.M. Ebdelatif), Swete & Zeitlinger, Lisse, The Netherlands, pp. 75–90.

Harvey, A.M. (2005) Differential effects of base-level, tectonic setting and climatic change on Quaternary alluvial fnas in the northern Great Basin, Nevada, USA, in *Alluvial Fans: Geomorphology, Sedimentology, Dynamics* (eds A.M. Harvey, A.E. Mather and M., Stokes), Geological Society, London, Special Publication 251, pp. 117–131.

Harvey, A.M. (2007) Differential recovery from the effects of a 100-year storm: significance of long-term hillslope-channel coupling; Howgill Fells, northwest England. *Geomorphology*, **84**, 192–208.

Harvey, A.M. and Wells, S.G. (1994) Late Pleistocene and Holocene changes in hillslope sediment supply to alluvial fan systems: Zzyzx, California, in *Environmental Change in Drylands: Biogeographical and Geomorphological Perspectives* (eds A.C. Millington and K. Pye), John Wiley & Sons, Ltd, Chichester, pp. 66–84.

Harvey, A.M. and Wells, S.G. (2003) Late Quaternary variations in alluvial fan sedimentologic and geomorphic processes, Soda Lake basin, eastern Mojave desert, California, in *Palaeoenviroments and Palaeohydrology of the Mojave and southern Great Basin Deserts*, (eds Y. Enzel, S.G. Wells and N. Lancaster), Geological Society of America, Special Paper 368, pp. 207–230.

Harvey, A.M., Alexander, R.W. and James, P.A. (1984) Lichens, soil development and the age of Holocene valley-floor landforms: Howgill Fells, Cumbria. *Geografisca Annaler*, **66A**, 353–366.

Harvey, A.M., Mather, A.E. and Stokes, M. (2005) Alluvial fans: geomorphology, sedimentology, dynamics – introduction. A review of alluvial fan research, in *Alluvial Fans: Geomorphology, Sedimentology, Dynamics* (eds A.M. Harvey, A.E. Mather and M. Stokes), Geological Society, London, Special Publication 251, pp. 1–7.

Harvey, A.M., Wigand, P.E. and Wells, S.G. (1999b) Response of alluvial fan systems to the late Pleistocene to Holocene climatic transition: contrasts between the margins of pluvial Lakes Lahontan ad Mojave, Nevada and California, USA. *Catena*, **36**, 255–281.

Harvey, A.M., Oldfield, F., Baron, A.F. and Pearson, G. (1981) dating of post-glacial landforms in the central Howgills. *Earth Surface Processes and Landforms*, **6**, 401–412.

Harvey, A.M., Silva, P., Mather, A.E. *et al.* (1999a) The impact of Quaternary sea-level and climatic change on coastal alluvial fans in the Cabo de Gata ranges, southeast Spain. *Geomorphology*, **28**, 1–22.

Harvey, A.M., Foster, G.C., Hannam, J. and Mather, A.E. (2003) The Tabernas alluvial fan and lake system, southeast Spain: applications of mineral magnetic and pedogenic iron oxide analyses towards clarifying the Quaternary sediment sequences. *Geomorphology*, **50**, 151–171.

Hooke, R.L. (1967) Processes on arid-region alluvial fans. *Journal of Geology*, **75**, 438–460.

Kostaschuk, R.A., Macdonald, G.M. and Putnam, P.E. (1986) Depositional processes and alluvial fan – drainage basin morphometric relationships near Banff, Alberta, Canada. *Earth Surface Processes and Landforms*, **11**, 471–484.

Leeder, M.R. and Mack, G.H. (2001) Lateral erosion ("toe cutting") of alluvial fans by axial rivers: implications for basin analysis and architecture. *Journal of the Geological Society, London*, **158**, 885–893.

Machette, M.N. (1985) Calcic soils of the southwestern United States, in *Soils and Quaternary geology of the Southwestern United States*, (ed. D.L. Weide), Geological Society of America, Special Paper 203, pp. 1–21.

Macklin, M.G. and Lewin, J. (2003) River sediments, great floods and centennial-scale Holocene climatic change. *Journal of Quaternary Science*, **18**, 102–105.

Mather, A.E., Harvey, A.M. and Stokes, M. (2000) Quantifying long term catchment changes of alluvial fan systems. *Geological Society of America Bulletin*, **112**, 1825–1833.

McFadden, L.D., Ritter, J.B. and Wells, S.G. (1989) Use of multiparameter relative-age methods for estimation and correlation of alluvial fan surfaces on a desert piedmont, eastern Mojave Desert, California. *Quaternary Research*, **32**, 276–290.

Nanson, G.C., Chen, X.Y. and Price, D.M. (1995) Aeolian and fluvial evidence of changing climate and wind patterns during the past 100 ka in the western Simpson Desert, Australia. *PalaeOgeography, Palaeoclimatology, Palaeoecology*, **113**, 87–102.

Oguchi, T. (1997) Late Quaternary sediment budjet in alluvial-fan – source-basin systems in Japan. *Journal of Quaternary Science*, **12**, 381–390.

Oguchi, T. and Ohmori, H. (1994) Analysis of relationships among alluvial fan area, source basin area, basin slope and sediment yield. *Zeitschrift fur Geomorphologie NF*, **38**, 405–420.

Pope, R.J.J. and Wilkinson, K.N. (2005) Reconciling the roles of climate and tectonics in Late Quaternary fan development on the Sparta piedmont, Greece, in *Alluvial Fans: Geomorphology, Sedimentology, Dynamics* (eds A.M. Harvey, A.E. Mather and M. Stokes), Geological Society, London, Special Publication 251, pp. 133–152.

Ritter, J.B., Miller, J.R. and Husek-Wulforst, J. (2000) Environmental controls on the evolution of alluvial fans in the Buena Vista valley, north central Nevada, during late Quaternary time. *Geomorphology*, **36**, 63–87.

Ritter, J.B., Miller, J.R., Enzel, Y. and Wells, S.G. (1995) Reconciling the roles of tectonism and climate in Quaternary alluvial fan evolution. *Geology*, **23**, 245–248.

Robinson, R.A.J., Spencer, J.Q.G., Strecker, M.R. *et al.* (2005) Luminescence dating of alluvial fans in intramontane basins of NW Argentina, in *Alluvial Fans: Geomorphology, Sedimentology, Dynamics* (eds A.M. Harvey, A.E. Mather M. Stokes), Geological Society, London, Special Publication 251, pp. 153–168.

Ryder, J.N. (1971) The stratigraphy and morphology of paraglacial fans in south central British Columbia. *Canadian Journal of Earth Sciences*, **8**, 279–298.

Schmidt, J.C. (1990) Recirculating flow and sedimentation in the Colorado River in the Grand Canyon, Arizona. *Journal of Geology*, **98**, 709–724.

Schmidt, J.C. and Graf, J.B. (1990) Aggradation and degradation of alluvial sand deposits 1965–1986, Colorado River, Grand Canyon National Park, Arizona. *U.S. Geological Survey Professional Paper*, **1493**, 74.

Schmidt, J.C. and Rubin, D.M. (1995) Regulated streamflow fine-grained depositsand effective discharges in canyons with abundant debris fans, in *Natural and Anthropogenic Influences in Fluvial Geomorphology* (eds J.E. Costa, A.J. Miller, K.W. Potter and P.R. Wllcock), Geophysics Monograph 89, American Geophysical Union, Washington, DC, pp. 177–195.

Schumm, S.A. and Litchty, R.W. (1965) Time, space and causality in geomorphic processes. *American Journal of Science*, **255**, 161–174.

Silva, P.G., Harvey, A.M., Zazo, C. and Goy, J.L. (1992) Geomorphology, depositional style and morphometric relationshipsof Quaternary alluvial fans in the Guadalentin depression (Murcia, Southeast Spain). *Zeitschrift fur Geomorphologie NF*, **36**, 325–341.

Silva, P.G., Bardaji, T., Calmel-Avila, M. *et al.* (2008) Transition from alluvial to fluvial systems in the Guadalentin depression (SE Spain): Lorca fan versus Guadalentin River. *Geomorphology*, **100**, 140–153.

Stanistreet, I.G. and McCarthy, T.S. (1993) The Okavango fan and the classification of subaerial fan systems. *Sedimentary Geology*, **65**, 115–133.

Stokes, M. (2008) Plio-Pleistocene drainage development in an inverted sedimentary basin: Vera basin, Betic Cordillera, SE Spain. *Geomorphology*, **100**, 193–211.

Wang, H., Harvey, A.M., Xie, S. *et al.* (2008) Tributary-junction fans of China's Yangtze Three-Gorges valley: morphological implications. *Geomorphology*, **100**, 131–139.

Wasson, R.J. (1979) Intersection point deposition on alluvial fans: an Australian example. *Geografiska Annaler*, **56A**, 83–92.

Weissmann, G.S., Bennett, G.L. and Lansdale, A.L. (2005) Factors controlling sequence development on Quaternary fluvial fans, San Joaquin Basin, California, USA, in *Alluvial Fans: Geomorphology, Sedimentology, Dynamics* (eds A.M. Harvey, A.E. Mather and M. Stokes), Geological Society, London, Special Publication 251, pp. 169–186.

Wells, S.G. and Gutierrez, A.A. (1982) Quaternary evolution of badlands in the southwestern Colorado Plateau, USA, in *Badland Geomorphology and Piping* (eds R.B. Bryan and A. Yair), Geobooks, Norwich, pp. 239–258.

Wells, S.G. and Harvey, A.M. (1987) Sedimentologic and geomorphic variations in storm generated alluvial fans, Howgill Fells, northwest England. *Geological Society of America Bulletin*, **98**, 182–198.

Wells, S.G., McFadden, L.D. and Dohrenwend, J.C. (1987) Influence of late Quaternary climatic changes on geomorphic and pedogenic processes on a desert piedmont, Eastern Mojave Desert, California. *Quaternary Research*, **27**, 130–146.

Whipple, K.X. and Dunne, T. (1992) The influence of debris-flow rheology on fan morphology, Owens Valley, California. *Geological Society of America Bulletin*, **104**, 887–900.

Whipple, K.X., Parkes, G., Paola, C. and Mohrig, D. (1998) Channel dynamics, sediment transport and the shape of alluvial fans: experimental study. *Journal of Geology*, **106**, 677–693.

Wolman, M.G. and Miller, J.P. (1960) Magnitude and frequency of forces in geomorphic processes. *Journal of Geology*, **68**, 64–74.

7

Overland Flow and Soil Erosion

Louise J. Bracken
Department of Geography, Durham University, Durham DH1 3LE, UK

7.1 Introduction

Soil erosion is a complex phenomenon involving the detachment and transport of soil particles, and the infiltration, ponding and runoff of rainfall. The magnitude and frequency of these processes is related to climate, rainfall characteristics, topography, soil characteristics, management practices, and scale. The inherent resistance of soil to erosional processes has been generally recognized and researched widely, alongside developments in our understanding of relationships between rainfall inputs and erosion. However, the full implications of dynamic properties of the soil and variability in inputs to the system have seldom been considered, despite issues of heterogeneity and spatial variability of surface properties providing a current focus for research. Recent thinking suggests that at the field scale interactions between climate, cultivation practices, and soil properties have the dominant influence on soil erosion, but at the catchment scale runoff appears to be the main factor influencing overland flow and erosion (Auzet, Poesen and Valentin, 2001). The complexity of soil erosion is highlighted in Table 7.1, which details a range of work attempting to unravel the factors promoting and resisting sediment entrainment. The table demonstrates some of the key debates such as the effect of surface roughness, with some proposing that surface roughness decreases erosion due to ponding of surface water (e.g. Onstad, 1984; Huang and Bradford, 1990; Hairsine, Moran and Rose, 1992), whereas others suggest that as roughness increases, flow concentration occurs, thereby increasing erosion (e.g. Abrahams and Parsons, 1990; Helming, Römkens and Prasad, 1998; Römkens, Helming and Prasard, 2001). This chapter covers a wide range of issues including overland flow and entrainment of sediment, rills and gullies as major routes of increased erosion, and sediment evacuation and the influence of soil properties. Badlands and soil erosion on agricultural land are discussed as two special cases of severe large-scale erosion. The chapter then outlines selected methods used to limit sediment transfer to help prevent soil erosion and closes with a brief conclusion.

Sediment Cascades: An Integrated Approach Edited by Timothy Burt and Robert Allison

Table 7.1 A summary of variables influencing soil erosion

Factors influencing soil erosion	Key research
Surface properties	
Soil erosion is size selective	Martinez-Mena Albaladejo and Castillo (1999)
Dry aggregates are more resistant to erosion	Nearing and Bradford (1985)
Slaking causes aggregates to be broken down more rapidly	Le Bissonnais (1990)
Surface roughness increases resistance to erosion	Farres (1978); Römkens and Wang (1987)
Surface roughness increases depression storage and decreases flow velocity	Hairsine, Moran and Rose (1992); Onstad (1984); Huang and Bradford (1990)
Surface roughness increases flow concentration	Abrahams and Parsons (1990); Helming, Römkens and Prasad (1998); Römkens, Helming and Prasard (2001)
Low soil moisture increases cohesion	Römkens and Wang (1987)
Low soil moisture increases slaking	Le Bissonnais (1990); Govers and Loch (1993); Bryan (1996)
Surface water increases splash and soil erosion	Palmer (1963); Mutchler and Larson (1971)
Surface water has no impact on splash	Moss (1991); Kinnell (1990)
Increased slope increases runoff and erosion	De Ploey (1990); Torri, Colica and Rockwell (1994); Sharma *et al.* (1993); Fox and Bryan (1999); Chaplot and Le Bissonnais (2000)
Increased slope decreases runoff by destroying the crust and increasing infiltration	Poesen (1984); Bryan and Poesen (1989); Slattery and Bryan (1992)
Slope has no effect on runoff or erosion	Lal (1990)
Crusting	
Different soil types produce crusts more readily, which encourages runoff and soil erosion	Singer and Le Bissonnais (1998)
Crust development results in increased runoff and erosion	Valentin and Bresson (1992); Casenave and Valentin (1992); Parsons *et al.* (2003)
Surface sealing protects the surface, decreasing erosion	De Ploey (1971); Govers (1991)
Erosion by wash and splash decrease during a runoff event due to surface sealing	Bradford, Remley and Ferris (1987a); Bradford, Ferris and Remley (1987b)
Pre-wetting reduces crust development	Yair (1990); Kidron and Yair (1997)
Vegetation and land management	
Bare areas produce more runoff	Wainwright (1996); Martinez-Mena, Albaladejo and Castillo (1998); Cerdà (1997); Cerdà *et al.* (1998); Leonard and Andrieux (1998); Lasanta *et al.* (2000); Parsons *et al.* (2003)
The change from grassland to shrubs results in rill development and increased soil erosion.	Wainwright, Parsons and Abrahams (2000)
No tillage decreases runoff soil erosion.	Castro *et al.* (1999)
Fire and burning increase runoff and erosion.	Imeson *et al.* (1992)
Rainfall	
The relationship between rainfall intensity and kinetic energy is complex due to variations in drop size	Parsons and Gadian (2000)

Table 7.1 *(Continued)*

Factors influencing soil erosion	Key research
Intensity is more important than duration for crusting	Morrison, Prunty and Giles (1985)
Decreasing rainfall intensity over a sequence of storms encourages rilling and hence increases soil erosion	Römkens, Helming and Prasard (2001)
Increased duration of rainfall increases soil erosion	Bryan (2000)
Repeated wetting and drying decreases aggregate stability and increases erosion	Tisdall, Cockcroft and Uren (1978); Shiel, Adey and Lodder (1988)
Higher rainfall intensities produce more soil erosion	Chaplot and Le Bissonnais (2000)
Wind	
Can increase or decrease erosion by influencing raindrop impact velocity and angle	Bryan (2000)

7.2 Overland Flow and Entrainment of Material

Overland flow usually occurs in small rills or as sheets of moderate depth over large surfaces; sometimes, larger gullies may erode, often along dry valley bottoms where surface runoff concentrates ('thalweg rills' – see Slattery, Burt and Boardman, 1994). For erosion to occur the rate of rainfall must be sufficient to produce runoff, and the shear stress produced by the moving water must exceed the resistance of the soil surface (Dietrich and Dunne, 1993; Bull and Kirkby, 1997). Erodibility is a function of the permeability of the surface, the physical and chemical properties that determine the cohesiveness of the soil, and the vegetation. The shear stress produced by overland flow depends on raindrop impact, slope, and depth. Depth is controlled by the relative rates of rainfall and infiltration, the velocity of flow, and by the length of slope. Raindrop impact can either dislodge particles or seal the surface to reduce infiltration and increase runoff (De Ploey, 1984; Bradford, Remley and Ferris, 1987a; Bradford, Ferris and Remley, 1987b; Govers, 1991).

There are two commonly accepted methods of surface runoff generation, Hortonian or infiltration-excess overland flow and saturation-excess overland flow (Anderson and Burt, 1990). The Variable Source Area (VSA) concept was developed for humid regions in the 1960s (Hewlett and Hibbert, 1967; Dunne and Black, 1970) and proposed that saturated areas produce most of the storm runoff as the water table rises to the soil surface over an expanding area as rainfall continues. Saturation-excess overland flow initially spreads up low-order tributaries, then up unchannelled swales and gentle footslopes of hillsides (Dunne, Moore and Taylor, 1975). The position and expansion of variable areas contributing runoff is related to geology, topography, soils, rainfall characteristics, and vegetation (Dunne and Black, 1970; Dunne, Moore and Taylor, 1975). Saturation-excess overland flow includes a contribution from rain falling on to saturated areas, but also exfiltration of subsurface flow, which can continue after rainfall has ceased, and is therefore unaffected by drop impact and rainsplash. Where exfiltration forms an important fraction of the overland flow, the shear force is augmented by a small lift force associated with the emergence of water from the soil (Kochel, Howard and MacLane, 1982). Saturation-excess overland flow usually occurs over surfaces stabilized by a dense ground cover and root mat. If soil transmissivity or gradient decreases downslope, or if topography converges in plan form, there is more likely to be some exfiltration. Kilinc and Richardson (1973) found that in thin flows entrainment efficiency increased with rainfall rate and that rainsplash was an important factor effecting particle entrainment in sheetflow. Savat (1979) then suggested that rainsplash has little effect on flow velocities and net drag force, but increased transport capacity and wash sediment concentrations, which in turn increased kinematic viscosity by up to 30%. This was confirmed by Torri, Sfalanga and Chisci (1987).

Recently, the VSA model itself has been the focus of critique, with calls for a new theory of runoff generation (e.g. McDonnell, 2003; Ambroise, 2004; Bracken and Croke, 2007). Hydrological connectivity is one possible concept of runoff generation and flood production that could provide a way forward, but currently there is much confusion in the literature about how the term is used and how it relates to existing research. There has been considerable research on aspects of hydrological 'connectivity', although not always referred to as such, including runoff generation at the patch, hillslope and catchment scale that makes it difficult to draw work together into a single theoretical model of runoff connectivity (e.g. Fitzjohn, Ternan and Williams, 1998; Cammeraat and Imeson, 1999; Ludwig, Wiens and Tongway, 2000). A consistent definition of the term, however, remains difficult to discern from published studies in hydrology and geomorphology.

In semi-arid and arid environments the variable source area concept does not hold because most runoff is produced by Hortonian overland flow (Bryan and Yair, 1982). There are no areas of saturation to initiate quick runoff because storms are infrequent and of short duration. Hortonian runoff production occurs where the rainfall intensity is greater than the infiltration capacity, and hence water cannot be absorbed into the soil and runs off over the landscape. Key runoff-producing areas tend to be found on steep slopes, either abandoned after agriculture or with sparse vegetation, and are composed of runoff-promoting soils such as marls (Bull *et al.*, 2000). These areas do not necessarily relate to the channel network and the mosaic pattern they form is key to producing floods in ephemeral channels. Infiltration-excess overland flow extends to the catchment divide, whereas saturation-excess overland flow is usually confined to slope-base concavities and hollows (Dunne and Black, 1970). The relationship between discharge and the distance from the divide is strongly affected by the soil profile at a particular elevation, soil depth, hydraulic conductivity and plan-form conductivity. Sediment transport in thin flows may be stimulated by raindrop impact (Moss, Walker and Hutka, 1979), but part of the transporting capacity may be satisfied by material entering the flow as rain splash. The Hortonian runoff model also applies to tilled fields where, as in many semi-arid situations, the balance between rainfall intensity and infiltration capacity favours surface-runoff production (Imeson and Kwaad, 1990; Burt and Slattery, 1996).

The rate of erosion and rate of deposition by flowing water are controlled by the difference between the transport capacity and the influx of sediment from upslope; erosion where the difference is positive and deposition where it is negative (Smith *et al.*, 1995; Abrahams *et al.*, 1998). Incision occurs where the total shear stress of the underloaded flow is (i) high enough to disrupt the root mat, (ii) is high enough to incise bare patches of soil between vegetation, (iii) where sheetwash transport through vegetation is available to evacuate sediment loosened by biogenic disruption, or (iv) where seepage erosion triggers the initial incision through the vegetated surface so the bare soil is exploited by saturation overland flow. The ability of overland flow to transport material is usually expressed in terms of flow power (Bagnold, 1966) or tractive shear stress (Horton, 1945; Yalin, 1971). Incision occurs at a critical shear stress (τ_{cr}) estimated by:

$$\tau_{cr} = \rho U_{*cr}^2 = (\rho g R S)_{cr} \tag{7.1}$$

where ρ is the density of water, U^* is shear velocity, g is the gravitational constant, R is hydraulic radius and S is slope.

This expression is used to link critical unit discharge to critical shear stress and slope at the point of incision (De Ploey, 1990). The initiation of motion is related to a critical threshold, determined by surface properties. On vegetation or fine sediment, the threshold is set by soil cohesion or vegetation mat strength. For unconsolidated sand or gravels, the threshold is related to the forces required to move the grain layer, often assuming equal mobility. The shape of the rill channel is strongly influenced by vertical differences in soil strength, caused by a tougher surface or a plough pan or other resistant layer at depth.

Texture, aggregate stability and soil shear strength are the main soil characteristics that are relevant for soil detachability (Arulanandan, Loganathan and Krone, 1975; Torri, 1987). De Ploey (1971) found that liquefaction in sandy sediments facilitates and even starts erosion, and was especially important during sheet erosion. However, the main property of clay governing its susceptibility to erosion is the ratio of dissolved sodium ions to other basic cations in pore water that helps to disperse material (Sherard, Ryker and Decker, 1972). Important changes in aggregation and crust characteristics can result from changes in the balance of the soil and pore fluid chemistry (Sargunam *et al.*, 1973; Gerits *et al.*, 1987). Yair *et al.* (1980) found that in the Zin Valley, Israel, due to the high stability and strong flocculation of clay-rich aggregates, rainsplash is ineffective in surface sealing. Infiltration capacities therefore remain high despite prolonged periods of rain. Rill system development may be closely related to regolith hydrological properties, and influenced by consistency, shrink–swell capacity and dispersibility (Gerits *et al.*, 1987; Imeson and Verstraten, 1988, 1989). Organic matter content also exerts a strong influence on aggregate stability (Imeson, Kwaad and Verstraten, 1982). However, Römkens, Helming and Prasard (2001) suggest that storms with initially high intensities, with the potential for early development of concentrated flow, have a greater likelihood of rill development, and those with low intensities early on in the storm, which are more likely to produce surface seals, tend to prevent erosion. Rainfall characteristics must therefore always be kept in mind when investigating soil-surface dynamics.

Runoff energy reflects flow discharge and hydraulics, which are strongly influenced by soil-surface properties such as micro-topography and vegetation (Govers, 1992; Bryan, 2000). Hydraulic conditions are important since they directly affect the erosive energy of overland flow. Studies of the applicability of deep channel hydraulics to overland flow have found significant differences (e.g. Horton, 1945; Savat, 1979). Deep channel flow is usually subcritical, hydraulically smooth and turbulent, with roughness elements submerged. However, overland flow on steep slopes is usually thin and discontinuous, and highly variable (Govers and Rauws, 1986; Govers, 1992; Bryan, 2000), which means flow is frequently supercritical and varies from rough to smooth over short distances (Govers and Rauws, 1986; Govers, 1992; Bryan, 2000). Rainsplash energy has little impact on deep flow, but can strongly influence shallow flows (Govers and Rauws, 1986; Bryan, 2000). Recent work suggests that the understanding of overland flow hydraulics necessary for physically based modelling of runoff and soil erosion has not been achieved due to the problem of modifying existing pipeflow equations rather than developing an individual theory of overland flow hydraulics consistent with the observations during fieldwork (Smith, Cox and Bracken, 2007).

A key soil characteristic that modifies overland flow is the surface roughness. Surface roughness is complex because it operates at a number of spatial scales and can be highly dynamic. For instance plough furrows are relatively large and influence flow pathways (form roughness) (Kirkby, Bracken and Reaney, 2002), individual stones and soil aggregates can be larger than the flow depth (grain roughness), yet on the other hand disintegration of aggregates, differential swelling, and surface sealing are also dynamic over the duration of a storm as small-scale roughness elements (Slattery and Bryan, 1992) (Figure 7.1). The hydraulic equations for the friction factor and laminar/turbulent flow are

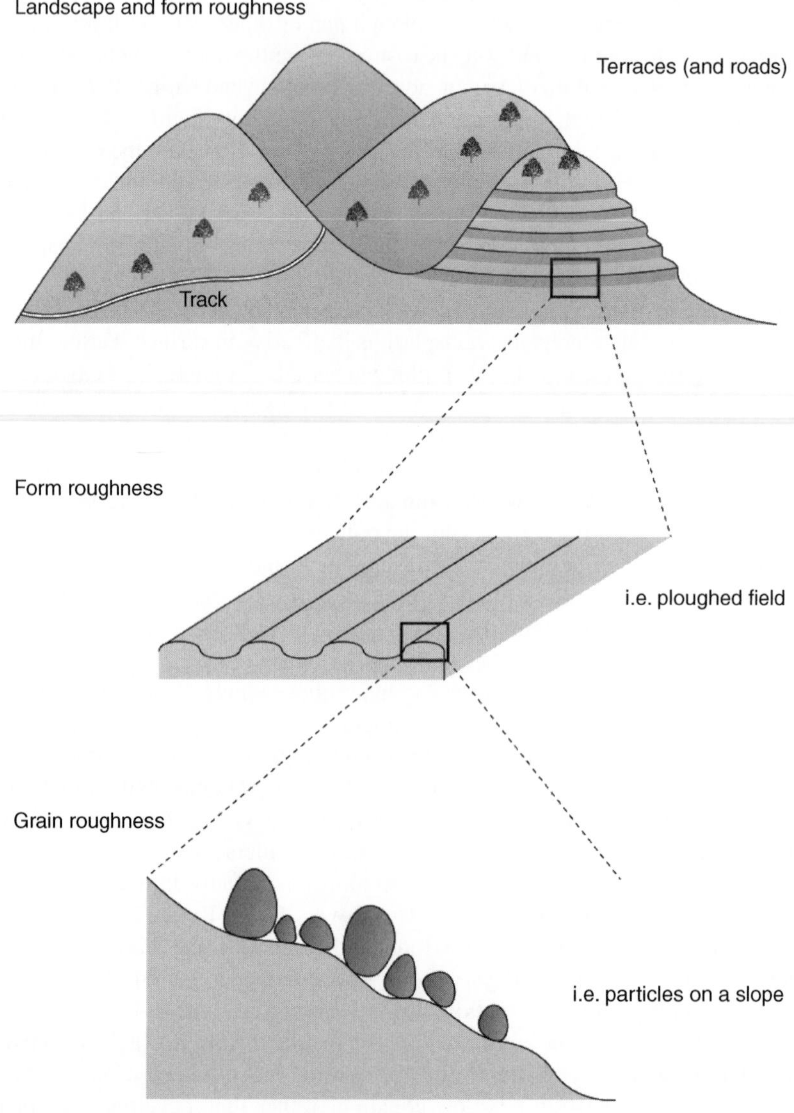

Figure 7.1 The different types of surface roughness

Table 7.2 Equations used to determine sediment transport in overland flow

Term	Equation	Used By
Bed shear stress	$\tau = \rho$ g d S	Slattery and Bryan (1992); Jayawardena and Bhuiyan (1999)
Specific flow power	$\omega = \tau$ u	Bagnold (1980), Abrahams *et al.* 1998; Slattery and Bryan (1992)
Unit flow power	u S	Govers (1985); Moore and Burch (1986); Huang (1995)
Excess flow power and flow depth	Log $T_w = -3.038 + 1.726$ log $(\omega - \omega_c) - 1.212$ log d	Abrahams *et al.* (2001)

Where τ is bed shear stress, ρ is water density (kg m^{-3}), g is acceleration due to gravity (m s^{-2}), d is flow depth (m), S is slope (°), ω is specific flow power, u is flow velocity (m s^{-1}), T_w is immersed sediment transport rate (kg s^{-3}).

used to determine the influence of roughness on surface flow. When roughness elements are fully submerged, hydraulic roughness declines with flow depth (an inverse relationship between the Darcy–Weisbach friction factor and the Reynolds number) (Savat, 1980; Govers, 1992; Bryan, 2000). However, when roughness elements equal or exceed flow depth, the Darcy–Weisbach friction factor and Reynolds number relationship becomes positive (Abrahams, Parsons and Luk, 1986; Abrahams and Parsons, 1991; Nearing *et al.*, 1998). This results in the decreasing importance of grain resistance (Govers and Rauws, 1986; Abrahams, Parsons and Luk, 1986) and sediment movement is controlled by form resistance. The transition between dominance by grain and form resistance varies with soil characteristics (Bryan, 2000). Different predictions of sediment transport in overland flow result from the different calculations (Govers and Rauws, 1986; Govers, 1985), and using grain shear stress may underpredict sediment transport capacity by overland flow (Abrahams and Parsons, 1994; Atkinson *et al.*, 2000) due to the impact of turbulent eddies (Abrahams *et al.*, 1998). Abrahams *et al.* (1998) therefore suggested the combined use of flow power and flow depth to capture the influence of roughness, and could be used to predict sediment transport in overland flow. A summary of equations used to predict sediment transport rate in overland is presented in Table 7.2. Table 7.3

Table 7.3 Equations used to determine sediment transport in rills

Term	Equation	Used By
Shear velocity	$u^* = g\, d\, S^{0.5}$	Govers (1985)
Stream power	$\omega = \rho$ q g S	Rose (1985)
Unit stream power	Y = r u	Govers and Rauws (1986); Moore and Burch (1986); Loch, Maroulis and Silburn (1989)
Froude number	$Fr = u/g\ d^{0.5}$	Savat (1976); Bryan (2000); Bryan and Oostwoud Wijdenes (1992)

Where u^* is shear velocity, g is acceleration due to gravity (m s^{-2}), S is slope (°), ω is flow power, Y is unit stream power, Fr is Froude number, u is flow velocity (m s^{-1}), d flow depth (m).

presents equations used to estimate sediment transport in rills and highlights the difference between processes included in the calculations for concentrated and unconcentrated flow.

7.3 Soil Properties

A chapter on overland flow and soil erosion is not complete without a brief discussion of soil properties that oppose erosion. The soil surface is not a smooth, rigid layer that just responds to action imposed on it, but is dynamic and interacts with applied stresses. Soil properties provide the main resistance to erosion, and are critical in determining spatial and temporal patterns of sediment transfer on hillslopes. Initially, it was thought that soil erodibility was a constant characteristic that could be accurately identified, but more recently it has been realized that this is far too simplistic and that soil erodibility is a highly complex response pattern, influenced by intrinsic soil characteristics and extrinsic environmental variables (Lal, 1990; Bryan, 2000; Knapen, Poesen and De Baets, 2007). Many studies have been undertaken to examine the influence of soil properties on erosion, and these have benefitted from a range of methodologies, environments, soil management techniques, soil types and rainfall conditions. This makes understanding soil erosion both exciting and challenging. In practice, there are a few key properties that influence soil erosion including shear strength, soil aggregation (Luk, 1979; Meyer and Harmon, 1984), and crusting and sealing (Bryan, 2000). However, infiltration rate, moisture content, and organic matter content are also important (Poesen and Savat, 1981). These properties influence the distribution of erosive forces, resistance to entrainment and water movement.

Soil strength varies with position in the soil and is influenced by the presence of aggregates and their behaviour during wetting. Raindrop impact during rainfall causes aggregates to break down (Panabooke and Quirk, 1957) but this can occur more rapidly if slaking takes place (Le Bissonnais, 1990). Hence dry aggregates are more resistant to erosion (Nearing and Bradford, 1985) but soil erodibility cannot be assessed on soil surface properties alone and there needs to be some consideration of the upper soil profile (Bryan, 2000). Aggregation influences soil strength, surface resistance and sealing. Aggregate formation involves physical stresses (e.g. frost action, root action, shrinkage and swelling), which can force particles together or apart and binding agents which cement particles together (e.g. humic acids, organic compounds, mineral deposits and electrostatic bonds; Gerits *et al.*, 1987; Bryan, 2000). Platelet building blocks are bonded together to form stable and resistant aggregates up to 90 μm diameter, although they can be up to 200 μm when combined with sand and silt particles (Bryan, 2000), and up to 250 μm when bound by roots, fungi and organic matter (Oades, 1993). According to Poesen (1981), aggregate size is the most important soil characteristic in terms of understanding soil erosion. Aggregates are affected by rainfall, and on wetting slaking can occur with maximum breakdown occurring when rain falls on dry soils (Le Bissonnais, Bruand and Jamagne, 1989). As water content increases, cohesion declines along with shear strength (Bryan, 2000).

Crust development results in a marked decrease in infiltration and possible protection against erosion by rainsplash and overland flow. There have been many studies investigating crusting, including rates and processes of formation, seal structure, micromorphometry, the relationship to rainfall (e.g. Farres, 1978; Römkens, Prasard and Whisler, 1990; Valentin and Bresson, 1992; Le Bissonnais, 1996; Assouline and Mualem, 1997) and the processes involved are raindrop compaction, clay mineral migration, vesicle formation, and clay skin orientation (Bryan, 2000). As rainfall hits the soil surface, pores can collapse (Moss, 1991) and the aggregates break down, providing particles for seal formation, which results in ponding and runoff generation. The degree of breakdown varies with soil characteristics and is related to soil moisture conditions (Le Bissonnais, 1990, 1996). Crusts tend to be classified as structural (formed by compaction and filtration of fines), depositional (caused by selective deposition of fines) (Arshad and Mermut, 1988; Sumner, 1995) or biotic (West, 1990; Eldridge, Zaady and Shachak, 2000) and each interacts differently with runoff. Different crusts have different effects on limiting soil erosion.

The soil's resistance to concentrated flow erosion is an important factor for predicting rill and gully erosion rates and hence soil erosion. Many experiments have been conducted to determine erosion resistance of different types of soils, but there has not been any attempt to establish trends across these data. Knapen, Poesen and De Baets (2007) collate all available data on the resistance of topsoil to erosion and, although they note that the heterogeneity of measurement methods and lack of standard definitions make direct comparability difficult, they suggest that channel erodibility and critical shear stress are the two most important variables to understand. They conclude that channel erodibility and critical shear stress are not related to each other, but both are needed to characterize the soil's resistance to overland flow. They propose that channel erodibility is the most appropriate parameter to represent the differences in soil erosion resistance under various soil and environmental conditions (Knapen, Poesen and De Baets, 2007).

7.4 Rill and Gully Development

There is considerable evidence to demonstrate that almost all cases of erosion involve rilling and/or gullying as the dominant process (Boardman, 2006). Rills are impermanent channels that vary in lateral position from year to year, over slopes of fine material (Bull and Kirkby, 1997). They may have a definable stream head which indicates the upstream extent of concentrated flow at a particular time (Dietrich and Dunne, 1993). Rills have a seasonal cycle of development and destruction (Schumm, 1956a, 1956b; Schumm and Lusby, 1963) and may be infilled by desiccated crust material, or by cultivation or other disturbance (Bull and Kirkby, 1997). Rill networks are important for soil erosion since sediment transport rates increase dramatically once rill incision occurs (Loch and Donnollan, 1983; Brunton and Bryan, 2000). Gullies are larger than rills and are more likely to be permanently incised channels, usually a V or U-shaped incision with side slopes close to the angle of rest of unconsolidated debris (Bull and Kirkby, 1997; Kirkby and Bracken, in press). Gullies are often defined for agricultural land in terms of channels too deep to easily remove with ordinary farm tillage equipment and typically range from 0.5 m to as much as 25–30 m deep (Poesen *et al.*, 2002). There have been many classifications of gullies based on form and geographical location (see

Imeson, Kwaad and Verstraten, 1982; Planchon, Fritsch and Valentin, 1987; Oostwoud Wijdenes and Bryan, 1991; Dietrich and Dunne, 1993; Poesen *et al.*, 2002). The transition from rill to gully represents a continuum, and any classification of hydraulically related erosion forms into separate classes, such as microrills, rills, megarills, ephemeral gullies and gullies, is, to some extent, subjective (Grissinger, 1996a, 1996b; Poesen *et al.*, 2002).

Rill and gully erosion are serious problems and are responsible for the destruction of agricultural land and structures such as roads, bridges and pipelines, and there is growing interest in off-site impacts of the sediment produced (Boardman *et al.*, 1994; Valentin, Poesen and Li, 2005). Off-site impacts include detrimental effects on water quality, reservoir capacity and floodplains (e.g. Oostwoud Wijdenes *et al.*, 2000; Vandekerckhove *et al.*, 2001; Wasson *et al.*, 2002; Krause *et al.*, 2003; De Vente, Poesen and Verstraeten, 2005). Estimates of the contribution of rill and gully erosion to overall soil erosion range from 70–80% (Boardman, 1998), 80% (Wasson *et al.*, 2002), to 90–98% Krause *et al.*, 2003). In any channel network, approximately half the total length of channels is in unbranched (first order) fingertip tributaries and environmental changes that promote channel extension can therefore have a large impact on the landscape. During discharge events channel heads may advance great distances upslope, or retreat downslope if the hollow refills. In extreme cases, gullies can grow in length by tens of metres per year, and may also incise their channels creating steep ravine banks (e.g. Nachtergaele *et al.*, 2002; Tucker *et al.*, 2006). Measurement of large gully heads in southeast Spain illustrates this well, with small rates of retreat occurring by erosion of flutes and prominences between large floods events, and more substantial retreat in conjunction with larger floods (Figure 7.2).

Traditionally there are two conceptual approaches to understanding channel initiation; the stability approach (Smith and Bretherton, 1972) and the threshold approach (Horton, 1945). The stability approach is derived from the conservation of sediment mass at the channel head and emphasizes that at this point in the landscape sediment

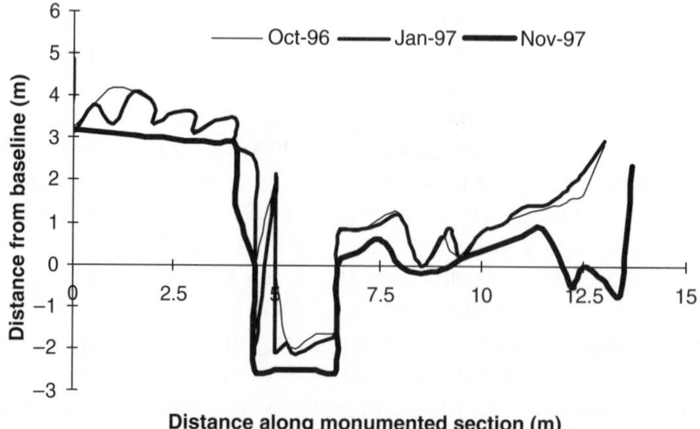

Figure 7.2 The episodic nature of gully head retreat for a box shaped gully head developed in marls in the Torrealvilla cathcment in southeast Spain. Lines represent the planform of the gully head

transport increases faster than linearly downslope. This usually requires wash processes to dominate. The threshold approach takes the view that the channel head represents a point at which processes not acting upslope become important. The balance of sediment determines whether the channel head becomes stable or migrates, but changing process domains drive incision. However, it is not clear if instability results from a change in process at the headcut, or a change in the intensity of the process operating, or a variation in the spatial distribution of processes that cause incision. More recently the two approaches have been combined in a single approach published by Kirkby (1994). The different approaches tend to be better suited to different environments and determine the two extremes of a range of factors that combine to produce channel heads. Current work has produced a family of continuous models that provide elementary theory of evolution of fluvial landscapes including the emergence of channelized flows, development of stable valleys, erosion of surfaces, relationships between surface forms and flow and landform variability (Smith, Birnir and Merchant, 1997; Smith, Merchant and Birnir, 1997). Two different approaches towards the areal modelling of gully system development agree in treating the ratio of advective (channel) to diffusive (sideslope) processes as a key determinant of the morphology of a gully system as it evolves (Kirkby and Bracken, in press).

Gullying is dependent on a wide range of factors including slope, critical drainage area, soil and lithology, piping, land-use change, construction of roads and buildings and climate change (see Valentin, Poesen and Li (2005) and Poesen *et al.* (2003) for more details). These factors have been extensively researched: however, few studies have attempted to model rill and gully development due to the rapidly changing hillslope morphology involved (Valentin, Poesen and Li, 2005). Willgoose, Bras and Rodriguez-Iturbe (1991) developed a coupled hillslope-channel model of catchment development based on the Hortonian threshold concept; highly non-linear interactions between channel head advance and contributing hillslope lead to network growth but the model was highly sensitive to initial conditions (Bull and Kirkby, 2003). Kirkby (1994) combined both the instability criteria and Horton threshold criteria in a model of stream head location which forecasts a constant relationship between slope and contributing area at stream head locations where instability is dominant (semi-arid environments), but a strong inverse relationship with the distance to the stream head where tractive thresholds dominate (temperate environments). The drainage basin model of Howard (1994) was also based on a combination of the threshold and mass conservation concepts. Both fluvial and slope processes can occur in each cell and the location of channel heads is defined by a morphometric criterion (Bull and Kirkby, 2003). Kirkby and Bull (2000) developed a model (see Figure 7.3 for the model outline) to generate simulated landscapes which has much in common with these other models. The model developed valley networks with plausible drainage densities, long profiles and hillslope profiles, even though the model formulation did not explicitly make a demarcation between 'hillslopes' and 'channels' (Figure 7.4). Smith, Birnir and Merchant (1997) and Smith, Merchant and Birnir (1997) developed the original concept proposed by Smith and Bretherton (1972) by unifying aspects of continuous, discrete and variational approaches to landscape modelling. More recently Sidorchuk (2005) proposed a new method of predicting detachment rate based on cohesive soil erosion for use in stochastic gully models, but also demonstrated that self-organized criticality could be used to understand gully development (Sidorchuk, 2006).

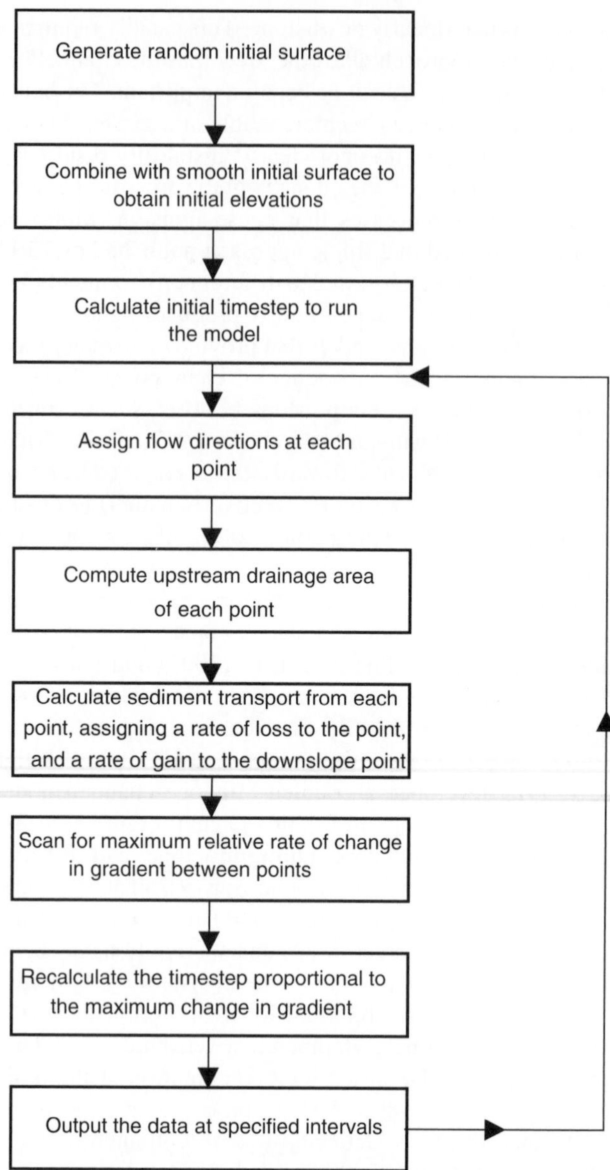

Figure 7.3 Outline flow diagram for the Kirkby and Bull (2000) two-dimensional gully simulation model

7.5 Badland Development

The term 'badland' is used to describe intensely dissected areas of unconsolidated sediments or poorly consolidated bedrock, with little or no vegetation, that are useless for agriculture (Gallart *et al.*, 2002; Faulkner, 2008). This is often the landscape produced when rill and gully erosion remains unchecked. Drainage density is very high, and is

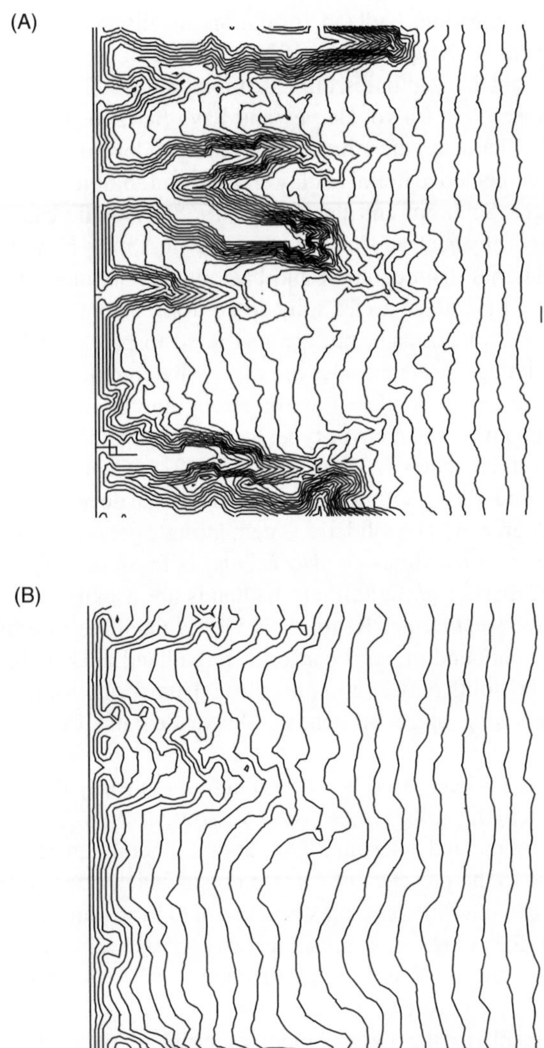

Figure 7.4 Example model outputs of simulated gullies from Kirkby and Bull (2000); (A) Effective bedload fraction (EBF) = 0.1 after 1000 years; (B) EBF = 1.0 after 10 000 years. Other parameters and initial surface are identical

characterized by steep V shaped channels. Badlands are different from rills and gullies in that they are not only linear erosive forms, but include hillslopes and divides usually carved in soft bedrock, although both forms may be closely related (Nogueras *et al.*, 2000; Gallart *et al.*, 2002; Faulkner, 2008). Badlands are frequently considered to be landscapes characteristic of dryland areas, but can also occur in wetter areas with high topographic gradients, bedrock weakness and high-intensity rainstorms (Schumm, 1956a). Usually, badlands consist of mosaics of physiographic units where all the stages can be found between the bare parent material at the bottom and/or sides of gullies, and soils with some kind of vegetation cover in more stable positions (Gallart *et al.*, 2002; Faulkner, 2008).

The main factors controlling badland formation are lithology, soils, mass movement processes and rilling and gullying. Lithology is a major factor for badland production, and is probably more important than tectonics, climate, topography or land use (Gerits *et al.*, 1987; Imeson and Verstraten, 1988; Calvo, Harvey and Paya-Serrano, 1991; Calvo *et al.*, 1991; Gallart *et al.*, 2002). Characteristics that favour badland development include unconsolidated or very poorly cemented material of clay and silt, and soluble minerals such as gypsum or halite (Scheidegger, Schumm and Fairbridge, 1968). Specific characteristics, such as structure, mineralogy, physical and chemical properties, may play either a primary or secondary role in material disintegration and badland development (Gallart *et al.*, 2002; Faulkner *et al.*, 2004). In badland areas, soils do not form a continuous three-dimensional body covering the entire landscape, but are restricted to patches where processes have allowed soil accumulation. These mosaics are especially complex in semi-arid Mediterranean landscapes. In Tabernas, southeast Spain, strong interrelationships have been found between soil development, ground cover type, hydrological and geomorphological behaviour, and topographical attributes (Alexander *et al.*, 1994; Cantón, 1999). The characteristics of the material underlying soils are thus crucial for the development of badlands, and these include the degree of consolidation, cementing agents, particle size range and distribution. Weathering of bedrock is also needed before erosion processes can act to produce badlands (Gallart *et al.*, 2002) and badlands are usually carpeted with regolith, produced by intensive weathering. Regolith is very vulnerable to erosion because it is usually composed of unbound mineral particles. In badland areas with moderate erosion rates, the evolution of regolith may be complex through the formation crusts, either physical or biological incorporating algae or lichens (Alexander and Calvo, 1990; Solé-Benet *et al.*, 1997).

Rainfall simulation has been used by several authors to study the response of badland surfaces to rainfall (Scoging, 1982; Imeson and Verstraten, 1988; Calvo *et al.*, 1991; Solé-Benet *et al.*, 1997; Bouma and Imeson, 2000; among others). On most badland surfaces, infiltration is low due to the presence of surface crusts which effectively seal the surface, although the response to rainfall is complex because of high spatial and temporal variability of regolith properties. Runoff response is usually fast and has been shown to be less than four minutes in some cases (Imeson, Kwaad and Verstraten, 1982; Calvo *et al.*, 1991; Solé-Benet *et al.*, 1997) (Figure 7.5). Microrelief patterns due to microrills, crustose lichens, pedestals and pinnacles can result in the development of efficient pathways which follow the microrelief (Imeson and Verstraten, 1988; Solé-Benet *et al.*, 1997). However, badland regoliths may allow high infiltration rates when there are open cracks or highly porous popcorn structures. On badland surfaces with deep cracks, true overland flow may be rare, and crack flow can feed rills and main channels with water and sediment during storms which are not long enough to lead to crack closure (Yair and Lavee, 1985). Therefore runoff during storms is a complex phenomenon that depends on lithology, antecedent conditions and rainfall characteristics (Hodges and Bryan, 1982; Regüés, Pardini and Gallart, 1995).

7.6 Soil Erosion on Agricultural Land

Agricultural land experiences almost constant degradation and some of the rates of erosion are high enough to give cause for concern (Boardman, 1990). Many studies have shown that

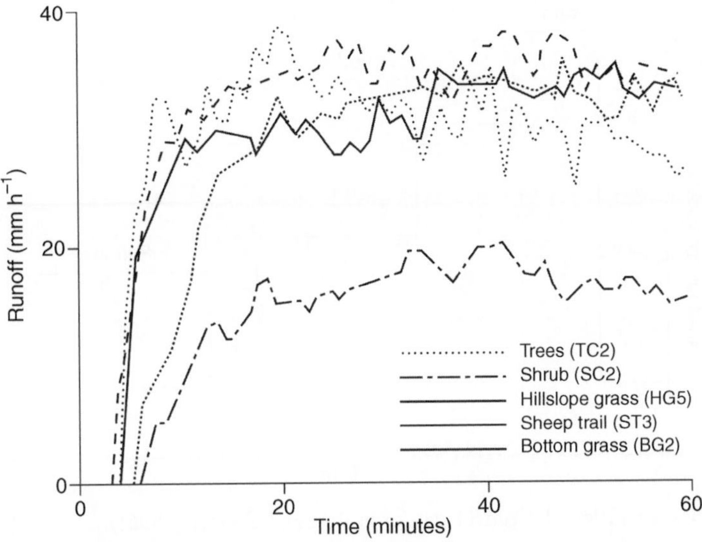

Figure 7.5 Runoff response from plot experiments (after Cerdà *et al.* (1998)).

changing from natural vegetation to cultivation can increase soil erosion rates by an order of magnitude or more (Morgan, 1986; Abernethy, 1990; Walling, 1999). Many agricultural soils are now at risk of erosion, not only due to climate change but due to a lack of awareness by farmers and policy makers (Boardman, 1990; Boardman *et al.*, 1994). Farmers are acutely influenced by economic incentives, react very quickly to changes in price and will hence adapt crops and practices to changes in market process and quota arrangements (Boardman, Poesen and Evans, 2003). Many factors affecting soil erosion are exacerbated on agricultural land. For instance, the use of heavy machinery causes compaction of the upper layers, farming practices encourage loss of organic matter, more and more crops are sown in autumn resulting in bare soil over the winter, and field sizes are increasing (Boardman, 1990). Subsidies are also encouraging cereals to be grown on unsuitable land, which is also likely to increase soil erosion (Burt, 2001). Within these general trends there are spatial and temporal variations, with some soils being more prone to erosion than others, and tillage cycles of different crops interacting in various ways with weather conditions (Burt, 2001). However, in general autumn-sown cereals and spring-sown maize are particularly problematic (Heathwaite and Burt, 1992; Boardman *et al.*, 1995; Burt and Slattery, 1996). Boardman (1992) describes the 'window of opportunity' as the important factor in promoting/preventing soil erosion caused by tillage–weather interactions, which is related to the time for which soil remains bare without a protective vegetation cover.

One of the main differences between agricultural soils and undisturbed areas is the use of repeated soil management practices such as ploughing. Imeson and Kwaad (1990) identified three periods in the evolution of a tilled layer from the starting point of a freshly ploughed field (Figure 7.6). Period 1: ploughing results in loose clods with low bulk density and high infiltration capacity. Once tillage is completed the soil starts to settle under its own weight and aggregates start to break up. The duration of this period is short and only lasts until the first rainfall. Period 2: during rainfall the surface is compacted more rapidly; raindrops

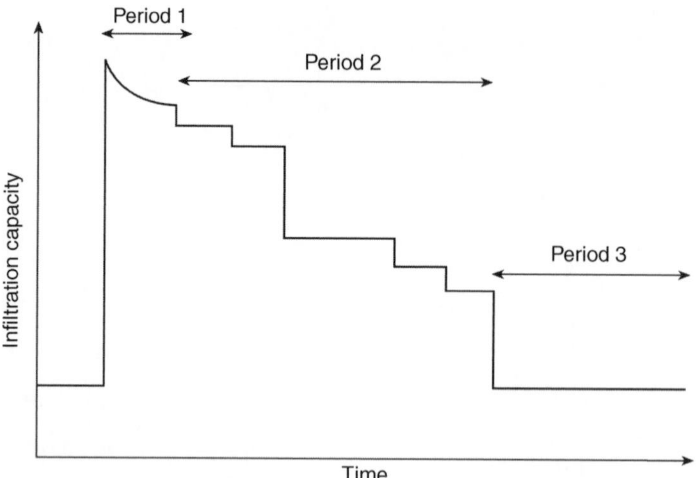

Figure 7.6 The evolution of a tilled layer (after Burt and Slattery (1996)).

effectively hammer the soil and clays swell to degrade the surface, resulting in a decrease in porosity and lower surface permeability. Degradation takes place discontinuously depending on the magnitude and frequency of rainfall events. Both structural and depositional crusts form. Hence infiltration capacity decreases and runoff and erosion increase, although desiccation and freeze–thaw may reverse these trends (Burt and Slattery, 1996) and soil strength and aggregate stability vary (Slattery, Burt and Boardman, 1994). Period 3: Imeson and Kwaad (1990) argued that there is no further change until tillage takes place once again after crusting and maximum compaction have occurred. However, Burt and Slattery (1996) disagreed and found that wetting and drying, freeze–thaw, microbial activity and cracking increased infiltration rates by breaking up the crust.

Due to the importance of soil erosion on agricultural land, many studies have focused on trying to measure and predict amounts of sediment removal and transfer. Middleton (1930) was one of the first to propose indices of soil erodibility combining properties affecting runoff and particle detachability. The Universal Soil Loss Equation (USLE) (Wischmeier and Smith, 1978) was one of the first methods of estimating soil erosion on agricultural land and is probably the most popular. The USLE is an empirical relationship derived from more than 8000 plot-years of erosion research data from agricultural land of low slope angle, to investigate long-term patterns and trends in soil erosion (Mitchell and Bubenzer, 1980; Sonneveld and Nearing, 2003). The USLE equation is as follows:

$$A = R \times K \times L \times S \times C \times P \tag{7.2}$$

where A represents soil loss (ha^{-1} yr^{-1}), R refers to the rainfall erosivity factor (a function of the kinetic energy of a storm and its 30 minute maximum intensity), K is the soil erodibility factor which reflects the susceptibility of a soil type to erosion (per unit of the R factor), L is an index of slope length, S is a slope gradient index, C is an index for the protective effect of the vegetation canopy and organic material in contact with the ground, and P is a factor that represents soil conservation measures that attempt to control erosion such as terraces and contour ploughing. Variables L, S, C and P are all expressed relative to original plot characteristics, that is a 22.6 m length slope, on a 9% slope, tilled land under clean tilled continuous fallow conditions, with up- and downslope ploughing (Sonneveld and Nearing, 2003).

The USLE has a number of advantages that has cemented its dominance as a method for estimating fluvial erosion and planning of soil conservation measures. The structure of Equation 7.2 is very simple and hence allows simple policy planning by changing land use (C and P) under given environmental conditions (R, K, L and S). The USLE also has relatively simple data requirements compared to contemporary physical-based models. Hence, the USLE has been applied at a number of levels from small watersheds up to continental scales (Sonneveld and Nearing, 2003).

However, there are a range of reasons why the application of the USLE is limited: (i) data from which the model was derived were solely based in the USA and application to all other environments is not necessarily correct (Morgan *et al.*, 1998); (ii) data on higher soil losses are scarce in the original data set so large errors can be expected for areas of high rainfall, in combination with steep slopes and high erodibility (Keyzer and Sonneveld, 1998; Sonneveld and Nearing, 2003); (iii) Nearing (1998) has shown that there may be problems with small soil losses consistently overestimated; (iv) models are scale dependent and were developed for the plot scale, so should only be applied at this scale due to changing dominance of processes as scale increases (Kirkby, 1999); (v) the model only describes the aspects of soil erosion that are well understood and the USLE is therefore incomplete (Römkens, Helming and Prasard, 2001); (vi) the multiplication of the six factors results in the magnification of errors when any factor is incorrect (Wischmeier, 1976); and (vii) the USLE only predicts the amount of soil moved on a field and not the sediment flux evacuated from a field (Trimble and Crosson, 2000).

Many attempts have been made at improving the predictive capability of the USLE. The most extensive work is probably that of Renard *et al.* (1998) in developing the Revised Universal Soil Loss Equation (RUSLE). The changes encompassed in the RUSLE include incorporation of new and more accurate data alongside more detailed consideration of specific processes (Sonneveld and Nearing, 2003). The RUSLE includes sub factors of vegetation cover, a new map for the rainfall erosivity factor, addition of seasonal variability in the soil erodibility factor, the inclusion of rill processes and a wider range of management practices (Renard *et al.*, 1991, 1998). These changes have increased the scope of application of the ULSE, but it is more difficult to assess the changes in accuracy (Sonneveld and Nearing, 2003), with a number of investigations suggesting the RUSLE is both more and less accurate than the USLE (e.g. Tiwari, Risse and Nearing, 2000). Other attempts to improve the USLE have included modifying the USLE to make it more applicable to a range of environments (e.g. the [modified] MUSLE (Glaetzer and Grierson, 1987), the use of fuzzy logic-based modelling (Tran *et al.*, 2002), using artificial neural networks to predict soil loss from natural plots (Lieznar and Nearing, in press), and mathematical model transformations (Sonneveld and Nearing, 2003). However, these modifications have not yet been tested in such a wide range of investigations as the USLE to be certain about their potential. It is important to note that as Trimble and Crosson (2000) suggest, the limitations of the USLE alongside new developments, leave us without an accurate idea of actual soil erosion occurring.

7.7 Transfers of Sediment Throughout Drainage Basins

Once soil is eroded, it can be redeposited locally or it can be moved off site. The pathways by which this occurs are determined by the nature of soil, bedrock and precipitation inputs, with the transport of soil roughly following the movement of flow. Many investigations into sediment movement have been undertaken at many scales and the importance of scaling is a focus for current research. Plots have been used to determine many of the variables influencing runoff production and soil erosion; however, at this scale storage of sediment and runoff is ignored (Boardman, 1998; Trimble and Crosson, 2000). Traditionally the link between hillslope and stream has been provided by the sediment delivery ratio (SDR):

$$\mathrm{SDR}\,(\%) = \mathrm{SY}/\mathrm{EROS} \tag{7.3}$$

where SY is the sediment yield ($\mathrm{t\,km^{-2}\,y^{-1}}$) and EROS is the catchment gross erosion ($\mathrm{t\,km^{-2}\,y^{-1}}$). It is generally accepted that the sediment delivery ratio decreases downstream as depositional opportunities increase with drainage basin area (Walling, 1983; Knighton, 1987; Trimble and Crosson, 2000). However, this view has been challenged by Church and Skaymaker (1989), who concluded that sediment yield increased with catchment area for rivers in British Columbia due to contributions from bank erosion and channel migration. More recent work at the event timescale for the upland UK indicated that suspended sediment concentrations increased downstream, but that sediment yield decreased in the headwaters, then increased towards the lower basin (Bull *et al.*, 1995). Bull *et al.* (1995) and Michaelides and Wainwright (2002) highlight the issues of changing connectivity with changing scale in drainage basins. In small low-order tributaries there is strong, direct coupling between hillslopes and channels, but as basin size increases there is an increasing lag between hillslope and channel response to storms and more available storage for both water and sediment fluxes. Although there is recognition of a threshold separating small and large drainage basins, the threshold area has never been identified (Phillips, 1995).

While the SDR is a useful concept it is generally agreed that it is a crude and unsatisfactory tool (Wolman, 1977; Walling, 1983; Trimble and Crosson, 2000). Most problems with using the sediment delivery ratio concept are concerned with the use of a simple relationship between yield and gross erosion and the necessary spatial and temporal lumping involved (Walling, 1983, 1988; Slattery and Burt, 1995). At the global scale estimates of changing sediment delivery ratio and sediment yield rely on calculations from representative basins. It is commonly accepted that sediment yields are increasing at all scales, from the catchment (Abernethy, 1990) to the global scale (Dedkov and Mozzherin, 1984; Milliman *et al.*, 1987) due to land disturbance, increasing population and increasing intensification of farming resulting in high rates of soil erosion. However, not all catchments are experiencing the same magnitude of impacts and all basins exhibit variability in response. For instance Alford (1992) reports a study of the Chao Phraya River in Thailand which has not seen an increase in sediment yields over the period 1950–1980. At smaller spatial scales is has also been proposed that sediment yields are unlikely to provide realistic estimates of erosion rates, especially on fields (Boardman and Favis-Mortlock, 1992; Evans, 1993).

The sediment budget was established as an alternative tool to elucidate and quantify the linkages between erosion processes operating and sediment yield evacuated from a catchment (e.g. Dietrich and Dunne, 1978; Trimble, 1983; Sutherland and Bryan, 1991), which is vital for investigating the impact of land-use change on erosion and sediment transfer (Walling, 1999). The precise form of a sediment budget varies with local physiographic conditions, and variations can occur over relatively small areas (Walling, 1999). Figure 7.7 presents the sediment budget for Coon Creek, Wisconsin (area of 360 km^2) over two different time periods and highlights changing amounts of sediment storage in the catchment with changing land use (from Trimble, 1983). Despite work of this nature, the sediment budget has remained essentially conceptual because it is difficult to assemble data on rates and fluxes involved for anything other than a small basin, but also a detailed understanding of sediment delivery processes cannot be achieved using aggregated or lumped data (Walling and Webb, 1992; Slattery and Burt, 1995; Walling, 1999). Fingerprinting of sediment sources has been an important area of research in trying to understand the sources of sediment and pathways within catchments (see Walling, 1988, 1999; Slattery and Burt, 1995; Walling and Woodward, 1995), but an alternative direction has been to investigate the mosaic of landscapes with different units having different responses to runoff and hence different rates and patterns of soil erosion. Further discussion of sediment budgets can be found in Chapter 11 (this volume).

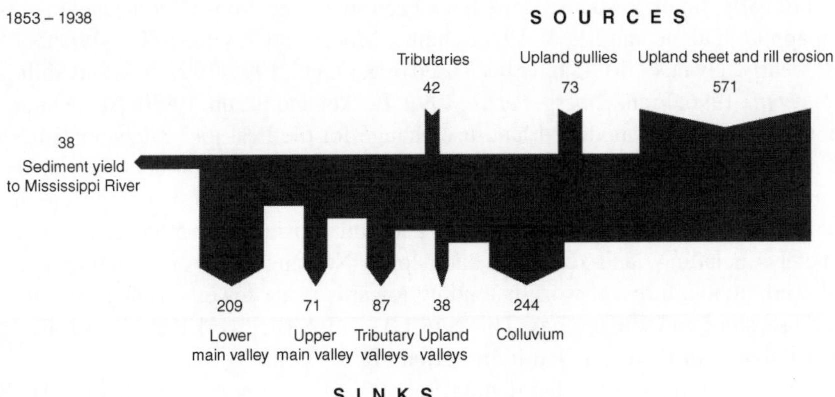

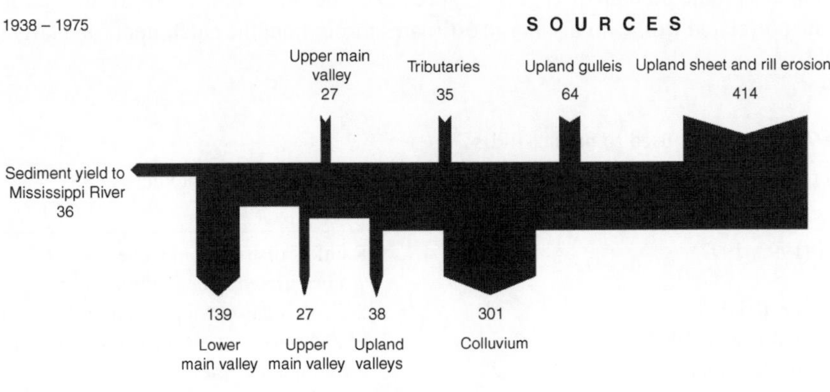

Figure 7.7 An example of a sediment budget (after Trimble (1983)).

In an attempt to spatially distribute runoff and sediment transfer, research has focused on portioning areas of the catchment producing runoff. For a storm to initiate catchment-scale runoff, it must overcome the spatial arrangement and threshold values of hydrologically similar surfaces at all scales. Previous research has concentrated on defining hydrological response units (HRUs): distributed, heterogeneously structured entities having a common climate, land use and underlying pedological–topographical–geological association controlling their hydrological dynamics (Flügel, 1995). These have had a variety of names including hydrologically similar units (HSUs) (Karnoven *et al.*, 1999), grouped response units (GRU) (Kouwen *et al.*, 1993), representative elementary areas (REAs) and hydrologically similar surfaces (HYSS) (Bull *et al.*, 2003). Response units have been found at a range of scales from plots (Bergkamp, Cammeraat and Mertinez-Fernandez, 1996) to hillslopes (Cerdà, 1995) to catchments (Imeson *et al.*, 1992; Yair, 1992). The mosaic pattern formed by these units is scale independent (Fitzjohn, Ternan and Williams, 1998), with the units found at one scale forming one level in a nested, hierarchical scalar system, which increases in complexity with scale (Fitzjohn, Ternan and Williams, 1998). It is assumed that sediment transfer will follow the pathways of water, yet this has not been explicitly investigated in these studies.

Traditionally, runoff response units have been identified from field measurements and field mapping (Dunne and Black, 1970; Dunne, Moore and Taylor, 1975; Myrabo, 1997); more recently, advances in GIS have been used (e.g. Flügel, 1995, 1997; Still and Shih, 1985; Sharma *et al.*, 1996; Frankenberger *et al.*, 1999; Becker and Braun, 1999). At the large scale Karnoven *et al.* (1999) modelled land-use change for the Lestijoki catchment in Finland (1290 km^2) using HRUs as inputs to a channel network model. A variety of variables have been used to estimate HRUs (Table 7.4), ranging from variables that are complex to monitor and estimate such as rainfall distribution and soil moisture, to more easily estimated parameters including land use, soil and slope. No single set of variables has been established, although recent work is tending towards using fewer variables and focusing on land use, slope and soil/geology. This contrasts with earlier work that tried to encompass more variables and more detailed information (Table 7.4).

Research in southeast Spain has demonstrated that is it possible to predict key HYSS by using GIS techniques to re-class the DEM along with maps of geology and land use into areas of high, medium and low potential runoff (Figure 7.8). These maps are then combined to produce a predicted map of HYSS (Figure 7.9). The predicted HYSS for the Rambla de Nogalte correlated well with discharge estimates throughout the catchment for a large flood

Table 7.4 Variables used to predict HRUs

Author	Number of variables	Variables used to predict HYSS
Flügel (1995, 1997)	6	Rainfall distribution, slope, aspect, land use, unsaturated zone, bedrock zone
Sharma *et al.* (1996)	4	Soil, vegetation, slope, rainfall distribution
Becker and Braun (1999)	3	Evaporation, runoff, groundwater recharge
Frankenberger *et al.* (1999)	3	Soil, land use, elevation
Karnoven *et al.* (1999)	3	Soil, land use, slope
Bull *et al.* (2003)	3	Geology, land use, slope

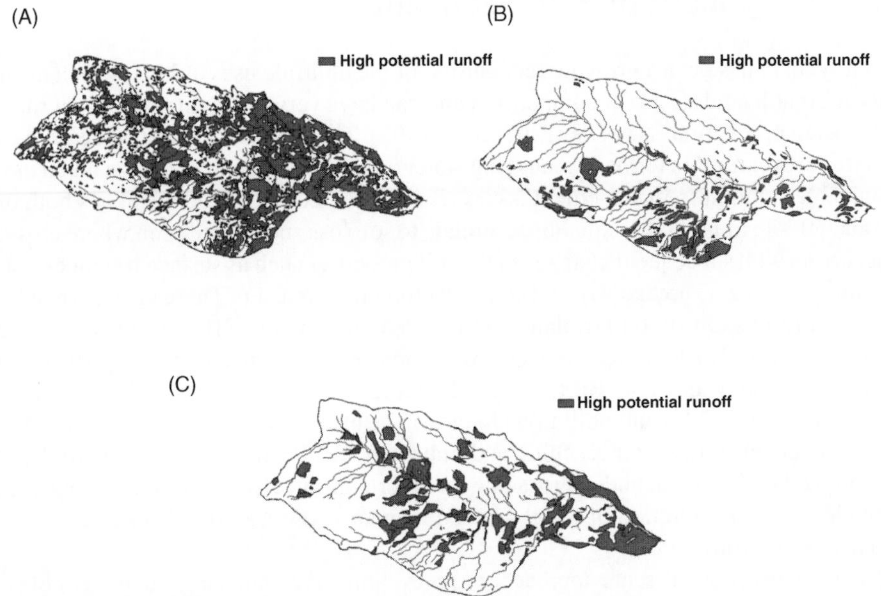

Figure 7.8 Maps of high runoff potential caused by characteristics of (A) geology, (B) slope and (C) land use

in September 1997. However, some areas identified as key source areas using GIS were not important in terms of flood generation. This was because of a lack of connectivity between the source area and the main channel. Disconnection was caused by long distances with high transmission losses, the construction of check dams or the existence of terraces.

Figure 7.9 The hydrologically similar surface (HYSS) classifications for the Rambla Nogalte catchment in southeast Spain

7.8 Management of Fine Sediment

Recently there has been a growing recognition of the multiple uses of land, which means that different land uses are in competition and can have very different impacts on runoff and erosion production (Newson, 1997; Burt, 1999). The move from an attitude of 'cure' towards 'prevention' in terms of managing water quantity and quality is also a much more sustainable solution to management issues (Burt, 1999, 2001). Soil erosion has both on-site and off-site impacts. On-site impacts relate to soil loss and degradation where erosion is occurring with subsequent changes in the soil properties such as surface roughness, soil chemistry, aggregate breakdown and crust development/break up. These changes result in loss of nutritious topsoil on farmland and reduced crop yields. Off-site impacts include transfer of sediment to water courses, deposition of sediment clogging up rivers and reservoirs (Heathwaite and Burt, 1992), flooding of properties by soil-laden runoff (Boardman *et al.*, 1994) and pollution (Johnes and Burt, 1993; Harris and Forster, 1997), which can be much more far reaching and impact on local communities. The problem is not confined to farmland, and overgrazing on moorland can also result in erosion of peat which leads to discoloration of water and sedimentation of reservoirs (Burt, Labadz and Butcher, 1997; Burt, 2001).

There are many small-scale approaches to preventing and managing small areas of soil erosion, such as using organic mulches to prevent further erosion (Smets, Poesen and Knapen, 2008), sediment trapping, grade stabilization, check dams and stone bunds. Moving up to the field scale, measures such as changing vegetative cover and practising conservation tillage (Knapen, Poesen and De Baets, 2007; Knapen *et al.*, 2008) have successfully been employed to limit soil loss. The most successful and most often used technique to prevent off-site impacts of soil erosion is the use of riparian buffer zones (RBZs). Research into buffer zones took off in the 1970s (Correll, 1997), and although RBZs may be better known for their function in nitrate and pollutant removal and transformation, the role of RBZs in trapping sediments has also been widely reported (e.g. Karr and Schlosser, 1978; Schlosser and Karr, 1981a, 1981b; Correl, 1997). (Detail on the use of RBZs for uses other then sediment removal can be found in Haycock *et al.* 1997)). Buffer zones function as a sediment trap and effectively reduce the coupling between hillslopes and channels and hence prevent eroded soil moving out of the catchment. Riparian vegetation facilitates the removal of suspended sediment from overland flow entering from upslope areas (Mitsch, Dorge and Wiemhoff, 1977; Peterjohn and Correll, 1984; Chescheir *et al.*, 1991). Once floodplains become sufficiently wide, they become an important source of runoff in their own right through saturation excess overland flow, as well as providing a pathway for slope runoff (Burt, 1997). The major sediment removal mechanisms associated with RBZs involve changes in flow hydraulics, which enhance the opportunity for infiltration of runoff and fine sediment into the soil, sediment deposition and filtration of sediment by the vegetation (Malanson, 1993; Dillaha and Inamdar, 1997). Hence, runoff needs to pass through the RBZ slowly to provide sufficient contact time for the removal mechanisms to function (Dillaha and Inamdar, 1997) (Figure 7.10).

The sediment trapping effects of riparian buffer zones (RBZs) are highly dependent on the volume and pathway of water moving through this zone (Burt, 1997; Correll, 1997). For example, overland storm flows may be effectively cleaned by forest riparian buffer zones when runoff is dominated by sheetflow from relatively small fields with slopes of less than 5% (Peterjohn and Correll, 1984), but when fields are larger and flow is concentrated

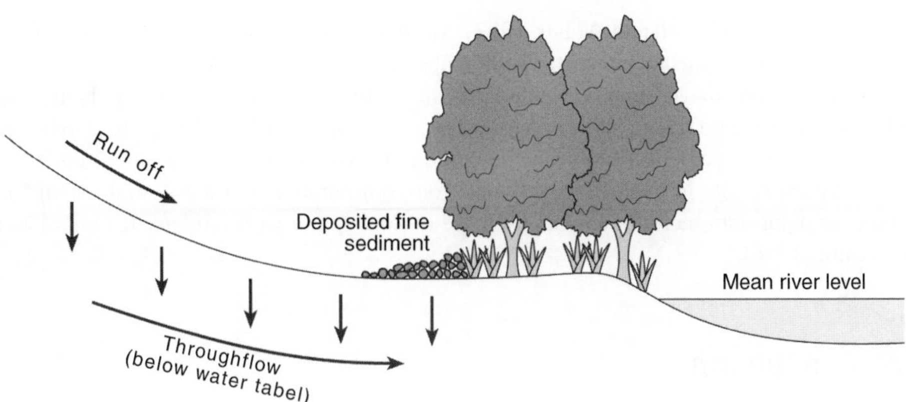

Figure 7.10 Riparian buffer zones used for managing water and sediment transfers

channels can be eroded through buffer zones (Jordan, Correll and Weller, 1993). There has been some success in treating rills and gullies by using level spreaders and grassed filter strips prior to forested buffers (Franklin, Gregory and Smolen, 1992). Riparian buffer zones for management of sediment transport are most effective in low-order tributaries due to greater interaction between upland runoff and the riparian zone (Lowrance, Vellidis and Hubbard, 1995), whereas RBZs for pollutant and nitrate removal have the most potential in high-order tributaries where a floodplain has formed and there is opportunity to decouple the hillslope–channel system (Burt, 2001). In low-order catchments, there is particular opportunity for deposition of eroded soil by the use of barriers to overland flow such as hedgerows (Burt, 2001).

Much work has looked at the efficiency of different types of RBZs. Grass or dense herbaceous vegetation has been found to be more effective at trapping particles from overland flow (Osborne and Kovacic, 1993; Dillaha and Inamdar, 1997), although the litter layer is also crucial in reducing runoff velocity (Correll, 1997). For instance Niebling and Alberts (1979) showed that a 5 m wide grass buffer reduced sediment discharge in sheetflow by 90% and high trapping efficiency of grass has been generally reported, although this decreases once submerged (Hayes, Barfield and Barnhisel, 1979; Young, Huntrods and Anderson, 1980; Tollner, Barfield and Hayes, 1982; Dillaha and Inamdar, 1997). Most sediment deposition occurs just upslope of the buffer due to the abrupt change in land use, and then a wedge-shaped deposit develops through the buffer over continued storms (Niebling and Alberts, 1979; Hayes, Barfield and Barnhisel, 1979; Tollner, Barfield and Hayes, 1982; Dillaha and Inamdar, 1997), with RBZ efficiency decreasing over time (Magette *et al.*, 1989). Importantly, the effectiveness of RBZs has shown to be less for concentrated flow (Dillaha *et al.*, 1989a) and questions have therefore been raised about using grass buffers in steep areas (Dillaha, Sherrard and Lee, 1989b).

Development and management of RBZs needs to be monitored otherwise the desired sediment sink can become a sediment source. Jordan *et al.* (1993) found that a buffer from a no-till cornfield was a net sediment source because of active erosion in the channel formed by the runoff as it incised through the buffer zone. This was probably caused by limited sediment load in the flow due to the no tillage and limited opportunity for deposition elsewhere (Jordan, Correll and Weller, 1993). Similarly Smith (1992) found that after eight years of growth in a forested buffer, the sediment eroded in the RBZ was more than twice the amount than from the pasture it was controlling. This was caused by the reduction in herbaceous plants and weeds beneath the canopy cover (Smith, 1992). There are also issues concerning the impact of runoff on vegetation growth. Research has suggested that sedimentation results in expansion of vegetation communities, but the spatial expansion of pioneer plants can be limited by an increase in flow velocity (Tsujimoto, Kitamura and Murakami, 1996).

7.9 Conclusion

Understanding overland flow and soil erosion is a key problem for the future, especially learning to ask the 'right' questions (Boardman, 2006). While long-term records provide evidence of the sediment loads in rivers both increasing and decreasing, land clearance, increasing intensification of land use and population growth tend to result in increased sediment delivery in river catchments (Abernethy, 1990), leading to problems of flooding, pollution and sedimentation. Cases where sediment loads have not increased are likely to be in areas where the fluvial system has been able to buffer effects of land-use change and hold sediment in storage (Walling, 1999), although, once these stores of sediment are effectively filled, the system will throughput higher amounts of sediment with the associated problems. Abernethy (1990) also suggested that problems of sediment delivery are particularly acute in developing countries where suspended sediment can be expected to double every 20 years due to population growth alone. Hence understanding sediment erosion and transport in the environment must remain a key direction for research, but must also be coupled with efficient management techniques and practices. The use of vegetation is generally accepted as an important tool in sediment management, offering both surface protection and sediment trapping capabilities. However, in some environments such as drylands, there is little vegetation, it tends to be scrubby and hence does not offer continuous surface protection and has limited effectiveness in trapping moving sediment. Therefore some of the geographical locations that are facing increased challenges of erosion and sediment control are areas where existing management techniques are of limited use. Even within temperate environments, we still know little about the actual pathways and time lags of sediment movement within fluvial systems. Hence we must continue to develop new methodologies and frameworks of understanding for investigating overland flow, soil erosion and sediment transport, and link these to more rigorous development of efficient management techniques for all environments.

References

Abernethy, C. (1990) The use of river and reservoir sedimentation data for the study of regional erosion rates and trends. International Symposium on Water Erosion Sedimentation and Resource Conservation, Dehradun, India, October 1990.

Abrahams, A.D. and Parsons, A.J. (1990) Determining the mean depth of overland flow in field studies of flow hydraulics. *Water Resources Research*, **26**, 501–503.

Abrahams, A.D. and Parsons, A.J. (1991) Relation between infiltration and stone cover on a semi-arid hillslope, S. Arizona. *Journal of Hydrology*, **122**, 49–59.

Abrahams, A.D. and Parsons, A.J. (1994) Hydraulics on interrill overland flow on stone covered desert surfaces. *Catena*, **23**, 111–140.

Abrahams, A.D., Parsons, A.J. and Luk, S.H. (1986) Resistance to overland flow on desert hillslopes. *Journal of Hydrology*, **88**, 343–363.

Abrahams, A.D., Li, G., Krishnan, C. and Atkinson, J.F. (1998) Predicting sediment transport by interrill overland flow on rough surfaces. *Earth Surface Processes and Landforms*, **23**, 1087–1099.

Abrahams, A.D., Li, G., Krishnan, C. and Atkinson, J.F. (2001) A sediment transport equation for interrill overland flow on rough surfaces. *Earth Surface Processes and Landforms*, **26**, 1443–1459.

Alexander, R.W. and Calvo, A. (1990) The influence of lichens on slope processes in some Spanish badlands, in *Vegetation and Erosion* (ed. J.B. Thornes), John Wiley and Sons, Chichester, pp. 385–398.

Alexander, R.W., Harvey, A.M., Calvo, A. *et al.* (1994) Natural stabilization mechanisms on badland slopes: Tabernas, Almeria, Spain, in *Environmental Change in Drylands: Biogeographical and Geomorphological Perspectives* (eds A.C. Millington and K. Pye), John Wiley and Sons, pp. 85–111.

Alford, D. (1992) Streamflow and sediment transport from mountain watersheds of the Chao Phraya basin, southern Thailand: a reconnaissance study. *Mountain Research and Development*, **12**, 257–268.

Ambroise, B. (2004) Variable 'active' versus 'contributing' areas or periods: a necessary distinction. *Hydrological Processes*, **18**, 1149–1155.

Anderson, M.G. and Burt, T.P. (1990) Subsurface runoff, in *Process Studies in Hillslope Hydrology* (eds M.G. Anderson and T.P. Burt) Wiley, Chichester, pp. 365–400.

Arshad, M.A. and Mermut, A.R. (1988) Micro morphological and physico-chemical characteristics of soil crust types in north-western Alberta, Canada. *American Journal of the Soil Science Society*, **52**, 724–729.

Arulanandan, K., Loganathan, P. and Krone, R.B. (1975) Pore and eroding fluid influences on surface erosion of soil. *Journal of Geotechnical Engineering Division, American Society of Civil Engineering*, **101**, 51–66.

Assouline, S. and Mualem, Y. (1997) Modelling the dynamics of seal formation and its effect on infiltration. *Water Resources Research*, **33**, 1527–1536.

Atkinson, J.F., Abrahams, A.D., Krishnan, C. and Li, G. (2000) Shear stress partitioning and sediment transport by overland flow. *Journal of Hydraulics Research*, **38**, 37–40.

Auzet, A.V., Poesen, J. and Valentin, C. (2001) Soil patterns as a key controlling factor of soil erosion. *Catena*, **46**, 85–87.

Bagnold, R.A. (1966) An approach to the sediment transport problem from general physics. *U.S. Geological Survey Professional Paper*, **422-J**.

Bagnold, R.A. (1980) An empirical correlation of bedload transport rates in flumes and natural rivers. *Proceedings of the Royal Society of London, Series A*, **372**, 453–473.

Becker, A. and Braun, P. (1999) Disaggregation, aggregation and spatial scaling in hydrological modelling. *Journal of Hydrology*, **217**, 239–252.

Bergkamp, G., Cammeraat, L.H. and Mertinez-Fernandez, J. (1996) Water movement and vegetation patterns on shrubland and an abandoned field in two desertification threatened areas in Spain. *Earth Surface Processes and Landforms*, **21**, 1073–1090.

Boardman, J. (1990) Soil erosion on the South Downs; a review, in *Soil Erosion on Agricultural Land* (eds J. Boardman, I.D.L. Foster and J.A. Nearing), John Wiley and Sons, pp. 87–106.

Boardman, J. (1992) Agriculture and erosion in Britain. *Geography Review*, **6**, 15–19.

Boardman, J. (1998) An average soil erosion rate for Europe: Myth or reality? *Journal of Soil Water Conservation*, **53** (1), 46–50.

Boardman, J. (2006) Soil erosion science: Reflections on the limitations of current approaches. *Catena*, **68**, 73–86.

Boardman, J. and Favis-Mortlock, D.T. (1992) Soil erosion and sediment loading of watercourses. *SEESOIL*, **7**, 5–29.

Boardman, J., Burt, T.P., Slattery, M.C. *et al.* (1995) Soil erosion and flooding as a result of a summer thunderstorm in Oxford and Berkshire, May 1993. *Applied Geography*, **16**, 21–34.

Boardman, J., Poesen, J. and Evans, R. (2003) Socio-economic factors in soil erosion and conservation. *Environmental science & policy*, **6**, 1–6.

Boardman, J., Ligneau, L., de Roo, A. and Vandaele, K. (1994) Flooding of property by runoff from agricultural land in north western Europe. *Geomorphology*, **10**, 183–196.

Bouma, N.A. and Imeson, A.C. (2000) Investigation of relationships between measured field indicators and erosion processes on badland surfaces at Petrer, Spain. *Catena*, **40**, 147–171.

Bracken, L.J. and Croke, J. (2007) The concept of hydrological connectivity and its contribution to understanding runoff dominated systems. *Hydrological Processes*, **21** (13), 1749–1763.

Brunton, D.A. and Bryan, R.B. (2000) Rill network development and sediment budgets. *Earth Surface Processes and Landforms*, **25**, 783–806.

Bradford, J.M., Remley, P.A. and Ferris, J.E. (1987a) Interrill soil erosion processes: I. Effect of surface sealing on infiltration, runoff and soil splash detatchment. *American Journal of Soil Science*, **51**, 1566–1571.

Bradford, J.M., Ferris, J.E. and Remley, P.A. (1987b) Interrill soil erosion processes: II. Relationahip of splash detatchment to soil properties. *American Journal of Soil Science*, **51**, 1571–1575.

Bryan, R. (1996) Erosional response to variations in interstorm weathering conditions. In M.G. Anderson and S.M. Brooks (Eds) *Advances in Hillslope Processes*, John Wiley and Sons, Chichester, 589–612.

Bryan, R.B. (2000) Soil erodibility and processes of water erosion on hillslope. *Geomorphology*, **32**, 385–415.

Bryan, R.B. and Poesen, J. (1989) Laboratory experiments on the influence of slope length on runoff, percolation and rill development. *Earth Surface Processes and Landforms*, **14**, 211–231.

Bryan, R.B. and Oostwoud Wijdenes, D. (1992) Field and laboratory experiments on the evolution of microsteps and scour channels on low angle slopes. *Catena Supplement Band*, **23**, 133–161.

Bryan, R.B. and Yair, A. (1982) Perspectives on studies of badland geomorphology, in *Badland Geomorphology and Piping* (eds R.B. Bryan and A. Yair), GeoBooks, Norwich, pp. 1–12.

Bull, L.J. and Kirkby, M.J. (1997) Gully processes and modelling. *Progress in Physical Geography*, **21**, 354–374.

Bull, L.J. and Kirkby, M.J. (2003) Channel heads and channel extension, in *Dryland Rivers: Hydrology and Geomorphology of Semi-Arid Channels* (eds L.J. Bull and M.J. Kirkby), John Wiley and Sons, Chichester, pp. 263–299.

Bull, L.J., Lawler, D.M., Leeks, G.J.L. and Marks, S. (1995) Downstream changes in suspended sediment fluxes in the River Severn, UK, in *Effects of Scale on Interpretation and Management of Sediment and Water Quality* (ed. W.R. Osterkamp), IAHS Publication 226, International Association of Hydrological Sciences, Wallingford, pp. 27–37.

Bull, L.J., Kirkby, M.J., Shannon, J. and Hooke, J.M. (2000) The impact of rainstorms on floods in ephemeral channels in southeast Spain. *Catena*, **38** (3), 191–209.

Bull, L.J., Kirkby, M.J., Shannon, J. and Dunsford, H. (2003) Predicting Hydrological Response Units (HRUs) in semi-arid environments. *Advances in Monitoring and Modelling*, **1**, 1–26.

Burt, T.P. (1997) *The hydrological role of buffer zones within the drainage basin system*, in *Buffer Zones: Their Processes and Potential in Water Protection* (eds N.E. Haycock, T.P. Burt, K.W.T. Goulding and G. Pinay), Quest Environmental, Herts, pp. 21–32.

Burt, T.P. (1999) Towards integrated management of rural drainage basins with particular reference to water quality issues, in *Water Quality; Processes and Policy* (eds S.T. Trudgill, D.E. Walling and B.W. Webb), pp. 257–259.

Burt, T.P. (2001) Integrated management of sensitive catchment systems. *Catena*, **42**, 275–290.

Burt, T.P. and Slattery, M.C. (1996) Time dependent changes in soil properties and surface runoff generation, in *Advances in Hillslope Processes* (eds M.G. Anderson and S.M. Brooks), John Wiley and Sons, Chichester, pp. 79–95.

Burt, T.P., Labadz, J.C. and Butcher, D.P. (1997) The hydrology and fluvial geomorphology of blanket peat: implications for integrated catchment management, in *Blanket Mire Degradation* (eds J.H. Tallis, R. Meade and P.D. Hulme), Macaulay Land Use Research Centre, Aberdeen, pp. 121–127.

Calvo, A., Harvey, A.M. and Paya-Serrano, J. (1991) Processes interactions and badland development in SE Spain, in *Soil Erosion Studies in Spain* (eds M. Sala, J.L. Rubio and J.M. García-Ruiz), Geoforma Ed, Logroño, pp. 75–90.

Calvo, A., Harvey, A.M., Paya-Serrano, J. and Alexander, R.W. (1991) Response of badland surfaces in SE Spain to simulated rainfall. *Cuaternario y geomorfología*, **5**, 3–14.

Cammeraat, L.H. and Imeson, A.C. (1999) The significance of soil-vegetation patterns following land abandonment and fire in Spain. *Catena*, **37**, 107–127.

Cantón, Y. (1999) Efectos hidrológicos y geomorfológicos de la cubierta y propiedades del suelo en paisaje de cárcavas. Unpublished PhD thesis. Universidad de Almeria. Spain. p. 394.

Casenave, A. and Valentin, C. (1992) A runoff capability classification system based on surface features in semi-arid areas of West Africa. *Journal of Hydrology*, **130**, 231–249.

Castro, N.M., Auzet, A.V., Chevallier, P. and Leprun, J.C. (1999) Land use change effects on runoff and erosion from plot to catchment scale on the basaltic plateau of southern Brazil. *Hydrological Processes*, **13**, 1621–1628.

Cerdà, A. (1995) Spatial distribution of infiltration on the matorral slopes in a Mediterranean environment, Genoves, Spain', in *Desertification in a European Context: Physical and Socio-economic Impacts* (eds R. Fantechi, D. Peter, P. Blabanis and J.L. Rubio), European Commission, Brussels, pp. 427–436.

Cerdà, A. (1997) Seasonal changes in the infiltration rates in a Mediteranean scrubland on limestone. *Journal of Hydrology*, **198**, 209–225.

Cerdà, A., Schnabel, S., Ceballos, A. and Gomez-Amelia, D. (1998) Soil hydrological response under simulated rainfall in the Dehesa land system, SW Spain under drought conditions. *Earth Surface Processes and Landforms*, **23**, 195–209.

Chaplot, V. and Le Bissonnais, Y. (2000) Field measurements of interrill erosion under different slopes and plot sizes. *Earth Surface Processes and Landforms*, **25**, 145–453.

Chescheir, C.M., Gilliam, J.W., Skaggs, R.W. and Broadhead, R.G. (1991) Nutrient and sediment removal in forested wetlands receiving pumped agricultural drainage water. *Wetlands*, **11**, 87–103.

Church, M. and Skaymaker, D. (1989) Disequilibrium of Holocene sediment yield in glaciated British Columbia. *Nature*, **337**, 452–454.

Correll, D.L. (1997) Buffer zones and water quality protection; basic principles, in *Buffer Zones: Their Processes and Potential in Water Protection* (eds N.E. Haycock, T.P. Burt, K.W.T. Goulding and G. Pinay), Quest Environmental, Herts, pp. 7–20.

Dedkov, A.P. and Mozzherin, V.T. (1984) *Eroziya I stock nanosov na zemle*, Izdatelstvo Kazanskogo Universiteta.

De Ploey, J. (1971) Liquefaction and rainwash erosion. *Zeitschrift für Geomorphologie*, **15**, 491–496.

De Ploey, J. (1984) Hydraulics of runoff and loess loam deposition. *Earth Surface Processes and Landforms*, **9**, 533–539.

De Ploey, J. (1990) Threshold conditions for thalweg gullying with special reference to Loess areas. *Catena Supplement*, **17**, 147–151.

De Vente, J., Poesen, J. and Verstraeten, G. (2005) The application of semi-quantitative methods and reservoir sedimentation rates for the prediction of basin sediment yield in Spain. *Journal of Hydrology*, **305** (1–4), 63–86.

Dietrich, W.E. and Dunne, T. (1978) Sediment budget for a small catchment in mountainous terrain. *Zeitschrift für Geomorphologie, Supplement Band*, **29**, 191–206.

Dietrich, W.E. and Dunne, T. (1993) The channel head, in *Channel Network Hydrology* (eds K. Beven and M.J. Kirkby), John Wiley and Sons, Chichester, pp. 175–219.

Dillaha, T.A. and Inamdar, S.P. (1997) Buffer zones as sediment traps or sources, in *Buffer Zones: Their Processes and Potential in Water Protection* (eds N.E. Haycock, T.P. Burt, K.W.T. Goulding and G. Pinay), Quest Environmental, Herts, pp. 33–42.

Dillaha, T.A., Reneau, R.B., Mostaghimi, S. and Lee, D. (1989a) Vegetative filter strips for agricultural non-point source pollution control. *Transactions of the ASCE*, **32** (2), 491–496.

Dillaha, T.A., Sherrard, J.H. and Lee, D. (1989b) Long term effectiveness and maintenance of vegetative filter strips. *Water Environment and Technology*, **1** (3), 418–421.

Dunne, T. and Black, R.D. (1970) Partial area contributions to storm runoff in a small New England watershed. *Water Resources Research*, **6**, 1296–1311.

Dunne, T., Moore, T.R. and Taylor, C.H. (1975) Recognition and prediction of runoff-producing zones in humid regions. *Hydrological Sciences Bulletin*, **20**, 305–327.

Eldridge, D.J., Zaady, E. and Shachak, M. (2000) Infiltration through three contrasting biological soil crusts in patterned landscapes in the Negev, Israel. *Catena*, **40**, 323–336.

Evans, R. (1993) On assessing accelerated erosion of arable land by water. *Soils and Fertilizers*, **56** (11), 1285–1293.

Farres, P. (1978) The role of time and aggregate size in the crusting process. *Earth Surface Proceedings*, **3**, 243–254.

Faulkner, H. (2008) Connectivity as a crucial determinant of badland morphology and evolution. *Geomorphology*, **100**, 91–103.

Faulkner, H., Alexander, R., Teeuw, R. *et al.* (2004) Variations in soil dispersivity across a gully head displaying shallow sub-surface pipes, and the role of shallow pipes in rill initiation. *Earth Surface Processes and Landforms*, **29**, 1143–1160.

Fitzjohn, C., Ternan, J.L. and Williams, A.G. (1998) Soil moisture variability in a semi-arid gully catchment: implications for runoff and erosion control. *Catena*, **32**, 55–70.

Fox, D.M. and Bryam, R.B. (1999) The relationship of soil loss to slope gradient for interrill erosion. *Catena*, **38**, 211–222.

Flügel, W.A. (1995) Delineating hydrological response units by geographical information system analyses for regional hydrological modelling using PRMS/MMS in the drainage basin of the River Bröl. *Germany. Hydrological Processes*, **9**, 423–436.

Flügel, W.A. (1997) Combining GIS with regional hydrological modelling using hydrological response units (HRUs): An application from Germany. *Mathematics and Computers in Simulation*, **43**, 297–304.

Frankenberger, J.R., Brooks, E.S., Todd-Walter, M. *et al.* (1999) A GIS-based variable source areas hydrology model. *Hydrological Processes*, **13**, 805–822.

Franklin, E.C., Gregory, J.D. and Smolen, M.D. (1992) Enhancement of the effectiveness of forested filter zones by dispersion of agricultural runoff. Report No. UNC-WRRI-92-270, Water Resources Research Institute, p. 28.

Gallart, F., Sole, A., Puidefabregas, J. and Lazaro, R. (2002) Badland systems in the Mediterranean, in *Dryland Rivers: Hydrology and Geomorphology of Semi-Arid Channels* (eds L.J. Bull and M.J. Kirkby), John Wiley and Sons, Chichester, pp. 299–326.

Gerits, J., Imeson, A.C., Verstraten, J.M. and Bryan, R.B. (1987) Rill development and badlands regolith properties. *Catena Supplement*, **8**, 141–160.

Glaetzer, B. and Grierson, I. (1987) *The Universal Soil Loss Equation. A Computer Package*, Roseworthy Agricultural College, Adelaide.

Govers, G. (1985) Selectivity and transport capacity of thin flows in relation to rill erosion. *Catena*, **12**, 35–49.

Govers, G. (1991) Rill erosion on arable land in central Belgium - rates, controls and predictability. *Catena*, **18** (2), 133–155.

Govers, G. (1992) Evaluation of transporting capacity formulae for overland flow, in *Overland Flow: Hydraulics and Erosion Mechanics* (eds A.J. Parsons and A.D. Parsons), UCL Press, London, pp. 140–273.

Govers, G. and Loch, R.J. (1993) Effects of initial water content and soil mechanical strength on the runoff and erosion resistance of clay soils. *Australian Journal of Soil Research*, **31**, 549–566.

Govers, G. and Rauws, G. (1986) Transporting capacity of overland flow on plane and on irregular beds. *Earth Surface Processes and Landforms*, **11**, 515–524.

Grissinger, E. (1996a) Rill and gullies erosion, in *Soil Erosion, Conservation, and Rehabilitation* (ed. M. Agassi), Marcel Dekker, New York, pp. 153–167.

Grissinger, E. (1996b) Reclamation of gullies and channel erosion, in *Soil Erosion, Conservation, and rehabilitation* (ed. M. Agassi), Marcel Dekker, New York, pp. 301–313.

Hairsine, P.B., Moran, C.J. and Rose, C.W. (1992) Recent developments regarding the influence of soil surface characteristics on overland flow. *Australian Journal of Soil Research*, **30**, 249–264.

Harris, G.L. and Forster, A. (1997) Pesticide contamination of surface waters – the potential role of buffer zones, in *Buffer Zones: Their Processes and Potential in Water Protection* (eds N.E. Haycock, T.P. Burt, K.W.T. Goulding and G. Pinay), Quest Environmental, Herts, pp. 62–69.

Hayes, J.C., Barfield, B.J. and Barnhisel, R.I. (1979) Filtration of sediment by simulated vegetation II; unsteady flow with non-homogeneous sediment. *Transactions of the ASCE*, **22** (5), 1063–1067.

Haycock, N.E., Burt, T.P., Goulding, K.W.T. and Pinay, G. (1997) *Buffer Zones: Their Processes and Potential in Water Protection*, Quest Environmental, Herts, p. 326.

Heathwaite, A.L. and Burt, T.P. (1992) The evidence for present and past erosion in the Slapton catchment, in *Soil Erosion, Past and Present*, vol. **22** (eds M. Bell and J. Boardman), Oxbow Monograph, pp. 89–100.

Helming, K., Römkens, M.J.M. and Prasad, S.N. (1998) Surface roughness related processes of runoff and soil loss: a flume study. *Journal of the Soil Science Society of America*, **62**, 243–250.

Hewlett, J.D. and Hibbert, A.R. (1967) Factors affecting the response of small watersheds to precipitation in humid areas, in *Forest Hydrology* (eds W.E. Sopper and H.W. Lull), Pergamon Press, New York, pp. 345–360.

Hodges, W.K. and Bryan, R.B. (1982) The influence of material behaviour on runoff initiation in the Dinosaur Badlands, Canada, in *Badland Geomorphology and Piping* (eds R.B. Bryan and A. Yair), Geo Books, Norwich, England, pp. 13–46.

Horton, R.E. (1945) Erosional development of streams and their drainage basins; hydrophysical approach to quantitative morphology. *Bulletin of the American Geological Society*, **56**, 275–370.

Howard, A.D. (1994) A detachment-limited model of drainage basin evolution. *Water Resources Research*, **30**, 2261–2285.

Huang, C. (1995) Empirical analysis of slope and runoff for sediment delivery from interill areas. *Soil Science Society of America Journal*, **59**, 982–990

Huang, C. and Bradford, J.M. (1990) Depressional storage for Markov–Gaussian surfaces. *Water Resources Research*, **26** (9), 2235–2242.

Imeson, A.C. and Kwaad, F.J.P.M. (1990) The response of tilled soils to wetting by rainfall and the dynamic character of soil erodibility, in *Soil Erosion on Agricultural Land* (eds J. Boardman, I.D.L. Foster and J.A. Dearing), John Wiley and Sons, Chichester, pp. 3–14.

Imeson, A.C. and Verstraten, J.M. (1988) Rills on badlands slopes; a physico-chemically controlled phenomenon. *Catena Supplement*, **12**, 139–150.

Imeson, A.C. and Verstraten, J.M. (1989) The microaggradation and erodibility of some semi arid and Mediterranean soils. *Catena Supplement*, **14**, 11–24.

Imeson, A.C., Kwaad, F.J.P.M. and Verstraten, J.M. (1982) The relationship of soil chemical and physical properties to the development of badlands in Morocco, in *Badlands Geomorphology and Piping* (eds R.B. Bryan and A. Yair), GeoBooks, Norwich, pp. 47–70.

Imeson, A.C., Verstraten, J.M., van Mulligen, E.J. and Sevink, J. (1992) The effects of fire and water repellency on infiltration and runoff under Mediterranean type forest. *Catena*, **19**, 345–361.

Jayawardena, A.W. and Bhuiyan, R.R. (1999) Evaluation of an interrill soil erosion model using laboratory catchment data. *Hydrological Processes*, **13**, 89–100.

Johnes, P.J. and Burt, T.P. (1993) Nitrate in surface waters, in *Nitrate: Processes, Patterns and Management* (ed. T.P. Burt), John Wiley and Sons, Chichester, pp. 269–317.

Jordan, T.E., Correll, D.L. and Weller, D.E. (1993) Nutrient interception by a forest receiving inputs from adjacent cropland. *Journal of Environmental Quality*, **22**, 467–473.

Karnoven, T., Koivusalo, H., Jauhianinen, M. *et al.* (1999) A hydrological model for predicting runoff from different land use areas. *Journal of Hydrology*, **217**, 253–265.

Karr, J.R. and Schlosser, I.J. (1978) Water resources and the land water interface. *Science*, **201**, 229–234.

Keyzer, M.A. and Sonneveld, B.G.J.S. (1998) Using the mollifier method to characterise datasets and models: the case of the Universal Soil Loss Equation. *ITC Journal*, **3–4**, 263–272.

Kidron, G.J. and Yair, A. (1997) Rainfall–runoff relationships over encrusted dune surfaces, Nizzana, Western Negev, Israel. *Earth Surface Landforms and Processes*, **22**, 1169–1184.

Kilinc, M. and Richardson, E.V. (1973) Mechanics of soil erosion from overland flow generated by simulated rainfall. *Colorado State University Hydrology Paper*, **63**, 82.

Kinnell, P.I.A. (1990) Modelling erosion by rain impacted flow. *Zeitschrift für Geomorphologie, Supplement Band*, **17**, 55–66.

Kirkby, M.J. (1994) Thresholds and instability in stream head hollows: a model of magnitude and frequency for wash processes, in *Process Models and Theoretical Geomorphology* (ed. M.J. Kirkby), John Wiley and Sons, London, pp. 295–352.

Kirkby, M.J. (1999) Translating models from hillslope (1ha) to catchment (1000 km2) scales, in *Regionalization in Hydrology*, (eds B. Diekkrüger, M.J. Kirkby and U. Schröder), IAHS Publication 254, International Association of Hydrological Sciences, Wallingford. pp. 1–12.

Kirkby, M.J. and Bull, L.J. (2000) Factors controlling gully growth in fine-grained sediments: A model applied to southeast Spain. *Catena*, **40**, 127–146.

Kirkby, M.J., Bracken, L.J. and Reaney, S. (2002) The influence of Landuse, Soils and Topography on the delivery of hillslope runoff to channels in SE Spain. *Earth Surface Landforms and Processes*, **27**, 1459–1473.

Knapen, A., Poesen, J. and De Baets, S. (2007) Seasonal variations in soil erosion resistance during concentrated flow for a loess-derived soil under two contrasting tillage practices. *Soil & Tillage Research*, **94** (2), 425–440.

Knapen, A., Poesen, J., Govers, G. *et al.* (2008) The effect of conservation tillage on runoff erosivity and soil erodibility during concentrated flow. *Hydrological Processes*, **22**, 1497–1508.

Knighton, A.D. (1987) River channel adjustment – the downstream dimension, in *River Channels: Environment and Process* (ed. K.S. Richards), Blackwell, Oxford, pp. 95–128.

Kochel, R.C., Howard, A.D. and MacLane, C.F. (1982) Channel networks developed by groundwater sapping in fine-grained sediments: analogs in some Martian valleys, in *Models in Geomorphology* (ed. M.J. Woldenberg), Allen and Unwin, Boston, pp. 313–341.

Kouwen, N., Soulis, E., Pietroniro, A. and Donald, J. (1993) Grouped response units for distributed hydrological modelling. *ASCE Journal of Water Resources Planning and Management*, **119**, 289–304.

Krause, A.K., Franks, S.W., Kalma, J.D. *et al.* (2003) Multi-parameter fingerprinting of sediment deposition in a small gullied catchment in SE Australia. *Catena*, **53** (4), 327–348.

Lal, R. (1990) *Soil Erosion in the Tropics: Principles and Management*, McGraw-Hill, New York.

Lasanta, T., Garcia-Ruiz, J.M., Perez-Rontome, C. and Sancho-Marcen, C. (2000) Runoff and sediment yield in a semi-arid environment: the effect of land management after farmland abandonment. *Catena*, **38**, 265–278.

Le Bissonnais, Y. (1990) Crust micromorphology and runoff generation on silty soil materials during different seasons. *Zeitschrift für Geomorphologie, Supplement Band*, **17**, 11–16.

Le Bissonnais, Y. (1996) Aggregate stability and assessment of soil crustability and erodibility: I. Theory and methodology. *European Journal for Soil Science*, **47**, 425–437.

Le Bissonnais, Y., Bruand, A. and Jamagne, M. (1989) Laboratory and experimental study of soil crusting: relations between aggregate breakdown mechanisms and crust structure. *Catena*, **16**, 377–392.

Leonard, J. and Andrieux, P. (1998) Infiltration characteristics of soils in Mediterranean vineyards in southern France. *Catena*, **32**, 209–223.

Loch, R.J. and Donnollan, T.E. (1983) Field rainfall simulator studies of two clay soils of the Darling Downs, Queensland II. Aggregate breakdown, sediment properties and soil erodibility. *Australian Journal of Soil Research*, **27**, 535–542

Loch, R.J., Maroulis, J.C. and Silburn, D.M. (1989) Rill erosion of a self-mulching black earth. *Australian Journal of Soil Research*, **27**, 535–542

Lowrance, R., Vellidis, G. and Hubbard, R.K. (1995) Denitrification in a restored riparian forest wetland. *Journal of Environmental Quality*, **24**, 808–815.

Ludwig, J.A., Wiens, J. and Tongway, D.J. (2000) A scaling rule for landscape patches and how it applies to conserving soil resources in savannas. *Ecosystems*, **3**, 82–97.

Luk, S.H. (1979) Effects of soil properties on soil erosion by wash and splash. *Earth Surface Processes*, **4**, 241–255.

Magette, W.L., Brinsfield, R.B., Palmer, R.E. and Wood, J.D. (1989) Nutrient and sediment removal by vegetated filter strips. *Transactions of the ASCE*, **32** (2), 663–667.

Malanson, G.P. (1993) *Riparian Landscapes*, Cambridge University Press, Cambridge.

Martinez-Mena, M., Albaladejo, J. and Castillo, V.M. (1998) Factors influencing surface runoff generation in a Mediterranean semi-arid environment: Chicamo watershed, SE Spain. *Hydrological Processes*, **12**, 741–754.

Martinez-Mena, M., Albaladejo, J. and Castillo, V.M. (1999) Influence of vegetal cover on sediment particle size distribution in natural rainfall conditions in a semiarid environment. *Catena*, **38**, 175–190.

McDonnell, J.J. (2003) Where does water go when it rains? Moving beyond the variable source area concept of rainfall–runoff response. *Hydrological Processes*, **17**, 1869–1875.

Meyer, L.D. and Harmon, W.C. (1984) Susceptibility of agricultural soil to interrill erosion. *American Journal of Soil Science*, **48**, 1151–1157.

Michaelides, K. and Wainwright, J. (2002) Modelling the effects of hillslope-channel coupling on catchment hydrological response. *Earth Surface Processes and Landforms*, **27**, 1441–1457.

Middleton, H.E. (1930) Properties of soils which influence soil erosion. Technical Report 178, U.S. Department of Agriculture.

Milliman, J.D., Qin, Y.S., Ren, M.E. and Saito, Y. (1987) Man's influence on the erosion and transport of sediment by Asian rivers; The Yellow River. *Journal of Geology*, **95**, 751–765.

Mitchell, J.K. and Bubenzer, G.D. (1980) Soil loss estimation, in *Soil Erosion* (eds M.J. Kirkby and R.P.C. Morgan), John Wiley and Sons, Chichester, pp. 20–35.

Mitsch, W.J., Dorge, C.L. and Wiemhoff, J.R. (1977) Ecosystem dynamics and a phosphorous budget of an alluvial cypress swamp in southern Illinois. *Ecology*, **60**, 1116–1124.

Moore, I.D. and Burch, G.J. (1986) Sediment transport capacity of sheet and rill flow: application of unit stream power theory. *Water Resources Research*, **22**, 1350–1360

Morgan, R.P.C. (1986) *Soil Erosion and Conservation*, Longman, Harlow.

Morgan, R.P.C., Quinton, J.N., Smith, R.E. *et al.* (1998) The European Soil Erosion Model (EUROSEM): A dynamic approach for predicting sediment transport from fields and small catchments. *Earth Surface Processes and Landforms*, **23**, 527–544.

Morrison, M.W., Prunty, L. and Giles, J.F. (1985) Characterising strength of soil crusts formed by simulated rain. *Soil Science Society of America Journal*, **49**, 427–431.

Moss, A.J. (1991) Rain impact on soil crusts I: formation on a granite derived soil. *Australian Journal of Soil Science*, **21**, 443–450.

Moss, A.J., Walker, P.H. and Hutka, J. (1979) Raindrop simulated transportation in shallow water flows: an experimental study. *Sedimentary Geology*, **22**, 165–184.

Mutchler, C.K. and Larson, C.L. (1971) Splash amounts from water drop impact on a smooth surface. *Water Resources Research*, **7**, 195–200.

Myrabo, S. (1997) Temporal and spatial scale response area and groundwater variation in till. *Hydrological Processes*, **11**, 1861–1880.

Nachtergaele, J., Poesen, J., Wijdenes, D.O. *et al.* (2002) Medium-term evolution of a gully developed in a loess-derived soil. *Geomorphology*, **46** (3–4), 223–239.

Nearing, M.A. (1998) Why soil erosion models over-predict small soil losses and under-predict large soil losses. *Catena*, **32**, 15–22.

Nearing, M.A. and Bradford, J.M. (1985) Single waterdrop splash detatchmentand mechanical properties of soils. *American Journal of Soil Science*, **49**, 547–552.

Nearing, M.A., Norton, L.D., Bulgakov, D.A. *et al.* (1998) Hydraulics and erosion in eroding rills. *Water Resources Research*, **33**, 865–876.

Newson, M.D. (1997) *Land, Water and Development*, 2nd edn, Routledge.

Niebling, W.H. and Alberts, E.E. (1979) Composition and yield of soil particles transported through sod strips. ASAE Paper 79-2065. American Society of Agricultural Engineers, Michigan.

Nogueras, P., Burjachs, F., Gallart, F. and Puigdefàbregas, J. (2000) Recent gully erosion in the El Cautivo badlands (Tabernas, SE Spain). *Catena*, **40**, 203–215.

Oades, J.M. (1993) The role of biology in the formation, stabilization and degradation of soil structure. *Geoderma*, **56**, 377–400.

Onstad, C.A. (1984) Depressional storage on tilled soil surfaces. *Transactions of the American Society of civil Engineers*, **27**, 729–732.

Oostwoud Wijdenes, D.J. and Bryan, R.B. (1991) Gully development on the Njemps Flats, Baringo, Kenya. *Catena Supplement*, **19**, 71–90.

Oostwoud Wijdenes, D., Poesen, J., Vandekerckhove, L. and Ghesquiere, M. (2000) Spatial distribution of gully head activity and sediment supply along an ephemeral channel in a Mediterranean environment. *Catena*, **39**, 147–167.

Osborne, L.L. and Kovacic, D.A. (1993) Riparian vegetated buffer strips in water quality restoration and stream management. *Freshwater Biology*, **29**, 243–258.

Palmer, R.S. (1963) The influence of thin water layer on water drop impact forces. International Association of Hydrological Sciences, **68**, 141–148.

Panabooke, C.R. and Quirk, J.P. (1957) Effect of initial water content on stability of soil aggregates in water. *Soil Science*, **1983**, 185–195.

Parsons, A.J. and Gadian, A.M. (2000) Uncertainty in modelling the detachment of soil by rainfall. *Earth Surface Processes and Landforms*, **25**(7), 723–728.

Parsons, A.J., Wainwright, J., Schlesinger, W.H. and Abrahams, A.D. (2003) The role of overland flow in sediment and nitrogen budgets of mesquite dunefields, southern Mexico. *Journal of Arid Environments*, **53**, 61–71.

Peterjohn, W.T. and Correll, D.L. (1984) Nutrient dynamics in an agricultural watershed: observations on the roles of a riparian forest. *Ecology*, **65**, 1466–1475.

Phillips, J.D. (1995) Decoupling of sediment sources in large river basins, *Effects of Scale on Interpretation and Management of Sediment and Water Quality*, (ed. W.R. Osterkamp),

IAHS Publication 226, International Association of Hydrological Sciences, Wallingford, pp. 11–16.

Planchon, O., Fritsch, E. and Valentin, C. (1987) Rill development in a wet savannah environment. *Catena Supplement*, **8**, 55–70.

Poesen, J. (1981) Rainwash experiments on the erodibility of loose sediment. *Earth Surface Processes*, **6**, 285–307.

Poesen, J. (1984) The influence of slope gradient on infiltration rate and Hortonian overland flow volume. *Zeitschrift für Geomoprhologie Supplement Band*, **49**, 117–131.

Poesen, J. and Savat, J. (1981) Detachment and transport of loose sediment by raindrop splash. Part II: Detachability and transportability measurements. *Catena*, **8**, 19–41.

Poesen, J., Vandekerckhove, L., Nachtergaele, J. *et al.* (2002) Gully erosion in dryland environments, in *Dryland Rivers: Hydrology and Geomorphology of Semi-Arid Channels* (eds L.J. Bull and M.J. Kirkby), John Wiley and Sons, Chichester, pp. 229–262.

Poesen, J., Nachtergaele, J., Verstraeten, G. *et al.* (2003) Gully erosion and environmental change: importance and research needs. *Catena*, **50** (2–4), 91–133.

Regüés, D., Pardini, G. and Gallart, F. (1995) Regolith behaviour and physical weathering of clayey mudrock as dependent on seasonal weather conditions in a badland area at Vallcebre. *Eastern Pyrenees. Catena*, **25**, 199–212.

Renard, K.G., Foster, G.R., Weesies, G.A. and Porter, G.P. (1991) RUSLE; revised soil loss equation. *Journal of Soil and Water Conservation*, **46**, 30–33.

Renard, K.G., Foster, G.R., Weesies, G.A. *et al.* (1998) Predicting soil erosion by water: a guide to conservation planning with Revised Soil Loss Equation (RUSLE), *Agriculture Handbook, 703*, USDA-ARS, Washington, DC.

Römkens, M.J.M. and Wang, J.Y. (1987) Soil roughness changes from rainfall. *Trans ASCE*, **31**, 408–413.

Römkens, M.J.M., Helming, K. and Prasard, S.N. (2001) Soil erosion under different rainfall intensities, surface roughness and soil water regimes. *Catena*, **46**, 103–123.

Römkens, M.J.M., Prasard, S.N. and Whisler, F.D. (1990) Surface sealing and infiltration, in *Process Studies in Hillslope Hydrology* (eds M.G. Anderson and T.P. Burt), John Wiley and Sons, Chichester, pp. 127–172.

Rose, C.W. (1985) Developments in soil erosion and deposition models, in *Advances in Soil Science* (ed. B.A. Stewart), Springer-Verlag, Berlin, pp. 1–64.

Sargunam, A., Riley, P., Arulanandan, K. and Krone, R.B. (1973) Physico-chemical factors in erosion of cohesive soils. *Journal of the Hydraulics Division, American Society of Civil Engineers*, **99**, HY3, 555–558.

Savat, J. (1976) Discharge velocities and total erosion of calcareous loess: a comparison between pluvial and terminal runoff. *Revue de Geomorphologie Dynamique*, **24**, 113–122

Savat, J. (1979) Laboratory experiments on erosion and deposition of loess by laminar sheetflow and turbulent rill flow. Proceedings of the Seminar Agricultural Soil Erosion in Temperate Non-Mediterranean Climate, 20–23, pp. 139–143.

Savat, J. (1980) Resistance to flow in rough, supercritical sheetflow. *Earth Surface Processes*, **5**, 103–122.

Scheidegger, A.E., Schumm, S.A. and Fairbridge, R.W. (1968) Badlands, in *Encyclopaedia of Geomorphology* (ed. R. Fairbridge), Dowden, Hutchinson & Ross, Inc, USA, pp. 43–48.

Schlosser, I.J. and Karr, J.R. (1981a) Riparian vegetation and channel morphology impact on spatial patterns of water quality in agricultural watersheds. *Environmental Management*, **5**, 233–243.

Schlosser, I.J. and Karr, J.R. (1981b) Water quality in agricultural watersheds: impact of riparian vegetation during baseflow. *Water Resources Bulletin*, **17**, 233–240.

Schumm, S.A. (1956a) Evolution of drainage systems and slopes in badlands at Perth Amboy, New Jersey. *Bulletin of the American Geological Society*, **67**, 597–646.

Schumm, S.A. (1956b) The role of creep and rainsplash on the retreat of badlands slopes. *American Journal of Science*, **254**, 693–706.

Schumm, S.A. and Lusby, G.C. (1963) Seasonal variations in infiltration capacity and runoff on hillslopes of Western Colorado. *Journal of Geophysical Research*, **63**, 3655–3666.

Scoging, H. (1982) Spatial variations in infiltration, runoff and erosion on hillslopes in semi-arid Spain, in *Badland Geomorphology and Piping* (eds R.B. Bryan and A. Yair), Geo Books, Norwich, pp. 89–112.

Sharma, K.D., Dhir, R.P. and Murthy, J.S.R. (1993) Modelling soil erosion in arid zone drainage basins. *Sediment Problems: Strategies for Monitoring, Prediction and Control*, **217**, 269–276.

Sharma, K.D., Menenti, M., Huygen, J. and Fernandez, P.C. (1996) Distributed numerical rainfall-runoff modelling in an arid region using thematic mapper data and a Geographical Information System. *Hydrological Processes*, **10**, 1229–1242.

Shiel, R.S., Adey, M.A. and Lodder, M. (1988) The effect of successive wet/dry cycles on aggregate size distribution in a clay texture soil. *Journal of Soil Science*, **39**, 71–80.

Sherard, J.L., Ryker, N.L. and Decker, R.S. (1972) Piping in earth dams of dispersive clay. Proceedings of the Special Conference on the Performance of Earth and Earth Supported Structures, ASCE, pp. 150–161.

Sidorchuk, A. (2005) Stochastic components in the gully erosion modelling. *Catena*, **63**, 299–317.

Sidorchuk, A. (2006) Stages in gully evolution and self-organized criticality. *Earth Surface Processes and Landforms*, **31**, 1329–1344.

Singer, M.J. and Le Bissonnais, Y.L. (1998) Importance of surface sealing in the erosion of some soils from a Mediterranean climate. *Geomorphology*, **24**, 79–85.

Slattery, M.C. and Bryan, R.B. (1992) Hydraulic conditions for rill incision under simulated rainfall: a laboratory experiment. *Earth Surface Processes and Landforms*, **17**, 127–146.

Slattery, M.C. and Burt, T.P. (1995) Size characteristics of sediment eroded from agricultural soil: dispersed versus non-dispersed, ultimate versus effective, in *River Geomorphology* (ed. E.J. Hickin), John Wiley and Sons, Chichester, pp. 1–17.

Slattery, M.C., Burt, T.P. and Boardman, J. (1994) Rill erosion along the thalweg of a hillslope hollow: a case study from the Cotswold Hills, Central England. *Earth Surface Processes and Landforms*, **19**, 377–385.

Smets, T., Poesen, J. and Knapen, A. (2008) Spatial scale effects on the effectiveness of organic mulches in reducing soil erosion by water. *Earth-Science Reviews*, **89**, 1–12.

Smith, C.M. (1992) Riparian afforestation effects on water yields and water quality in pasture catchments. *Journal of Environmental Quality*, **21**, 237–245.

Smith, M., Cox, N.J. and Bracken, L.J. (2007) Applying flow resistance equations to overland flows. *Progress in Physical Geography*, **31** (4), 363–387.

Smith, R.E., Goodrich, D.C., Woolheiser, D.A. and Unkrich, C.L. (1995) KINEROS – a kinematic runoff and erosion model, in *Computer Models of Watershed Hydrology* (ed. V.P. Singh), Water Resources Publications, Colorado, pp. 697–732.

Smith, T.R. and Bretherton, F.P. (1972) Stability and the conservation of mass in drainage basin evolution. *Water Resources Research*, **8**, 1506–1529.

Smith, T.R., Birnir, B. and Merchant, G.E. (1997) Towards an elementary theory of drainage basin evolution: I. The theoretical basis. *Computers and Geosciences*, **23**, 811–822.

Smith, T.R., Merchant, G.E. and Birnir, B. (1997) Towards an elementary theory of drainage basin evolution: II. Computational evaluation. *Computers and Geosciences*, **23**, 823–849.

Solé-Benet, A., Calvo, A., Cerdà, A. *et al.* (1997) Influence of micro-relief patterns and plant cover on runoff related processes in badlands from Tabernas (SE Spain). *Catena*, **31**, 23–38.

Sonneveld, B.G.J.S. and Nearing, M.A. (2003) A bobparametric/parametric abalysis of the Universal Soil Loss Equation. *Catena*, **52**, 9–21.

Still, D.A. and Shih, S.F. (1985) Using Landsat data to classify land use for assessing the basin wide runoff index. *Water Resources Bulletin*, **21**, 931–940.

Sumner, M.E. (1995) Soil crusting: chemical and physical processes. The view from Georgia, in *Sealing, Crusting and Hard Setting Soils: Productivity And Conservation* (eds H.B. Smith *et al.*), Australian Soil Science Society, Brisbane.

Sutherland, R.A. and Bryan, R.B. (1991) Sediment budgeting: a case study in Katiorin drainage basin, Kenya. *Earth Surface Processes and Landforms*, **16**, 383–398.

Tisdall, J.M., Cockcroft, B. and Uren, N.C. (1978) The stability of soil aggregates as affected by organic materials, microbial activity, and physical disruption. *Australian Journal of Soil Research*, **16**, 9–17.

Tiwari, A.K., Risse, L.M. and Nearing, M.A. (2000) Evaluation of WEPP model and its comparison with USLE and RUSLE. *Transactions of the American Society of Agricultural Engineers*, **43**, 1129–1135.

Tollner, E.W., Barfield, B.J. and Hayes, J.C. (1982) Sedimentology of erect vegetative filters. *Proceedings of the Hydraulics Division, ASCE*, **108** (12), 1518–1531.

Torri, D. (1987) A theoretical study of soil detachability. *Catena Supplement*, **10**, 15–20.

Torri, D., Colica, A. and Rockwell, D. (1994) Preliminary study of the erosion mechanisms in a biancana badland (Tuscany, Italy). *Catena*, **23**, 281–294.

Torri, D., Sfalanga, M. and Chisci, G. (1987) Threshold conditions for incipient rilling. *Catena Supplement*, **8**, 94–106.

Tran, L.T., Ridgley, M.A., Nearing, M.A. *et al.* (2002) Using fuzzy logic-based modelling to improve the performance of the revised Universal Soil Loss Equation, in *Sustaining the Global Farm* (eds D.E. Stott *et al.*), International Soil Conservation Organisation, pp. 919–923.

Trimble, S.W. (1983) Changes in sediment storage in Coon Creek basin, Driftless Area, Wisconsin, 1853–1975. *Science*, **214**, 181–183.

Trimble, S.W. and Crosson, P. (2000) US soil erosion rates – myth and reality. *Science*, **289**, 248–250.

Tsujimoto, T., Kitamura, T. and Murakami, S. (1996) Basin morphological processes due to deposition of suspended sediment affected by vegetation. 2nd International Symposium on Habitat Hydraulics, Quebec, INRS & IAHR.

Tucker, G.E., Arnold, L., Bras, R.L. *et al.* (2006) Headwater channel dynamics in semiarid rangelands, Colorado high plains, USA. *Geological Society of America Bulletin*, **118** (7–8), 959–974.

Valentin, C. and Bresson, L.M. (1992) Morphology, genesis and classification of surface crusts in loamy and sandy soils. *Geoderma*, **55**, 225–245.

Valentin, C., Poesen, J. and Li, Y. (2005) Gully erosion: Impacts, factors and control. *Catena*, **63** (2–3), 132–153.

Vandekerckhove, L., Poesen, J., Wijdenes, D.O. and Gyssels, G. (2001) Short-term bank gully retreat rates in Mediterranean environments. *Catena*, **44** (2), 133–161.

Wainwright, J. (1996) Infiltration, runoff and erosion characteristics of agricultural land in extreme strom events, SE France. *Catena*, **26**, 27–47.

Wainwright, J., Parsons, A.J. and Abrahams, A.D. (2000) Plot-scale studies of vegetation, overland flow, and erosion interactions: case studies form Arizona and New Mexico. *Hydrological Processes*, **14**, 2921–2943.

Wasson, R.J., Caitcheon, G., Murray, A.S. *et al.* (2002) Sourcing sediment using multiple tracers in the catchment of Lake Argyle, Northwestern Australia. *Environmental Management*, **29** (5), 634–646.

Walling, D.E. (1983) The sediment delivery problem. *Journal of Hydrology*, **69**, 209–237.

Walling, D.E. (1988) Measuring sediment yield from river basins, in *Soil Erosion Research Methods* (ed. R. Lal), Soil and Water Conservation Society, Iowa, pp. 39–73.

Walling, D.E. (1999) Linking land user, erosion and sediment yields in river basins. *Hydrobiologia*, **410**, 223–240.

Walling, D.E. and Webb, B.W. (1992) Water quality I: physical characteristics, in *The Rivers Handbooks*, vol. **1** (eds P. Calow and G.E. Petts), Blakwell, Oxford, pp. 48–72.

Walling, D.E. and Woodward, J.C. (1995) Tracing suspended sediment sources in river basins; a case study of the River Culm, Devon, UK. *Journal of Marine and Freshwater Research*, **46**, 327–336.

West, N.E. (1990) Structure and formation of microphytic soil crusts in wildland ecosystems of arid and semi-arid regions. *Advances in Ecological Research*, **20**, 179–223.

Willgoose, G., Bras, I. and Rodriguez-Iturbe, I. (1991) A coupled channel network growth and hillslope evolution model 1: Theory. *Water Resources Research*, **27**, 1671–1684.

Wischmeier, W.H. (1976) Use and misuse of the universal soil loss equation. *Journal of Soil and Water Conservation*, **31**, 5–9.

Wischmeier, W.H. and Smith, D.D. (1978) Predicting rainfall erosion losses, in *Agricultural Handbook, 537*, U.S. Department of agriculture, Washington, DC.

Wolman, M.G. (1977) Changing needs and opportunities in the sediment field. *Water Resources Research*, **13**, 50–59.

Yair, A. (1990) The role of topography and surface cover upon soil formation along hillslopes in arid climates. *Geomorphology*, **3**, 287–299.

Yair, A. (1992) The control of headwater area on channel runoff in a small arid watershed, in *Overland Flow: Hydraulics and Erosion Mechanics* (eds A.J. Parsons and A.D. Abrahams), UCL Press, London, pp. 53–68.

Yair, A. and Lavee, H. (1985) Runoff generation in arid and semiarid zones, in *Hydrological Forecasting* (eds M.G. Anderson and T.P. Burt), John Wiley and Sons, Chichester, pp. 183–220.

Yair, A., Bryan, R.B., Lavee, H. and Adar, E. (1980) Runoff and erosion processes and rates in the Zin Valley northern Negev, Israel. *Earth Surface Processes and Landforms*, **5**, 205–225.

Yalin, M.S. (1971) *Mechanics of Sediment Transport*, Pergamon Press.

Young, R.A., Huntrods, T. and Anderson, W. (1980) Effectiveness of vegetative buffer strips in controlling pollution from feedlot runoff. *Journal of Environmental Quality*, **9**, 483–487.

8

Erosional Processes and Sediment Transport in Upland Mires

Martin G. Evans[1] and Timothy P. Burt[2]

[1]Upland Environments Research Unit, School of Environment and Development, University of Manchester, Manchester, M13 9PL, UK

[2]Department of Geography, Durham University, Durham, DH1 3LE, UK

8.1 The Role of Upland Peatlands in the Sediment Cascade

Upland mires play a distinctive role in the wider upland sediment cascade. Extensive cover of friable organic soil means that both the quantity and quality of sediment flux differs markedly from similar catchments where mire development is absent.

Peat accumulations are found in many upland areas of the humid temperate zone (Charman, 2002). In areas of high relief, steep slopes and rapid drainage may limit peat to local topogenous accumulations, for example on cirque floors. However, in areas of low slope and in more maritime locations, extensive areas of blanket peat may accumulate (Lindsay, 1995). Blanket peatlands are areally extensive and can be the dominant land-cover type defining the sediment cascade across entire ranges of hills and plateaux. Blanket peatlands are found across the globe, including Tasmania, South Island, New Zealand, southern South America and the western fringes of northern North America. Extensive blanket peatland is found in the amphi-Atlantic region from Newfoundland to Scandinavia. Blanket peatlands within this zone in the UK contain 15% of the total global area of this cover type (Tallis, Meade and Hulme, 1997). However, blanket peatlands of the UK and Ireland are, uniquely, extensively eroded. Peatland erosion mobilizes large stores of organic

Sediment Cascades: An Integrated Approach Edited by Timothy Burt and Robert Allison

sediment, dramatically increases drainage density, and increases the efficiency of slope–channel linkage within the sediment cascade (Evans and Warburton, 2007). Consequently, eroding peatlands have very high sediment yields and the erosional status of a mire system has a major influence on its functioning within the wider sediment cascade. This is of particular significance as peatlands are climatically determined landforms and their extent and erosional status are likely to be highly sensitive to predicted climate change (Lindsay, 1995; Bragg and Tallis, 2001; Evans and Warburton, 2007). Over the past 20 years significant progress has been made in understanding the sediment system of upland peatlands. This chapter aims to synthesize the key advances and assess implications for the role of peatland systems in the wider sediment cascade. The geographical focus of the review is largely on the UK, reflecting the location of the authors' peatland research and the locus of most of the recent work on peatland geomorphology. It is intended, however, that analysis of peatland sediment systems developed here should be applicable to upland peatlands more widely.

8.2 Timescales of Erosion and Initiation

Appropriate timescales for consideration of the role of peatlands in upland systems are defined by the period of peatland formation. Most northern peatlands lie at latitudes which were glaciated during the last glacial period so that peatland ages are constrained to Holocene timescales.

8.2.1 Timescales for Peat Initiation

Iverson (1964) proposed a predictable pattern of interglacial soil development where progressive leaching of soils leads in the final oligocratic phase to podzolization and peat formation. However, dating of basal peat layers suggests that, where climate is favourable, blanket peat cover became established very early in the Holocene (Bell and Walker, 1992). In England, upland peat deposition began around 8 ka in the north Pennines (Johnson and Dunham, 1963; Evans, unpublished data). Similarly, early dates are reported for North-West Scotland (Moore, Merryfield and Price, 1984). Solem (1986) reports basal peat dates as early as 7800 yr BP for Norwegian blanket bogs, and Korhola (1995) cites peat formation as early as 10 700 yr BP in Finland. In North America peat initiation appears to have been largely climatically controlled and is associated with cooler, wetter conditions in the mid-Holocene. Recorded basal peat dates across much of northern North America fall in the range 6–3 ka (Ovenden, 1990). Ovenden suggests that, unlike maritime sites in Europe where precipitation totals are an important control on rates of peat accumulation, initiation of peat cover in more continental North American sites has been strongly influenced by temperature changes. In contrast, dates for peat initiation in the Southern Pennines (UK) cluster around 5000 yr BP and coincide with the elm decline. Moore (1975) suggests that this indicates an association between peat initiation and forest clearance by Mesolithic humans, and that human impacts may have been sufficient to trigger irreversible initiation of peat formation in climatically marginal locations (Moore, Merryfield and Price, 1984).

It is clear that, although there may be local variation associated with climate and human impacts, blanket peat has been a feature of upland landscapes for most of the Holocene. Consideration of the role of upland peat bogs in the sediment cascade will therefore necessarily entail some consideration of past as well as present impacts.

8.2.2 Peat Erosion – Timescales

In most upland areas of the UK and Ireland, extensive gullying of upland blanket peats provides evidence for historic or contemporary erosion of the peat mass (Bradshaw and McGee, 1988; Large and Hamilton, 1991; Grieve, Hipkin and Davidson, 1994; Tallis, 1997; Wishart and Warburton, 2001). Dating the onset of erosion is problematic for two reasons. First, the physical properties of peat tend to preclude formation of dateable depositional sequences associated with the erosion phase. Second, because the peat consists of organic material which has been preserved for periods of the order of 10^3 years, it is hard to be certain that organic matter for radiocarbon dating is contemporaneous with deposition. Lacustrine sediments are one depositional environment where sedimentary records of peat erosion appear to be preserved. Stevenson, Jones and Battarbee (1990), Rhodes and Stevenson (1997) and Bradshaw and McGee (1988) report lake-sediment sequences downstream of areas of eroding peat where the organic content of the deposit undergoes a stepped increase. This horizon is interpreted as representing the onset of erosion. The onset of peat erosion in Wicklow is dated to 3000 yr BP and in Donegal to 1500 yr BP (Bradshaw and McGee, 1988). These dates are the uppermost dates in a series of radiocarbon dates which are normally superposed. Dates from further up the cores, above the inferred onset of erosion, are inverted due to the in-wash of preserved (i.e. older) organic material. Stevenson, Jones and Battarbee (1990) date the onset of erosion in a series of Scottish and Irish lakes to between AD 1300 and 1800. These dates are based on the extrapolation of lead-210 and caesium-137 dating as well as radiocarbon dating. An alternative approach to dating erosion phases has been developed by Tallis (e.g. 1985, 1995). Tallis has identified erosion phases in the Southern Pennines, UK by studying the biostratigraphy of intact peat sequences adjacent to eroded gullies. Shifts to a drier peat surface (inferred from change-related shifts in the proportion of *Sphagnum* and *Racomitrium* moss) which are not correlated with climatically forced drying of the bog surface (as recorded at uneroded sites) are interpreted as being the result of increased local drainage due to gullying. The dates for these events span AD 1250–1800, clustering around AD 1450.

Attempts to date phases of peat erosion have been associated with research into the causes of peat erosion. A large number of possible triggers for peat erosion phases have been suggested. These can be categorized as the onset of peat erosion by the crossing of either an intrinsic threshold or an extrinsic threshold. Bower (1961) considered a range of possible triggers for peat erosion in the Pennines including: mass movements triggered by expansion of peat on to steeper slopes and mass movements triggered by an increase in peat depth (intrinsic thresholds); and, fire, overgrazing, climate change, and atmospheric pollution (extrinsic thresholds). Bower argued, because the distribution of peat erosion appeared to correlate with climatic variables and with topography, that these were also likely to be important controls on initiation of erosion. Subsequent studies have demonstrated, at least locally, that extrinsic changes have placed stress on the bog-surface vegetation, reducing resistance to erosion and triggering change. Shimwell (1974) demonstrated that seventeenth and eighteenth century erosion of Peak District blanket peat occurred at a time when sheep stocking levels were high, displacing grazing on to the bog surfaces. Similarly, there are many documented cases of local erosion after burning of the vegetation (Anderson, 1997).

Rhodes and Stevenson (1997) demonstrated that of nine lake basins studied the sedimentary record of peat erosion was statistically associated with the charcoal deposition record (a proxy for fire history) in only one case. These lakes in Scotland and Ireland were distant from pollution sources, and the dated erosion phases pre-date major expansion of grazing. Because of the temporal correlation of initiation of erosion across widely spaced sites, and the coincidence of the dating with the Little Ice Age (LIA), Rhodes and Stevenson (1997) concluded that the increased storminess in the LIA induced erosion. In this scenario peat erosion was initiated by the crossing of an extrinsic threshold, since climatic change increased the available erosive energy. Tallis (1997) has also argued that climate change is an important driver of erosion; however, his model is rather more complex. In Tallis' view the particular sequencing of climate change has played an important role in triggering erosion. He argues that widespread drying of the bog surfaces during the medieval climatic optimum caused cracking and decreased infiltration capacity of the surface so that it became particularly vulnerable to erosive forces during the subsequent wetter period of the LIA.

A complete understanding of the initiation of peat erosion is elusive; in part, dating problems and the difficulty of inferring causation from correlation hinder progress. However, it may be that the concept of a single erosional phase with a single cause is excessively simplistic. Tallis (1997) describes erosion of peat at Holme Moss in the Southern Pennines as a result of a combination of historical causes and suggests that this may be a model for blanket peat more generally. The bog-surface vegetation is stressed by a range of pressures, reducing resistance to erosion, at the same time that climate change is increasing the efficacy of erosive forces. It is clear that peatlands are not a static blanket over the landscape, rather they are a dynamic component of the sediment cascade. The following sections explore the nature of that dynamism.

8.3 Upland Mires as Components of the Landscape Sediment System – A Functional Typology

Classification of mires has represented a major direction in mires research, and a large number of schemes exist (e.g. Hughes, 1995; Scott and Jones, 1995; Charman, 2002). Classifications by hydrology and topography (e.g. Seminiuk and Seminiuk, 1995) are of particular relevance here in considering the role of upland mires in the wider sediment cascade. The following section takes the topographic location and hydrological connectivity of mires as the basis for development of a functional typology of upland mires based on their functioning within the wider landscape sediment system. The intention is not to add to the existing morass of mire classification but rather to highlight the important and varying contribution of upland mires to the sediment budget of many upland regions. What is developed is a conceptual classification of mire function within the sediment cascade.

The functioning of mires within the sediment cascade may be divided into those mires which are net sources of sediment and those which are net sinks. As a first approximation, these categories might be regarded as equivalent to the 'eroded' and 'intact' mires which are common reference points in the ecological literature. The large accumulations of organic sediment stored within mire systems mean that it is helpful to subdivide this classification to consider sources and sinks of mineral and organic sediment separately. A further useful subdivision is based on sediment delivery mechanisms which vary according to the position of the mire in the sediment cascade and so are closely related to topographically based classifications of mires. Table 8.1 outlines a classification of upland mires on this basis and highlights the dominant processes in the organic and mineral sediment systems. The basis for the classification is outlined briefly below; details of the geomorphology of eroded and intact mires are described in the subsequent sections.

The classification emphasizes that the contingent position of a mire within the sediment cascade as well as the erosional status of the mire will determine the functioning of the mire within the wider sediment system. Inevitably, such a scheme is a considerable simplification; in reality, there is considerable spatial and temporal variability in mire type and function. Table 8.2 illustrates a situation where bare eroding peat, intact peat, and eroded but revegetated peat all occur within the 22 km^2 of the Moor House National Nature Reserve in the Northern Pennines (Garnett and Adamson, 1997).

The extent of peat erosion in the British Isles is such that most areas are a mosaic of intact and eroded peat. Within the mosaic, patterns of sediment supply are highly variable as the different component areas play different roles in the sediment budget. In addition to spatial variation, the role of mires in the wider sediment cascade changes over time. In the British Isles widespread upland mire formation occurred over the period 8–5 ka. The period of mire formation would have seen upland sites changing from a source of mineral sediment to a net sink as the mineral surface was covered and the mire surface begins to trap colluvial and aeolian deposition. Johnson and Dunham (1963) report extensive clay laminae in the lower peats at Moor House associated with colluvial delivery of mineral material to incipient blanket peat downslope. In some locations mires continue to trap mineral sediment periodically; for example, Ashmore *et al.* (2000) report major mineral horizons within blanket bog stratigraphies on the Hebrides caused by upslope mass movement events.

Blanket peat is commonly underlain by clay-rich glacial and periglacial deposits. In locations where peat formation occurred as early as 8 ka, the reduction in mineral sediment yields may have been particularly marked as the damping effect of the peat blanket was superimposed on elevated paraglacial yields. There are several examples in the literature of expanding peatlands burying pre-existing channels with the effect of reducing local drainage density and, by implication, sediment flux (Vardy, Aitken and Bell, 2000; Thorp and Glanville, 2003)

Widespread erosion of UK blanket peatland dates to the last millennium (Tallis, 1997; Higgitt, Warburton and Evans, 2001; Evans and Warburton, 2007). At this stage, upland mires would have become a major source of organic sediment, and eventually of inorganic sediment as erosion reached the mineral substrate. Recent work has highlighted extensive *Sphagnum* and *Eriophorum* revegetation of eroded gullies at sites across the Pennine range (Clement, 2005; Crowe, Evans and Allott, 2008) so that peat is actively accumulating under contemporary conditions. Timescales for complete revegetation of eroded gully

Table 8.1 Classification of principal sediment transfers in eroded and intact peatlands

Mire status	Topographic position	Typical mire type	Mineral sediment flux		Organic sediment flux	
			Principal inputs	*Principal outputs*	*Principal inputs*	*Principal outputs*
Intact	*Summit*	Blanket peat	Aeolian	Solutes	Peat accumulation	Dissolved organic matter
	Mid-slope	Blanket/hollow	Colluvial	Solutes	Peat accumulation	Dissolved organic matter
	Valley	Blanket/basin/raised	Fluvial/colluvial	Solutes	Peat accumulation	Dissolved organic matter
Eroded	*Summit*	Blanket peat	Aeolian	Fluvial where erosion has reached the mineral substrate	Peat accumulation	Wind/rain action on bare peat/gully erosion
	Mid-slope	Blanket/hollow	Colluvial	Fluvial where erosion has reached the mineral substrate	Peat accumulation	Fluvial gully erosion
	Valley	Blanket/basin/raised	Fluvial/colluvial	Fluvial where erosion has reached the mineral substrate	Fluvial	Wind/rain action and bank erosion

Table 8.2 Percentage cover types in the 22 km^2 Moor House National Nature Reserve, Northern Pennines (data from Garnett and Adamson, 1997).

Mire type	Area (ha)	Percentage cover
Eroding	173	7.8
Revegetated	207	9.4
Pristine	1381	62.8
Other soils	439	19.9

systems under favourable conditions are as short as 10 years (Crowe, 2007) It appears therefore that eroded peat landscapes may experience cyclic 'cut and fill' (Figure 8.1). Blanket peat therefore forms a dynamic temporal and spatial mosaic across upland surfaces. The role of mire systems in the wider landscape sediment cascade is strongly influenced by the erosional status of mire systems. The characteristics of mire erosional states are considered below.

Figure 8.1 Images of gully erosion and revegetation at Shiny Brook in the Southern Pennines, UK. (a) Active gully erosion in blanket peat, 1981. (b) Vegetated gully infill at the same site, 2002.

8.4 Intact Mires

The focus of the limited geomorphological work on upland mires has been on degraded systems. In part this is due to the fact that eroding mires are perceived as sites of environmental degradation, but is also related to the fact that in intact mires surface morphology is controlled by hydrological rather than geomorphological processes. Ingram (1982) demonstrated that the profile of a raised mire in Scotland conformed to that predicted for a groundwater mound with flow controlled by Darcy's Law. The characteristic domed water table of raised mire systems could be described by the mean hydraulic conductivity of the peat mass and the rate of lateral discharge. Water table is the ultimate control on the ground surface level in peat bogs through its control on peat decomposition rates and peat shrinkage. The long-term bog profile is therefore controlled by the hydrological characteristics of the peat mass. This groundwater mound hypothesis has been validated for simple mire forms (Ingram, 1987) although the assumption of constant hydraulic conductivity (K) in the peat mass does not hold in all cases and numerical modelling solutions using variable K are required to describe more complex mire forms (Clymo, 2004). Almost by definition, the drainage density of intact mires is low. However, areas of intact peat are dissected by stream systems which most likely pre-date the onset of peat formation (although demonstrating this unequivocally is difficult). This is particularly so in blanket peat as opposed to raised mires.

Intact mires are sites of sediment accumulation. Mineral sediment fluxes are low because of the mantle of peat. In blanket peatlands there may be downslope delivery of mineral material from steeper slopes where peat cover has not formed. This is particularly common in cirque-floor situations where peat cover lies downslope of bare talus slopes. Surface sediments are, however, typically protected from wind and rain action by dense moorland vegetation so that sediment movement over the surface is minimal. Where intact mires are cut by pre-existing river channels, there may be mineral export from the base of the peat profile by bank erosion. In intact mire systems, organic sediment (peat) accumulates at a rate determined by the balance between the input of dead plant material and the rate of decomposition. The cumulic nature of intact peat soils means that these are by definition important sites for the accumulation of soil carbon (Worrall *et al.*, 2003), which will remain sequestered so long as the mire remains intact (see also Section 8.8).

The principal export of organic matter from intact peatlands is in dissolved form, dissolved organic matter (DOM) gives peatland waters their characteristic brown hue. Export of organic sediment from intact mires can occur in three main ways: first, net transport of peat by downslope creep; second, by bank collapse into fluvial channels (Evans and Warburton, 2001, 2005); and third, by rapid mass movement. Bower (1960) recognized that periodic mass movements were active agents of erosion in blanket peat. Peat slides or 'bog bursts' have been reported and described from several areas (e.g. Alexander, Coxon and Thorn, 1986; Carling, 1986; Dykes and Kirk, 2001; Warburton, Higgitt and Mills, 2003; Dykes and Warburton, 2007). Carling (1986) describes a spectrum of such phenomena from bog bursts, where a slurry of wet peat flows downhill, through to slides of intact material. Dykes and Warburton (2007) formalized definitions of six main classes of peat mass movement (Table 8.3). Peat slides are typically associated with failure at or just below the peat–mineral interface due to high pore-water pressures. Slides are commonly associated with either natural or artificial peat drainage lines. Most failures appear to occur during summer convective storms, although winter events due to prolonged heavy rain or

Table 8.3 Classification of peat mass movements (Dykes and Warburton, 2007).

Failure type	Characteristics
Bog bursts	Flow failure of raised bogs
Bog flows	Flow failure of blanket bogs
Bog slides	Shear failure in blanket bogs leading to sliding
Peat slides	Shear failure at the peat–mineral interface in blanket bogs
Peaty debris flows	Shear failure in the mineral substrate beneath blanket peat
Peat flows	Flow failures of other peat deposits

rain-on-snow have been recorded (Warburton, Holden and Mills, 2004). Peat slides constitute low-frequency high-magnitude organic sediment transfer events. Warburton and Higgitt (1998) assembled data showing eight major recorded peat slides in the Northern Pennines over a period of 65 years. Volumes ranged from 1800 to 38 500 m^3 (180–3850 t). Given the otherwise quiescent nature of sediment export from intact mires, these figures represent a major, though spatially discrete, contribution to the total sediment budget of blanket peat catchments. In addition to the immediate sediment transport associated with the peat slide event, there may be chronic erosion of the devegetated landslide scar. This includes mineral sediment eroded by flow concentrated in the slide scar, and organic sediment weathered from the bare peat edges of the slide scar. As Bower (1960) noted, peat slides and consequent chronic sediment production are a possible mechanism for the initiation of mire degradation so that rather than being a process of intact mires *per se*, peat slides may mark the transition from intact to eroded status.

8.5 Eroded Mires

8.5.1 Erosion Processes in Blanket Peat Catchments

Characteristic patterns of peat erosion were outlined in detail by Bower (1960, 1961). Bower identified two modal patterns of gullies in peat: Type I, a closely spaced dendritic pattern characteristic of lower angled slopes; and, Type II, more widely spaced linear gullies found on slopes typically above 5°. Bower hypothesized that the primary cause of gully erosion was fluvial incision as water flowed between bog pools at high flow.

The first detailed work on erosion processes (although quantitative measurements of erosion rates were undertaken earlier by Crisp (1966) and Tallis (1973)) in peat were two studies of eroding peatland catchments in the UK. Labadz, Burt and Potter (1991) examined sediment yield and sediment delivery in the Shiny Brook catchment, an eroding blanket peat catchment in the Southern Pennines. Francis (1990) studied erosion in a peat catchment at Plynlimon in central Wales. In both cases intrastorm patterns of sediment yield tended to exhibit positive hysteresis with respect to discharge, suggesting that sediment exhaustion during storms is characteristic of eroding blanket peat catchments. Freshly exposed peat is massive and fibrous and relatively resistant to erosion. Peat exposed at the surface is weathered to a friable consistency which is more easily mobilized. Three processes appear to be important: desiccation of surface peat, frost action, particularly formation of needle ice, and raindrop impact (Yang, 2005). Two factors then contribute to the supply-limited nature of peatland streams. First, sediment production by these processes is to some extent limited since the mantle of weathered peat protects the underlying fresh peat from further physical weathering; and second, the highly erodible nature of the weathered layer means that it is rapidly removed in the early stages of a storm. Sediment exhaustion has been observed at a range of temporal scales in peat catchments (Labadz, Burt and Potter, 1991). Progressive exhaustion may also occur over a series of storms, or through the annual cycle (Figure 8.2).

The rate of removal of the weathered layer of surface peat has been extensively monitored through the use of erosion pins in a range of environments. Evans and Warburton (2007) reviewed measured rates and showed that surface recession of bare peat is typically in the range 5–45 mm yr^{-1}. Surface retreat measured by erosion pins records removal of peat by wind and water erosion as well as by oxidation and chemical decomposition. Wind erosion in peats is an understudied area; the best available data come from work by Foulds and Warburton (2007a, 2007b) who demonstrated that maximum rates of wind erosion occur during wind-blown rain events. They suggest that, although wind erosion may be locally important on areas of low slope (summit areas), overall erosion rates are dominated by water erosion of peat. Sediment trap data suggest that *c.* 50% of recorded surface retreat is water erosion and hence delivered to the fluvial system (Yang, 2005).

Seasonal patterns of sediment production are closely linked to the relative importance of frost action and desiccation as weathering processes. Francis (1990) showed that sediment flux in gully systems in mid-Wales was at a maximum in the autumn due to the mobilization of weathered peat produced by summer desiccation. In contrast, sediment trap data from the Northern Pennines shows a winter peak in sediment production associated with frost action (Figure 8.3).

Overall, although there is good evidence on the rates of sediment production from eroding peatlands and the resulting forms have been well characterized, detailed knowledge of the processes of erosion is at a relatively early stage, particularly in comparison with the weight of soil erosion research on mineral soil systems.

8.5.2 Sediment Delivery in Peatland Catchments

Burt, Labadz and Butcher (1997) suggest that the sediment delivery ratio for eroded peat will approach unity: peat is primarily removed from eroding catchments as fine particulate suspended load which, once entrained, is unlikely to be redeposited. In actively eroding systems, sediment delivery ratios are high (i.e. a high percentage of eroded soil is exported from the basin; cf. Figure 1.1) but work on eroded and revegetated peatlands indicates that the process of revegetation exerts a considerable influence over the efficiency of sediment delivery through the fluvial system. Evans, Warburton and Yang (2006) contrast the sediment budget of two systems in the Northern and Southern Pennines (Figure 8.4). Rough Sike in the Northern Pennines is a small catchment in eroded blanket peatland where there has been extensive revegetation so that the gully floors are fully vegetated and sediment delivery to the channel system is across a vegetated slope–channel interface, which is a zone of fine sediment deposition. In contrast, Upper North Grain is an actively eroding blanket peat catchment in the Southern Pennines characterized by extensive gullying and limited revegetation. Figure 8.4 shows that organic sediment yield from the revegetated system is reduced sixfold compared to the actively eroding system where sediment yields are high (195 t km^{-2} yr^{-1}). The effect of revegetation is to dramatically

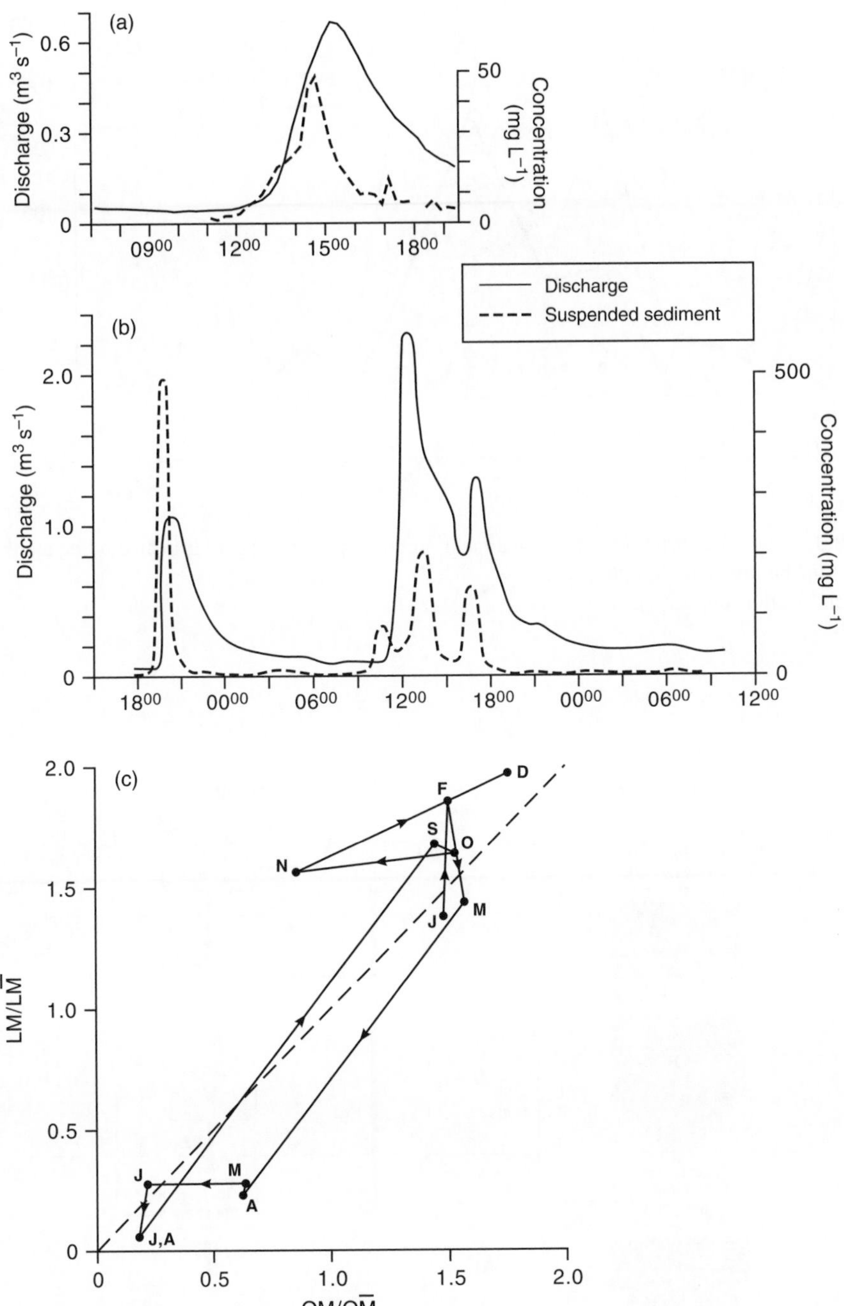

Figure 8.2 Three types of hysteresis in runoff sediment load relation in blanket peat catchments. (a) Intrastorm hysteresis, on Rough Sike an $0.83\,km^2$ catchment in the Northern Pennines. (b) Interstorm hysteresis on 'Wes 1' a $0.0042\,km^2$ catchment in the Southern Pennines (after Labadz, Burt and Potter, 1991).
(c) Annual hysteresis in the upper Severn catchment, Plynlimon, mid-Wales (area $0.94\,km^2$) (after Francis, 1990).

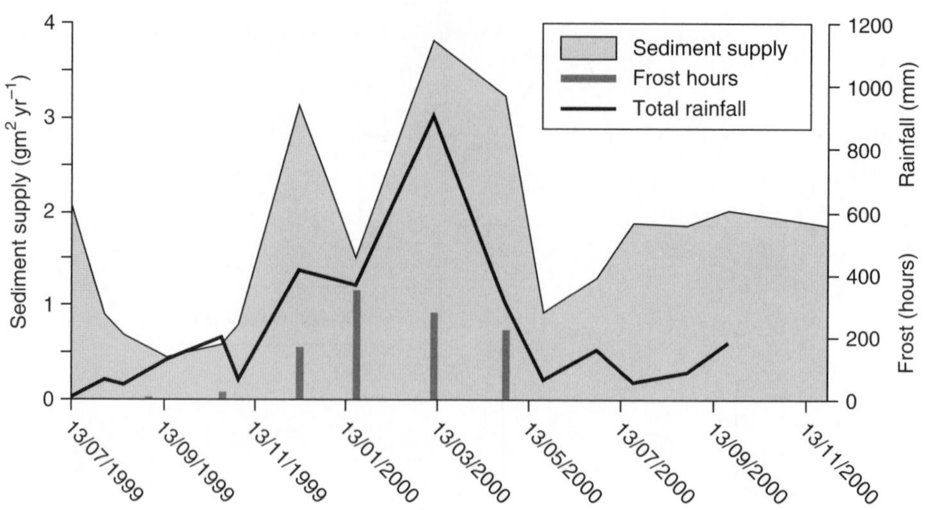

Figure 8.3 Climatic controls on rates of sediment production from gully walls in an eroding blanket peatland in the Northern Pennines, UK
(Evans and Warburton, 2007).

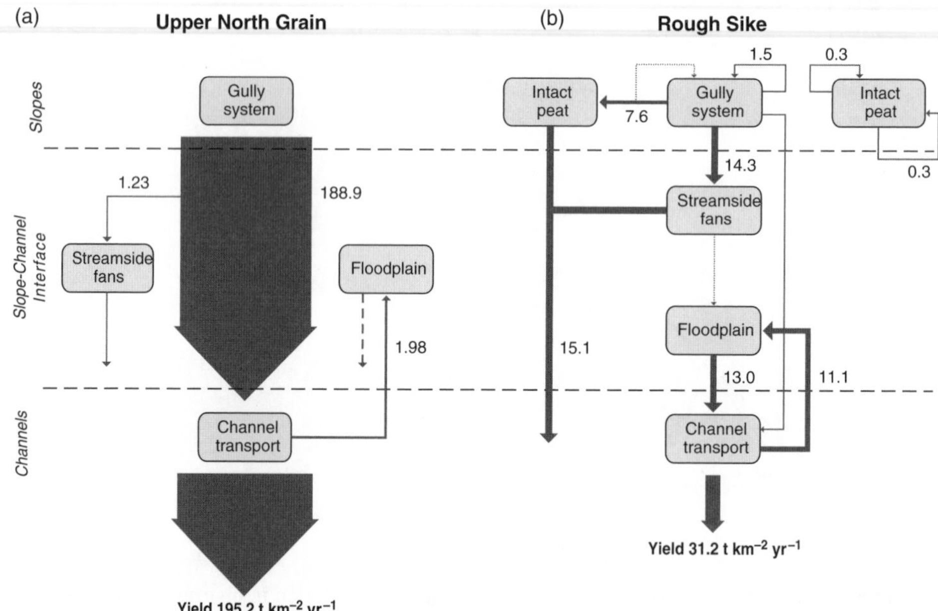

Figure 8.4 Estimated organic sediment budgets for (a) Upper North Grain, an actively eroding blanket peat catchment in the Southern Pennines and (b) Rough Sike, an eroded and substantially re-vegetated blanket peatland in the Northern Pennines (Evans et al 2006).

reduce connectivity between the hillslope gullies and the main channel system so that in the revegetated system the reduced sediment flux is sourced largely from channel rather than hillslope sediment sources.

One important source of channel-derived sediments is through cantilever bank failures (Evans and Warburton, 2001). Bank collapse provides a source of coarse peat to the system but appears also to influence the pattern of suspended sediment transport. Carling (1983) reported frequent 'spikes' of suspended sediment concentration on the falling limb of the hydrograph in Great Eggleshope Beck in the Northern Pennines. He associated these spikes with bank collapse events. Similar events have been observed on Rough Sike in the Northern Pennines. Bank collapse on the falling limb is a common characteristic of rivers with flashy runoff regimes (Frydman and Beasley, 1976; Lawler, 1992). The frequent occurrence of bank collapse events on the falling limb of the storm hydrograph promotes deposition of blocks on bars. Deposition of boulder-scale peat blocks on bar tops and in smaller streams as block jams across the channel has a significant influence on channel form in peatland streams (Evans and Warburton, 2001, 2007). Peat blocks are characteristically deposited in large flood events and consequently form storage elements in the fluvial system at timescales of years to decades. Rapid abrasion of peat blocks during transport is a source of fine sediment to the system when peat blocks are remobilized and abraded during transport (Evans and Warburton, 2001). One significant gap in current knowledge of peatland sediment systems is the quantitative extent of the contribution of peat blocks to the sediment budgets of peatland catchments.

8.6 Valley Mires

The focus of this chapter so far has been on the sediment cascade of upland landscapes; consequently it has focused on blanket peats. Valley mires can occur within blanket peat systems, in particular along floodplain sections of main drainage lines from the peatland. Valley mire systems can also exist downstream of the main peatland system where local valley constriction forms alluvial reaches within upland river systems. Work on valley mires has been largely divorced from studies of upland peat erosion. Consequently, we know relatively little of the dynamics of sediment transfer between mire types. As is clear from Table 8.1, the sediment budget of valley mires is a balance of sediment delivery from upstream and from upslope. Further work on the impact of valley mires on the efficiency of slope-channel linkages within peatlands is required. The role of valley mires in regulating sediment transport in upland valleys has been more widely studied. It is generally accepted that valley wetlands 'filter' sediment from river flows although actual evidence for this is sparse (Kleiss, 1996). Kleiss studied a large turbid river system (Cache River, Arkansas) and demonstrated the importance of floodplain deposition. Large floodplain wetlands effected a 14% downstream reduction in sediment load. In a survey of wetlands in Pennsylvania, Wardrop and Brooks (1998) have shown maximum sedimentation rates of 2200 g m^{-2} (mineral) and 1130 g m^{-2} (organic). The highest rates were recorded in three mire types: headwater floodplains, riparian depressions and impoundments. Phillips (1989) reviewed evidence for sediment storage in wetlands and showed that, for 10 study catchments, 14–58% of total upland sediment production was stored in alluvial wetlands.

Deposition on floodplain wetlands is in part due to normal processes of floodplain sedimentation (see Chapter 11); dense wetland vegetation cover can significantly enhance these processes. Newall and Hughes (1995) demonstrated that flow velocities are significantly reduced in dense vegetation stands due to flow diversion around the plants and that turbulence was reduced where flow traversed vegetation. Klove (1997) showed that the settling efficiency of peat in ponds decreases as water depth increases due to increased turbulence, and suggested that shallow flow on artificial floodplains would provide effective removal of peat from waters draining peat mines. All these studies are consistent with the observation that well-vegetated floodplain mires can be sinks for organic sediment. On Rough Sike in the Northern Pennines, coarse peat deposition is commonly seen just upstream and downstream of linear patches of sedge, consistent with local flow diversion and reduction in velocity (Evans and Warburton, 2001). Figure 8.4 illustrates the importance of floodplain deposition within the wider sediment budget of this Northern Pennine system. Overall, it is clear that there is a significant research need for investigation of the transfers of sediment between upland mire types and the long-term role of valley mire systems as buffers between upland and lowland sediment systems. In addition, of course, if the valley mire is intact, the mire itself will continue to develop, accumulating organic matter *in situ* as well as trapping sediments from upstream. Both processes will be important in ensuring that the catchment system continues to sequester carbon. Erosion of valley mires, on the other hand, will minimize sediment trapping and may limit mire growth. This is discussed further below (Section 8.8).

8.7 Peatland Sediment Yields

Most of the available data on sediment yield from upland blanket peat catchments relate to the UK. Typical measured sediment yields for all UK rivers are in the range 1–500 t $km^{-2} yr^{-1}$ with most rivers in the range 1–50 t $km^{-2} yr^{-1}$ and upland yields approaching 100 t $km^{-2} yr^{-1}$ (Walling and Webb, 1981). Sediment yields recorded from areas of upland peat vary across the whole range: Labadz, Burt and Potter (1991) reviewed published rates of sediment yield from catchments with extensive peat cover and identified yields ranging from 3 to 66 t $km^{-2} yr^{-1}$ in undisturbed catchments with higher rates in disturbed catchments. In the Northern Pennines, Crisp (1966) reported rates of 112 t $km^{-2} yr^{-1}$ from the Rough Sike catchment in Upper Teesdale, whereas remeasurement of sediment yield from this system after an interval of 40 years showed sediment yield reduced to 39 t $km^{-2} yr^{-1}$ because of significant revegetation of the system (Evans and Warburton, 2005). Higher long-term average rates have been indicated by surveys of sediment accumulation in reservoirs. Labadz, Burt and Potter (1991) reported published rates ranging from 4 to 205 t $km^{-2} yr^{-1}$. In the Southern Pennines, Evans, Warburton and Yang (2006) reported yields of up to 265 t $km^{-2} yr^{-1}$ from small actively eroding peatland catchments. What is clear is that sediment yields from eroding blanket peat can be at least as high as from upland catchments dominated by mineral sediment sources. Additionally, as noted by Labadz, Burt and Potter (1991), the low density of eroded peat means that the volumetric erosion rate, and consequently the rate of morphological change is much higher in peatland catchments.

Peatlands are sensitive systems (Bragg and Tallis, 2001) existing within a relatively narrow envelope of hydrological, ecological and geomorphological conditions. The systems are closely coupled so that the initiation of erosion is clearly linked to anthropogenic and/or hydroclimatic impacts on peatlands, but where conditions favour revegetation there are dramatic impacts on the sediment system. Figure 8.5 is a conceptual model of the

Figure 8.5 Conceptual model of sediment delivery in intact, eroded and revegetated blanket peatlands

peatland sediment system under conditions of intact peat, erosion and revegetation. It is based on a model presented in Evans and Warburton (2007) and expanded here to include intact peatlands. It emphasizes the transition from hillslope to channel sediment sources as eroded catchments revegetate (discussed above) and provides indicative sediment yields for the three system states. These indicate an order of magnitude reduction in typical sediment yields between actively eroding and eroded but revegetated catchments, and another order of magnitude reduction if intact peatlands are considered. In summary, sediment yield from peatland systems is strongly controlled by the erosional status of the system (expressed as vegetation cover), which influences both the production of sediment (largely from bare surfaces) and the delivery of sediment to the channel system.

8.8 Peatland Sediment Cascades and Biogeochemical Cycling

The dynamic nature of sediment flux from peatland systems also has significant implications for the role that peatlands play in biogeochemical cycling. The impacts of peat erosion on biogeochemical systems can be divided into two main classes: first, those directly associated with changes in sediment flux from eroding or recovering systems; and second, the effects on peatland function of morphological change in the peatland surface driven by erosional processes.

8.8.1 Sediment-Related Mechanisms

Perhaps the most significant impact of changing sediment flux from peatland systems is the impact on peatland carbon budgets. Soil organic carbon is the largest part of terrestrial carbon storage and in the UK, for example, over 50% of soil carbon is stored in upland peatlands (Milne and Brown, 1997). Intact peatlands are important carbon sinks. The carbon balance for a typical upland bog shows fixation of gaseous carbon at rates on the order of $50\,g\,C\,m^{-2}\,yr^{-1}$ (Worrall *et al.*, 2003). Removal of carbon from catchments as part of either the dissolved (DOC) or particulate (POC) fluvial organic carbon flux reduces this optimum rate of sequestration. In actively eroding systems where annual particulate carbon losses are on the order of $50–100\,g\,C\,m^{-2}\,yr^{-1}$, the erosional losses of carbon alone may be sufficient to shift a peatland catchment system from net carbon sink to net carbon source (Evans, Warburton and Yang, 2006). In terms of feedbacks between the sediment system and climate change, it is important to know what proportion of the erosional flux becomes atmospherically active carbon through chemical and biological transformation to the dissolved or gaseous phases. Indicative data presented by Pawson (2007) suggests that on the order of 30% of the particulate carbon flux may be oxidized within the fluvial system, which indicates a significant feedback. Detailed work on the fate of particulate carbon in transit between source and sink (the sea) is one of the urgent research needs in peatland carbon science.

In addition to their high carbon content and consequent role in the carbon cycle, organic sediments from eroding peatlands play an important role as vectors of peatland contaminants. Peat soils are highly effective in binding metal species so that intact peatlands are zones of storage of atmospherically deposited metals. In areas with a history of industrial contamination this can lead to the accumulation of potentially hazardous concentrations of metals in surface peats (Rothwell *et al.*, 2007a). Mobilization of these polluted organic sediments due to erosion releases contaminated sediment into the fluvial system and to a lesser degree leads to elevated dissolved metal concentrations in stream water (Rothwell, Evans and Allott, 2007; Rothwell *et al.*, 2007b).

8.8.2 Morphologically-Driven Biogeochemical Change

The effects of peatland erosion are expressed not only in the nature of the sediment flux associated with erosional activity but also through the effects of the consequent erosional morphology, particularly gully formation, on peatland function. Water table is the key control on intact peatland function and incision of gully systems through intact peatlands leads to marked reduction of water table in the zone adjacent to the gully (Figure 8.6). Water table controls a range of peatland processes and lowering of the water table should theoretically increase DOC flux from peatlands and has been shown to increase CO_2 flux from decomposition of acrotelm peats (Silvola *et al.*, 1996). Worrall and Burt (2005) successfully modelled DOC flux from an upland peatland using water table and temperature data. Field data on the relation between DOC flux and water table is highly variable and not straightforward to interpret (Worrall, Burt and Adamson, 2004, 2008); laboratory manipulations show increased DOC and CO_2 fluxes but with complex lagged responses to water table change (Blodau and Moore, 2003). Under certain circumstances in polluted peatlands, oxidation of sulphur associated with lower water tables reduces DOC solubility so that gullied systems are observed to have lower DOC flux (Daniels *et al.*, 2008). Patterns of carbon flux are therefore strongly influenced by changing water table but the net effect on the carbon balance is variable in space associated with local soil conditions (Holden, 2005).

The observation by Holden (2005) that detailed understanding of peatland hydrology and geomorphology is central to accounting for carbon cycling in peatlands is a conclusion

Figure 8.6 Modelled water table depth for a gullied peatland catchment in the Southern Pennines. Water table modelled using wetness index and proximity to gullies shows the significant effect of morphological modification of the catchment by erosion on patterns of water table (Allott et al 2009).

which can be extended more generally to peatland ecosystem functions. The organic nature of the sediment and the importance of near-surface water tables are the defining characteristics of peatland systems. The erosional status of peatlands and the role they play in the sediment cascade are important controls on peatland material which demand detailed study alongside the ecological controls on peatland function.

8.9 Peatlands in the Sediment Cascade

Significant progress in research on peatland sediment systems over the past 20 years means that the major controls on sediment flux from peatland systems and the internal sediment dynamics within peatlands in varying states of erosion are now reasonably well understood. The major research needs in this area are to develop modelling to provide quantitative and predictive applications of the current conceptual models. However, understanding of the wider role of peatlands in the sediment cascade is less well advanced. As discussed above, sediment yields from upland peatlands are broadly comparable to the range of values from all upland catchments. Locally, small catchment yields from mineral systems may be close to an order of magnitude greater (e.g. Warburton, Evans and Johnson, 2003) but the extensive nature of the gully erosion observed in upland Britain means that, in the wider landscape, peat erosion is a major component of upland erosion (McHugh, Harrod and Morgan, 2002). Over Holocene timescales, the shift from widespread intact blanket peatlands across the uplands of the UK to the current degraded pattern is likely to have entailed significant shifts in the relative magnitude of organic and mineral sediment fluxes. For much of the Holocene, the land surface has been sealed by thick organic accumulations. Erosion in the last millennium (Higgitt, Warburton and Evans, 2001) initially increased organic sediment flux but increasingly, as erosional gullies incised the mineral substrate, mineral sediment yields will have increased too. The magnitude of these effects at the landscape scale, together with the dynamic role that upland mires play as a buffer between upland and lowland sediment systems at long timescales is poorly understood. Progress in these areas is urgently required, not just to understand the sediment system, but because the fate of sediment sourced from upland peatlands is an important determinant of both carbon losses and pollutant release from organic sediment fluxes.

This chapter has focused on the mechanisms and effects of the erosion of upland mires. It has drawn examples largely from the uniquely eroded upland peatlands of the UK. However, because peatlands are climatically determined landforms, the significant climate changes predicted for the next century will cause climatic stress to wide areas of global peatlands ranging from the melting of permafrost to desiccation of some peatland surfaces. Although there has been considerable work on the potential effects of these changes on biological and chemical cycling, particularly of carbon in peatlands (e.g. Gorham, 1991; Moore, Roulet and Waddington, 1998; Freeman *et al.*, 2004; Belyea and Malmer, 2004), the potential changes associated with physical instability have not yet been sufficiently considered.

Evidence from the UK system indicates that timescales for the physical removal of peat are on the order of several centuries so that, although in some locations the metastable equilibrium condition may be a shift from peatlands to mineral soils, the transient condition of degraded peatlands will be a medium-term reality. Peatlands provide a range of valued ecosystem services (Bonn, Rebane and Reid, 2009; Hubacek *et al.*, in press) ranging from carbon sequestration through hydrological regulation and regulation of lowland sediment flux to recreational spaces. While in an intact peatland these services are regulated by interactions between the ecological and hydrological systems, in degrading systems geomorphological controls become significant. Predicting the impact of physical disturbance of peatlands on delivery of ecosystem services under future climate conditions remains an important challenge for biogeochemical science in which the role of peatlands in the sediment cascade is a key part.

References

Alexander, R.W., Coxon, P. and Thorn, R.H. (1986) A bog flow at Straduff Townland, County Sligo. *Proceedings of the Royal Irish Academy*, **86B** (4), 107–120.

Allott, T.E.H., Evans, M.G., Lindsay, J.L., Agnew, C.T., Freer, J.E., Jones, A., and Parnell, M. (2009) Water tables in Peak District blanket peatlands Report to Moors for the Future/ Environment Agency, Edale.

Anderson, P. (1997) Fire damage on blanket mires, *Blanket Mire Degradation: Causes and Consequences: Proceedings of the Mires Research Group of the British Ecological Society*, (eds J.H. Tallis, R. Meade and P.D. Hulme), pp. 16–27.

Ashmore, P., Brayshay, B.A., Edwards, K.J. *et al.* (2000) Allochthonous and autochthonous mire deposits, slope instability and palaeoenvironmental investigations in the Borve Valley, Barra, Outer Hebrides, Scotland. *Holocene*, **10** (1), 97–108.

Bell, M. and Walker, M.J.C. (1992) *Late Quaternary Environmental Change*, Longman.

Belyea, L.R. and Malmer, N. (2004) Carbon sequestration in peatland: patterns and mechanisms of response to climate change. *Global Change Biology*, **10** (7), 1042–1052.

Blodau, C. and Moore, T.R. (2003) Experimental response of peatland carbon dynamics to a water table fluctuation. *Aquatic Science*, **65**, 47–62.

Bonn, A., Rebane, M. and Reid, C. (2009) Ecosystem services a new rationale for the conservation of upland environments, in *Drivers of Environmental Change in the Uplands* (eds A. Bonn, T.E.H. Allott, K. Hubacek and J. Stewart), Routledge, Abingdon, p. 498.

Bower, M.M. (1960) Peat erosion in the Pennines. *Advancement of Science*, **64**, 323–331.

Bower, M.M. (1961) The distribution of erosion in blanket peat bogs in the Pennines. *Transactions of the Institute of British Geographers*, **29**, 17–30.

Bradshaw, R. and McGee, E. (1988) The extent and time-course of mountain blanket peat erosion in Ireland. *New Phytologist*, **108**, 219–224.

Bragg, O.M. and Tallis, J.H. (2001) The sensitivity of peat-covered upland landscapes. *Catena*, **42** (2–4), 345–360.

Burt, T.P., Labadz, J.C. and Butcher, D.P. (1997) The hydrology and fluvial geomorphology of blanket peat: implications for integrated catchment management, *Blanket Mire Degradation: Causes and Consequences: Proceedings of the Mires Research Group of the British Ecological Society* (eds J.H. Tallis, R. Meade and P.D. Hulme), pp. 121–127.

Carling, P.A. (1983) Particulate dynamics, dissolved and total load in two small basins, Northern Pennines, UK. *Hydrological Sciences Journal*, **28** (3), 355–375.

Carling, P.A. (1986) Peat slides in Teesdale and Weardale, Northern Pennines, July 1983: Description and failure mechanisms. *Earth Surface Processes and Landforms*, **11**, 193–206.

Charman, D. (2002) *Peatlands and Environmental Change*, John Wiley and Sons, Chichester, 301.

Clement, S. (2005) The future stability of upland blanket peat following historical erosion and recent re-vegetation. Unpublished PhD thesis, Durham University, p. 387.

Clymo, R.S. (2004) Hydraulic conductivity of peat at Ellergower Moss, Scotland. *Hydrological Processes*, **18** (2), 261–274.

Crisp, D.T. (1966) Input and output of minerals for an area of Pennine moorland: the importance of precipitation, drainage, peat erosion and animals. *Journal of Applied Ecology*, **3**, 327–348.

Crowe, S.K. (2007) Natural revegetation of eroded blanket peat: implications for blanket bog restoration. Unpublished PhD thesis, School of Environment and Development, University of Manchester, p. 338.

Crowe, S.K., Evans, M.G. and Allott, T.E.H. (2008) Geomorphological controls on the re-vegetation of erosion gullies in blanket peat: implications for bog restoration. *Mires and Peat*, **3**, 1–14.

Daniels, S.M., Evans, M.G., Agnew, C.T. and Allott, T.E.H. (2008) Sulphur leaching from headwater catchments in an eroded peatland, South Pennines, UK. *Science of the Total Environment*, **407**, 481–496.

Dykes, A.P. and Kirk, K.J. (2001) Initiation of a multiple peat slide on Cuilcagh Mountain, Northern Ireland. *Earth Surface Processes and Landforms*, **26** (4), 395–408.

Dykes, A.P. and Warburton, J. (2007) Mass movements in peat: A formal classification scheme. *Geomorphology*, **86** (1–2), 73–93.

Evans, M. and Warburton, J. (2001) Transport and dispersal of organic debris (peat blocks) in upland fluvial systems. *Earth Surface Processes and Landforms*, **26** (10), 1087–1102.

Evans, M.G. and Warburton, J. (2005) Sediment budget for an eroding peat-moorland catchment in Northern England. *Earth Surface Processes and Landforms*, **30** (5), 557–577.

Evans, M. and Warburton, J. (2007) *The Geomorphology of Upland Peat: Process, Form and Landscape Change*, Blackwell, Oxford, p. 262.

Evans, M., Warburton, J. and Yang, J. (2006) Sediment Budgets for Eroding Blanket Peat Catchments: Global and local implications of upland organic sediment budgets. *Geomorphology*, **79** (1–2), 45–57.

Foulds, S.A. and Warburton, J. (2007a) Significance of wind-driven rain (wind-splash) in the erosion of blanket peat. *Geomorphology*, **83** (1–2), 183–192.

Foulds, S.A. and Warburton, J (2007b) Wind erosion of blanket peat during a short period of surface desiccation (North Pennines, Northern England). *Earth Surface Processes and Landforms*, **32** (3), 481–488.

Francis, I.S. (1990) Blanket peat erosion in a mid-Wales catchment during two drought years. *Earth Surface Processes and Landforms*, **15**, 445–456.

Freeman, C., Fenner, N., Ostle, N.J. *et al.* (2004) Export of dissolved organic carbon from peatlands under elevated carbon dioxide levels. *Nature*, **430**, 195–198.

Frydman, S. and Beasley, D.H. (1976) Centrifugal modelling of riverbank failure. *Journal of Geotechnical Engineering Division. Proceedings of the American Society of Civil Engineers*, **102**, 395–409.

Garnett, M. and Adamson, J.K. (1997) Blanket mire monitoring and research at Moor House National Nature Reserve, Blanket Mire Degradation: Causes and Consequences: Proceedings of the Mires Research Group of the British Ecological Society (eds J.H. Tallis, R. Meade and P.D. Hulme), pp. 116–117.

Gorham, E. (1991) Northern Peatlands: Role in the carbon cycle and probable responses to climatic warming. *Ecological Applications*, **1**, 182–195.

Grieve, I.C., Hipkin, J.A. and Davidson, D.A. (1994) Soil erosion sensitivity in upland Scotland, *Scottish Natural Heritage Research Survey and Monitoring Reports 24*, Scottish Natural Heritage, Edinburgh.

Higgitt, D.L., Warburton, J. and Evans, M.G. (2001) Sediment transfer in upland environments, in *Geomorphological Processes and Landscape Change: Britain in the Last 1000 Years* (eds D.L. Higgit and E.M. Lee), Blackwell, Oxford, pp. 190–214.

Holden, J. (2005) Peatland hydrology and carbon release: why small-scale process matters. *Philosophical Transactions of the Royal Society of London Series A: Mathematical Physical and Engineering Sciences*, **363**, 2891–2913.

Hubacek, K., Beharry, N., Bonn, A. *et al.* (in press) Ecosystem services in dynamic and contested landscapes: the case of UK uplands, in *Land Use and Management: The New Debate* (eds M. Winter and M. Lobley), Earthscan, London.

Hughes, J.M.R. (1995) The current status of European wetland inventories and classifications. *Vegetatio*, **118** (1–2), 17–28.

Ingram, H.A.P. (1982) Size and shape in raised mire ecosystems – a geophysical model. *Nature*, **297** (5864), 300–303.

Ingram, H.A.P. (1987) Ecohydrology of Scottish Peatlands. *Transactions of the Royal Society of Edinburgh Earth Sciences*, **78**, 287–296.

Iverson, J. (1964) Retrogressive vegetational succession in the post-glacial. *Journal of Ecology*, **52**, 59–70.

Johnson, G.A.L. and Dunham, K. (1963) *The Geology of Moor House, Monograph 2, Nature Conservancy Council.*, HMSO, London.

Kleiss, B.A. (1996) Sediment retention in a bottomland hardwood wetland in eastern Arkansas. *Wetlands*, **16** (3), 321–333.

Klove, B. (1997) Settling of peat in sedimentation ponds. *Journal of Environmental Science and Health A*, **32** (5), 1507–1523.

Korhola, A. (1995) Holocene climatic variations in southern Finland reconstructed from peat-initiation data. *The Holocene*, **5** (1), 43–57.

Labadz, J.C., Burt, T.P. and Potter, A.W.R. (1991) Sediment yields and delivery in the blanket peat moorlands of the southern Pennines. *Earth Surface Processes and Landforms*, **16**, 265–271.

Large, A.R.G. and Hamilton, A.C. (1991) The distribution, extent and causes of peat loss in Central and Northwest Ireland. *Applied Geography*, **11**, 309–326.

Lawler, D.M. (1992) Process dominance in bank erosion systems, in *Lowland Floodplain Rivers* (eds P.A. Carling and G.E. Petts), John Wiley and Sons, Chichester.

Lindsay, R. (1995) *Bogs: The Ecology, Classification and Conservation of Obrotrophic Mires*, Scottish Natural Heritage, Edinburgh, p. 119.

McHugh, M., Harrod, T. and Morgan, R. (2002) The extent of soil erosion in Upland England and Wales. *Earth Surface Processes and Landforms*, **27** (1), 99–107.

Milne, R. and Brown, T.A. (1997) Carbon in the vegetation and soils of Great Britain. *Journal of Environmental Management*, **49**, 413–433.

Moore, P.D. (1975) Origin of blanket mires. *Nature*, **256**, 267–269.

Moore, P.D., Merryfield, D.L. and Price, M.D.R. (1984) The vegetation and development of blanket Mires, in *European Mires* (ed. P.D. Moore), Academic Press, London, pp. 203–235.

Moore, T.R., Roulet, N.T. and Waddington, J.M. (1998) Uncertainty in Predicting the Effect of Climatic Change on the Carbon Cycling of Canadian Peatlands. *Climatic Change*, **40** (2), 229–245.

Newall, A.M. and Hughes, J.M.R. (1995) Microflow environments of aquatic plants in flowing water wetlands, in *Hydrology and Hydrochemistry of British Wetlands* (eds J.M.R. Hughes and A.L. Heathwaite), John Wiley and Sons, Chichester.

Ovenden, L. (1990) Peat accumulation in northern wetlands. *Quaternary Research*, **33**, 377–386.

Pawson, R.R. (2007) Assessing the role of particulates in the fluvial organic carbon flux from eroding peatland systems. Unpublished PhD thesis, School of Environment and Development, University of Manchester, p. 304.

Phillips, J. (1989) Fluvial sediment storage in wetlands. *Journal of the American Water Resources Association*, **25** (4), 867–873.

Rhodes, N. and Stevenson, T. (1997) Palaeoenvironmental evidence for the importance of fire as a cause of erosion of British and Irish blanket peats, *Blanket Mire Degradation: Causes and Consequences: Proceedings of the Mires Research Group of the British Ecological Society* (eds J.H. Tallis, R. Meade and P.D. Hulme), pp. 64–78.

Rothwell, J.J., Evans, M.G. and Allott, T.E.H. (2007) Lead contamination of fluvial sediments in an eroding blanket peat catchment. *Applied Geochemistry*, **22** (2), 446–459.

Rothwell, J.J., Evans, M.G., Lindsay, J.B. and Allott, T.E.H. (2007a) Scale-dependent spatial variability in peatland lead pollution in the southern Pennines, UK. *Environmental Pollution*, **145**, 111–120.

Rothwell, J.J., Evans, M.G., Daniels, S.M. and Allott, T.E.H. (2007b) Baseflow and stormflow metal concentrations in streams draining contaminated peat moorlands in the Peak District National Park. *Journal of Hydrology*, **341**, 90–104.

Scott, D.A. and Jones, T.A. (1995) Classification and inventory of wetlands – A global overview. *Vegetatio*, **118** (1–2), 3–16.

Seminiuk, C.A. and Seminiuk, V. (1995) A geomorphic approach to global classification for inland mires. *Vegetatio*, **118** (1–2), 103–124.

Shimwell, D. (1974) Sheep grazing intensity in Edale, 1692–1747, and its effect on blanket peat erosion. *Derbyshire Archaeological Journal*, **94**, 35–40.

Silvola, J., Alm, J., Ahlholm, U. *et al.* (1996) CO^2 fluxes from peat in boreal mires under varying temperature and moisture conditions. *Journal of Ecology*, **84**, 219–228.

Solem, T. (1986) Age, origin and development of blanket mires in Sor-Trondelag, central Norway. *Boreas*, **15**, 101–115.

Stevenson, A.C., Jones, V.J. and Battarbee, R.W. (1990) The cause of peat erosion: a palaeolimnological approach. *New Phytologist*, **114**, 727–735.

Tallis, J.H. (1973) Studies on Southern Pennine peats V. Direct observations on peat erosion and peat hydrology at Featherbed Moss, Derbyshire. *Journal of Ecology*, **61** (1), 1–22.

Tallis, J.H. (1985) Erosion of blanket peat in the Southern Pennines: new light on an old problem, in *The Geomorphology of North-West England* (ed. R.H. Johnson), Manchester University Press.

Tallis, J.H. (1995) Climate and erosion signals in British blanket peats: the significance of *Racomitrium lanuginosum* remains. *Journal of Ecology*, **83**, 1021–1030.

Tallis, J.H. (1997) The Southern Pennine experience: an overview of blanket mire degradation, *Blanket Mire Degradation: Causes and Consequences: Proceedings of the Mires Research Group of the British Ecological Society* (eds J.H. Tallis, R. Meade and P.D. Hulme), pp. 7–15.

Tallis, J.H., Meade, R. and Hulme, P.D. (1997) Introduction to blanket mire degradation: causes and consequences, *Blanket Mire Degradation: Causes and Consequences: Proceedings of the Mires Research Group of the British Ecological Society* (eds J.H. Tallis, R. Meade and P.D. Hulme), pp. 1–2.

Thorp, M. and Glanville, P. (2003) Mid-Holocene sub-blanket-peat alluvia and sediment sources in the Upper Liffey Valley, Co. Wicklow, Ireland. *Earth Surface Processes and Landforms*, **28** (9), 1013–1024.

Vardy, S.R., Aitken, A.E. and Bell, T. (2000) Mid-Holocene palaeoenvironmentqal history of Easter Axel Heiberg Island: Evidence from a rapidly accumulating peat deposit. Geo-Canada 2000: The Millennium Geoscience Summit, Calgary, Canadian Society of Exploration Physicists.

Walling, D.E. and Webb, B.W. (1981) The reliability of suspended sediment load data. in *Erosion and Sediment Transport Measurement*, IAHS publication 133, pp. 79–88.

Warburton, J. and Higgitt, D. (1998) Hart Hope peat slide: an example of slope channel coupling, in *Geomorphological Studies in the North Pennines* (ed. J. Warburton), BGRG, Durham, p. 104.

Warburton, J., Evans, M.G. and Johnson, R.M. (2003) Discussion on 'the extent of soil erosion in upland England and Wales'. *Earth Surface Processes and Landforms*, **28** (2), 219–223.

Warburton, J., Higgitt, D. and Mills, A. (2003) Anatomy of a Pennine peat slide. *Northern England Earth Surface Processes and Landforms*, **28** (5), 457–473.

Warburton, J., Holden, J. and Mills, A.J. (2004) Hydrological controls of surficial mass movements in peat. *Earth-Science Reviews*, **67** (1–2), 139–156.

Wardrop, D.H. and Brooks, R.P. (1998) The occurrence and impact of sedimentation in central Pennsylvania wetlands. *Environmental Monitoring and Assessment*, **51** (1–2), 119–130.

Wishart, D. and Warburton, J. (2001) An assessment of blanket mire degradation and peatland gully development in the Cheviot Hills, Northumberland. *Scottish Geographical Journal*, **117** (3), 185–206.

Worrall, F. and Burt, T.P. (2005) Predicting the future DOC flux from upland peat catchments. *Journal of Hydrology*, **300** (1–4), 126–139.

Worrall, F., Burt, T.P. and Adamson, J.K. (2004) Can climate change explain increases in DOC flux from upland peat catchments? *The Science of the Total Environment*, **326**, 95–112.

Worrall, F., Burt, T.P. and Adamson, J.K. (2008) Long-term records of dissolved organic carbon flux from peat-covered catchments: evidence for a drought effect? *Hydrological Processes*, **22**, 3181–3193.

Worrall, F., Reed, M., Warburton, J. and Burt, T.P. (2003) Carbon budget for British upland peat catchment. *Science of the Total Environment*, **312**, 133–146.

Yang, J. (2005) *Monitoring and Modelling Sediment Flux from a Blanket Peat Catchment in the Southern Pennines, Geography*, Unpublished PhD Thesis, School of Environment and Development, University of Manchester.

9

Gravel-Bed Rivers

Michael Church

Department of Geography, University of British Columbia, Vancouver, British Columbia, Canada V6T 1Z2

9.1 Introduction: Where are Gravel-Bed Rivers Found?

The term gravel is used as a catch-all term for clastic material coarser than 2 mm in diameter. It includes pebbles, cobbles, and even an admixture of boulders. In some engineering systems of soil texture designation, gravel is formally defined as the pebble class. In rivers, there is always some finer material present. In many gravel deposits, sand and finer material may constitute up to 30% or more by weight of the total material present. Yet it is the larger material that dominates the character of the deposit. In rivers, this means that the bed and lower banks are composed of the larger clasts, often with little visible evidence of the sand that is sequestered in the interstices of the gravel beneath. Sand may also occur in the upper banks, occupying bar-top positions and forming the surficial sediment of the floodplain (see Section 9.4).

Gravel is derived from physical rock weathering and commonly enters the stream system in montane or upland regions where bedrock is apt to be exposed at the surface and where sufficiently steep slopes may deliver rock directly to the streams. Most gravel derives from point sources in the landscape, such as rock outcrops or landslides, or from linear sources along stream banks, such as undercut slopes. Weathered rock, unconsolidated colluvium or glacial deposits are the principal source materials. In steep mountains, recurrent debris flows may be the dominant means by which coarse material is delivered from hillslopes into stream channels.

Consequently, gravel-bed rivers are characteristic of montane and upland valleys (Figure 9.1b) and of mountain forelands (Figure 9.1c). Gravel-bed channels are also common in arctic and subarctic regions where the frost climate readily promotes mechanical weathering of rock, and in glacial forefields (Figure 9.1a) and formerly glaciated regions where glacial action has initially created rock debris that is later re-entrained by running water and deposited along stream channels.

Sediment Cascades: An Integrated Approach Edited by Timothy Burt and Robert Allison

Figure 9.1 Characteristic settings for gravel-bed streams. (a) Glacial forefield: Lewis Glacier, Baffin Island, Canada. (b) Montane valley: Iskut River, northwest British Columbia, Canada (photograph by J. M. Ryder). (c) Mountain foreland, Waimakiriri River, Canterbury plains, New Zealand. (d) Alluvial fan, Ekalugad Fjord, Baffin Island, Canada. All views upstream

Gravel is often deposited where relatively steep, competent tributaries flow into trunk channels on substantially lower gradients. The gravel is deposited in an alluvial fan – hence alluvial fans, typical of upland environments, are characteristically gravelly (Figure 9.1d). Finally, steep, rock-bound channels may have intermittent and/or partial gravel cover where conditions permit sediment to settle. In this circumstance, the presence or absence of a gravel cover has a profound effect on the capacity of the flows to further erode the rock.

The competence of a stream to move gravel is determined according to Shields (1936) relation

$$\tau_* = \tau / g(\rho_s - \rho)D \tag{9.1}$$

in which τ_* is the Shields number, a non-dimensional representation of shear stress exerted on the stream bed, τ is the actual shear stress, g is the acceleration of gravity, ρ_s and ρ are the sediment and fluid density, respectively, and D is the diameter of sediment that is mobilized. The relation was arrived at by dimensional analysis and experimentation. In gravel mixtures, the critical value of τ_* to initiate sediment movement is commonly found to take a value in the range 0.040–0.045 (Shields' original evaluation gave $\tau_{*c} = 0.06$, but he conducted his experiments on very narrowly graded materials which would naturally have assumed relatively tight packing. Recent estimates for sediment mixtures vary between 0.036 and 0.047) and the grain size of the mixture is indexed by D_{50}, the median size.

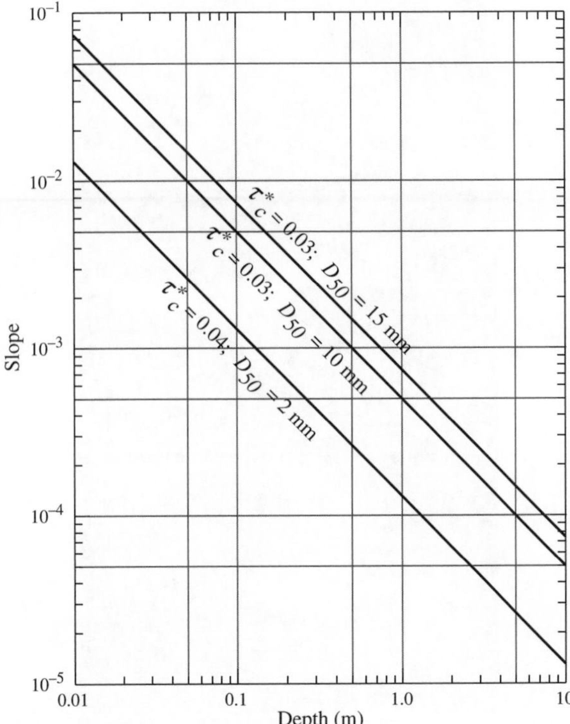

Figure 9.2 Existence diagram for gravel-bed streams, based on ideal and threshold criteria discussed in the text. Mobile gravel occurs above the limit relation; below it, flows are incompetent

In steady, uniform flow $\tau = \rho g dS$, in which d is flow depth and S is the energy gradient of the stream (equivalent to the water surface slope for the specified condition). Hence, for a first approximation of the domain of gravel-bed streams we find, from these formulae, that $dS > 0.066D_{50}$ (adopting $\tau_{*c} = 0.040$) or, substituting for D_{50} the minimum value 2 mm (0.002 m), $dS > 0.00\,013$. Figure 9.2 illustrates the consequent limit regime in stream depth and gradient for mobile gravel to be found.

In fact, this first calculation is practically biased since sediments, once entrained, can be maintained in motion at lower Shields numbers than those required for entrainment. A more realistic kinematic limit value of τ_* is probably closer to 0.030, or even lower. On the other hand, it is rare to find a streambed dominated by gravel with median size smaller than 10 or 15 mm, so the limit value falls in the range $0.00050 < dS < 0.00075$.

9.2 The Character of Gravel-Bed Streams

Headwater channels in steep terrain receive clastic material with a wide range of grain sizes from rock outcrop, either by mass wasting processes or by direct quarrying along the channel. On gradients steeper than about 0.05 (*c.* 3°), material must be very large or it must be constrained in some way if it is to remain in the channel. Unconstrained material on

Figure 9.3 Characteristic morphologies of gravel-bed streams: (a) boulder cascade, Cascade Mountains, Washington State; (b) elementary riffle–pool sequence, Jordan River, Vancouver Island; (c) pool–riffle–bar structure, Lardeau River, southeast British Columbia

such gradients is swept away by flows as deep as the material is large (see Chapter 2). Consequently, the channel takes the form of a continuous cascade (Figure 9.3a) or a step–pool cascade in which key rocks are jammed against each other and against the banks, creating structures sufficiently strong to withstand strong fluid forces (Figure 9.2). Finer material may be retained behind these boulder or cobble structures so that sediment transport during normal flows remains small (giving rise to the common vision of crystal clear mountain streams). However, the rare flows that are sufficient to break the key structures may mobilize much or all of the accumulated sediment in the channel, giving rise to a debris flood or debris flow (see Chapter 2). Such channels are, then, episodically purged of sediment in dangerously violent events.

Water in the landscape seeks the steepest path of descent, so stream gradients in general decline downstream. On the other hand, streamflow and depth increase. Referring, again, to the Shields equation, we find that the largest clast that a stream may transport has $D \approx 20dS$ and this limits stream competence – that is, the size of the largest clasts that can be mobilized. Downstream, S declines faster than d increases. Consequently, sediments steadily become finer and so better sorted as they are transported downstream onto lower gradients by competence limited flows. However, the fining sequence is often interrupted at tributary confluences, where coarser sediment may be injected into the trunk channel (Rice, 1998; Rice and Church, 1998).

A second important criterion that affects the sediment transporting character of the stream and the consequent morphology of the channel is the relative scale of the flow and the sediment, expressed as the relative roughness, D/d, in which D is some relatively large grain size (often D_{84}, the size at which 84% of the material in the streambed is finer; this value is selected because it falls one standard deviation above the mean in a normal distribution). On gradients lower than about 0.05 and with $D/d < 0.3$, approximately, sediments begin to be stacked into aggregate deposits, so that rudimentary structures, such as riffles and bars appear.

Initially, such features may have little expression (Figure 9.3b), the channel bed remains relatively plane and the flow uniformly rapid. One may, in fact, define such channels as 'rapids' (they have also been called 'plane bed' channels; cf. Montgomery and Buffington, 1997). As relative roughness declines, structures become better expressed, issuing in the well-known riffle–bar–pool triplet (Figure 9.3c) that characterizes the morphology of most gravel-bed channels. In the range $0.07 \geq S \geq 0.02$, step-pools, rapids and riffle–pool features often alternate along the channel according to the local variations in gradient.

Gravel-bed rivers on gradients below about 0.02 and with $D/d < 0.1$ (which we will call 'large rivers', for convenience) express well-developed riffle–pool–bar triplets (Figures 9.3c and 9.4) in which the riffles and bars represent many grains-deep aggregations of gravel. The individual clasts have now become so small in comparison with the scale of the flow that they no longer dominate the organization of the flow over them. The channel itself now directs flow so that systematic secondary flows develop that in turn mould the channel by directing bed material sediment transport out of the pools and onto the riffles, or onto the superposed bars (Figure 9.5). The secondary flows direct water toward one side or the other of the river, where flow convergence creates strong currents that scour the pools. These flows are organized into successively opposed secondary circulations along the channel so that the riffles take up successively opposed oblique orientations and the bars alternate (hence 'alternate bars') along the channel. In larger rivers, this organized morphology becomes recognizable as the – usually somewhat irregular – equivalent of meanders in finer grained channels.

The sequence of morphologies briefly described above, determined as it is by gradient and bed material grain size, occurs systematically along the river system, as has been described by Grant, Swanson and Wolman (1990), and by Montgomery and Buffington (1997). Both of those accounts emphasize features in steeper, montane channels. A summary of these features is given in Table 9.1.

The channel geometry of rivers is strongly influenced by the character of the bank materials, which determines the aspect ratio (ratio of width to depth) of the channel. In small channels, including relatively steep ones, the large material lining the banks, out of the way of the direct current, effectively defends the banks and maintains a relatively narrow

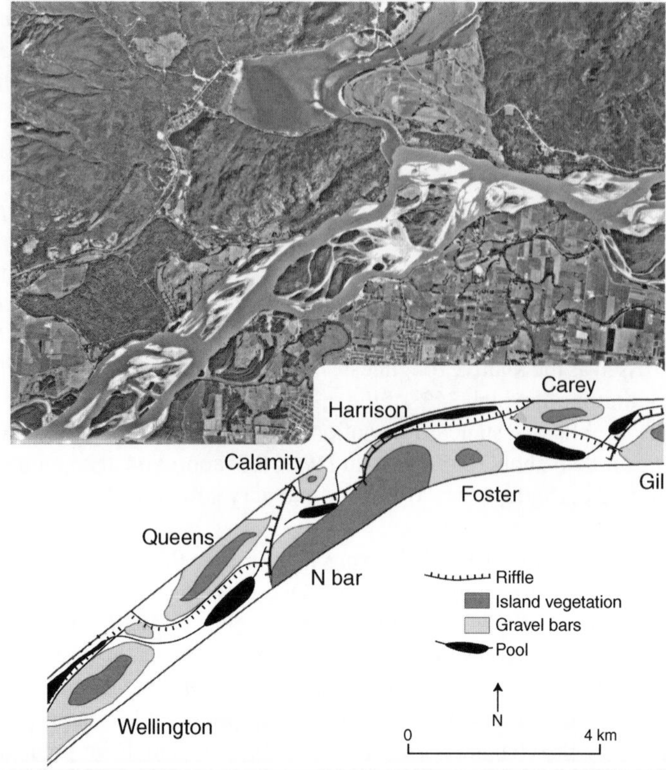

Figure 9.4 Aerial photograph and idealized pattern of pool–riffle–bar structures in a large gravel-bed river; lower Fraser River, British Columbia. Principal bars are named (photo reproduced by permission of the Government of British Columbia).

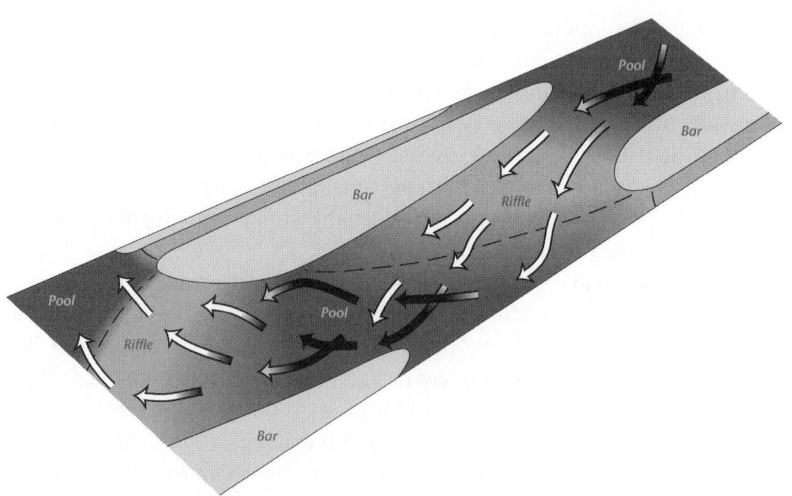

Figure 9.5 Anatomy and circulation of a riffle–pool–bar unit

Table 9.1 Morphological characteristics of gravel-bed channels. Summarized from several studies; details given in Church (2002)

Reach type	Gradient	Relative roughness	Comments
Cascade	> 0.04	≥1	Rock-bound, or jammed cobbles or boulders; may incorporate large wood in forest environments; includes the distinctive step-pool type
Rapid	0.02–0.05	0.3–1.0	Also called 'plane bed'; a feature with relatively high gradient, high velocity and significant downstream extension
Riffle	0.01–0.02	0.1–0.5	Major repeating element in gravel-bed channels, usually in alternately opposed forms; locally high gradient and velocity
'Glide'	< 0.02	0.1–0.5	An extended riffle, often a sediment-filled pool; informal term introduced by fisheries ecologists
Pool	< 0.01	< 0.1	Closed depression, often exhibiting zero gradient at low flow

channel. In the presence of heavy vegetation, the penetration of tree roots below channel depth also creates relatively strong, stable banks (Millar, 2000). In large rivers, however, gravelly banks that are often the earlier deposits of the river itself, have no greater strength than the bed material, are often not protected by effective root penetration, and are subject as well to gravitational failure. Hence they are relatively weak, so that the channels characteristically become relatively wide and shallow, with important consequences for sediment transport (see Section 9.3.1).

River bars constitute the accumulation of temporarily stored sediment in transit through the river system. Hence, the overall morphology of the channel is influenced, as well, by the volume of sediment delivered to the stream channel and moved downstream. As the sediment supply increases and the channel becomes more congested, the conveyance of water over or around the accumulated sediment creates an increasingly wide channel and more and more extensive bars, ultimately issuing in the division of flow into a number of subchannels around the bars. This is the braided state, long recognized as a distinctive channel morphology (though probably not so distinctive as has been imagined: see Section 9.4.2).

9.3 Sediment Transport in Gravel-Bed Rivers

9.3.1 Fundamental Aspects

Clastic sediment moving through a river system is commonly classified, according to the mode of transport, as either suspended load or bedload. The basis for the classification is the mechanics of the transport process and measurement procedures are designed accordingly. A more fundamental distinction, from the perspective of river morphology, is between washload and bed material load (Church, 2006). The latter is the material resident on the bed and lower banks of the channel that give the channel form and character. The former is finer material that, once entrained, can be moved in continuous suspension a relatively long way through the system and, consequently, is not commonly found in the local bed except in interstitial positions within the bed where it may have been filtered out of the subsurface flow, and in backwaters or overbank. In gravel-bed streams, washload is medium sand and finer material. Bed material moves almost exclusively as bedload, with the exception of coarser sand which may intermittently go into suspension.

Most of the sediment that moves through gravel-bed channels is wash material (see Chapter 10). In montane streams, all of the transported sediment may be wash material for periods of time that in some channels may extend to years or decades. When a gravel mobilizing flood finally occurs, however, the total yield is rebalanced so that the long-term yield of sediment is equivalent to the sediment input, which may amount to as much as 50% or more of coarse material. Farther down the system, as competence limitation and sediment abrasion act to reduce the size of the resident material, the fraction of fine material in the sediment load increases to the point that, in large gravel-bed rivers, the proportion of coarse material in transport may be as low as the order of 1% of the total load. The systematic mismatch between the proportion of the load that is sand size or finer, and the proportion that is gravel, indicates that the finer material is sourced in larger proportion and moves through the system more quickly than the coarse material. But despite the modest proportion of bed material transport in larger gravel-bed streams, it is this material (by definition) that dominates the morphology of the channel.

How, given the dramatic change downstream in the proportion of coarse material that can be moved through the channel network, is a sediment balance maintained? It is evident that coarse material must be broken down to finer material within the channel. It has long been supposed that this effect must be achieved by abrasion during transport. The rapidity with which initially angular rocks become edge rounded is evidence that this process is somewhat effective. However, the cobble and boulder materials that may be fed into stream headwaters probably are not primarily worn in this way. Most sediment attrition in river channels occurs while the grains are sequestered in stream deposits. The wet or wet–dry conditions of the channel bed and floodplain create highly effective weathering environments. In northern and alpine environments, freeze–thaw physical weathering dominates whereas in tropical environments chemical weathering promotes the breakdown of mineral materials. Jones and Humphrey (1997) have demonstrated the effectiveness of *in situ* weathering of riverbed materials, some of which may be sequestered for centuries or millennia in floodplain deposits.

Sediment transport in gravel-bed channels has been divided into three stages (Ashworth and Ferguson, 1989). In stage 1 transport, movement is limited to overpassing of grains delivered from upstream that are generally finer than the local bed, which takes very limited or no part in the transport. In stage 2, episodic movements of individual grains occur and entrainment is, in general, restricted to the grains exposed on the surface of the bed or to a few-grains-deep layer that does not, in aggregate, exceed in thickness twice the diameter of the larger grains exposed on the bed (Wilcock and McArdell (1997) specified $2D_{90}$). In stage 3, movement of the bed becomes general and extends to several grain diameters depth. In stage 2, not all exposed grains are in motion – indeed, most of the grains on the bed remain static most of the time. The transport is said to be 'partial' in the sense that only some fraction of the grains of a given size resident on the bed are mobile at any one time. Therefore, through the course of a flood, the characteristics of the population of grains on the bed changes little (Wilcock and DeTemple, 2005). Furthermore, since Shields'

Equation 9.1 posits a linear relation between applied stress and the competence of the flow, we expect that, in stage 2 transport, the transport may be size-selective, with the probability for motion of smaller grains exceeding that for larger ones.

Reformulating these conditions in terms of the Shields number, we may define stage 1 as occurring when $\tau_{*c} < 0.04$, or less than the value required to entrain material from the local bed but greater than necessary to maintain already entrained grains in motion. Stage 2 transport is found to occur when $0.04 < \tau_* < 0.1$ or, more generally, when τ_* is no more than about twice the threshold value for local entrainment. Thereafter, general movement of bed material occurs. It is found that bed material transport in gravel-bed streams is almost always restricted to stages 1 and 2; that is, that τ_* almost never exceeds $2\ \tau_{*c}$. This occurs because, when the shear force exerted on the bed exceeds this limit, the corresponding forces on the non-cohesive banks is sufficient to erode them, leading to widening of the channel, reduction in flow depth, and reduction in shear force (and the Shields number) to within the customarily observed range. Accordingly, such channels can be defined as 'threshold channels' (Church, 2006), signifying that the dynamic condition of the channel almost always remains near or below the threshold for motion of the bed material. When an extreme flood does impose forces that exceed this limit on the bed and banks very rapid erosion may occur, leading to dramatic reorganization of the channel in the attempt to accommodate the large flow (Figure 9.6). This is the mechanism for the sometimes catastrophic changes created by major floods along gravel-bed rivers.

Figure 9.6 Rapid enlargement of a gravel-bed channel by an unusually large flood. Highland Creek, Toronto, Canada, after a 500 $m^3\ s^{-1}$ flood in August 2005. Shields number for the flow was on the order of 0.3 (for $D = 100$ mm bed grain). The black outline indicates the well-vegetated pre-flood river channel banks
(photograph reproduced by courtesy of the City of Toronto Engineering Department and Professor Joseph Desloges).

9.3.2 Self-Organization of the Channel Bed

Bed material transport in gravel-bed channels induces, via grain–grain interactions, additional phenomena that act to modify the criteria given above for transport. Because of the inherently size-selective character of entrainment, finer grains are preferentially moved away. They may not move far since, as they move over the bed, they encounter voids into which they may fall between the static larger grains – a process termed 'spontaneous percolation'. For static grain D_s and mobile grain D_m, if $D_m/D_s < 0.15$ the smaller mobile grain may fall through the smallest opening in a tightly packed array of spherical grains; on the other hand, if $D_m/D_s > 0.22$, the mobile grain may fail to pass the openings in the most open packing of such grains. While these limits may vary for the mixtures of irregular grains found in streambeds, the net effect is still to clear finer material from the bed surface, leaving a layer, one grain deep, that is coarser than the material beneath. This is the characteristic 'armour layer' found on the surface of gravel-bed streams (Figure 9.7a). The percolated fine material commonly collects in a congested layer in the first 10 cm below the surface, however (Wooster *et al.*, 2008), where it may create significant problems for streambed organisms.

Figure 9.7 Sediment organization in gravel-bed rivers. (a) Lightly armoured surface, Quesnel River, British Columbia: direction of flow is left to right. (b) Heavily imbricated cobble armour, Coquihalla River, British Columbia: direction of flow is left to right. (c) Gravel patches, Peace River, British Columbia: flow is from left to right – the finer material is overrunning coarser material. (d) Stone lines in Gunnerside Gill, North Yorkshire, UK

The result of this process is to increase the relative exposure and likelihood to be entrained of coarse material, while reducing the exposure of finer grains. Consequently, the likelihood for grains of any given size to be transported is shifted away from simple inverse relation to particle size (the prediction of Shields criterion) toward equal likelihood for any grain to be entrained; a condition termed 'equal mobility'. If the condition of equal mobility were realized, then all grains would move off at the same value of τ and we would find, from Shields' relation, that τ_{*c} is inversely proportional to grain size. This phenomenon is called the 'hiding effect' (in reference to the fact that the smaller grains are 'hidden from the flow' under or in the lee of larger grains) and, in the simple functional form presented here, can be called a 'complete hiding effect'. In this circumstance, no transport-associated sediment sorting would occur.

Finer gravels and sand may also be transported into areas where the current slackens and be deposited there, constituting a lateral segregation mechanism. More generally, because of the relation between fluid force and competence, the transport of sediment by running water, once mobilized, is inherently a sorting process. Different sized material is found at different places on the channel bed. Accordingly, sediment is sorted into 'patches' (Paola and Seal, 1995) with characteristic surface dimensions of order metres to tens of metres, or even hundreds of metres in the largest channels (Figure 9.7c). These vertical and lateral sorting mechanisms act to condition the bed to withstand the ambient fluid forces and, again, to reduce the propensity for sediment to be moved. Grain size is locally adjusted to bring τ_* closer to the critical value.

Another mechanism that promotes grain sorting on the bed surface is the interaction between mobile grains and static ones. A mobile grain much larger than the local substrate grains will simply roll or slide over the bed, while a mobile grain much smaller is apt to be swept around protruding static grains by the force of the deflected flow. Grains of similar size, however, are apt to be blocked and to stop against or immediately adjacent to the static one. This promotes the homogeneity of patches. It also leads to a stone-on-stone structural arrangement within patches known as imbrication (Figure 9.7b). This arrangement significantly increases bed stability because the entrainment of an individual clast is no longer a matter simply of overcoming its inertia, but entails prying the clast loose from its 'locked' position in the imbricate structure.

Grain–grain interaction produces more complex structures. In flows very close to threshold over a gravel bed, the largest stones on the bed do not move at all. They form key obstacles that block other large clasts when they do move. The result is the development of stone clusters (Brayshaw, 1984), stone lines and cells (Church, Hassan and Wolcott, 1998) consisting of the largest stones in the streambed (Figure 9.7d). The incorporated clasts commonly are larger than D_{70} of the surface material. The structures that they create may occupy up to 20% of the bed surface and carry a large proportion of the applied fluid force. This, again, substantially increases the stability of the bed surface (Church, Hassan and Wolcott, 1998).

The sedimentary effects described above constitute self-organization of the bed that results from grain movements and interactions with static grains. The summary effect is to increase the critical Shields number for the entrainment of grains from a structured gravel bed by a factor of two or more. The Shields number, which traditionally (and in the discussion to this point) has been interpreted as an index of grain stability on the streambed is, in fact, an index of *bed* stability: it discriminates the chance for a grain to be entrained from the bed, account taken not just of grain inertia, but also of its packing arrangement in relation to adjacent grains and to the overall bed structure in its neighbourhood. These circumstances mediate both the fluid force that actually impinges on the grain and the force necessary to move it. As data have accumulated, it has become clear that the critical shear stress to entrain sediment, τ_{*c}, indeed varies over the 2× range posited above (Figure 9.8). Values greater than 0.1 have been reported from the field (Mueller, Pitlick and Nelson, 2005).

Taken altogether, the emergent properties of the bed represent a response to the imposed flows that increases the stability of the bed against those flows by reducing the propensity for grains to be entrained. Of course, absolute stability is never reached because of the continual variation of the flows and the introduction of additional bed material from upstream.

As an interesting comparison, similar arrangements do not occur in sand beds because the individual grains are sufficiently small that a wide range of flows can entrain any size. Consequently, full mobilization of the bed is common and stage 3 transport is normal in sands. Indeed, the strongest flows can entrain fine gravel in a similar manner, and exceptionally large flows move gravel in the manner of sand, so that sandy-type bedforms,

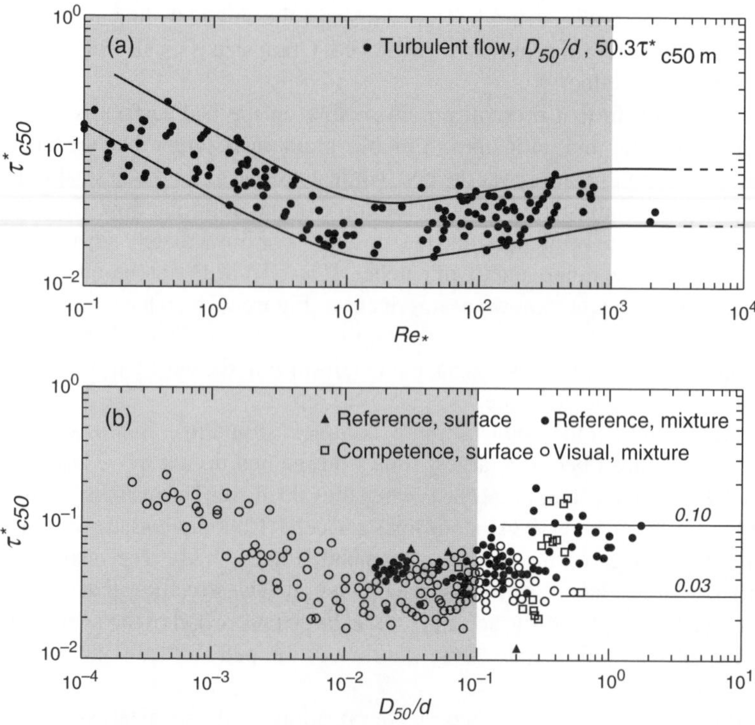

Figure 9.8 Variation of τ_{*c}, the critical Shields number (a) Shields' criterion: data for high relative roughness excluded. The abscissa is the grain Reynolds number, introduced by Shields. (b) Critical Shields number versus relative roughness; the range of τ_{*c} for gravel, encompassing also the range for stage 2 transport, is delimited. In both diagrams, the domain for gravel is highlighted (from Buffington and Montgomery, 1997, their figure 3, somewhat simplified; reproduced by permission of the American Geophysical Union).

such as dunes, are known in gravel (e.g. Dinehart, (1992) reported values of τ_* in the range $0.1 < \tau_* \leq 0.3$ in the North Fork Toutle River, in Washington state). There is no reason to suppose that there is a physically mediated sharp break between these contrasting modes of sedimentary behaviour, which depend on the physically continuous variation in the scales of flows and of sediment properties. There is, however, something of a practically imposed break. It is rather unusual, except in quite small channels, to find streambeds dominated by fine gravel. There is often a relative paucity of grains within the size range 0.5 to as much as 8 mm in widely graded fluvial gravels (Figure 9.9). To find the median grain size (the size customarily used to gauge the transport of sediment mixtures) in this range is quite rare. The reason for the gap is not entirely resolved. It has been proposed that clasts in this size range rapidly disintegrate into their constituent mineral grains. Alternatively, it is supposed that such grains – the smallest and most inherently mobile of gravels – outrun larger gravels and continue in motion over sand downstream from the gravel–sand transition. It is also possible that these sizes outrun others simply because they fall within the range of sizes that, although relatively highly mobile, are nevertheless excluded from percolation below the surface by limited pore openings, hence cannot effectively be hidden and do not show up in significant quantities in bed material samples collected from the streambed. That gravel–sand transitions are abrupt is well-known (Yatsu, 1955; Sambrook Smith and Ferguson, 1995: Figure 9.10). These circumstances demarcate distinctive sedimentary behaviour between sands and gravels.

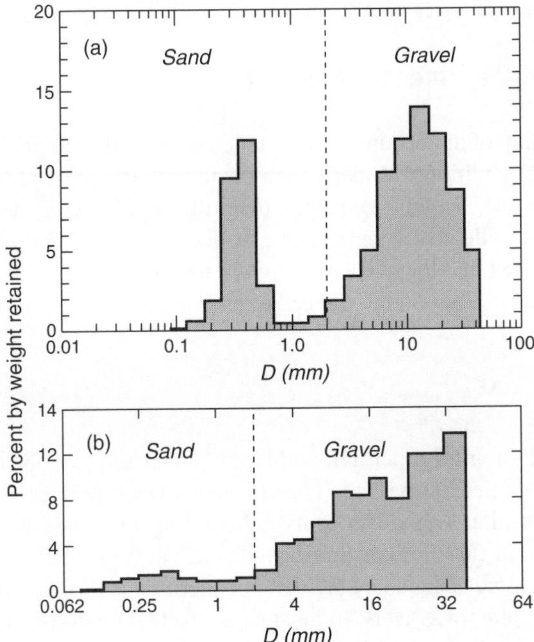

Figure 9.9 Size analyses of fluvial gravels. (a) Bar gravel from lower Fraser River, British Columbia, a relatively extreme example (reproduced by permission of John Wiley & Sons Ltd). (b) Bar gravel from Harris Creek, British Columbia (reproduced by permission of the American Geophysical Union).

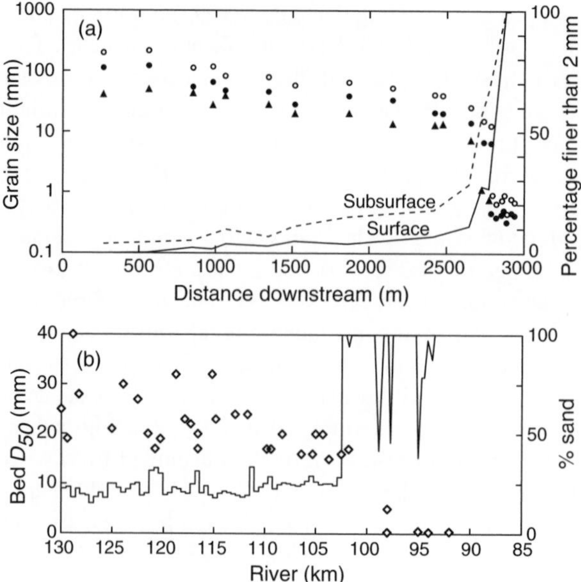

Figure 9.10 Downstream fining of gravel stream beds culminating in the gravel-sand transition. (a) Allt Dubhaig, Scotland (Sambrook Smith and Ferguson, 1995; reproduced by permission of SEPM Society for Sedimentary Geology). (b) Lower Fraser River, British Columbia: positions are given as distances upstream from the river mouth

9.3.3 Predicting Sediment Transport

The definitive measure of understanding in science is the ability to predict the phenomena, given the governing conditions. Modern attempts to predict sediment transport on the bed began with du Boys (1879), who conceptualized the process as a sheet-like progression of successive layers collectively several grains deep, over the bed (in essence, stage 3 transport). He proposed that the mass of sediment transported was a somewhat non-linear function of the fluid force applied to the bed in excess of some threshold force required to initiate movement:

$$i_b = C\tau(\tau - \tau_c) \tag{9.2}$$

wherein i_b is the mass transport per unit width of channel and τ_c is the critical shear stress required to initiate motion of the bed. The constant C incorporates sediment properties. Shields adopted a similar formalism, only substituting his non-dimensional quantities. A substantial number of variations on this formula were developed throughout the twentieth century and mostly tested using data obtained in experimental flumes. Tests have variously been made for sands and for gravels. In most cases, following du Boys' original intuition, the tests were conducted at relatively high transport rates.

In a notable variation on this approach Bagnold (1966, 1980) proposed that the mass of sediment transported by a stream ought to depend on the power of the stream (that is, on its ability to do work, subject to an adjustment to account for the efficiency with which momentum might be transferred from the water to the sediment grains). Hence, he adopted the predictor variable ρqS, in which $q = Q/w$ is the flow per unit width of channel and w is channel width: $\rho qS = \rho vdS = \rho v\tau/g$, which factors water velocity into the practical calculation.

Most of the candidate formulae put forward have adopted a 'bulk' approach to predicting total sediment transport; that is, it has been assumed that the character of the sedimentary material can be adequately described by a single representative size. In most cases, the chosen size is the median grain size (D_{50}) of either the subsurface bed material or of the surface bed material. The reasoning behind the former choice is that, in the long run, this must represent the material that is actually transported by the stream; the reasoning for the latter is that it represents the material exposed to the flow and available to be entrained. Two critical assumptions underlie these decisions. First, for a single size to successfully represent the sediment mixture, it must be assumed that the condition of equal mobility is actually achieved and that transport (at least locally) is not a sorting process. Second, it is assumed that the sediment in (or on) the bed is representative of the sediment that is transported (again, no sorting). In this case, sediment transport is a three parameter problem comprising the flow, the sediment character, and the amount transported. The sedimentary phenomena described above, however, indicate that neither assumption is correct. Sediments are to a greater or lesser extent differentially transported according to size and the problem involves four parameters, since the resident population of sediment and the transported population may be different. But it is reduced to a three parameter problem in a flume that is run to equilibrium with the imposed sediment feed (or circulation, in the case of a recirculating arrangement).

The most significant essay in the early approach to the prediction of gravel transport was that of Meyer-Peter and Müller (1948) based on three extensive series of flume tests, mainly using gravel (mean grain sizes varied between 0.38 and 28.65 mm; gradients varied between 0.0004 and 0.023 (the data are recorded in Smart and Jaeggi, 1983)). They proposed

$$q_{b*} = 8[(q'/q)(K_b/K_r)^{3/2}\tau_* - 0.047]^{3/2} \tag{9.3}$$

wherein $q_{b*} = q_b/\sqrt{g}(\rho_s - \rho/\rho)D_m^{3/2}$ is the dimensionless volumetric bedload transport per unit width, q' is the sidewall-corrected specific discharge, K_b and K_r are the total and grain-roughness Manning–Strickler coefficients, respectively, τ_* is the Shields number of the flow and 0.047 is the value experimentally arrived at for τ_{*c}. (The formula presented here is in non-dimensional form to conform with Shields' work, but was originally presented in dimensioned form.) Reanalysis in the light of modern hydraulic principles (Wong and Parker, 2006) has yielded an amended form

$$q_{b*} = 3.97(\tau_{b*} - 0.0495)^{3/2} \tag{9.3a}$$

in which τ_{b*} is the shear stress applied to the bed (i.e. sidewall corrected). This reduces the predicted values of sediment transport by about a factor of two. The revised formula appears to predict gravel transport in some simple river sections reasonably well, while the value of τ_{*c} falls about mid-way in the observed range, and is similar to the frequently accepted value 0.045 for mixed gravels without developed structures.

A notable feature of the way in which prediction formulae of this type have been applied is that the shear stress (or stream power) is customarily based on average values of the hydraulic variates for the channel, or portion thereof for which the calculation is being made. It seems, at face value, unlikely that sediment entrainment in a turbulent water flow from a structured granular bed would physically be so simply explained. Another perspective is to notice that the arguments in the shear stress or stream power formulations are all scaled according to the flow. On this basis, it would seem that du-Boys-type formulae may simply be scale relations of the flow. An alternative approach, then, would be to scale sediment transport by flow directly. This is what Schoklitsch (1962; summarizing a long development) did, finally presenting

$$q_b = 0.93(q - q_c)S^{3/2} \tag{9.4}$$

There are a number of variants of this equation (see further discussion in Chapter 2).

The direct measurement of bedload transport in rivers has for a long time remained a difficult exercise because most samplers disturb the natural process and because it may be dangerous to deploy instruments during significant sediment transporting flows. In recent years, however, an increasing number of field data have become available, mostly – though by no means entirely – from smaller streams. These data show that the relation between shear stress (or stream power) and sediment transport in gravel-bed channels customarily is extremely sensitive (Figure 9.11) – the exponent in the transport relation is much greater than the value 3/2 found by Meyer-Peter and Müller and by many other experimental investigators. This outcome can be anticipated from the phenomenological observations reported in Section 9.3.1. When transport begins, grain movements are isolated and sporadic. As flow increases transport increases sharply not because of rapidly increasing intensity of transport, but because more and more of the bed begins to contribute to the overall transport. Hence, stage 2 transport is very non-linearly sensitive to river stage. Only in stage 3, when further increase in the transport is obtained by deepening the active layer, which involves transferring momentum by the very inefficient process of grain–grain collisions, might the relation asymptotically approach the high transport limit of 3/2 power sensitivity (Figure 9.11e).

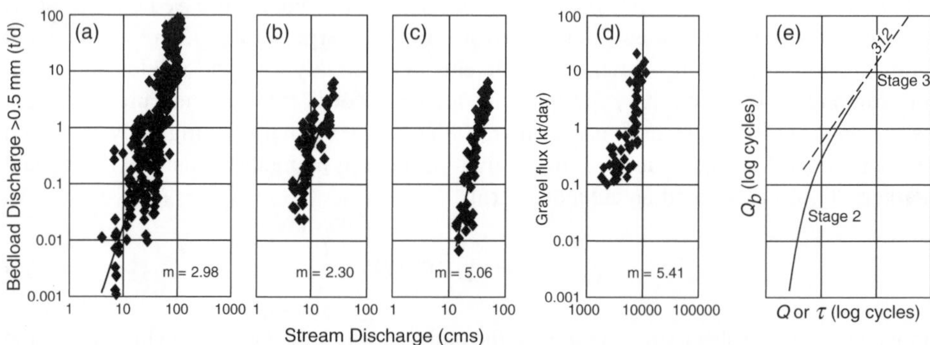

Figure 9.11 Example relations between bedload transport and streamflow in some gravel-bed streams. (a) Little Granite Creek near Bondurant, WY ($A = 56.4\ \text{km}^2$). (b) Valley Creek near Stanley, ID ($A = 381\ \text{km}^2$). (c) Selway River near Lowell, ID ($A = 4950\ \text{km}^2$) (all the foregoing from Emmett and Wolman, 2001; Reproduced by permission of Wiley-Blackwell). (d) Fraser River at Agassiz, British Columbia ($A = 218\,000\ \text{km}^2$); m is the slope of a fitted power relation. (e) general concept of the variation of bedload transport with discharge or shear stress.

The entrainment of any particular grain is, in fact, a probabilistic outcome of the chance impingement of turbulent eddies on grains with varying constraint and varying effective admittance to the impulsive force. In light of this, it has been asked whether there is any finitely definable threshold for motion. In view of the extreme sensitivity of the transport at low rates, an expedient answer to this question is that it really does not matter – functional approximations of the transport rapidly decline to very small numbers as flows and transport decrease below the practically detectable threshold of motion. These considerations led Barry, Buffington and King (2004), in the spirit of Schoklitsch, to propose a relatively simple power function for sediment transport, which they tested with reasonable success against field data from montane channels in the western USA. Their equation (amended 2005) has the form

$$q_b = \alpha(Q/Q_2)^{\beta} \tag{9.5}$$

wherein $\alpha = 0.0008A^{1/2}$, with A the contributing drainage area, and $\beta = 3.56 - 2.45q_*$. These two factors adjust the calculation for the conditions expected to constrain bedload transport within the catchment and within the channel. Variable α expresses the rate of increase of sediment supply downstream and is proposed to increase approximately linearly with distance. In light of the variety of land surface condition and source materials, this is unlikely to represent a general result. $q_* = [(\tau_{Q2} - \tau_{D50s})/(\tau_{Q2} - \tau_{D50ss})]^{3/2}$ wherein τ_{D50s} $0.03g(\rho_s - \rho)D_{50s}$ for the D_{50} (median) size of sediment on the bed surface, and τ_{D50ss} is the equivalent expression for the subsurface sediment. 0.03 is the selected critical Shields number, τ_{*c}, for entrainment of these sediments. This factor is supposed to adjust for the surface armouring and structural effects described in Section 9.3.2 and is based on an observation of Dietrich *et al.* (1989) that, as a gravel-bed channel becomes more starved of sediment, the armour ratio, D_{50s}/D_{50ss} increases (the range is 1.0 to approximately 4), expressing the increasingly heavy winnowing of finer sediments from the surface. The empirical range of armour ratios is further investigated by Pitlick *et al.* (2008). Whether the present formulation adequately covers all structural effects remains untested. In light of this concern and concern for the coefficient α, it would be prudent to anchor formulaic prediction of sediment transport in any particular channel by a number of field measurements obtained at workably moderate flows (Wilcock, 2001).

It was remarked above that sediment transport is, in fact, a four parameter problem in view of the distinction between the population of sediment grains resident on the bed and the population that is transported. This is the consequence of the size-selective exchange of sediments between the bed and the actively transported sediment. The effect is evident in diagrams that portray the distribution of transported sediment by individual size class, normalized by the distribution of sizes resident on the bed (Figure 9.12). To tackle this issue, it is necessary to formulate sediment transport by individual size classes. This problem has been most successfully pursued by G. Parker and P. R. Wilcock. The most recent and most comprehensive formulation from this work was presented by Wilcock and Crowe (2003) following formalisms earlier established by Parker:

$$W_i^* = \begin{matrix} 0.002(\tau/\tau_{ri})^{7.5} & \tau/\tau_{ri} < 1.35 \\ 14[1 - 0.894(\tau/\tau_{ri})^{-0.5}]^{4.5} & \tau/\tau_{ri} \geq 1.35 \end{matrix} \tag{9.6}$$

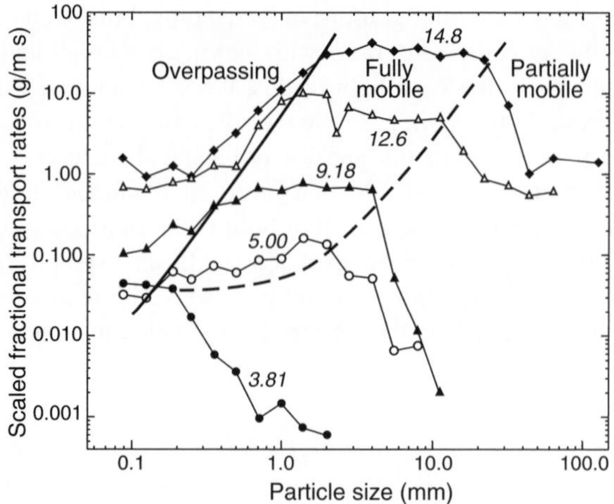

Figure 9.12 Fractional transport diagram from Harris Creek, British Columbia. The ordinate represents the fraction of material of size i present in the transported material, p_i, normalized by f_i, the proportion of that size that is present in the bed, and is scaled by the bedload transport rate per unit width, q_b, Fully mobile material has the ratio $p_i/f_i = 1$ and plots as a straight line equal to q_b for that flow. Figures give flows in $m^3 s^{-1}$
(from Church and Hassan, 2002; their figure 2b; reproduced by permission of the American Geophysical Union).

wherein $W_i^* = q_{bi}/F_i \cdot g(\rho_s - \rho/\rho)/(\tau/\rho)^{3/2}$ is a non-dimensional representation of volumetric transport per unit width of channel; q_{bi} is the volumetric transport per unit width of sediment of size class i, and F_i is the proportion of size i *on the bed surface* (this non-dimensional form is different to that introduced in Equation 9.3). τ_{ri} is a 'reference shear stress' for size class i; it is analogous to the critical shear stress but differs in that it is chosen for some arbitrarily small transport rate (here $W_i^* = W_r^* = 0.002$) on reasoning that the actual threshold of motion is difficult to detect. Equation 9.6 is an empirical fit to experimental data (Wilcock, Kenworthy and Crowe, 2001) of form similar to results previously arrived at by Parker (1990); there is no simple physical explanation for the form, but the great sensitivity to τ, especially for small shear rates, conforms with field observations. It remains unclear whether the formula is asymptotically consistent with stage 3 data.

A notable feature of the Wilcock and Crowe analysis is that τ_{ri} is found to vary systematically with the sand content of the sediment (Figure 9.13a), ranging from about 0.036 when no sand is present to about 0.021 for sand content beyond about 25% (when the interstitial space in framework gravel approaches its capacity to hold sand). The consequence of this outcome is that gravel transport is increased in the presence of sand (Figure 9.13b), a result previously established in the field (Ferguson, Prestegaard and Ashworth, 1989). A second notable feature is that their hiding factor, incorporated through the specification of a particular value of τ_{ri} for each grain size follows a variable function (Figure 9.14)

$$\tau_{ri}/\tau_{rs50} = (D_i/D_{s50})^b \tag{9.6a}$$

in which quantities are normalized by those for the median size of the surface sediment and b varies between 0.12 and 0.67. This indicates that small grains are almost completely hidden (in the sense introduced in Section 9.3.2) while large grains are not strongly hidden at all. The

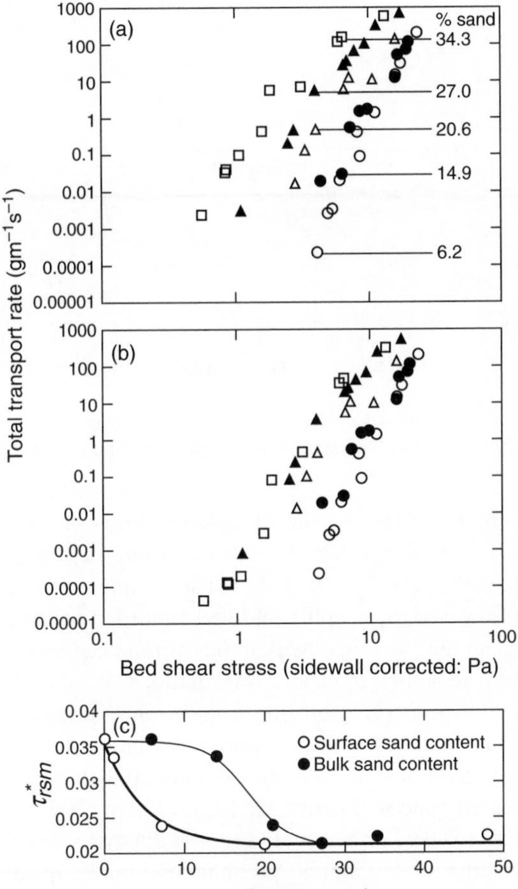

Figure 9.13 Effect of sand content on gravel transport rate: (a) variation of total transport with sand content and (b) variation of gravel fraction transport with sand content. (From Wilcock, Kenworthy and Crowe, 2001, their figure 4b and d; reproduced by permission of the American Geophysical Union). (c) Variation of reference shear stress with sand content of the bed. τ_{rsm} is the reference shear stress for the mean grain size on the bed surface (Wilcock and Crowe, (2003); their figure 4, with minor modifications; with permission from American Society of Civil Engineers).

result confirms commonsense expectation. Altogether, this formulation incorporates a more realistic appraisal of bed state than any preceding one, but actual structural constraint still remains unparameterized.

It often happens that the bed material grain-size distribution, necessary to effect the prediction of sediment transport by individual size classes, is not available. But it remains that the contrasting mobility of sand and gravel presents a serious handicap to bulk calculations. To mitigate this problem Wilcock and Kenworthy (2002) have presented a procedure, analogous to that just described, which makes use of only two fractions: sand and gravel.

All of the above formulation does not arrive at a transport prediction for a river. Formulations based on flume results represent at best the transport for specific conditions

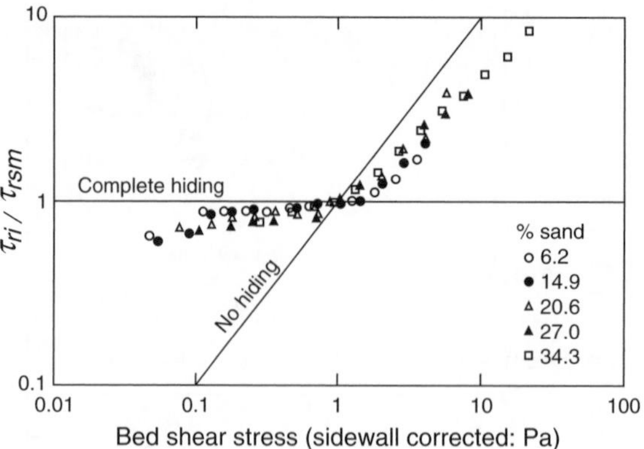

Figure 9.14 The reference shear stress for various grain sizes (Wilcock and Crowe, 2003; their figure 5, with permission from American Society of Civil Engineers).

that may hold at a particular place on the streambed. But river channels exhibit variable geometry both across and along stream, and variations in the bed sediments. These conditions create spatially varying transport (Lisle *et al.*, 2000), part of the continual adjustment amongst flow, sediment influx and bed condition leading to local erosion and sedimentation and to cumulative adjustment of the overall channel morphology. In the past it has been customary to apply transport formulations to whole channel averages of geometric and hydraulic properties, but the non-linear character of the formulation is bound to lead to bias in such cases. A transport formulation, properly applied to a river (as opposed to a flume) must be embedded in a computational model that accounts for the variations in river geometry and sediments (see Li and Millar (2007) for an example using a two-dimensional river model). Furthermore, a model that performs continual calculations must be a morphodynamic model; that is, one that continually updates the morphological changes induced by transport. A difficult problem in such exercises, however, is that the initial spatial variability of bed sediments is almost never known and remains extremely difficult to map for any considerable distance along a stream channel.

Ferguson (2003) has considered this problem in the context of the much more frequently employed one-dimensional models for sediment transport and morphodynamic change. In this case, one must overcome the problem that cross-sectional channel geometry is not represented. Ferguson proposed geometrical idealizations of actual cross-sectional geometries (most of which are roughly trapezoidal or triangular) in order to study the variations in shear stress that may be expected across a channel and the effect that it may have on transport estimates. Careful formulation of the model can lead to reasonable reach-length predictions of sediment transport (see Ferguson and Church, in press), but estimation of local variations in transport – the variations that determine the local morphology of river channels – remains generally beyond reach.

9.3.4 The Sediment Budget in Gravel-Bed Rivers

The somewhat extended foregoing discussion is focused on the estimation of gravel transport at a single point along a river. This reflects the historical focus of interest. The subject developed in that way probably because the underlying hydraulic variates – flow, water depth and velocity – also are customarily determined at a single point, or cross-section, by stream gauging. However, engineers are more often interested in the cumulated transport over intervals of time and the distribution of the transported sediments along the river, and this temporally and spatially extended perspective is essential in order to understand the sedimentology and morphological development of rivers. These *desiderata* open a different perspective on bed material transport – to infer the transport from observations of morphological change.

The fundamental requirements in this 'inverse approach' to estimate sediment transport are the compilation of detailed surveys of an extended reach of the channel at selected times and knowledge of the transport obtained by some independent means (such as direct measurements) at one place in the reach (or more, if checks are desired). The method has not heretofore enjoyed prominence because the surveying requirements have been prohibitive, given the former state of survey technology. That, however, has changed within the past two decades, making inverse approaches feasible. The method is particularly attractive for application in larger gravel-bed channels, where a significant lapse of time may occur within which trends of local scour and fill persist, so that periodic surveys can capture a reasonable approximation of the true changes effected by sediment transport. Inverse methods and associated issues have recently been reviewed by the present writer (Church, 2006), so that only an outline will be given here.

The basis for estimating sediment transport from changes in channel morphology is sediment continuity:

$$\partial q_{\mathrm{b}}/\partial x + \partial q_{\mathrm{b}}/\partial y + (1-p)\partial z/\partial t + \partial C_{\mathrm{b}}/\partial t = 0 \tag{9.7}$$

wherein q_{b} is volumetric bed material transport per unit width of channel, p is the porosity of the sediment deposits, C_{b} is the volumetric concentration per unit bed area of sediment in motion, x and y indicate downstream and cross-stream directions, and z is bed elevation. The problem is usually presented in one dimension by averaging cross-sectional changes to give

$$\partial Q_{\mathrm{b}}/\partial x + (1-p)\partial A_{\mathrm{b}}/\partial t = 0 \tag{9.7a}$$

in which $A_b = wdz$ is the net deposition or scour (dz) in the cross-section of width w and Q_b is the transport integrated across the channel. For purposes of survey, this equation is represented in finite difference form:

$$\Delta Q_{\mathrm{b}}/\Delta x + (1-p)\Delta A_{\mathrm{b}}/\Delta t = 0 \tag{9.7b}$$

that is

$$(1-p)\Delta V + (Q_{\mathrm{bo}} - Q_{\mathrm{bi}})\Delta t = 0 \tag{9.7c}$$

$$\Delta V = V_{\mathrm{i}} - V_{\mathrm{o}} \tag{9.7d}$$

$\Delta V = \Delta A_{\mathrm{b}}\Delta x$ is the net change in sediment volume in a reach of length Δx during the intersurvey period Δt and V_{i} and V_{o} are the volumetric input and output during that interval for the area of channel bed within the reach.

While cross-section surveys were initially exploited to provide measurements, contemporary surveys yield distributed arrays of elevation determinations that are applied to construct digital elevation models (DEMs) of the stream bed, whence DEMs of difference define the volumetric changes (Lane, Richards and Chandler, 1995; McLean and Church, 1999). Ashmore and Church (1998) and Lane, Westaway and Hicks (2003) have examined issues of bias and precision arising from the survey, while the critical issue of survey frequency has been considered by Lindsay and Ashmore (2002).

9.4 Sediments and Morphodynamics of Gravel-Bed Rivers

9.4.1 Sediments

The sedimentology and morphology of alluvial rivers is the consequence of the bed material transport. The continual mutual adjustment between the flows and the morphology of the river is effected by the transport. At the sedimentary level this determines bed sediment texture and stratigraphy (Bridge, 2003).

Because, in gravel-bed rivers, bed material is preponderantly moved as bedload, and because for most flows transport remains near threshold and involves only a single grain layer, most sediment deposits are built of successive emplacements of individual grains, leading to single or few-grains-thick layers that are conformable with the local bed attitude – that is, mostly plane. Because grains are mainly stopped by encounter with an effective obstacle grain, the deposit is relatively well sorted and exhibits imbrication. Transport frequently occurs in only a restricted portion of the bed (Lisle *et al.*, 2000) and the cumulative effect is visualized as a creeping sheet of material, one or two grains deep and with limited lateral extent (as in Figure 9.7c). Where transport is sufficiently vigorous, the feature is recognizable as a physical 'bedload sheet' (Drake *et al.*, 1988; Venditti, Nelson and Dietrich, 2008; Rice *et al.*, 2009). Secondary currents direct transport toward the channel margins and onto bar flanks (see Figure 9.5) where most persistent deposition occurs. Continued transport along the same trajectory leads to the build-up of successive sheets into a tabular gravel body of metre-order thickness which may have an avalanching distal face directed away from the channel. Ahead of this face there is often a back-channel or chute channel that runs 'behind' the bar. The imbricate structure of the bed sediments, which reveals the direction of the depositing flow, is preserved in the stratigraphy, which may otherwise possess few features apart from avalanche-front foresets.

Because of the sorting effects of competence and interception, individual sheets are well sorted. Gravels may be deposited first in flows still competent to transport sands. Hence, even in the presence of a dominantly sandy load, gravels are deposited in grain–grain contact, forming a so-called 'framework' gravel. Under reduced flows later, sand infiltrates into the interstices, ultimately to comprise about 20% of the deposit (and suggesting critical reference Shields numbers for re-entrainment of sediment as low as 0.025). On avalanching fronts, however, sand and gravel may be deposited simultaneously into quieter water. Sand content may then dominate the deposit, gravel clasts lose mutual contact, and we find a 'matric' gravel (referring to the matrix of sand). In bar-tail backwaters, pure sand deposits may accumulate.

Sheets often form recognizable patches, but size gradation occurs within sheets, especially when deposited on an upward incline, such as across a bar flank. The increasing bed elevation controls flow depth and competence so that larger materials are deposited near the head and in the lower part of the deposit. Similar size gradation may also occur along a bar flank as currents slacken from the bar-head riffle into the adjacent pool (see Figure 9.5). Along the length of a bar, sediment size gradation is often as great as mean grain-size change down many kilometres of the river.

Gravel deposits build to median flood level, the limitation in bar height being the outcome of the inability of the river currents to continue to transport gravel in the weaker currents over the developed bar top. Consequently, bar surfaces are customarily flooded in high but still normal floods. Sands are then deposited on bar tops and, once some vegetation becomes established, may build to a considerably higher level that defines the floodplain surface. The result is a widespread characteristic stratigraphy consisting of tabular, flat-lying or laterally dipping gravels, with local tabular foresets, to about two-thirds full channel depth, overlain by massive or laminated sands, fining upward (Figure 9.15). Montane channels

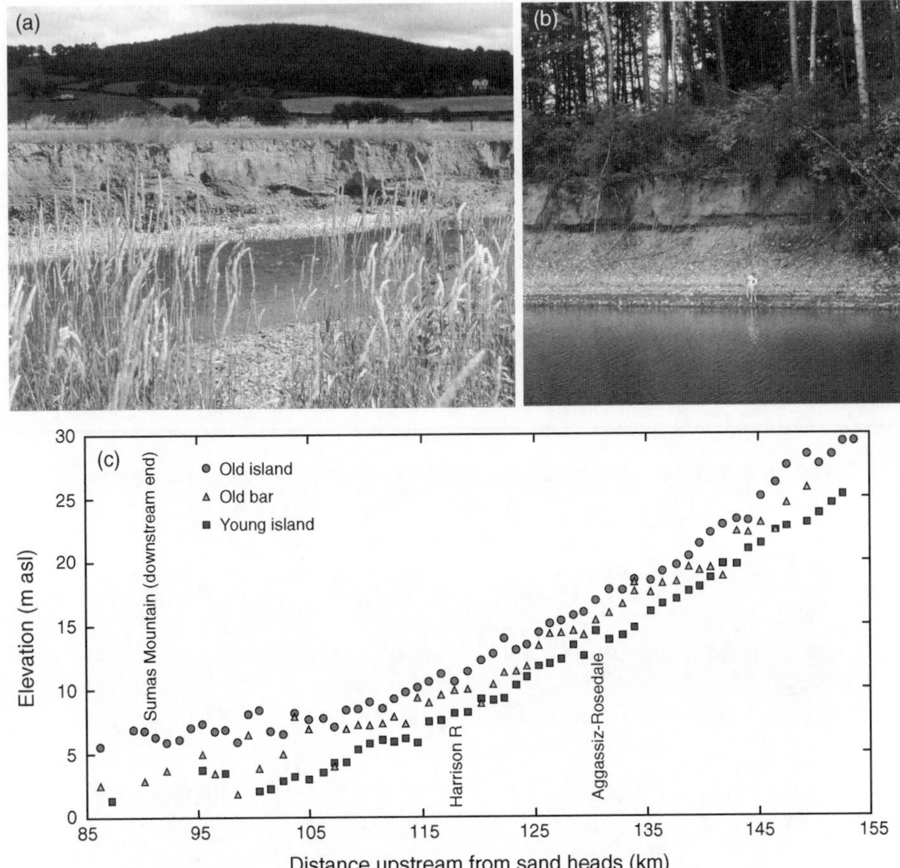

Figure 9.15 Photographs to illustrate characteristic gravel stratigraphy. (a) River Severn near Caersws, Wales, a thick sand member over channel-bed gravel. (b) Lower Fraser River, British Columbia, 3 m sand member over bar gravels in a large gravel-bed river. (c) Variation of sand member thickness along lower Fraser River: three levels are shown, for old-established island surfaces (floodplain level); recently established islands (immature vegetation); and for old bars. Distances upstream from river mouth

often lack the sandy member. Valley and foreland rivers typically exhibit it but its thickness is variable, depending locally upon the time elapsed since the inception of the deposit. The time to fully develop the floodplain surface may be centuries on large rivers.

9.4.2 Morphodynamics

The principal sediment deposits in alluvial river channels are the river bars, features that collectively define the morphology of the channel. Because of the limited 'lifting' capacity of rivers for gravel moved at near-threshold flows, sediments are deposited along bar flanks with limited vertical development. The consequence is a lateral style of channel instability. Developing river bars relatively rapidly approach their full height and length – the former scaled by channel depth and the latter by the channel width, hence, ultimately, by the scale of the flows. They then grow by lateral accretion of successive sediment lobes, accumulations of bedload sheets accreted to the bar flank (Figure 9.16). Bar growth is, accordingly, allometric and the bar a palimpsest of successively accreted lobes. Growth continues until the river channel reaches confining valley walls or the evolving channel alignment causes sedimentation to cease locally. Not infrequently, the channel may become so extended around the bar that it avulses into the chute channel, effectively cutting the bar off. Flood flows rework and dissect the bar surface, increasing the complexity of the sedimentary pattern on a mature bar surface (Bluck, 1976, 1979; Rice *et al.*, 2009).

Lateral bar growth forces the channel to erode the opposite bank, on the deep edge of the adjacent pool (Figure 9.5), in order to maintain conveyance. The eroded sediments are then moved to the next downstream bar. This establishes a characteristic sediment transport distance that corresponds with bar spacing (Pyrce and Ashmore, 2003).

Overall bar style corresponds with the channel morphology, the two major variants being lateral (bank attached) bars and medial (unattached) bars. The former type, in the

Figure 9.16 Photograph to illustrate the lateral style of gravel bar growth: Calamity Bar on lower Fraser River. Delineated areas represents sequential major lobate additions to the outboard flank of the bar. A high-flow chute channel persists behind the bar and the bar core has become vegetated. See Rice *et al.* (2009) for discussion of the history of this bar

form of a succession of alternate (left and right bank) bars is characteristic of single-thread channels, while the latter is the basic unit of braided channels. In transitional (wandering) channels one type may transform into the other, depending upon the activity of the chute channel.

The timescale for bar development has not, in general, been investigated. In Fraser River, a large gravel-bed river in western Canada, it is 100 years (Church and Rice, 2009), but on active braidplains, it may be on the order of only days or weeks. Development time must be related to the mean local bed material flux (or bed material concentration) but, while this is often assumed, the character of the association has not been systematically investigated.

Because gravels are deposited in-channel, deposits reduce channel conveyance, which must be recovered by compensating bank erosion. Where that erosion cannot occur, or cannot occur sufficiently quickly, the channel is prone to avulsion – that is, to diversion outside the channel boundary to form an entirely new channel. Avulsion is common in unconfined montane channels, where sediments may be large and channel banks may be well defended by large materials so that rapid channel adjustment to extreme flows is difficult. Alluvial fans – where there is no topographic confinement while gradient and competence are declining rapidly – are particularly prone to this process. Within-channel avulsions are also common on braidplains, which has led to a common perception of braided channels as distinctive, multithread channel systems. Ashmore (2001) and Egozi and Ashmore (2009) have, however, built a case that, at any one time, there is one dominant channel carrying most of the sediment load in the braided system and that this channel exhibits many of the characteristics of single-thread channels. Braided systems have frequently been represented as sediment-choked systems but the channel type may be as much a consequence of competence limitation in a laterally unconfined setting.

It is evident that channels are less stable where sedimentation occurs. In montane drainage basins, a convergence of tributaries often occurs near the head of a master or trunk valley and available gradient characteristically declines here as well. Sediment delivery, focused in this position, creates notably unstable reaches in this vicinity, where gradient-mediated competence declines quickly (Figure 9.17). One has, in effect, a valley-head, confined alluvial fan.

9.5 Conclusion

The transport of gravel in fluvial systems is surely one of the most complex problems in classic mechanics. One is considering the transport of large numbers of clasts of varying size and shape by a turbulent flow. Entrainment involves mobilization from a bed with complex structure and varying constraints. The moving grains interact with a static bed of complex geometry. Much of the time, the intensity of transport is too low to permit an easy continuum approximation of the process. The result is a modern 130-year history of mainly empirical attempts to characterize the phenomenon. Mostly, it has been studied in the extremely reduced situation presented by laboratory flumes. Parts of the problem – in particular, bed structure – remain unparameterized.

Recently developed fractional formulae (that is, ones that compute the transport for individual size classes), although still empirical, appear able to arrive at reasonable approximations of sediment transport (that is, well within an order of magnitude, which is the range often exhibited by none too reliable field measurements) when the governing

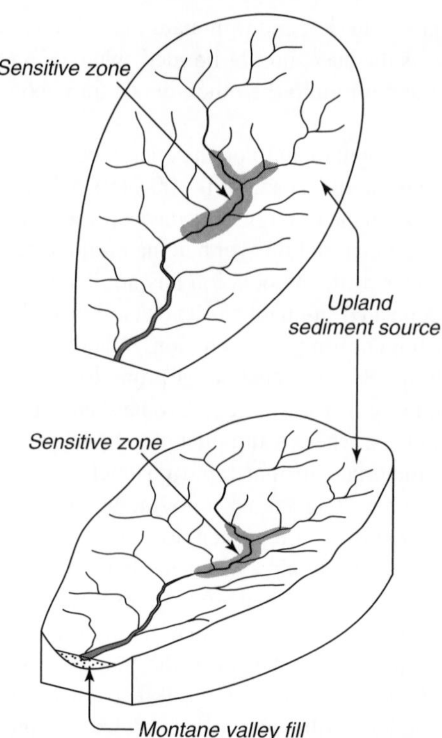

Figure 9.17 Focus of sedimentation and channel instability in montane channel networks

conditions, sediments and channel geometry are sufficiently carefully specified. We understand qualitatively how the transport creates the morphology and sediments of the channel, and its deformation style. There is no doubt, however, that there remains a great deal yet to accomplish.

Acknowledgement

The diagrams were created on very short notice by Eric Leinberger in the University of British Columbia Department of Geography.

References

Ashmore, P.E. (2001) Braiding phenomena: statics and kinetics, in *Gravel-Bed Rivers V* (ed. M.P. Mosley), New Zealand Hydrological Society, Welllington, pp. 95–114.

Ashmore, P.E. and Church, M. (1998) Sediment transport and river morphology: a paradigm for study, in *Gravel-Bed Rivers in the Environment* (eds P.C. Klingeman, R.L. Beschta, P.D. Komar and J.B. Bradley), Water Resources Publications, Highland Ranch, CO, pp. 115–148.

Ashworth, P.J. and Ferguson, R.I. (1989) Size-selective entrainment of bed load in gravel bed streams. *Water Resources Research*, **25**, 627–634.

Bagnold, R.A. (1966) An approach to the sediment transport problem from general physics. *U.S. Geological Survey Professional Paper*, **422-I**, 37.

Bagnold, R.A. (1980) An empirical correlation of bedload transport rates in flumes and natural rivers. *Royal Society of London, Proceedings*, **A372**, 453–473.

Barry, J.J., Buffington, J.M. and King, J.G. (2004) A general power equation for predicting bedload transport rates in gravel bed rivers. *Water Resources Research*, **40**, 22. W10401. doi: 10.1029/2004WR003190, Reply to comment in Vol. **41**, W07016. doi: 10.1029/2005WR004172, p. 2.

Bluck, B.J. (1976) Sedimentation in some Scottish rivers of low sinuosity. *Royal Society of Edinburgh Transactions*, **69**, 425–456, + 1 plate.

Bluck, B.J. (1979) Structure of coarse grained braided stream alluvium. *Royal Society of Edinburgh Transactions*, **70**, 181–221.

Brayshaw, A.C. (1984) Characteristics and origin of cluster bedforms in coarse-grained alluvial channels, in *Sedimentology of Gravels and Conglomerates*, vol **10** (eds E.H. Koster and R.J. Steel), Canadian Society of Petroleum Geologists, Memoir, pp. 77–85.

Bridge, J.S. (2003) *Rivers and Floodplains: Forms, Processes and Sedimentary Record*, Blackwell Science, Oxford, p. 491.

Buffington, J.M. and Montgomery, D.R. (1997) A systematic analysis of eight decades of incipient motion studies, with special reference to gravel-bedded rivers. *Water Resources Research*, **33**, 1993–2029.

Church, M. (2002) Geomorphic thresholds in riverine landscapes. *Freshwater Biology*, **47**, 541–557.

Church, M. (2006) Bed material transport and the morphology of alluvial river channels. *Annual Review of Earth and Planetary Sciences*, **34**, 325–354.

Church, M. and Hassan, M. (2002) Mobility of bed material in Harris Creek. *Water Resources Research*, **38**, 1237. doi: 10.1029/2001WR000753, p. 12.

Church, M. and Rice, S.P. (2009) Form and growth of bars in a wandering gravel-bed river. *Earth Surface Processes and Landforms*, **34**, 1422–1432. doi: 10.1002/esp.1831.

Church, M., Hassan, M. and Wolcott, J.F. (1998) Stabilizing self-organized structures in gravel-bed stream channels: field and experimental observations. *Water Resources Research*, **34**, 3169–3179.

Dietrich, W.E., Kirchner, J., Ikeda, H. and Iseya, F. (1989) Sediment supply and the development of the coarse surface layer in gravel-bedded rivers. *Nature*, **340**, 215–217.

Dinehart, R.L. (1992) Evolution of coarse gravel bed forms: field measurements at flood stage. *Water Resources Research*, **28**, 2667–2689.

Drake, T.G., Shreve, R.L., Dietrich, W.E. *et al.* (1988) Bedload transport of fine gravel observed by motion-picture photography. *Journal of Fluid Mechanics*, **192**, 193–217.

Du Boys, M.P. (1879) Etudes du régime et l'action exercé par les eaux sur u lit à fond de graviers indéfiniment affouiable. *Annales des Ponts et Chaussées*, **5** (18), 141–195.

Egozi, R. and Ashmore, P.E. (2009) Experimental analysis of braided channel pattern response to increased discharge. *Journal of Geophysical Research – Earth Surface*, **114**, F02012. doi: 10.1029/2008JF001099.

Emmett, W.W. and Wolman, M.G. (2001) Effective discharge and gravel-bed rivers. *Earth Surface Processes and Landforms*, **26**, 1369–1380.

Ferguson, R.I. (2003) The missing dimension: effects of lateral variation on 1-D calculations of fluvial bedload transport. *Geomorphology*, **56**, 1–14.

Ferguson, R.I. and Church, M. (2009) A critical perspective on 1D modelling of river processes: gravel load and aggradation in lower Fraser River. *Water Resources Research*. doi:10.1029/2009WR007740

Ferguson, R.I., Prestegaard, K.L. and Ashworth, P.J. (1989) Influence of sand on hydraulics and gravel transport in a braided gravel bed river. *Water Resources Research*, **25**, 635–643.

Grant, G.E., Swanson, F.J. and Wolman, M.G. (1990) Pattern and origin of stepped-bed morphology in high gradient streams, Western Cascades, Oregon. *Geological Society of America Bulletin*, **102**, 340–352.

Jones, L.S. and Humphrey, N.F. (1997) Weathering-controlled abrasion in a coarse-grained, meandering reach of the Rio Grande: implications for the rock record. *Geological Society of America Bulletin*, **109**, 1080–1088.

Lane, S.N., Richards, K.S. and Chandler, J.H. (1995) Morphological estimation of the time-integrated bed load transport rate. *Water Resources Research*, **31**, 761–772.

Lane, S.N., Westaway, R.M. and Hicks, D.M. (2003) Estimation of erosion and deposition volumes in a large, gravel-bed, braided river using synoptic remote sensing. *Earth Surface Processes and Landforms*, **28**, 249–271.

Li, S.S. and Millar, R.G. (2007) Simulating bed-load transport in a complex gravel-bed river. *Journal of Hydraulic Engineering*, **133**, 323–327.

Lindsay, J.B. and Ashmore, P.E. (2002) The effects of survey frequency on estimates of scour and fill in a braided river model. *Earth Surface Processes and Landforms*, **27**, 27–43.

Lisle, T.E., Nelson, J.M., Pitlick, J. *et al.* (2000) Variability of bed mobility in natural, gravel-bed channels and adjustments to sediment load at local and reach scales. *Water Resources Research*, **36**, 3743–3755.

McLean, D.G. and Church, M. (1999) Sediment transport along lower Fraser River. 2. Estimates based on the long-term gravel budget. *Water Resources Research*, **35**, 2549–2559.

Meyer-Peter, E. and Müller, R. (1948) Formulas for bed-load transport. International Association for Hydraulic Research, 2nd Meeting, Stockholm. Proceedings, pp. 39–64.

Millar, R.G. (2000) Influence of bank vegetation on alluvial channel patterns. *Water Resources Research*, **36**, 1109–1118.

Montgomery, D.R. and Buffington, J.M. (1997) Channel-reach morphology in mountain drainage basins. *Geological Society of America Bulletin*, **109**, 596–611.

Mueller, E.R., Pitlick, J. and Nelson, J.M. (2005) Variations in the reference Shields stress for bed load transport in gravel-bed streams and rivers. *Water Resources Research*, **41**. W04006. doi: 10.1029/WR003692.

Paola, C. and Seal, R. (1995) Grain size patchiness as a cause of selective deposition and downstream fining. *Water Resources Research*, **31**, 1395–1407.

Parker, G. (1990) Surface-based bedload transport relation for gravel rivers. *Journal of Hydraulic Research*, **28**, 417–436.

Pitlick, J., Mueller, E.R., Segura, C. *et al.* (2008) Relation between flow, surface-layer armoring and sediment transport in gravel-bed rivers. *Earth Surface Processes and Landforms*, **33**, 1192–1209.

Pyrce, R. and Ashmore, P.E. (2003) The relation between particle path length distributions and channel morphology in gravel-bed streams: a synthesis. *Geomorphology*, **56**, 167–187.

Rice, S.P. (1998) Which tributaries interrupt downstream fining along gravel-bed rivers? *Geomorphology*, **22**, 39–56.

Rice, S.P. and Church, M. (1998) Grain size along two gravel-bed rivers: statistical variation, spatial pattern and sedimentary links. *Earth Surface Processes and Landforms*, **23**, 345–363.

Rice, S.P., Church, M., Wooldridge, C.L. and Hickin, E.J. (2009) Morphology and evolution of bars in a wandering gravel-bed river: lower Fraser River, British Columbia *Sedimentology*, **56**, 709–736. doi: 10.1111/j.1365-3091.2008.00994.x.

Sambrook Smith, G.H. and Ferguson, R.I. (1995) The gravel–sand transition along river channels. *Journal of Sedimentary Research*, **A65**, 423–430.

Schoklitsch, A. (1962) *Handbuch des Wasserbaues*, 3rd edn, Springer, Vienna.

Shields, A. (1936) *Anwendung der Aenlichkeitsmechanik und der Turbulenzforschung auf die Geschiebebewegung*, Preussische Versuchsanstalt fur Wasserbau und Schiffbau, Berlin, Mitteilungen. Heft 26 (Translated by W.P. Ott and J.C. van Uchelen, United States Department of Agriculture, Soil Conservation Service, Cooperative Laboratory, California Institute of Technology, Pasadena, p. 43.

Smart, G.M. and Jaeggi, M.N.R. (1983) Sediment transport on steep slopes. Eidgenössischen Technischen Hochschule Zürich, *Mitteilunger der Versuchsanstalt für Wasserbau, Hydrologie und Glaziologie*, **64**, 191.

Venditti, J.G., Nelson, P.A. and Dietrich, W.E. (2008) The domain of bedload sheets, Marine Sandwave and River Dune Dynamics III. International Workshop Proceedings, 1–3 April, Leeds University, (eds D.R., Parsons, T., Garlan and J.L., Best), pp. 315–321.

Wilcock, P.R. (2001) Toward a practical method for estimating sediment-transport rates in gravel-bed rivers. *Earth Surface Processes and Landforms*, **26**, 1395–1408.

Wilcock, P.R. and Crowe, J.C. (2003) Surface-based transport model for mixed-size sediment. *Journal of Hydraulic Engineering*, **129**, 120–128.

Wilcock, P.R. and DeTemple, B.T. (2005) Persistence of armor layers in gravel-bed streams. *Geophysical Research Letters*, **32**, L08402. doi: 10.1029/2004GL021722, p. 4.

Wilcock, P.R. and Kenworthy, S.T. (2002) A two-fraction model for the transport of sand/gravel mixtures. *Water Resources Research*, **38**. doi: 10.1029/2001WR000684.

Wilcock, P.R. and McArdell, B.W. (1997) Partial transport of a sand/gravel sediment. *Water Resources Research*, **33**, 235–245.

Wilcock, P.R., Kenworthy, S.T. and Crowe, J.C. (2001) Experimental study of the transport of mixed sand and gravel. *Water Resources Research*, **37**, 3349–3358.

Wong, M. and Parker, G. (2006) Reanalysis and correction of bed-load relation of Meyer-Peter and Müller using their own database. *Journal of Hydraulic Engineering*, **132**, 1159–1168.

Wooster, J.K., Dusterhoff, S.R., Cui, Y. *et al.* (2008) Sediment supply and relative size distribution effects on fine sediment infiltration into immobile gravels. *Water Resources Research*, **44**, W03424. doi: 10.1029/2006WR005815, p. 18.

Yatsu, E. (1955) On the longitudinal profile of the graded river. *American Geophysical Union Transactions*, **36**, 655–663.

10
The Fine-Sediment Cascade

Pamela S. Naden
Centre for Ecology and Hydrology, Wallingford, UK

10.1 Introduction

Fine sediment is arguably the most important component of the sediment cascade – whether this is measured in terms of flux, its role in transporting nutrients and contaminants, or its importance to aquatic ecosystems. This chapter focuses on the transport and storage of fine sediment in the river system, and assesses our knowledge at a range of scales from point processes, through individual river reaches and the dynamic behaviour of fine sediment, to consideration of the large-catchment scale and the role of the entire river network in the fine-sediment cascade. Interactions with channel sources and sinks, as well as the links with other chapters, are indicated in the schematic shown in Figure 10.1. First, the importance of the fine-sediment cascade and its relationship to its geomorphological setting and the time and space scales of interest are examined.

10.1.1 Importance of Fine Sediment

Globally, the fluvial system contributes an estimated 13.5–22 billion tonnes of fine sediment per year to the coastal oceans (Holeman, 1968; Panin, 2004; Syvitski *et al.*, 2005; Walling, 2006), making up an estimated 90–95% of the total sediment flux (Syvitski *et al.*, 2003). This overall flux is affected by two dominant anthropogenic impacts which act in opposite ways. Land-use change – particularly deforestation and intensive agriculture – has led to increases in soil erosion and delivery to the river system, particularly for small- to medium-sized catchments (Owens and Walling, 2002; Dearing and Jones, 2003; Walling *et al.*, 2003a). However, at the large basin scale, the building of reservoirs means that an estimated 25–30% of the global fine-sediment flux is retained within some 45 000 reservoirs (Vörösmarty *et al.*, 2003). Much of the global picture relating to sediment flux is dominated by a few very large basins which are explored in detail in Chapter 13. For Europe, sediment production is estimated to be $1800 \times 10^6\,t\,yr^{-1}$, with $540 \times 10^6\,t\,yr^{-1}$ stored in rivers and on

Sediment Cascades: An Integrated Approach Edited by Timothy Burt and Robert Allison

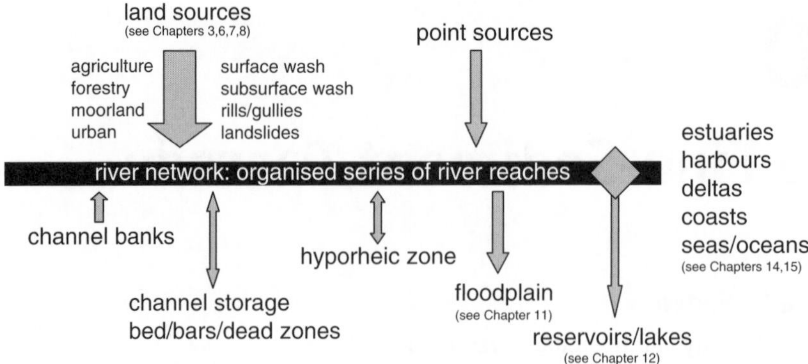

Figure 10.1 Schematic of the fine-sediment cascade showing dominant interactions and links with other chapters

floodplains, 346×10^6 t yr stored in reservoirs and $200 \times 10^6\,\mathrm{t\,yr^{-1}}$ mined from fluvially active areas, leaving a flux of $714 \times 10^6\,\mathrm{t\,yr^{-1}}$ (i.e. 40% production) which is deposited in lowland zones (estuaries, harbours and deltas) or discharged into the surrounding seas and oceans (Owens, 2007).

In addition to its dominance with regard to sediment flux, fine sediment is a major player in the transport of nutrients and contaminants within fluvial systems. The silt and clay fraction of fine sediment is chemically active, due to its large surface area and ionic charge. More than 90% of the total riverine flux of P, Ni, Si, Rb, U, Co, Mn, Cr, Th, Pb, V and Cs, and almost all Fe, Al and rare earth elements are carried as particulates (Martin and Meybeck, 1979). In addition, some 55% nitrogen flux (Meybeck, 1982) and 27% carbon flux (Meybeck, 1982; Degens, Kempe and Richey, 1991) from rivers to the world's oceans may be linked to particulates, although this percentage varies substantially between catchments (Lal, 2002; Alvarez-Cobelas, Angeler and Sánchez-Carrillo, 2008). Within river systems, it is now established that fine sediment is a key factor in the transport, storage and recycling of both nutrients (Bowes and House, 2001; Clarke and Wharton, 2001; House, 2003; Evans, Johnes and Lawrence, 2004; Bowes, Leach and House, 2005; Jarvie *et al.*, 2005) and contaminants (Singh *et al.*, 1997; Macklin, Hudson-Edwards and Dawson, 1997; Rees *et al.*, 1999; Kronvang *et al.*, 2003; Collins, Walling and Leeks, 2005). Hence, knowledge of the transport and behaviour of fine sediment is highly important for understanding the eutrophication of water bodies and impacts on ecosystem and human health.

Fine sediment is also fundamentally important for a range of aquatic habitats. Here, there are issues of sediment quantity as well as quality. High concentrations of suspended sediment have been shown to be detrimental to fish and a comprehensive review of the effects on a range of fish species dependent on both the concentration of suspended sediment and the duration of exposure is presented by Newcombe and Jensen (1996). Evidence for the detrimental impact of both surface and subsurface deposition of fine sediment on periphyton, invertebrates and fish species comes from a range of field and experimental studies (e.g. Quinn *et al.*, 1992; Soulsby *et al.*, 2001; Yamada and Nakamura, 2002; Pedersen, Friberg and Larsen, 2004; Rabeni, Doisy and Zweig, 2005; Matthaei *et al.*, 2006; Niyogi *et al.*, 2007). In particular, the settling and infiltration of fine sediment into gravel interstices impedes water flow within the hyporheic zone and reduces oxygen levels, adversely affecting both benthic organisms and fish spawning grounds (Beschta and Jackson, 1979;

Acornley and Sear, 1999; Soulsby *et al.*, 2001). Here, the effect of particle size and organic content has been shown to be particularly important (Greig, Sear and Carling, 2005; Heywood and Walling, 2007). At the other end of the spectrum, supply of fine sediment is critical to the maintenance of habitats on floodplains, mud flats and deltas – problems with too little sediment are reported by Batalla (2003).

Fine sediment is also responsible for reservoir siltation, significantly reducing the life of high capital cost impoundments for water supply and irrigation, and for river aggradation, leading to problems for flood defence and navigation (Owens *et al.*, 2005). Dredging is an expensive option and problematic for contaminated sediments (Köthe, 2003). All these impacts underline the importance of fine sediment and the need to understand the fine-sediment cascade in order to provide effective management strategies (see Chapters 1 and 7).

10.1.2 Geomorphological Setting, Time and Space Scales

In describing the fine-sediment cascade, it is essential to be aware of both the geomorphological setting and the time and space scales of relevance to fine-sediment transport. Fine sediment, which is sourced from the land, input from external point sources (e.g. construction sites, industrial or sewage effluent) or derived from bank erosion, can be moved through all types of river reach including bedrock, cohesive-bed, boulder-bed, gravel-bed and sand-bed rivers. To encapsulate transport in these settings, three terms are useful. *Washload* consists of material which moves through the river system but has no interaction with the river bed. Silt and clay generally fall into this category, particularly in the context of sand-bed rivers where they are present in the bed in only negligible quantities. *Throughput load* is a term which has been coined to define the transport of fine material which has some interaction with the bed. For example, in gravel-bed rivers, sand, silt and clay move as a type of washload but may be deposited in the interstices of gravel or in limited areas of the channel such as bars, benches, pools and dead zones. This fine sediment is insufficient to cover the bed and does not determine the bed morphology. However, it is of prime concern for aquatic ecosystems and is the subject of a great deal of current research. This deposited fine sediment may be mobilized as part of the active bed layer, but may subsequently move either as bedload, by saltation or rolling, or in suspension. *Bed material load* is that portion of the transported load which is found on the bed of the stream and undergoes exchange between the bed and the water column. Bed material load may move as bedload or in suspension but is an integral part of the channel morphology. The movement of sand as dunes or sand sheets and the movement of sediment in alluvial channels fall into this category.

In structuring our knowledge of the fine-sediment cascade, it is helpful to consider the time and space scales of interest. Loosely based on Church (1996), three domains can be identified: these have different modes of explanation and are described in Figure 10.2. The first of these is the domain of hydraulics in which the physics of sediment transport is used to determine the flux of fine sediment through river cross-sections or short river reaches. Here the explanation is deterministic, with space–time averaging of turbulent eddy scales and short duration burst–sweep cycles which are responsible for the movement of individual grains but where flux is determined by the steady-state average behaviour of the fluid. Expanding both time and space scales, the domain of hydrology is entered where the important characteristics of sediment behaviour are the dynamics over hydrological

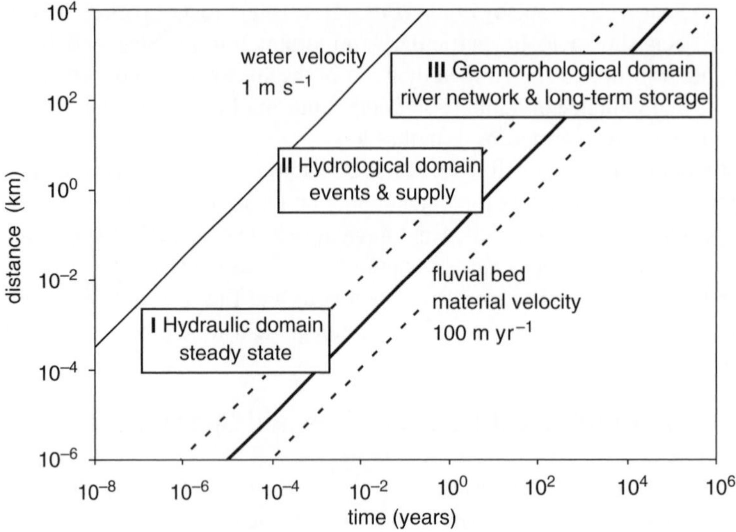

Figure 10.2 Time-space scales in the sediment cascade showing domains of interest. Also shown are the virtual velocities for water (thin black line) and fluvial bed material (thick black line and dashed lines showing possible range of values) as given in Church (1996); suspended sediment takes a trajectory somewhere between the velocities of water and bed material which may itself depend on scale

events. Here, the fine-sediment flux is described by the interplay between the hydrology and sediment storage, with the role of channel processes and sediment routing becoming more important and the emergence of hysteretic behaviour in the observed dynamics. For larger catchments and longer time scales, the role of geomorphology becomes increasingly important and the flux now becomes contingent on the structure of the river network and the history of sediment erosion and transfer throughout the entire basin.

This chapter takes each of the three domains in turn and reviews recent research which has contributed significantly to improving our understanding of the fine-sediment cascade. It predominantly focuses on fine sediment that is moved as washload or throughput load and transported in suspension. The interaction of fine sediment with floodplain and channel morphology is covered in Chapter 11.

10.2 Fine-Sediment Transport

The first domain of interest is the hydraulic domain in which the focus is on the physics of sediment transport and how this controls fine-sediment conveyance through a cross-section or a short river reach. First, the nature of fine sediment needs to be understood.

10.2.1 The Nature of Fine Sediment

Fine sediment is normally taken to comprise sand, silt and clay particles, that is those grain sizes with a diameter <2 mm. Grain size may also be quoted in phi units (defined as

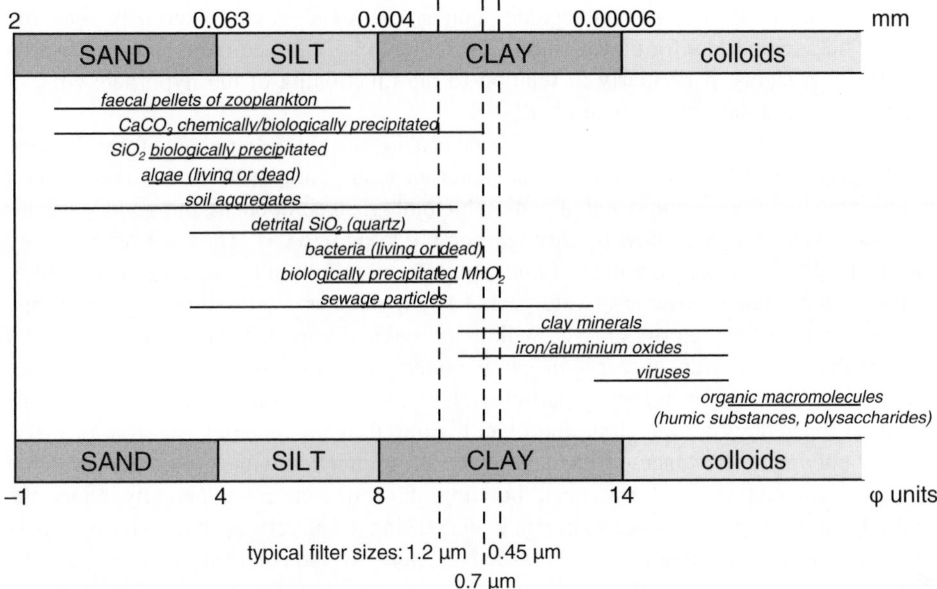

Figure 10.3 Definition of fine-sediment size characteristics, including organic and colloidal material using size ranges given in Tipping (1988)

$\varphi = \log_2[D/D_0]$ where D is the grain size in mm and D_0 is 1 mm). Figure 10.3 gives the standard definition of sediment size fractions in terms of particle diameter in both millimetres (on a logarithmic scale) and phi units. It also shows typical filter sizes (1.2, 0.7 and 0.45 μm) used in the measurement of suspended sediment. While once the term sediment strictly referred to mineral grains, the definition of fine riverine sediment has gone through something of a revolution in recent years (Droppo, 2001) and it is now accepted that it includes both organic and flocculated material. This is important for understanding both the flux of material and its transport characteristics. Accordingly, Figure 10.3 includes the particle size ranges of the common organic materials, flocs and, for completeness, colloids found in aquatic systems (Tipping, 1988).

As particle size decreases, the surface area per unit volume of the particle increases and interparticle, or cohesive, forces dominate particle behaviour. There is no clear size boundary between cohesive and non-cohesive sediment. Sand-sized material is generally non-cohesive. Clay-sized material is cohesive and exhibits very strong inter-particle forces due to its surface ionic charge. Silt may be cohesive, dependent on the presence of clay and organic material, and is generally classed as such. Cohesive sediments consist of both inorganic minerals and organic material such as plant and animal detritus and bacteria. The term *seston* is used in the ecological community to refer to any particulate matter such as plankton or other living organisms as well as organic detritus and inorganic particles. Thus, an additional component of the fine-sediment cascade in its broadest sense is autochthonous material derived from biological components within the river system. An example of this is the faecal pellets of blackfly larvae which can make up some 40–60% of the fine material deposited on the bed of chalk streams (Wharton *et al.*, 2006). Knowledge of the characteristics (particle size and organic content) of this biogenic material is generally lacking but is important for understanding both the flux of fine sediment and its effect on aquatic ecosystems, particularly in relation to the functioning of the hyporheic zone or permeable region beneath a stream bed.

Cohesive sediment is commonly transported in a flocculated or aggregated form (Droppo and Ongley, 1992). Flocs or aggregates are heterogeneous, composite structures composed of an active biological component, a non-viable biological component, inorganic particles and water held within or flowing through pores (Droppo, 2001). They are never purely inorganic. They are formed either within the water column or in the surface of the bed or transported to aquatic systems as water-stable soil aggregates. While some large flocs may readily break up, the majority of riverine flocs are stable (Droppo, 2004). Liao *et al.* (2002) showed that van der Waal and/or hydrophobic interactions were important in the development phase of flocs by bringing particles close enough to form specific interactions. However, once flocs are initiated, microbial binding through bacterial secretion of extracellular polymeric substances (EPS) takes over, in conjunction with further strengthening via ionic interactions and hydrogen bonding. Flocculation fundamentally alters the hydrodynamic properties of the sediment by increasing the effective particle size by orders of magnitude and by changing the shape, density, porosity and composition of the particle. This changes the downward flux of fine sediment and is a fundamental factor in the deposition of fine material both in channels (Tipping, Woof and Clarke, 1993) and on floodplains (Nicholas and Walling, 1996).

The effect on particle size is shown in Figure 10.4. The pictogram in Figure 10.4a shows an image of flocculated sediment in suspension and a disaggregated suspension following sonification. The change in particle size is clearly visible. Flocs are found to comprise about 10–27% total number of particles but represent about 92–97% of the total volume of suspended sediment. Grain sizes between 1 and 2.6 μm are entirely incorporated into flocs (Droppo and Ongley, 1992). Figure 10.4b shows example distributions of both effective particle size and absolute (disaggregated) particle size for both suspended sediment and bed sediments taken from Phillips and Walling (1999). It clearly shows the shift in size with the median size of the flocculated sediment lying between 20 and 80 μm compared with an absolute value of about 5 μm. Phillips and Walling (1999) also show a seasonal variation in particle size, with the size of flocculated particles being higher in the summer when biological activity is greatest but with the absolute particle size being higher in winter when erosion is more active.

10.2.2 Settling Velocity of Fine Sediment

The primary characteristic of fine sediment which dictates its transport and storage within river systems is its settling velocity. Very small particles (<1 μm) and colloids are subject to Brownian diffusion and do not settle. For larger particles where gravitational forces exceed the effects of diffusion, the settling velocity of small individual particles travelling through static water may be calculated, assuming spherical particles, from Stokes' Law

$$V_g = \frac{RgD^2}{C_1 \nu} \tag{10.1}$$

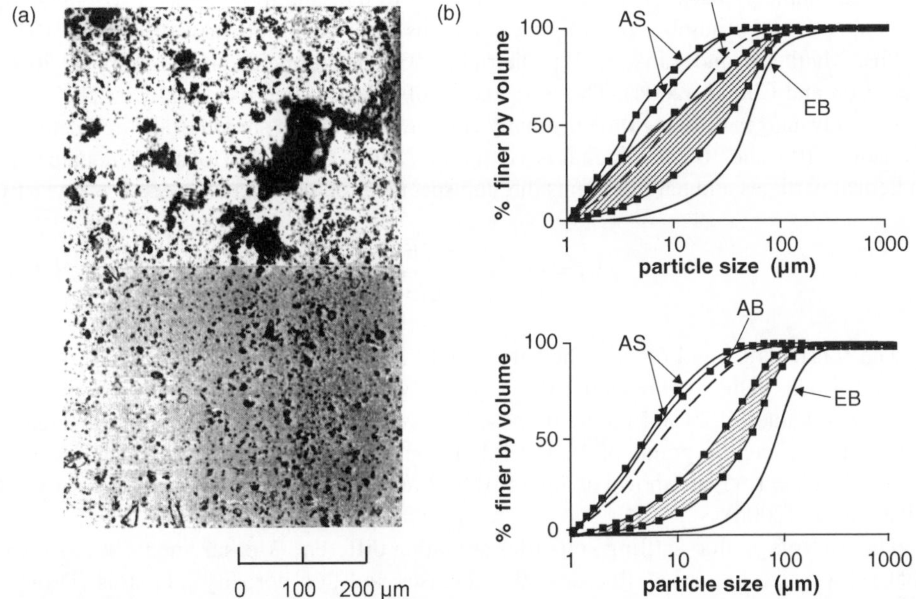

Figure 10.4 Effect of flocculation on particle size.: (a) Pictogram of flocculated suspension and disaggregated suspension following sonification
(reprinted, with permission, from Droppo, I.G. and Ongley, E.D. (1992) The state of suspended sediment in the freshwater fluvial environment: a method of analysis. Water Research, 26, 65–72. © Elsevier.); (b) Example distributions of effective and absolute particle size: AB and EB are absolute and effective size of bed material respectively; AS is absolute size of suspended material; shading denotes range of effective size of suspended material.
(reprinted, with permission, from Phillips, J.M. and Walling, D.E. (1999) The particle size characteristics of fine-grained channel deposits in the River Exe Basin, Devon, UK. Hydrological Processes, 13, 1–19. © John Wiley & Sons Ltd).

where V_g is settling velocity, R is submerged specific gravity or $(\rho_s - \rho)/\rho$, ρ_s is particle density, ρ is density of fluid, g is acceleration due to gravity, D is particle diameter, ν is kinematic fluid viscosity ($10^{-6}\,\mathrm{m}^{-2}\,\mathrm{s}^{-1}$ for water at 20 °C) and C_1 is a constant with a theoretical value of 18. This gives accurate results for spherical particles provided that the particle Reynolds number ($V_g D/\nu$) is less than 1.0, indicating that viscous forces dominate fluid resistance. At large particle Reynolds numbers, when inertial forces dominate, boundary-layer separation occurs behind the particle such that its rapid settling is resisted predominantly by the turbulent drag of the wake behind the particle. In this case, the settling velocity is given by

$$V_g = \sqrt{\frac{4}{3}\frac{RgD}{C_2}} \tag{10.2}$$

where C_2 is the constant asymptotic value of the drag coefficient for particle Reynolds numbers between 10^3 and 10^5. Experiments have shown C_2 to be 0.4 for smooth spheres and approximately 1 for natural grains (Cheng, 1997). For the transitional range of particle

Reynolds numbers between 1 and 10^3, equivalent to the sand to fine gravel size range of $0.1 < D < 4$ mm, a number of different equations have been put forward (Rubey, 1933; Gibbs, Matthews and Link, 1971; Dietrich, 1982; Hallermeier, 1981; Ahrens, 2000; Ferguson and Church, 2004). The most recent of these by Ferguson and Church (2004) has the advantages of being dimensionally correct, providing a smooth transition between Equations 10.1 and 10.2 to which it is asymptotic, and of being able to accommodate both spherical particles and natural grains through specification of its parameters. It is given by

$$V_g = \frac{RgD^2}{C_1\nu + \sqrt{0.75C_2RgD^3}} \qquad (10.3)$$

The constants C_1 and C_2 take the values of 18 and 0.4 for smooth spheres; for typical natural sands, values of 18 and 1.0 are proposed for use with size specified in sieve diameters. Values of 20 and 1.1 are proposed for use with nominal diameters, with a likely limit for very angular grains of 24 and 1.2. The proposed curves for spheres and natural sediments ($R = 1.65$) are shown in Figure 10.5a. Methods for measuring particle density are given in Lal (2006).

The controls on floc-settling velocities are rather different. Measurements have shown that as floc size increases, floc density decreases and floc porosity increases (Droppo *et al.*, 1997). Thus, although the primary control on settling velocity is still floc size, there is much more scatter in the relationship. Floc densities are typically between the density of water and 1.4 g cm^{-3}, with the majority being around 1.1 g cm^{-3} (Droppo, Walling and Ongley, 2002). Settling rates are, therefore, about an order of magnitude lower than a similar size of solid particle. Some typical relationships reported in the literature are shown in Figure 10.5b. It should be noted that, compared to the exponent of 2 in the Stokes'

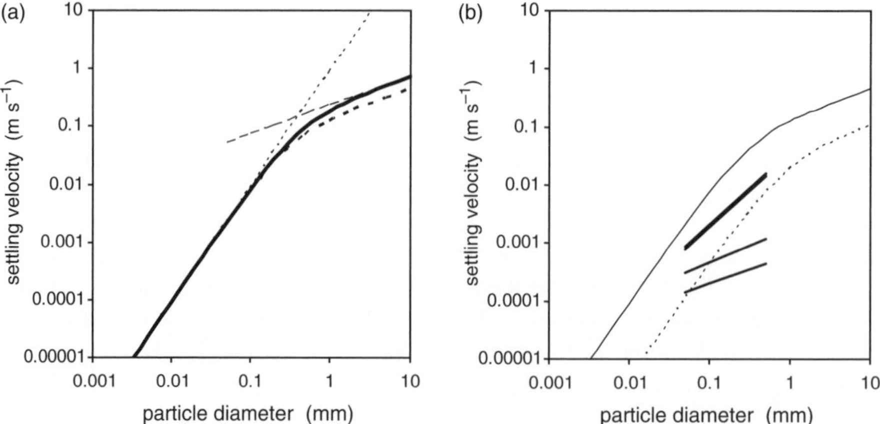

Figure 10.5 Settling velocity as a function of particle size. (a) Stokes Law (black dotted line), wake law for smooth spheres (black dashed line), Ferguson and Church (2004) for spheres (heavy black solid curve) and natural sediments (heavy black dashed curve) with $R = 1.65$. (b) Typical settling velocities for flocs, for example soil aggregates (upper heavy black line) and suspended solids (lower heavy black lines) after Williams, Walling and Leeks (2008) with Ferguson and Church (2004) settling velocities for spheres with $R = 1.65$ (black solid curve) and $R = 0.11$ (black dotted curve)

equation, these relationships have exponents of particle diameter in the range 0.5 to 1.0 for flocs and about 1.2–1.3 for soil aggregates. Williams, Walling and Leeks (2008) identify the other controls on the settling velocity of flocs and aggregates as the porosity of the particle and its fractal dimension. Although these relationships plot well below the Stokes' Law indicating lower settling velocities for the given size, it should be remembered that the flocs are composed of particles which may be an order of magnitude or more smaller in size and therefore this represents an increase in settling velocity for those particles, which enhances their propensity to deposit. Thus, the process of flocculation and the transport of fine sediment as flocs and aggregates are clearly fundamental for understanding the dynamics of the fine-sediment cascade. For further details on the transport of cohesive sediment, the reader is referred to the excellent review in Krishnappan (2007) and the applications given in Haralampides, Corquodale and Krishnappan (2003) and Bungartz and Wanner (2004).

For very high sediment concentrations, mutual particle hindrance and increased drag cause a reduction in the settling velocity compared to that for individual grains. This effect may be represented as

$$V'_g = V_g(1-C)^n \tag{10.4}$$

where V'_g is settling velocity in a dispersion of other grains, V_g is settling velocity of a single grain, C is *volume* concentration and n is an exponent which varies between 2.32 and 4.65 dependent on grain Reynolds number (Richardson and Zaki, 1958). Typical high values of suspended sediment concentrations are of the order of $1000\,\mathrm{mg\,L^{-1}}$ which, assuming a particle density of $2650\,\mathrm{kg\,m^{-3}}$, gives a volume concentration of 10^{-4} implying a negligible 0.001% reduction in settling velocity. However, hyperconcentrated flows have also been reported in Chinese rivers with concentrations of $300\,\mathrm{g\,L^{-1}}$ (Xu, 2002). Applying the same calculation gives a reduction in settling velocity of about 30%. Hyperconcentration also occurs near the river bed, giving rise to 'fluffy' surficial deposits with a high water content which may subsequently be readily entrained (Droppo and Stone, 1994).

10.2.3 Conveyance of Fine Sediment

Settling velocity determines both the transport mode of fine sediment as well as its propensity for deposition. Silt, clay and flocculated material will normally travel in suspension while sand may be transported either in suspension or as bedload, by rolling or saltation, dependent on the flow conditions. Of key importance for calculating the flux of suspended material is the distribution of sediment concentration with depth.

Suspended sediment includes all grains kept aloft by fluid turbulence such that the weight of suspended material is balanced by an upward momentum transfer from fluid eddies (Leeder, 1982). Thus, the quantity of particles in suspension depends on the fluid velocity and turbulence as well as on the settling velocity of the sediment. Typical measured concentration profiles are illustrated graphically in Figure 10.6. This clearly shows that the higher the settling velocity of the particle, the higher is the concentration near the bed, while particles with a very low settling velocity are evenly distributed throughout the water column.

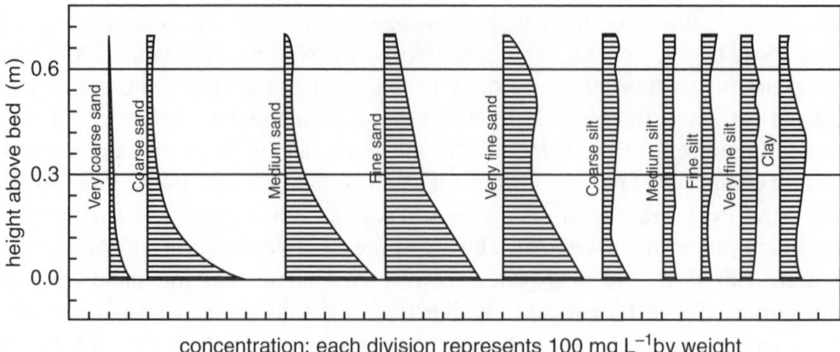

Figure 10.6 Typical concentration profiles for suspended sediment of a range of particle sizes/settling velocities in the Missouri River at Kansas City, USA
(reprinted from Brakensiek, D.L., Osborn, H.B. and Rawls, W.J. (coords.) 1979 Field Manual for Research in Agricultural Hydrology. United States Department of Agriculture, Agricultural Handbook 224. 550 pp. Courtesy of the United States Department of Agriculture).

Assuming a logarithmic velocity distribution, particles in suspension will exhibit a vertical concentration profile which can be calculated from diffusivity theory (Rouse, 1937) as

$$C_y/C_a = [(d-y)/y.\, a/(d-a)]^z \tag{10.5}$$

where C_y is concentration of suspended sediment at height y above the bed, C_a is the reference concentration at height a above the bed, d is total flow depth and $z = V_g/\beta\kappa u_*$ where V_g is settling velocity, β is the ratio of the diffusivity of suspended sediment to that of water and is often taken to be 1, κ is Von Karman's constant and u_* is bed shear velocity. The Rouse equation has been found to agree with concentration profiles in the lower part of stream flows but tends to underestimate concentrations in the upper parts of the flow (e.g. Bennett, Bridge and Best, 1998). Various criticisms have been made of the Rouse equation which are summarized in Bridge (2003). The main problem lies in the calculation of the sediment diffusivity and alternative formulations for β have been cited (e.g. van Rijn, 1984). In practice, C_a and a must also be measured or calculated and there are various methods for doing this (e.g. van Rijn, 1984; Cao, 1999). An alternative to the convection–diffusion approach is through an explicit two-phase flow model (e.g. Greimann and Holly, 2001) but these methods are as yet relatively undeveloped.

In order to calculate the suspended sediment conveyance or transport capacity, the velocity and suspended sediment profiles must be determined for the full width of the channel and their product integrated for the whole cross-section. Alternatively, there is a plethora of empirical functions and sediment transport models available which are reviewed by Papanicolaou *et al.* (2008).

Of particular interest for the fine-sediment cascade is the conveyance of suspended sediment through vegetated channels where the assumption of a logarithmic velocity profile

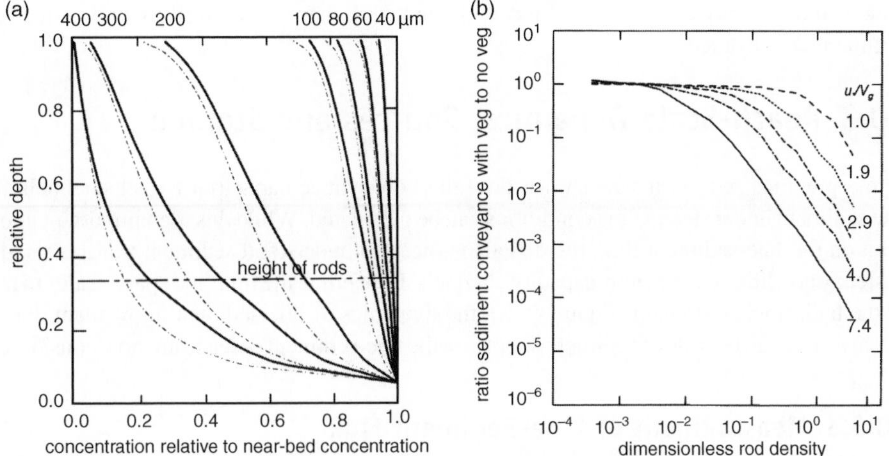

Figure 10.7 Effect of vegetation on sediment conveyance.: (a) Vertical concentration profiles with vegetation (solid lines) compared to Rouse model (dot-dash lines) for a range of grain sizes indicated along the top of the graph.; (b) Reduction in sediment conveyance in the vegetated case for a range of grain sizes as indicated by the ratio of shear velocity (u*) to settling velocity (V_g) (reproduced from López, F. and Garcia, M. (1998) Open-channel flow through simulated vegetation: suspended sediment transport modelling. Water Resources Research, 34, 2341-2352. Courtesy of the American Geophysical Union).

no longer holds and a double-averaging methodology needs to be adopted. An example using vertical rods to represent rigid vegetation is provided by López and Garcia (1998). They use a modified k–ε turbulence model with additional drag terms to close the Reynolds-averaged Navier–Stokes equations for continuity and momentum to give velocity profiles for clear water in vegetated channels. This is then applied in conjunction with the equation for the vertical diffusion of sediment to give concentration profiles in the presence of vegetation. Figure 10.7a shows that, compared to the Rouse distributions, the presence of vegetation has a tendency to produce a more uniform distribution of fine sediment – particularly for the coarser size fractions. The transport capacity is calculated by numerically integrating the product of the double-averaged downstream velocity and sediment concentration. Figure 10.7b shows the impact on transport capacity of vegetation at a range of densities for sediment with different settling velocities. The rapid reduction in transport capacity with increasing vegetation density is readily apparent, particularly at high particle Reynolds numbers.

In practice, vegetation is not a uniform field of rigid rods but provides a varied hydraulic environment with both marginal vegetation and in-channel submerged vegetation. In the case of large trailing macrophytes, fine sediment may be found in patches in the lee of stems (Cotton *et al.*, 2006). This vegetation also creates a spatially organized flow pattern or pseudo-braiding which may actually increase the overall sediment transport capacity. The exchange of fine sediment with marginal habitats (Gurnell *et al.*, 2006) is another interesting feature of natural channels in which lateral diffusivity is important for modelling conveyance and deposition (Sharpe and James, 2006). Further research is needed to understand and model the flow patterns in vegetated channels as well as their effect on sediment conveyance.

10.3 Reach-Scale Dynamics, Sources and Storage

In the previous section it was shown how the conveyance capacity of river channels for fine sediment under steady flow conditions can be calculated. While this is an important upper limit on the fine-sediment flux, there is also a need to understand sediment transport under unsteady conditions and in non-capacity channels. Expanding the time and space scales to that of the hydrological domain (Figure 10.2), the dynamics of fine-sediment movement during storm events and the role of sediment supply on the fine-sediment cascade are now considered.

10.3.1 Measurement of Fine-Sediment Flux

The flux of suspended sediment is usually derived from measurements of water discharge and suspended sediment concentrations. This assumes that the sediment travels at the same speed as the water. This assumption has recently been brought into question by experimental work of Muste *et al.* (2005) and neglect of this may lead to an overestimate of flux (Aziz, Prasad and Bhattacharya, 1992). However, perhaps more significant is the need for effective temporal resolution in flux monitoring. It is well known that $> 90\%$ of fine-sediment flux is carried in $< 10\%$ time. An example time series of flow and suspended sediment concentration is shown in Figure 10.8. It is evident from this that fine sediment may vary by several orders of magnitude over a single flow event. For this reason, good flux estimates rely heavily on event sampling coupled with continuous turbidity measurements, generally using optical backscatter, and there are a number of well-established instrumentation techniques and protocols available (Evans, Wass and Hodgson, 1997; Wass and Leeks, 1999; Old *et al.*, 2006).

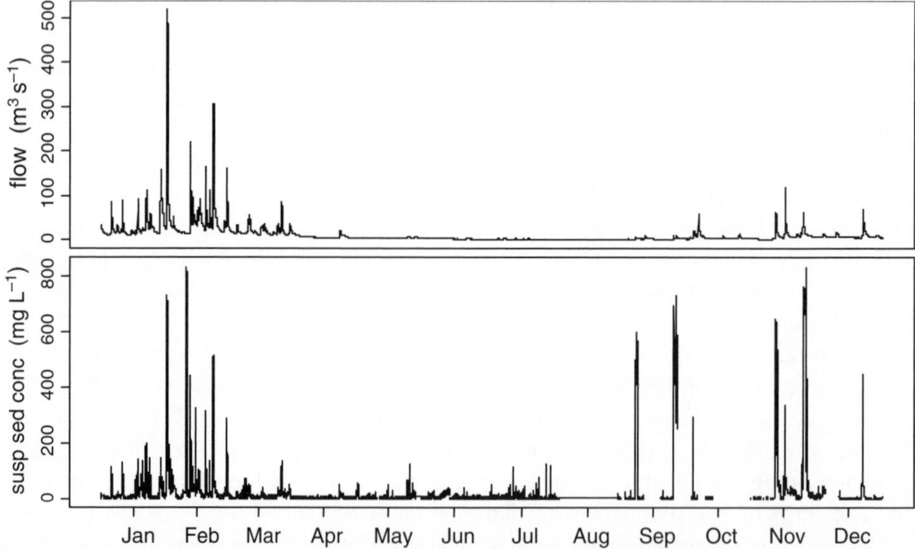

Figure 10.8 Example time series of flow and suspended sediment concentration for 1995 from the River Swale, UK based on continuous turbidity and storm event sampling

As these protocols reveal, the reliability of determining suspended sediment from turbidity depends on rigorous weekly or fortnightly cleaning of the turbidity sensor, regular calibration of the probe using standards and the stability of the relationship between suspended sediment and turbidity. A number of factors, such as particle size, shape, composition and water colour, have been identified as intervening in this relationship and a comprehensive review is given in Gippel (1995). The uncertainty associated with these variables may preclude the use of turbidity at very low suspended sediment concentrations ($<10\ \text{mg L}^{-1}$) so that calculation of flux is best achieved using a median value of suspended sediment concentration from grab samples taken say weekly to indicate the background suspended solids during low flows, coupled with event sampling.

Even in small catchments, the relationship between suspended sediment and turbidity can be quite scattered (Figure 10.9a). In large catchments with multiple sources, the relationship between suspended sediment and turbidity may vary between each storm. This is shown in Figure 10.9b for the River Swale over a series of storm events sampled during 1994 and 1995; the relationship during an individual storm is often extremely good but the overall scatter is high. This emphasizes the importance of using *both* turbidity and automatic sediment sampling in tandem in order to obtain good continuous data and therefore good flux measurements. Newer approaches using acoustic or fibre optic technologies may have potential for development in the future.

10.3.2 Fine-Sediment Dynamics over Hydrological Events

The part of the fine-sediment cascade which is dominated by washload or throughput load may be characterized as a 'high flux–low storage' system. One of the reasons that it is so

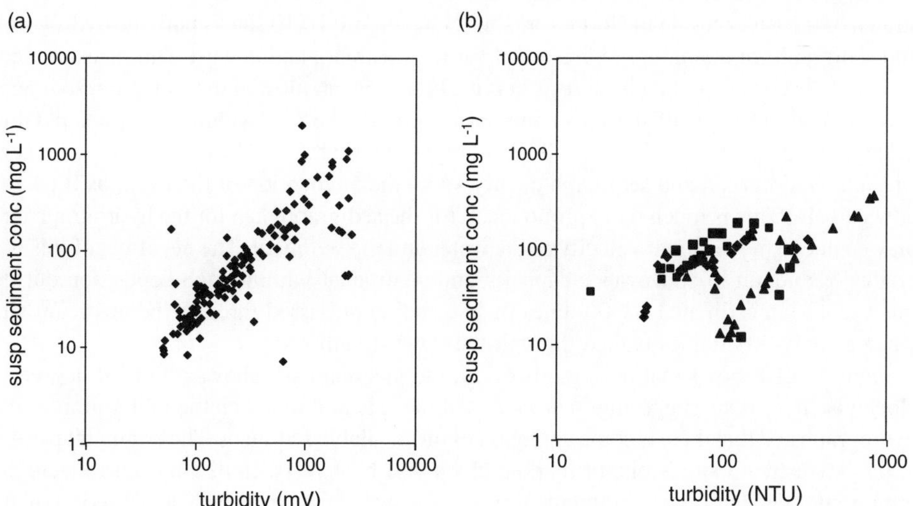

Figure 10.9 Example relationships between turbidity and suspended sediment. (a) Small catchment with well-defined relationship (Denbrook, Devon, UK – data collected by Barnaby Smith, CEH and Patricia Butler, North Wyke Research used with permission). (b) River Swale – each symbol represents a different event
(after Smith *et al.*, 2003a).

difficult to predict and model in detail is that much of the time fine-sediment flux is limited by its availability, that is the flow has the capacity to convey much more fine sediment than is available to be transported. At other times, fine sediment is transport-limited which leads to deposition. In order to understand the dynamics of fine-sediment storage in the channel, it is therefore important to recognize the times, places and conditions under which these different conditions occur.

If fine sediment were not limited by storage, then sediment rating curves would show a tightly defined relationship between fine-sediment concentration and river discharge. However, sediment rating curves in general show a very high degree of scatter. This has variously been attributed to variations with season, sediment source area and the rising and falling limbs of the hydrograph. In this last case, patterns of hysteresis can provide a useful tool for exploring sediment dynamics. Many catchments, particularly small ones, exhibit clockwise or positive hysteresis, that is sediment concentration increases more rapidly than river discharge, often peaks before discharge and shows much lower concentrations at the same discharge on the falling limb of the hydrograph. The cause of this has generally been attributed to depletion of the store of available sediment (Bogen, 1980; Walling and Webb, 1982) or an increased proportion of subsurface flow during the recession limb of a hydrograph (Wood, 1977; Becht, 1989). Anticlockwise or negative hysteresis occurs less frequently and has been associated with cases where sediment sources are distant from the point of measurement (Heidel, 1956) or where sediment is derived from bank collapse following the flood peak (Sarma, 1986; Ashbridge, 1995) or mobilized from the bed following biofilm break-up (Lawler *et al.*, 2006). For larger catchments, other factors need to be considered such as the downstream propagation of the flood wave (Petts *et al.*, 1985; Bull, 1997), the timing of water and sediment inputs from tributary streams (Smith *et al.*, 2003a), size of temporary storage within the river channel (Asselman, 1999) and the relative velocity of the water and the sediment.

Many of these points can be illustrated by reference to measurements taken over a 55 km stretch of the River Swale in North Yorkshire, UK. Figure 10.10 shows both the hydrograph and sedigraph for a simple in-bank event for the upstream and downstream measurement sites and the corresponding hysteresis curves. In this event, most of the water and sediment was sourced upstream of the upper measuring point. The total volume of water passing through the upstream site is $7.4 \times 10^6\,m^3$ compared with $8.5 \times 10^6\,m^3$ for the downstream site. The hydrograph and sedigraph clearly show the attenuation of the event as it passes downstream. This is much more pronounced for the sedigraph than for the hydrograph and may be due to the different velocity of the water and the sediment. The net effect of this is a gradual reduction in conveyance capacity and significant within-reach deposition during this event – an estimated 2230 tonnes of fine sediment passed through the upstream site whereas only 901 tonnes passed through the downstream site.

Figure 10.10 shows that during this event, the upstream site shows a limited degree of clockwise hysteresis suggesting that more sediment is mobilized on the rising limb of the hydrograph and that there is some depletion of the available sediment. The dominant pattern at the downstream site is one of marked clockwise hysteresis. In this case, it reflects the progressive deposition of sediment during its travel down the reach and a consequent reduction in available sediment. Later in the season, similar events show anticlockwise hysteresis at the upstream site which is consistent with the hypothesis that in-channel sources have largely been exhausted and material is being sourced from more distal sources. This effect is less pronounced at the downstream site due to the supply of in-channel

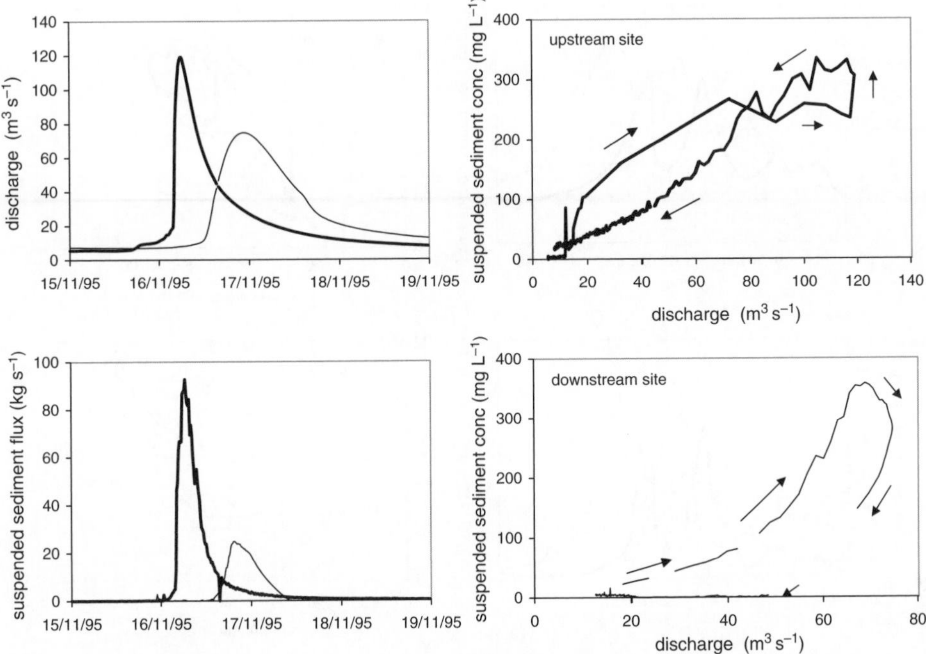

Figure 10.10 Passage of water and sediment downstream for River Swale, North Yorkshire, UK, during an event largely sourced in the headwaters: thick line denotes upstream site of Catterick; thin line denotes downstream site of Leckby Grange

sediment deposited in the intervening reach by events such as that described in Figure 10.10. Thus, over an annual period and particularly in events where both water and sediment is largely sourced from above the upstream site, a slug of fine sediment is seen as being progressively cascaded down through the river reach.

Figure 10.11 shows a contrasting type of in-bank event. Here, the entire catchment is contributing with both water and sediment input from tributaries joining the main channel throughout the monitored reach. In this case, although the timing of the peak at the downstream site is later than that at the upstream site, both the hydrograph and the sedigraph peaks increase as tributary rivers bring in more sediment and water. The water flux through the upstream site is 15.8×10^6 m^3 and through the downstream site is 23.3×10^6 m^3; the associated sediment flux is 516 tonnes and 1017 tonnes respectively. It should also be noted that this event has much lower sediment concentrations compared to the earlier event suggesting that available sediment is more limited. For the upstream site, both sedigraphs of this double peaked event show anticlockwise hysteresis with the first event having higher concentrations of fine sediment compared to the second event. The anticlockwise hysteresis suggests that material comes from more distal sources; the progressive reduction in concentration over the two events suggests a reduction in the availability of sediment with each successive event. At the downstream site, there is evidence for increased mobility of sediment for flows greater that 50 m^3 s^{-1} which may indicate mobilisation from within the channel. There is only limited anticlockwise hysteresis but a pronounced drop in concentration for a given flow during the second of the two events, implying a reduction in sediment availability.

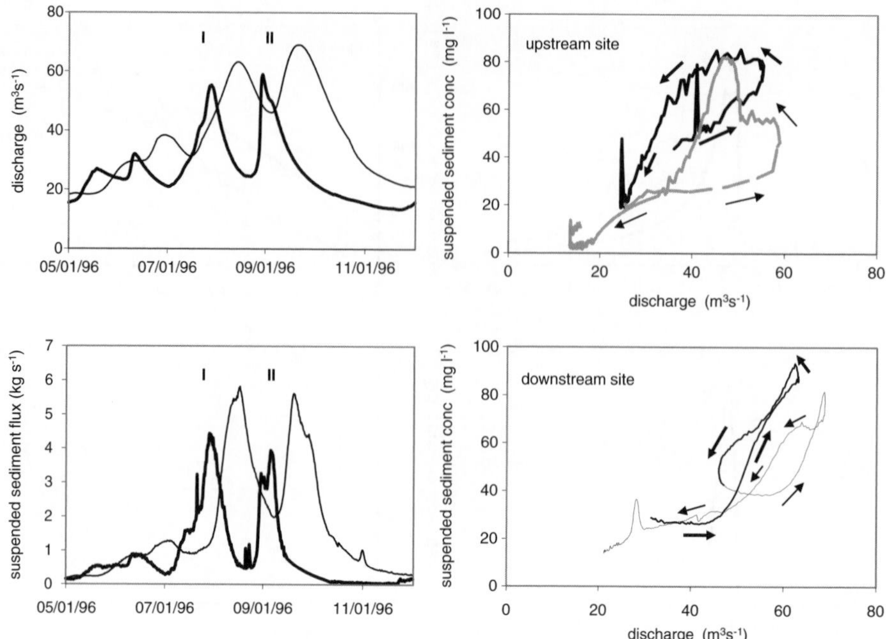

Figure 10.11 Passage of water and sediment downstream for River Swale, North Yorkshire, UK, during a double event sourced throughout the larger catchment: thick line denotes upstream site of Catterick; thin line denotes downstream site of Leckby Grange. Hysteresis patterns for the two major events (labelled I and II) are shown in solid and dashed lines respectively

These findings for the River Swale are similar to those of Asselman (1999) for the River Rhein in Germany where the mixing and routing of water from different sources as well as the amount of sediment in storage were also found to be important drivers of the sediment flux during events. A full analysis of all the events on the Swale over the period 1994–1996 also points to the importance of a small number of overbank events and to changing sediment storage within the reach (Smith *et al.*, 2003a, 2003b).

10.3.3 In-Channel Storage

Following on from the dynamics illustrated above, it is clear that in order to understand the fine-sediment cascade, it is important to be able to quantify in-transit storage, that is the amount of sediment that is temporarily stored *en route* to the sea. Floodplain storage is dealt with in Chapter 11 and may account for between 10 and 50% of the annual flux (Walling, Owens and Leeks, 1999). The focus here is on in-channel storage, particularly in relation to gravel-bed rivers in which the storage of fine sediment may be relatively short-term and unimportant in terms of volume but may be a substantial proportion of the annual fine-sediment flux and is critical for both biota and biogeochemistry or water quality.

Fine sediment may be stored on the river bed in discrete patches around vegetation (Cotton *et al.*, 2006) or in vegetated channel margins (Gurnell *et al.*, 2006), in the lee of cobbles or pebble clusters (Brayshaw, Frostick and Reid, 1983), as sand ribbons (Bridge and Jarvis, 1976) or drapes in pools (Lisle and Hilton, 1999), in dead zones (Tipping, Woof and Clarke, 1993) or on channel bars (Gurnell, Blackall and Petts, 2008). These stores may respond to individual flood events (Krein, Petticrew and Udelhovem, 2003); they may be seasonal in character such as those associated with vegetation growth; or they may reflect the passage of sediment waves downstream. Examples of the latter include short-term effects such as the reservoir release of *c*. 7000 m^3 of silt to pebble-sized sediment reported by Wohl and Cenderelli (2000). Here, fine sediment was deposited in downstream pools but then reworked in successive events such that the fine sediment passed almost entirely through the 12 km reach over a single year. Passage of a sediment wave downstream on the River Piddle is also reported by Walling and Amos (1999) who attributed it to the reworking of sediment deposited in the upper and middle reaches of this small stream following erosion of cultivated areas during the winter months. In this case, fine sediment was only slowly transmitted downstream due to the low energy of this groundwater-fed chalk stream. Longer-term impacts due to catchment disturbance (e.g. Bartley and Rutherfurd, 2005) are considered in the next section.

In addition to surface deposits, fine sediment may also be deposited within the hyporheic zone of gravel-bed rivers in the interstices between individual clasts. This may happen through simultaneous deposition of the gravel framework and fine matrix material (Andrews and Parker, 1987). Alternatively, fine sediment may subsequently be deposited within clean gravels (Schälchli, 1995) or move down into the subsurface as the gravel framework dilates at flows near the threshold of motion (Allan and Frostick, 1999). The importance of intergravel flows, rather than just surface infiltration, has also been demonstrated (Petticrew, Krein and Walling, 2007). This fine sediment fundamentally affects the permeability of the bed and the oxygenation of the gravels vital to fish-egg survival (Acornley and Sear, 1999; Greig, Sear and Carling, 2005) and the benthic and hyporheic invertebrate communities (Wood and Armitage, 1997). It is difficult to be precise about the depth to which fine sediment may be deposited within a gravel bed and the depth to which the active bedload layer, required to remobilize the contained fine sediment, may extend. In gravel-bed rivers which exhibit armouring, the depth of infiltration of fines will depend on the grain size of the armour layer, water depth and velocity. Beschta and Jackson (1979), Frostick, Lucas and Reid (1984) and Lisle (1989) found that this depth typically ranged between 5 and 10 cm, although, in very high flows, it may be more than twice the median grain size of the armour layer.

There are various techniques used for determining fine-sediment storage. The prime method for determining the content of fines at depth is freeze-coring (Petts *et al.*, 1989). This provides a measure of the fine-sediment content as a percentage of the total core by weight, and figures for the chemically/ecologically important silt and clay fractions vary between 0 and 20%. Milan, Petts and Sambrook (2000) give typical values for the upper 30 cm gravel from three types of rivers in the south of the UK (Table 10.1). Type I rivers are on impermeable geology while Type II rivers are chalk and Type III rivers have either a limestone or sandstone geology.

Freeze-coring is an expensive and disruptive method of measurement and, for surface and near-surface (upper 5–10 cm) interest, the disturbance technique of Lambert and Walling (1988) refined by Collins and Walling (2007a) has been used to determine channel storage of fine sediment. In this method, a metal cylinder is carefully lowered into the channel until it rests on the channel bed. The cylinder is then slowly rotated and pushed

Table 10.1 Typical values for fine sediment content in the upper 30 cm of gravel-bed rivers in southern UK (after Milan, Petts and Sambrook, 2000); mean percentage by weight, range given in brackets

	Type I stream impermeable ($n = 20$)	Type II stream chalk ($n = 11$)	Type III stream sandstone/limestone ($n = 20$)
D_{50} of armour layer (mm)	55 (31–73)	42 (20–84)	38 (13–60)
D_{50} substrate (mm)	26.6 (14–48)	5.6 (0.5–12)	10.1 (1.1–29)
Sand (0.063–2 mm) fraction %	11 (6.5–16.5)	42 (28.0–64.1)	21.5 (9.5–43.0)
Silt (0.004–0.062 mm) fraction %	3.5 (0.6–7.3)	4.9 (0.9–8.1)	7.4 (2.0–18.0)
Clay (<0.004) fraction %	0.6 (<0.1–1.9)	0.6 (<0.1–1.2)	1.7 (0.3–5.2)

downwards in order to create a seal with the bed. The contained area of bed is then disturbed to release the fine sediment. Disturbance of the water within the cylinder will simply release surface sediment; disturbance of the bed will release fine sediment stored down to the disturbance depth (usually about 5 cm). Representative samples of the water within the cylinder are taken and its sediment content measured. The measurements provide a snapshot in time; assuming that fine sediment is mobilized during high flows, the highest values of storage might be expected during prolonged low flows. Aside from this time dimension, estimates of channel storage are only as good as representative sampling of sediment patches and procedures for scaling up to river reaches (multiplying by the width of the channel and the length of the reach) allow. However, they provide a useful indicator of the amount of storage and its relation to estimates of the annual flux of sediment. Published values for rivers in the UK are given in Table 10.2 which includes the range of measured point storage values.

Sampled values of storage are similar to those quoted for Ontario rivers by Droppo and Stone (1994). Some indication of the variation between site measurements for each main river is indicated. All figures relating to both storage and flux in Table 10.2 should be taken as estimates with broad error bands which have yet to be fully quantified. The high percentage of the annual load seen in the cases of the Tern, Pang and Lambourn are consistent with the low flushing capacity of these predominantly groundwater-fed rivers which experience large amounts of macrophyte growth during the summer months. In other rivers, the low percentage storage is relative to a high flux but the amount of fine sediment per unit bed area stored within these rivers can be just as high as that of chalk streams.

The aggregated estimates for individual rivers given in Table 10.2 mask a number of important variations noted in the literature. An important variation is the increase in storage with distance downstream; the ranges given in Table 10.2 for the large catchments in Yorkshire (Swale, Ure, Nidd and Wharfe) illustrate the difference between the most upstream and the most downstream reach which is partly a function of the width of the channel and partly a reflection of changes in point measurements per unit bed area. Downstream effects were also noted in the Frome and Piddle catchments (Collins and Walling, 2007c). Seasonal patterns have been observed at individual sites but no consistent pattern has been found across sites, which implies that local factors are important in determining fine-sediment storage (Collins and Walling, 2007b). Summer maxima (Adams and Beschta, 1980; Frostick, Lucas and Reid, 1984; Walling, Owens and Leeks, 1998; Walling *et al.*, 2003b) may reflect both flow regime, land use and macrophyte growth.

Table 10.2 Sediment storage in channel bed (fine sediment in upper 5 cm of gravel-bed rivers)

River	Catchment area, km^2	Point storage range, $g\ m^{-2}$	Reach storage, t	Storage, $t\ km^{-1}$	Annual sediment flux, $t\ yr^{-1}$	Channel store as % annual flux	Reference
Lowland groundwater-dominated rivers in the UK							
Tern	231	860–5500	639	13	1733	17–61	Collins and Walling, 2007b
Pang	166	470–2290	191	5	498	7–92	Collins and Walling, 2007b
Lambourn	234	770–1760	229	7	1076	4–34	Collins and Walling, 2007b
Frome	437	410–2630	795	2–29	4370	11–39	Collins and Walling, 2007c
Piddle	138	260–4340	730	4–19	1281	29–97	Collins and Walling, 2007c
Large UK catchments with moorland headwaters and agricultural areas in the lower reaches							
Tweed	4390	120–960	4329		66 010	7	Owens, Walling and Leeks, 1999
Swale	1457	170–3730	3798	3–93	42 352	9	Walling, Owens and Leeks, 1998
Ure	983	180–850	3710	6–21	28 887	13	Walling, Owens and Leeks, 1998
Nidd	549	580–2410	1333	9–36	7719	17	Walling, Owens and Leeks, 1998
Wharfe	818	510–2670	1866	8–53	10 816	17	Walling, Owens and Leeks, 1998

Winter maxima have also been observed (Sear, 1993; Acornley and Sear, 1999; Walling and Amos, 1999) which tie in with increased sediment delivery. The source of these sediments is now explored.

10.3.4 Source Apportionment

A fundamental consideration when describing the fine-sediment cascade is the source of the sediment. Except in very small catchments (e.g. Bartley *et al.*, 2007), the individual monitoring of all the components is precluded by the intensive and expensive spatial and temporal sampling that would be required. Accordingly, the past two decades have seen the continued development and application of tracing and fingerprinting techniques for apportioning sediment sources. Specific advances have been made in composite fingerprinting and some attempt, albeit not generally implemented as yet, to quantify uncertainty.

Many papers report the proportion of suspended sediment or the proportion of fine material found on the river bed from various sources based on these techniques so it is important to understand the method, the various assumptions made and the potential uncertainty in these estimates. The key assumption in fingerprinting is that one or more properties of the transported sediment reflect those of the sources involved. Early work looked at single tracer groups, for example mineralogy, mineral-magnetic, radiometric, organic, chemical, isotopic or physical, but no single property was found that would reliably distinguish different sources and reliance shifted to composite signatures (Walling, Woodward and Nicholas, 1993; Collins, Walling and Leeks, 1997). Figure 10.12 shows the basic steps in source apportionment through composite fingerprinting based on Collins, Walling and Leeks (1997), Rowan, Goodwill and Franks (2000), Motha *et al.* (2002) and

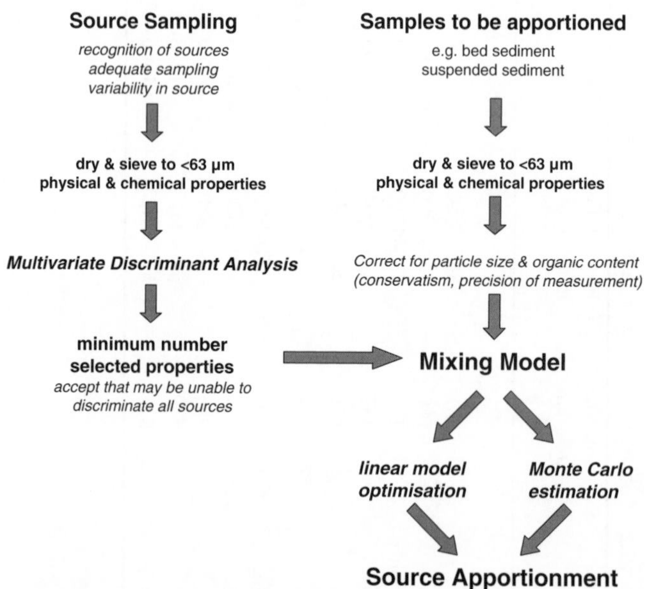

Figure 10.12 Annotated outline of composite fingerprinting method

Motha *et al.* (2003). Also shown in italics are some of the important issues encountered at each step.

Two sets of samples need to be collected: a selection of source materials needed to characterize the material likely to be eroded and the set of suspended sediment or bed sediment samples for which source apportionment is required. The collection of source materials is clearly a key determinant of the results and should both focus on material likely to be eroded (top 2 cm surface, eroding banks, poached areas, eroding tracks, etc.) and encompass variations in surface types (e.g. different land uses) and different geology as appropriate. Less obvious or hidden sources, such as subsurface soils in areas where underdrainage of clay soils promotes subsurface erosion (Chapman *et al.*, 2005; Bilotta *et al.*, 2008), should not be overlooked. A wide range of sediment chemical and physical properties is then needed to identify those properties which may be able to distinguish between different sources. Which ones are effective will vary from catchment to catchment. Once these have been identified using some form of multivariate discriminant analysis, a simple mixing model (Equation 10.6) is applied to measurements of these properties in each of the samples to be apportioned. This should use the minimum possible number of properties as the model is likely to be overparameterized. If different sources are indistinguishable, they will need to be amalgamated. Corrections for both particle size and organic content should be applied as these parameters fundamentally affect the ability of the sediment to absorb chemistry (Collins, Walling and Leeks, 1997). Corrections for conservatism (Motha *et al.*, 2002) and weights to reflect precision of measurement (Collins, Walling and Leeks, 1997) may also be applied. At its simplest, the mixing model can be expressed for each sample to be apportioned as

$$\hat{C}_i = \sum_{j=1}^{n} (p_j \, s_{ij} \, k) \text{ subject to} \sum_{j=1}^{n} p_j = 1 \text{ and } 0 \leq p_j \leq 1 \qquad (10.6)$$

where $\hat{C}_i$ is the predicted concentration of property i in the sample, p_j is the proportion of source type j, s_{ij} is a representative concentration of property i for source j and k is a correction factor for the sample.

There are two basic approaches to fitting the model: optimization using constrained linear programming (Collins, Walling and Leeks, 1997) or some form of Monte Carlo simulation which recognizes that there is unlikely to be a single optimum solution (Rowan, Goodwill and Franks, 2000; Motha *et al.*, 2002; Collins and Walling, 2007a). Results from Monte Carlo simulations may indicate relatively wide uncertainty bands for the derived proportions but generally give distinct upper limits and the same order of importance of the different sources as more traditional optimization techniques. Thus, although precise values may need to be treated with caution, the overall relative importance of sources is generally reliable. Example surveys of suspended sediment samples in the UK (Walling, 2005) clearly separate those catchments where channel bank or subsurface erosion dominates from those where surface erosion dominates (Figure 10.13a). Regional contrasts (Figure 10.13b) have also been identified in the source of bed-sediment samples (Walling, Collins and McMellin, 2003c).

In addition to the dominant climate, soil type, land management and bank erodibility factors which clearly have a role in the observed pattern, distance downstream is also important in the cascade of fine sediment. Work using radionuclides within the Yellowstone

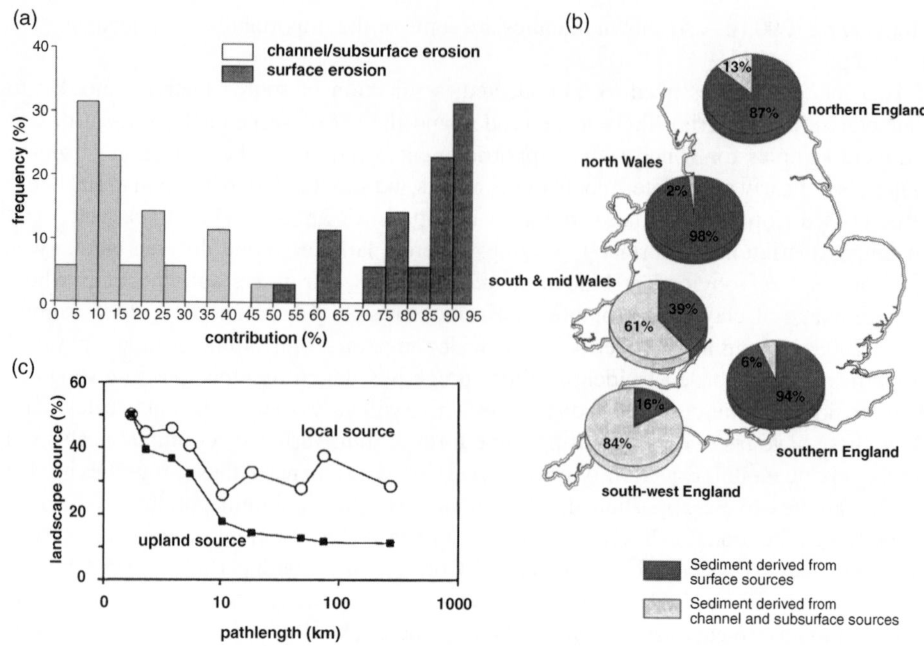

Figure 10.13 Source apportionment.: (a) Suspended sediment in UK rivers (reproduced, with permission, from Walling, D.E. (2005) Tracing suspended sediment sources in catchments and river systems. Science of the Total Environment, 344, 159–184. © Elsevier). (b) Provenance of interstitial sediment in the UK (With kind permission from Springer Science + Business Media: Walling, D.E., Collins, A.L. and McMellin, G. (2003) A reconnaissance survey of the source of interstitial fine sediment recovered from salmonid spawning gravels in England and Wales. Hydrobiologia, 497, 91–108.) (c) Downstream trends in provenance from Yellowstone (reproduced from Whiting, P.J., Matisoff, G. and Fornes, W. (2005) Suspended sediment sources and transport distances in the Yellowstone River basin. Geological Society of America Bulletin, 117, 515-529. Courtesy of the Geological Society of America).

River basin clearly demonstrates that the dominance of upstream sources diminishes with distance downstream (Figure 10.13c) as transport distances increase with catchment area and the relative importance of river banks increases due to lateral erosion and interaction between the river and its floodplain (see Chapter 11).

10.4 A Whole-Catchment View

The key issues in the fine-sediment cascade highlighted above are the source of the sediment and the importance of sediment routing and channel storage. It has also been seen that fine-sediment transport is highly intermittent and shows behaviour which is both threshold-dependent and hysteretic. In this section, it is shown that assembling this information at the large basin scale requires both proper identification of time and space scales, appropriate simplification of process representation and a spatially distributed approach.

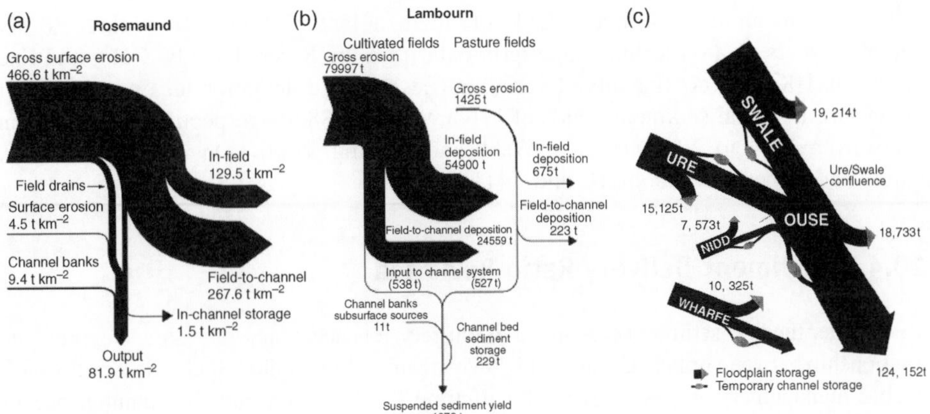

Figure 10.14 Sediment budgets for two contrasting small catchments in the UK and for the River Ouse, North Yorkshire, UK
(reproduced, with permission, from Walling, D.E., Russell, M.A., Hodgkinson, R.A. and Zhang, Y. (2002) Establishing sediment budgets for two small lowland agricultural catchments in the UK. Catena, 47, 323–353. © Elsevier; from Walling, D.E., Collins, A.L., Jones, P.A., Leeks, G.J.L. and Old, G. (2006) Establishing fine-grained sediment budgets for the Pang and Lambourn LOCAR study catchments. Journal of Hydrology, 330, 126–141. © Elsevier; Walling, D.E., Owens, P.N. and Leeks, G.J.L. (1998) The role of channel and floodplain storage in the suspended sediment budget of the River Ouse, Yorkshire, UK. Geomorphology, 22, 225–242. © Elsevier).

10.4.1 Examples of Fine-Sediment Budgets

Once sources are apportioned and storage assessed, an overall fine-sediment budget can be constructed. Examples of the range of contemporary fine-sediment budgets are given in Figure 10.14. They have been constructed to represent the annual sediment yield on the basis of short-term records and an array of fingerprinting and other tracing techniques. Figures 10.14a and 10.14b are for contrasting types of small catchment. Both show the importance of within-field and field-to-channel deposition, with relatively small amounts of fine sediment being conveyed through the channel system. This is especially the case for the River Lambourn which is a chalk catchment, fed predominantly by groundwater. Figure 10.14b also highlights the important contribution of cultivated compared to pasture fields. In underdrained clay soils and in wetter areas with steeper slopes, the contribution from pasture fields will be much higher (Bilotta *et al.*, 2008). Figure 10.14c shows an example of the much larger catchment of the River Ouse which shows the relative importance of flux, floodplain and channel storage. Figures for in-channel storage are given in Table 10.2 and amount to 10–20% of the total flux; Figure 10.14c gives the figures for floodplain losses which account for 10–50% of the overall fine-sediment flux.

The budgets shown in Figure 10.14 represent the annual flux, based on a relatively short record of flux data. However, another characteristic of sediment flux, occasioned not only by its relationship with flow but also by depletion of available catchment storage, is its high variability from year to year. In the UK, measurements made on the Swale and Calder in Yorkshire (Wass and Leeks, 1999), the Tweed in Southern Scotland (Bronsdon and Naden, 2000) and the Cyff and Tanllwyth in Wales show differences in annual sediment yields of up to a factor of five, from three to six complete years of monitoring. Long-term data from the Rivers Creedy, Culm and Exe in Devon, UK, give coefficients of variation (i.e. standard deviation as a percentage of the mean) in annual sediment yields of 116.3, 96.4 and 88.8% respectively over the ten years from 1994 to 2003 (Harlow, Webb and Walling, 2006). On the River Creedy, yields varied between about 10 and 74 $\mathrm{t\,km^{-2}\,yr^{-1}}$.

10.4.2 Sediment Delivery Ratio Revisited

To consider the fine-sediment cascade in its entirety, it is necessary to be able to express the relationship between upland erosion and downstream sediment flux. Such a description will enable management issues such as the lifetime of reservoirs and the maintenance of downstream habitat to be addressed. The sediment delivery ratio (SDR), that is the ratio of the sediment yield of a catchment compared to the average sediment erosion across the catchment, has in the past been used as a lumped descriptor of the fine-sediment cascade. However, there are a number of conceptual and practical problems associated with this lumped measure which have recently been highlighted. Parsons *et al.* (2006) point to issues relating to quoted hillslope erosion rates and the definition of area used in the delivery ratio, while Lu and Richards (2008) emphasize issues relating to the time periods over which erosion rates and sediment yields produce meaningful averages – a factor which is not helped by the lack of long-term data on sediment flux through river networks.

For landscape units in steady state or equilibrium, catchment yield should be approximately equal to the amount eroded from upstream slopes with no net change in storage, and the sediment delivery ratio should equal 1. If $\mathrm{SDR} \neq 1$ then it implies aggradation or degradation over the chosen time period. For larger catchments, it is therefore clear that the SDR, as calculated over observational time periods, is a function of time and the catchment condition as the system is not stationary over these time periods but contingent on its erosional history as dictated by both anthropogenic and geomorphological change. Human colonization, the expansion of agriculture and the associated increase in soil erosion have been widely documented in the USA (Trimble, 1975, 1983; Meade, Yuzyk and Day, 1990), Australia (Wasson, Olive and Rosewell, 1996; Wasson *et al.*, 1998; Rutherfurd, 2000) and the UK (Macklin, 1999; Macklin and Lewin, 2008). Timescales for the passage of sediment slugs following human disturbance can be several centuries dependent on volumes and calibre of material. Examples of recovery from human disturbance are given in Wohl and Cenderelli (2000) and Bartley and Rutherfurd (2005). The legacy of geomorphological change following glaciation is also evident in examples from Canada (Church *et al.*, 1999; Church, 2002) where the characteristic time for sediment transfer in large basins has been estimated to be of the order of 10^2–10^6 years.

It is also clear that as spatial scale increases, the system moves from a condition in which the hillslope is dominant to a system in which the channel network dominates and the importance of travel distance and in-channel storage as well as the link to geomorphology and floodplain storage (see Chapter 11) becomes paramount. The scale at which this happens depends on the relative travel times. Lu, Moran and Sivapalan (2005) present a simple lumped two-store linear model which recognizes the roles of hillslope and channel storage and the processes of redeposition and reworking of sediment (Figure 10.15). It is analogous to the hydrological model of Sivapalan, Jothityangkoon and Menabde (2002).

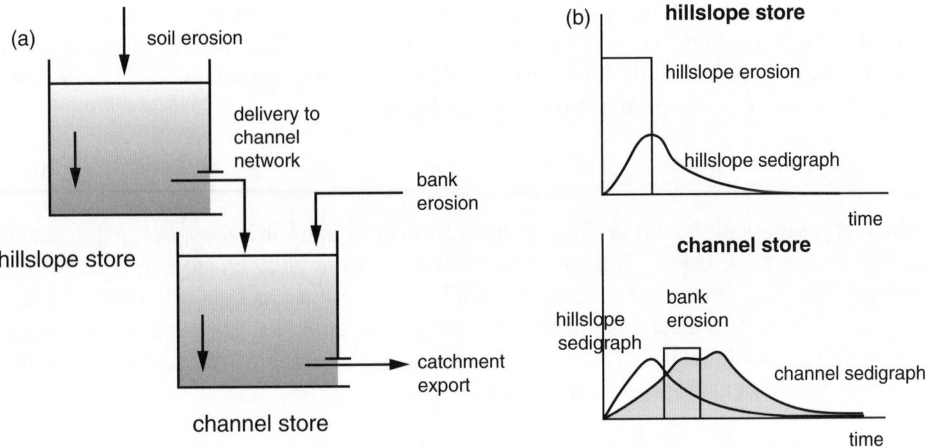

Figure 10.15 Simple two store linear model
(reproduced from Lu, H., Moran, C.J. and Sivapalan, M. (2005) A theoretical exploration of catchment-scale sediment delivery. Water Resources Research, 41, Article no. W09415. Courtesy of the American Geophysical Union).

The SDR for an event may be derived as

$$\gamma_e = \frac{t_{er}}{t_t}\frac{1}{B_n}\left(\frac{1}{A_h} + \frac{e_n t_i}{e_h t_{er}}\right) \quad \text{where} \quad A_h = 1 + \lambda_h t_h \quad \text{and} \quad B_n = 1 + \lambda_n t_n \qquad (10.7)$$

where γ_e is event SDR, t_{er} is effective storm duration, t_t is event time, e_n is channel-network erosion rate, t_i is duration of channel erosion, e_h is upland erosion rate, λ_h is proportion of hillslope sediment redeposited per unit time, t_h is mean residence time within the hillslope store, λ_n is proportion of channel erosion redeposited per unit time and t_n is mean residence time within the channel store. This model has been shown to fit the SDR data of Roehl (1962) and, with a suitable choice of parameters, can simulate all the observed behaviour in SDR with catchment area. The shift from hillslope to channel dominance may be calculated as

$$\frac{1}{A_h} < \frac{e_n t_t}{e_h t_{er}} \qquad (10.8)$$

While the lumped approach given in Figure 10.15 is one way of putting together the hillslope and network routing components of the fine-sediment cascade, for large catchments it is often necessary to link specific upstream inputs to downstream impacts in order to provide a basic understanding of the system for the purpose of managing fine sediment. This requires a spatially distributed approach which is exemplified by the SedNet model presented by Prosser *et al.* (2001), Lu *et al.* (2004), Lu, Moran and Prosser (2006) and Ding and Richards (2009). This takes a similar approach to that shown in Figure 10.15 in that it separates out hillslope and channel components of the system and treats individual sediment sources separately but it is implemented within a spatially distributed GIS framework. Individual processes are approximated in a simple way and the linkage between source and sea is essentially treated as a chained set of reservoirs in which there is the potential for deposition. One application is to the Murray–Darling basin in Australia which has an area of about $1.1 \times 10^6\,\text{km}^2$. The model is set up using spatial units of 50–100 km^2. Annual sediment yields from each unit are given by

$$Y_i = T_i + B_i + G_i + \gamma_i E_i - D_i \tag{10.9}$$

where Y_i is sediment yield (t yr^{-1}), T_i is input from upstream tributaries, B_i is mean annual supply from bank erosion, G_i is mean annual supply from gully erosion, γ_i is a hillslope delivery ratio, E_i is mean annual input from hillslope erosion and D_i is a deposition or loss term which includes channel deposition and losses to floodplains, lakes and reservoirs. For an individual unit the SDR is $Y_i/(Y_i + D_i)$ and for any downstream link, k, the mean annual delivery of sediment from upstream link i is given by

$$\lambda_{ik} = I_i \prod_{j=1}^{M_{ik}} \gamma_j \tag{10.10}$$

where I_i is the sediment supply from within the link i, M_{ik} is the number of river links along the route from link i to location k, and γ_j are the successive link element SDRs which represent the probability of sediment passing through river link j as determined by the amount of deposition in that river link. The total sediment yield at the downstream point is given by

$$T_k = \sum_{i}^{N} \lambda_{ik} \tag{10.11}$$

where N is the total number of link elements contributing to location k.

An example of the results from this model is given in Figure 10.16 for the sediment delivery ratio for both the clay and silt fractions of fine sediment in the Murray–Darling

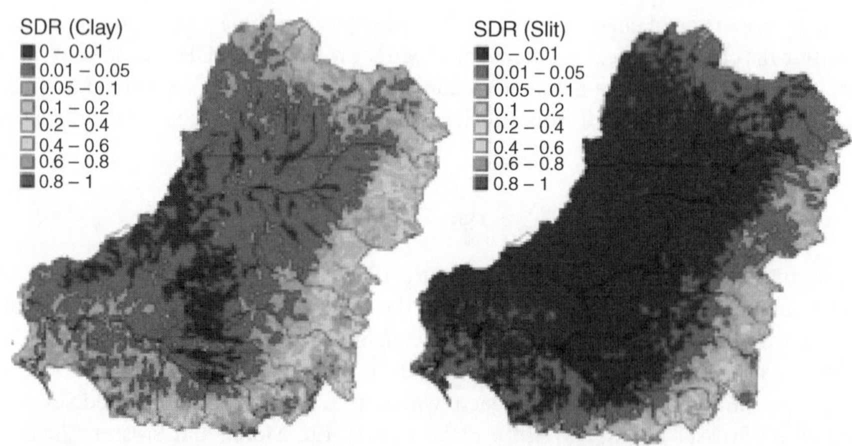

Figure 10.16 Results of SDR for clay and silt fractions in the Murray–Darling basin (reproduced, with permission, from Lu, H., Moran, C.J., Prosser, I.P. (2006) Modelling sediment delivery ratio over the Murray Darling Basin. Environmental Modelling and Software, 21, 1297–1308. © Elsevier).

Basin. The calibre of the sediment affects the residence time of the sediment via the settling velocity. It is apparent from Figure 10.16 that very little of the sediment generated in the uplands reaches the basin outlet on an annual timescale and, in this particular example, the main management concern in the region is the delivery of fine sediment to lowland rivers and the impact on aquatic ecosystems.

Lu *et al.* (2004) use the model to examine the cost-effectiveness of various management scenarios to reduce the delivery of fine sediment generated by upland soil erosion to the channels of the lowland area. The model could also be used to look at longer term impacts of climate, land use or other environmental change based on a range of scenarios. The approach focuses on an appropriate level of detail for the scale of the problem. Further work is needed to develop simplified formulations of the governing processes in line with time-integration of our physical understanding. Perhaps this, together with the basic data collection and testing of model outputs, is one of most important challenges for the future.

10.5 Conclusion – the Need for Future Research

This chapter has sought to highlight areas of the fine-sediment cascade in which significant progress has been made in the past few decades. It has spanned the three major disciplines of hydraulics, hydrology and geomorphology in order to bring out the key advances and identify some of the remaining issues. The fundamental influence of time and space scales on current thinking in these areas has also been demonstrated. Looking to the future, a number of areas for future research can be identified.

In the hydraulics field, there is still a need to improve estimates of fine-sediment conveyance capacity and move away from empirical relationships. Given the complex flow fields found in natural channels, this will require the development of two-phase space–time averaged models. A particular requirement is parameterization of the trapping of fine sediment in gravel beds and vegetation. Both unsteady and non-capacity flows need to be considered. In developing these models, it is important to recognize the process of flocculation and the specific characteristics of flocculated and biogenic sediments, and to develop sound protocols for their measurement.

A great deal of progress has been made in the monitoring of fine-sediment fluxes and collecting data on both in-channel and floodplain storage. However, there is still very limited long-term monitoring of sediment flux through entire river networks. This is a crucial requirement for the effective management of fine sediment and provision of test data for GIS models at the larger scale. New technologies such as acoustics or fibre optics need to be explored as to their applicability. It is also increasingly necessary to provide proper estimates of uncertainty in flux measurements, source apportionment and volumes of channel and floodplain storage.

At the whole-catchment scale, the development and testing of simplified process-based models, a recognition of time-dependence in sediment transfer, the development of better relationships between sediment transport and storage, and estimates of sediment residence times are fundamental requirements. In parallel, it is also necessary to quantify the timing, amounts, types and quality of sediment which are detrimental to ecosystems and human well-being and infrastructure. This information then needs to be coupled with an understanding of the fine-sediment cascade in order to provide a sound basis for sediment management, policy and application of mitigation measures.

10.6 Acknowledgements

The author gratefully acknowledges CEH colleagues Dr Gareth Old for comments on an earlier draft and Dr Barnaby Smith for work on the sediment dynamics of the River Swale which was undertaken as part of the Land Ocean Interaction Study funded by the Natural Environment Research Council of the UK. Flow data for the River Swale was supplied by the UK Environment Agency.

References

Acornley, R.M. and Sear, D.A. (1999) Sediment transport and sedimentation of Brown trout (*Salmo trutta* L) spawning gravels in chalk streams. *Hydrological Processes*, **13**, 447–458.

Adams, J.N. and Beschta, R.L. (1980) Gravel composition in Oregon coastal streams. *Canadian Journal of Fisheries and Aquatic Sciences*, **37**, 1514–1521.

Ahrens, J.P. (2000) A fall-velocity equation. *Journal of Waterway, Port, Coastal, and Ocean Engineering, American Society of Civil Engineers*, **126**, 99–102.

Allan, A. and Frostick, L.E. (1999) Framework dilation, winnowing and matrix particle size: the behaviour of some sand-gravel mixtures in a laboratory flume. *Journal of Sedimentary Research*, **69**, 21–26.

Alvarez-Cobelas, M., Angeler, D.G. and Sánchez-Carrillo, S. (2008) Export of nitrogen from catchments: a worldwide analysis. *Environmental Pollution*, **156**, 261–269.

Andrews, E.D. and Parker, G. (1987) Formation of a coarse surface layer as the response to gravel mobility, in *Sediment Transport in Gravel-Bed Rivers* (eds C.R. Thorne, J.C. Bathurst and R.D. Hey), John Wiley and Sons, Chichester, pp. 269–325.

Ashbridge, D. (1995) Processes of river bank erosion and their contribution to the suspended sediment load of the River Culm, Devon, in *Sediment and Water Quality in River Catchments* (eds I.D.L. Foster, A.M. Gurnell and B.W. Webb), John Wiley and Sons, Chichester, pp. 229–245.

Asselman, N.E.M. (1999) Suspended sediment dynamics in a large drainage basin: the River Rhine. *Hydrological Processes*, **13**, 1437–1450.

Aziz, N.M., Prasad, S.N. and Bhattacharya, S.K. (1992) Suspended sediment concentration profiles using conservation laws. *Journal of Hydraulic Research*, **30**, 539–554.

Bartley, R. and Rutherfurd, I. (2005) Re-evaluation of the wave model as a tool for quantifying the geomorphic recovery potential of streams disturbed by sediment slugs. *Geomorphology*, **64**, 221–242.

Bartley, R., Hawdon, A., Post, D.A. and Rothe, C.H. (2007) A sediment budget for a grazed semi-arid catchment in the Burdekin basin, Australia. *Geomorphology*, **87**, 302–321.

Batalla, R.J. (2003) Sediment deficit in rivers caused by dams and instream gravel mining. A review with examples from NE Spain. *Cuaternario y Geomorfología*, **17**, 79–91.

Becht, M. (1989) Suspended load yield of a small alpine drainage basin in upper Bavaria, in *Landforms and Landform Evolution in West Germany* (ed. F. Ahnert). *Catena (Supplement)*, vol. **15**, pp. 329–342.

Bennett, S.J., Bridge, J.S. and Best, J.L. (1998) The fluid dynamics of upper-stage plane beds. *Journal of Geophysical Research*, **103**, 1239–1274.

Beschta, R.L. and Jackson, W.L. (1979) The intrusion of fine sediments into a stable gravel bed. *Journal of the Fisheries Research Board of Canada*, **36**, 204–210.

Bilotta, G.S., Brazier, R.E., Haygarth, P.M. *et al.* (2008) Rethinking the contribution of drained and undrained grasslands to sediment-related water quality problems. *Journal of Environmental Quality*, **37**, 906–914.

Bogen, J. (1980) The hysteresis effect of sediment transport systems. *Norsk Geografiska Tidsskrift*, **34**, 45–54.

Bowes, M.J. and House, W.A. (2001) Phosphorus and dissolved silicon dynamics in the River Swale catchment, UK: a mass balance approach. *Hydrological Processes*, **15**, 261–280.

Bowes, M.J., Leach, D.V. and House, W.A. (2005) Seasonal nutrient dynamics in a chalk stream: the River Frome, Dorset, UK. *Science of the Total Environment*, **336**, 225–241.

Brakensiek, D.L., Osborn, H.B. and Rawls, W.J. (coords.) (1979) Field Manual for Research in Agricultural Hydrology, United States Department of Agriculture, Agricultural Handbook 224. 550 pp.

Brayshaw, A.C., Frostick, L.E. and Reid, I. (1983) The hydrodynamics of particle clusters and sediment entrainment in coarse alluvial channels. *Sedimentology*, **30**, 137–143.

Bridge, J.S. (2003) *Rivers and Floodplains. Forms, Processes and Sedimentary Record*, Blackwell, Oxford. 491 pp.

Bridge, J.S. and Jarvis, J. (1976) Flow and sedimentary processes in meandering River South Esk, Glen Clova, Scotland. *Earth Surface Processes and Landforms*, **1**, 303–336.

Bronsdon, R.K. and Naden, P.S. (2000) Suspended sediment in the Rivers Tweed and Teviot. *Science of the Total Environment*, **251/252**, 95–113.

Bull, L.J. (1997) Relative velocities of discharge and sediment waves for the River Severn, UK. *Hydrological Sciences Journal*, **42**, 649–660.

Bungartz, H. and Wanner, S.C. (2004) Significance of particle interaction to the modelling of cohesive sediment transport in rivers. *Hydrological Processes*, **18**, 1685–1702.

Cao, Z. (1999) Equilibrium near-bed concentration of suspended sediment. *Journal of Hydraulic Engineering ASCE*, **125**, 1270–1278.

Chapman, A.S., Foster, I.D.L., Lees, J.A. and Hodgkinson, R. (2005) Sediment sources and transport pathways in the Rosemaund experimental catchment, Herefordshire, UK. *Hydrological Processes*, **19**, 2875–2897.

Cheng, N.-S. (1997) Simplified settling velocity formula for sediment particle. *Journal of Hydraulic Engineering ASCE*, **123**, 149–152.

Church, M. (1996) Space, Time and the Mountain – how do we order what we see?, in The Scientific Nature of Geomorphology: Proceedings of the 27th Binghampton Symposium in Geomorphology (eds B.L. Rhoads and C.E. Thorn), pp. 147–170.

Church, M. (2002) Fluvial sediment transfer in cold regions, in *Landscapes of Transition: Landform Assemblages and Transformations in Cold Regions* (eds K. Hewitt, M.-L. Byrne, M. English and G. Young), Kluwer Academic, Netherlands, pp. 93–117.

Church, M., Ham, D., Hassan, M. and Slaymaker, O. (1999) Fluvial clastic sediment yield in Canada: scaled analysis. *Canadian Journal of Earth Sciences*, **36**, 1267–1280.

Clarke, S.J. and Wharton, G. (2001) Sediment nutrient characteristics and aquatic macrophytes in lowland English rivers. *Science of the Total Environment*, **266**, 103–112.

Collins, A.L. and Walling, D.E. (2007a) Sources of fine sediment recovered from the channel bed of lowland groundwater-fed catchments in the UK. *Geomorphology*, **88**, 120–138.

Collins, A.L. and Walling, D.E. (2007b) The storage and provenance of fine sediment on the channel bed of two contrasting lowland permeable catchments, UK. *River Research and Applications*, **23**, 429–450.

Collins, A.L. and Walling, D.E. (2007c) Fine-grained bed sediment storage within the main channel systems for the Frome and Piddle catchments, Dorset, UK. *Hydrological Processes*, **21**, 1448–1459.

Collins, A.L., Walling, D.E. and Leeks, G.J.L. (1997) Source type ascription for fluvial suspended sediment based on a quantitative composite fingerprinting technique. *Catena*, **29**, 1–27.

Collins, A.L., Walling, D.E. and Leeks, G.J.L. (2005) Storage of fine-grained sediment and associated contaminants within the channels of lowland permeable catchments in the UK, in *Sediment*

Budgets I, IAHS Publication 291 (eds D.E. Walling and A.J. Horowitz), IAHS Press, Wallingford, UK, pp. 259–268.

Cotton, J.A., Wharton, G., Bass, J.A.B. *et al.* (2006) The effects of seasonal changes to in-stream vegetation cover on patterns of flow and accumulation of sediment. *Geomorphology*, **77**, 320–334.

Dearing, J.A. and Jones, R.T. (2003) Coupling temporal and spatial dimensions of global sediment flux through lake and marine sediment records. *Global and Planetary Change*, **39**, 147–168.

Degens, E.T., Kempe, S. and Richey, J.E. (1991) Summary: biogeochemistry of major world rivers, in *Biogeochemistry of Major World Rivers* (eds E.T. Degens, S. Kempe and J.E. Richey), John Wiley and Sons, Chichester, pp. 323–347.

Dietrich, W.E. (1982) Settling velocity of natural particles. *Water Resources Research*, **18**, 1615–1626.

Ding, J. and Richards, K.S. (2009) Preliminary modelling of sediment production and delivery in the Xihanshui River basin, Gansu, China. *Catena*. doi: 10.1016/j.catena.2009.05.014.

Droppo, I.G. (2001) Rethinking what constitutes suspended sediment. *Hydrological Processes*, **15**, 1551–1564.

Droppo, I.G. (2004) Structural controls on floc strength and transport. *Canadian Journal of Civil Engineering*, **31**, 569–578.

Droppo, I.G. and Ongley, E.D. (1992) The state of suspended sediment in the freshwater fluvial environment: a method of analysis. *Water Research*, **26**, 65–72.

Droppo, I.G. and Stone, M. (1994) In-channel surficial fine-grained sediment laminae. Part I: Physical characteristics and formational processes. *Hydrological Processes*, **8**, 101–111.

Droppo, I.G., Walling, D.E. and Ongley, E.D. (2002) Suspended sediment structure: implications for sediment transport/yield modelling, in *Modelling Erosion, Sediment Transport and Sediment Yield*, UNESCO International Hydrological Programme Technical Documents in Hydrology no. 60 (eds W. Summer and D.E. Walling), UNESCO, Paris, pp. 205–228.

Droppo, I.G., Leppard, G.G., Flannigan, D.T. and Liss, S.N. (1997) The freshwater floc: a functional relationship of water and organic and inorganic floc constituents affecting suspended sediment properties. *Water, Air and Soil Pollution*, **99**, 43–53.

Evans, D.J., Johnes, P.J. and Lawrence, D.S. (2004) Physico-chemical controls on phosphorus cycling in two lowland streams. Part 2 – the sediment phase. *Science of the Total Environment*, **329**, 165–182.

Evans, J.G., Wass, P.D. and Hodgson, P. (1997) Integrated continuous water quality monitoring for the LOIS river programme. *Science of the Total Environment*, **194/195**, 111–118.

Ferguson, R.I. and Church, M. (2004) A simple universal equation for grain settling velocity. *Journal of Sedimentary Research*, **74**, 933–937.

Frostick, L.E., Lucas, P.M. and Reid, I. (1984) The infiltration of fine matrices into coarse-grained alluvial sediments and its implications for stratigraphical interpretation. *Journal of the Geological Society*, **141**, 955–965.

Gibbs, R.J., Matthews, M.D. and Link, D.A. (1971) The relationship between sphere size and settling velocity. *Journal of Sedimentary Petrology*, **41**, 7–18.

Gippel, C.J. (1995) Potential of turbidity monitoring for measuring the transport of suspended solids in streams. *Hydrological Processes*, **9**, 83–97.

Greig, S.M., Sear, D.A. and Carling, P.A. (2005) The impact of fine sediment accumulation on the survival of incubating salmon progeny: implications for sediment management. *Science of the Total Environment*, **344**, 241–258.

Greimann, B.P. and Holly, F.M. (2001) Two-phase flow analysis of concentration profiles. *Journal of Hydraulic Engineering ASCE*, **127**, 753–762.

Gurnell, A.M., Blackall, T.D. and Petts, G.E. (2008) Characteristics of freshly deposited sand and finer sediments along an island-braided gravel-bed river. *Geomorphology*, **99**, 254–269.

Gurnell, A.M., Van Oosterhout, M.P., De Vlieger, B. and Goodson, J.M. (2006) Reach-scale interactions between aquatic plants and physical habitat: River Frome, Dorset. *River Research and Applications*, **22**, 667–680.

Hallermeier, R.J. (1981) Terminal settling velocity of commonly occurring sand grains. *Sedimentology*, **28**, 859–865.

Haralampides, K., Corquodale, J.A. and Krishnappan, B.G. (2003) Deposition properties of fine sediment. *Journal of Hydraulic Engineering ASCE*, **129**, 230–234.

Harlow, A., Webb, B. and Walling, D. (2006) *Sediment yields in the Exe Basin: a longer-term perspective, in Sediment dynamics and the hydromorphology of fluvial systems*, IAHS Publication 306 (eds J.S. Rowan, R.W. Duck and A. Werrity), IAHS Press, Wallingford, UK, 12–20.

Heidel, S.G. (1956) The progressive lag of sediment concentration with flood waves. *Transactions of the American Geophysical Union*, **37**, 56–66.

Heywood, M.J.T. and Walling, D.E. (2007) The sedimentation of salmonid spawning gravels in the Hampshire Avon catchment, UK: implications for the dissolved oxygen content of intragravel water and embryo survival. *Hydrological Processes*, **21**, 770–788.

Holeman, J.N. (1968) The sediment yield of major rivers of the world. *Water Resources Research*, **4**, 737–747.

House, W.A. (2003) Geochemical cycling of phosphorus in rivers. *Applied Geochemistry*, **18**, 739–748.

Jarvie, H.P., Jurgens, M.D., Williams, R.J. *et al.* (2005) Role of bed sediments as sources and sinks of phosphorus across two major eutrophic UK river basins: the Hampshire Avon and Herefordshire Wye. *Journal of Hydrology*, **304**, 51–74.

Köthe, H. (2003) Existing sediment management guidelines: an overview. *Journal of Soils and Sediments*, **3**, 139–143.

Krein, A., Petticrew, E. and Udelhovem, T. (2003) The use of fine sediment fractal dimensions and colour to determine sediment sources in a small watershed. *Catena*, **53**, 165–179.

Kronvang, B., Laubel, A., Larsen, S.E. and Friberg, N. (2003) Pesticides and heavy metals in Danish streambed sediment. *Hydrobiologia*, **494**, 93–101.

Krishnappan, B.G. (2007) Recent advances in basic and applied research in cohesive sediment transport in aquatic systems. *Canadian Journal of Civil Engineering*, **34**, 731–743.

Lal, R. (2002) Soil erosion and the global carbon budget. *Environment International*, **29**, 437–450.

Lal, R. (2006) *Encyclopedia of Soil Science*, 2nd edn, CRC Press. 1600 pp.

Lambert, C.P. and Walling, D.E. (1988) Measurement of channel storage of suspended sediment in a gravel-bed river. *Catena*, **15**, 65–80.

Lawler, D.M., Petts, G.E., Foster, I.D.L. and Harper, S. (2006) Turbidity dynamics during spring storm events in an urban headwater. *Science of the Total Environment*, **360**, 109–126.

Leeder, M.R. (1982) *Sedimentology. Process and Product*, George Allen and Unwin, London. 344 pp.

Liao, B.Q., Allen, D.G., Leppard, G.G. *et al.* (2002) Interparticle interactions affecting the stability of sludge flocs. *Journal of Colloid and Interface Science*, **249**, 372–380.

Lisle, T.E. (1989) Sediment transport and resulting deposition in spawning gravels, north coastal Carolina. *Water Resources Reseach*, **26**, 1303–1319.

Lisle, T.E. and Hilton, S. (1999) Fine bed material in pools of natural gravel bed channels. *Water Resources Research*, **35**, 1291–1304.

López, F. and Garcia, M. (1998) Open-channel flow through simulated vegetation: suspended sediment transport modelling. *Water Resources Research*, **34**, 2341–2352.

Lu, H. and Richards, K.S. (2008) Sediment delivery: new approaches to modelling an old problem, in *River Confluences, Tributaries and the Fluvial Network* (eds S.P. Rice, A.G. Roy and B. Rhoads), John Wiley and Sons, Chichester, 337–366.

Lu, H., Moran, C.J. and Prosser, I.P. (2006) Modelling sediment delivery ratio over the Murray Darling Basin. *Environmental Modelling and Software*, **21**, 1297–1308.

Lu, H., Moran, C.J. and Sivapalan, M. (2005) A theoretical exploration of catchment-scale sediment delivery. *Water Resources Research*, **41**, W09415, doi: 10.1029/2005WR004018

Lu, H., Moran, C.J., Prosser, I.P. and DeRose, R. (2004) Investment prioritisation based on broadscale spatial budgeting to meet downstream targets for suspended sediment loads. *Water Resources Research*, **40**, W09501, doi: 10.1029/2003WR002966

Macklin, M.G., Hudson-Edwards, K.A. and Dawson, E.J. (1997) The significance of pollution from historic metal mining in the Pennine orefields on river sediment contaminant fluxes to the North Sea. *Science of the Total Environment*, **194**, 391–397.

Macklin, M.G. (1999) Holocene river environments in prehistoric Britain: human interaction and impact. *Journal of Quarternary Science*, **14**, 521–530.

Macklin, M.G. and Lewin, J. (2008) Alluvial response to the changing Earth system. *Earth Surface Processes and Landforms*, **33**, 1374–1395.

Martin, J.-M. and Meybeck, M. (1979) Elemental mass-balance of material carried by major world rivers. *Marine Chemistry*, **7**, 173–206.

Matthaei, C.D., Weller, F., Kelly, D.W. and Townsend, C.R. (2006) Impacts of fine sediment addition to tussock, pasture, dairy and deer farming streams in New Zealand. *Freshwater Biology*, **51**, 2154–2172.

Meade, R.H., Yuzyk, T.R. and Day, T.J. (1990) Movement and storage of sediment in rivers of the United States and Canada, in *Surface Water Hydrology* (eds M.G. Wolman and H.C. Riggs), Geological Society of America, Boulder, Colorado, pp. 255–280.

Meybeck, M. (1982) Carbon, nitrogen and phosphorus transport by world rivers. *American Journal of Science*, **282**, 401–450.

Milan, D.J., Petts, G.E. and Sambrook, H. (2000) Regional variations in the sediment structure of trout streams in southern England: benchmark data for siltation assessment and restoration. *Aquatic Conservation*, **10**, 407–420.

Motha, J.A., Wallbrink, P.J., Hairsine, P.B. and Grayson, R.B. (2002) Tracer properties of eroded sediment and source material. *Hydrological Processes*, **16**, 1983–2000.

Motha, J.A., Wallbrink, P.J., Hairsine, P.B. and Grayson, R.B. (2003) Determining the sources of suspended sediment in a forested catchment in southeastern Australia. *Water Resources Research*, **39**, 1056, doi: 10.1029/2004WR000794

Muste, M., Yu, K., Fujita, I. and Ettema, R. (2005) Two-phase versus mixed-flow perspective on suspended sediment transport in turbulent channel flows. *Water Resources Research*, **41**, W10402, doi: 10.1029/2004WR003595

Newcombe, C.P. and Jensen, J.O.T. (1996) Channel suspended sediment and fisheries: a synthesis for quantitative assessment of risk and impact. *North American Journal of Fisheries Management*, **16**, 693–727.

Nicholas, A.P. and Walling, D.E. (1996) The significance of particle aggregation in the overbank deposition of suspended sediment on river floodplains. *Journal of Hydrology*, **186**, 275–293.

Niyogi, D.K., Koren, M., Arbuckle, C.J. and Townsend, C.R. (2007) Stream communities along a catchment land-use gradient: subsidy-stress responses to pastoral development. *Environmental Management*, **39**, 213–225.

Old, G.H., Leeks, G.J.L., Packman, J.C. *et al.* (2006) River flow and associated transport of sediments and solutes through a highly urbanised catchment, Bradford, West Yorkshire. *Science of the Total Environment*, **360**, 98–108.

Owens, P.N. (2007) Introduction. Background and Summary of this issue on Sediment linkages. *Journal of Soils and Sediments*, **7**, 273–276.

Owens, P.N. and Walling, D.E. (2002) Changes in sediment sources and floodplain deposition rates in the catchment of the River Tweed, Scotland, over the last 100 years: the impact of climate and land use change. *Earth Surface Processes and Landforms*, **27**, 403–423.

Owens, P.N., Walling, D.E. and Leeks, G.J.L. (1999) Deposition and storage of fine-grained sediment within the main channel system of the River Tweed, Scotland. *Earth Surface Processes and Landforms*, **24**, 1061–1076.

Owens, P.N., Batalla, R.J., Collins, A.J. *et al.* (2005) Fine-grained sediment in river systems: environmental significance and management issues. *River Research and Applications*, **21**, 693–717.

Panin, A. (2004) Land-ocean sediment transfer in palaeotimes, and implications for present-day natural fluvial fluxes, in *Sediment transfer through the fluvial system*, IAHS Publication 288 (eds V. Golosov, V. Belyaev and D.E. Walling), IAHS Press, Wallingford, UK, 115–124.

Papanicolaou, A.N.T., Elhakeem, M., Krallis, G. *et al.* (2008) Sediment transport modelling review – current and future developments. *Journal of Hydraulic Engineering ASCE*, **134**, 1–14.

Parsons, A.J., Wainwright, J., Brazier, R.E. and Powell, D.M. (2006) Is sediment delivery a fallacy? *Earth Surface Processes and Landforms*, **31**, 1325–1328.

Pedersen, M.L., Friberg, N. and Larsen, S.E. (2004) Physical habitat structure in Danish lowland streams. *River Research and Applications*, **20**, 653–669.

Petticrew, E.L., Krein, A. and Walling, D.E. (2007) Evaluating fine sediment mobilization and storage in a gravel-bed river using controlled reservoir releases. *Hydrological Processes*, **21**, 198–210.

Petts, G.E., Foulger, T.R., Gilvear, D.J. *et al.* (1985) Wave-movement and water quality variations during a controlled release from Kielder Reservoir, North Tyne River, UK. *Journal of Hydrology*, **80**, 371–389.

Petts, G.E., Thoms, M.C., Brittan, K. and Atkin, B. (1989) A freeze-coring technique applied to pollution by fine sediments in gravel-bed rivers. *Science of the Total Environment*, **84**, 259–272.

Phillips, J.M. and Walling, D.E. (1999) The particle size characteristics of fine-grained channel deposits in the River Exe Basin, Devon, UK. *Hydrological Processes*, **13**, 1–19.

Prosser, I.P., Rutherfurd, I.D., Olley, J.M. *et al.* (2001) Large-scale patterns of erosion and sediment transport in river networks, with examples from Australia. *Marine and Freshwater Research*, **52**, 81–99.

Quinn, J.M., Davies-Colley, R.J., Hickey, C.W. *et al.* (1992) Effects of clay discharge on streams: 2. benthic invertebrates. *Hydrobiologia*, **248**, 235–247.

Rabeni, C.F., Doisy, K.E. and Zweig, L.D. (2005) Stream invertebrate community functional responses to deposited sediment. *Aquatic Sciences*, **67**, 395–402.

Rees, J.G., Ridgeway, J., Knox, R.W.O.B. *et al.* (1999) Sediment-borne contaminants in rivers discharging into the Humber estuary, UK. *Marine Pollution Bulletin*, **37**, 316–329.

Richardson, J.F. and Zaki, W.N. (1958) Sedimentation and fluidisation. *Transactions of Institute of Chemical Engineers*, **32**, 35–53.

Roehl, J.E. (1962) Sediment source areas, delivery ratios and influencing morphological factors. *International Association of Hydrological Sciences Publication*, **59**, 202–213.

Rouse, H. (1937) Modern conceptions of the mechanics of turbulence. *Transactions of American Society of Civil Engineers*, **102**, 463–543.

Rowan, J.S., Goodwill, P. and Franks, S.W. (2000) Uncertainty estimation in fingerprinting suspended sediment sources, in *Tracers in Geomorphology* (ed. I.D.L. Foster), John Wiley and Sons, Chichester, UK, pp. 279–290.

Rubey, W. (1933) Settling velocities of gravel, sand and silt particles. *American Journal of Science*, **225**, 325–338.

Rutherfurd, I.D. (2000) Some human impacts on Australian stream channel morphology, in *River Management: the Australasian Experience* (eds S. Brizga and B. Finlayson), John Wiley and Sons, Chichester, UK, pp. 11–50.

Sarma, J.N. (1986) Sediment transport in the Burhi Dihing River, India, in *Drainage Basin Sediment Delivery*, IAHS Publication 159 (ed. R.F. Hadley), International Association of Hydrological Sciences, Wallingford, pp. 199–215.

Schälchli, U. (1995) Basic equations for siltation of riverbeds. *Journal of Hydraulic Engineering*, **121**, 274–287.

Sear, D.A. (1993) Fine sediment infiltration into gravel spawning beds within a regulated river experiencing floods – ecological implications for salmonids. *Regulated Rivers – Research and Management*, **8**, 373–390.

Sharpe, R.G. and James, C.S. (2006) Deposition of sediment from suspension in emergent vegetation. *Water SA*, **32**, 211–218.

Singh, M., Ansari, A.A., Muller, G. and Singh, I.B. (1997) Heavy metals in freshly deposited sediments of the Gomati River (a tributary of the Ganges River) – effects of human activities. *Environmental Geology*, **29**, 246–252.

Sivapalan, M., Jothityangkoon, C. and Menabde, M. (2002) Linearity and non-linearity of basin response as a function of scale: discussion of alternative definitions. *Water Resources Research*, **38**, 1012, doi: 10.1029/2001 WR000482

Smith, B.P.G., Naden, P.S., Leeks, G.J.L. and Wass, P.D. (2003a) The influence of storm events on fine sediment transport, erosion and deposition within a reach of the River Swale, Yorkshire, UK. *Science of the Total Environment*, **314–316**, 451–474.

Smith, B.P.G., Naden, P.S., Leeks, G.J.L. and Wass, P.D. (2003b) Characterising the fine sediment budget of a reach of the River Swale, Yorkshire, UK during the 1994 to 1995 winter season. *Hydrobiologia*, **494**, 135–143.

Soulsby, C., Youngson, A.F., Moir, H.J. and Malcolm, I.A. (2001) Fine sediment influence on salmonid spawning habitat in a lowland agricultural stream: a preliminary assessment. *Science of the Total Environment*, **265**, 295–307.

Syvitski, J.P.M., Peckham, S.D., Hilberman, R. and Mulder, T. (2003) Predicting the terrestrial flux of sediment to the global ocean: a planetary perspective. *Sedimentary Geology*, **162**, 5–24.

Syvitski, J.P.M., Vörösmarty, C.J., Kettner, A.J. and Green, P. (2005) Impact of humans on the flux of terrestrial sediment to the global coastal ocean. *Science*, **308**, 376–380.

Tipping, E. (1988) Colloids in the aquatic environment. *Chemistry and Industry*, **15**, 485–490.

Tipping, E., Woof, C. and Clarke, K. (1993) Deposition and resuspension of fine particles in a riverine 'dead zone'. *Hydrological Processes*, **7**, 263–277.

Trimble, S.W. (1975) Denudation studies: can we assume stream steady state? *Science*, **188**, 1207–1208.

Trimble, S.W. (1983) A sediment budget for Coon Creek basin in the Driftless Area, Wisonsin, 1853–1977. *American Journal of Science*, **283**, 454–474.

Van Rijn, L.C. (1984) Sediment transport, Part II: Suspended load transport. *Journal of Hydraulic Engineering*, **110**, 1613–1641.

Vörösmarty, C.J., Meybeck, M., Fekete, B. *et al.* (2003) Anthropogenic sediment retention: major global impact from registered river impoundments. *Global and Planetary Change*, **39**, 169–190.

Walling, D.E. (2005) Tracing suspended sediment sources in catchments and river systems. *Science of the Total Environment*, **344**, 159–184.

Walling, D.E. (2006) Human impact on land-ocean sediment transfer by the world's rivers. *Geomorphology*, **79**, 192–216.

Walling, D.E. and Amos, C.M. (1999) Source, storage and remobilisation of fine sediment in a chalk stream. *Hydrological Processes*, **13**, 323–340.

Walling, D.E. and Webb, B.W. (1982) Sediment availability and the prediction of storm-period sediment yields, in *Recent developments in the explanation and prediction Of erosion and sediment yield*, IAHS Publication 137 (ed D.E. Walling), IAHS Press, Wallingford, UK, pp. 327–337.

Walling, D.E., Collins, A.L. and McMellin, G. (2003) A reconnaissance survey of the source of interstitial fine sediment recovered from salmonid spawning gravels in England and Wales. *Hydrobiologia*, **497**, 91–108.

Walling, D.E., Owens, P.N. and Leeks, G.J.L. (1998) The role of channel and floodplain storage in the suspended sediment budget of the River Ouse, Yorkshire, UK. *Geomorphology*, **22**, 225–242.

Walling, D.E., Owens, P.N. and Leeks, G.J.L. (1999) Rates of contemporary overbank sedimentation and sediment storage on the floodplains of the main channel systems of the Yorkshire Ouse. *Hydrological Processes*, **13**, 993–1009.

Walling, D.E., Woodward, J.C. and Nicholas, A.P. (1993) A multi-parameter approach to fingerprinting suspended sediment sources, in *Tracers in Hydrology*, IAHS Publication 215 (eds N.E. Peters, E. Hoehn, Ch. Leibundgut, N. Tase and D.E. Walling), IAHS Press, Wallingford, UK, pp. 329–337.

Walling, D.E., Russell, M.A., Hodgkinson, R.A. and Zhang, Y. (2002) Establishing sediment budgets for two small lowland agricultural catchments in the UK. *Catena*, **47**, 323–353.

Walling, D.E., Owens, P.N., Foster, I.D.L. and Lees, J.A. (2003a) Changes in the sediment dynamics of the Ouse and Tweed basins in the UK, over the last 100–150 years. *Hydrological Processes*, **17**, 3245–3269.

Walling, D.E., Owens, P.N., Carter, J. *et al.* (2003b) Storage of sediment-associated nutrients and contaminants in river channel and floodplain systems. *Applied Geochemistry*, **18**, 195–220.

Walling, D.E., Collins, A.L., Jones, P.A. *et al.* (2006) Establishing fine-grained sediment budgets for the Pang and Lambourn LOCAR study catchments. *Journal of Hydrology*, **330**, 126–141.

Wass, P.D. and Leeks, G.J.L. (1999) Suspended sediment fluxes in the Humber catchment, UK. *Hydrological Processes*, **13**, 935–953.

Wasson, R.J., Olive, L.J. and Rosewell, C.J. (1996) Rates of erosion and sediment transport in Australia, in *Variability in Stream Erosion and Sediment Transport*, IAHS Publication 224 (eds L.J. Olive, R.J. Loughran and J.A. Kesby), IAHS Press, Wallingford, UK, pp. 269–279.

Wasson, R.J., Mazari, R.K., Starr, B. and Clifton, G. (1998) The recent history of erosion and sedimentation on the Southern Tablelands of southeastern Australia: sediment flux dominated by channel incision. *Geomorphology*, **24**, 291–308.

Wharton, G., Cotton, J.A., Wotton, R.S. *et al.* (2006) Macrophytes and suspension-feeding invertebrates modify flows and fine sediments in the Frome and Piddle catchments, Dorset (UK). *Journal of Hydrology*, **330**, 171–184.

Whiting, P.J., Matisoff, G. and Fornes, W. (2005) Suspended sediment sources and transport distances in the Yellowstone River basin. *Geological Society of America Bulletin*, **117**, 515–529.

Williams, N.D., Walling, D.E. and Leeks, G.J.L. (2008) An analysis of the factors contributing to the settling potential of fine fluvial sediment. *Hydrological Processes*, **22**, 4153–4162.

Wohl, E.E. and Cenderelli, D.A. (2000) Sediment deposition and transport patterns following a reservoir sediment release. *Water Resources Research*, **36**, 319–333.

Wood, P.A. (1977) Controls of variation in suspended sediment concentration in Rover Rother, West Sussex, England. *Sedimentology*, **24**, 437–445.

Wood, P.J. and Armitage, P.D. (1997) Biological effects of fine sediment in the lotic environment. *Environmental Management*, **21**, 203–217.

Yamada, H. and Nakamura, F. (2002) Effect of fine sediment deposition and channel works on periphyton biomass in the Makomanai River, northern Japan. *River Research and Applications*, **18**, 481–493.

Xu, J. (2002) Implication of relationships among suspended sediment size, water discharge and suspended sediment concentration: the Yellow River basin, China. *Catena*, **49**, 289–307.

11

Streams, Valleys and Floodplains in the Sediment Cascade

Stanley W. Trimble
Department of Geography, 1255 Bunche Hall, UCLA, Los Angeles CA, 90095-1524, USA

11.1 Introduction

In the cascade of sediment from hillsides to the sea, streams, valleys and floodplains are the 'zone of transfer' (Schumm, 1977). Although there is still much to learn, geomorphological research over the past century and especially over the past few decades has begun to demonstrate how complex these transfers are. This chapter considers morphology, processes and the intricacies of sediment budgets at various scales of space and time.

11.2 Classification of Streams, Valleys and Floodplains

Many excellent studies examine the range of morphologies and inherent processes in streams, valleys and floodplains (Wolman and Leopold, 1957; Nanson, 1986; Nanson and Croke, 1992; Bridge, 2003). To introduce all these types and their permutations would require a treatise several times longer than this book. Discussing sediment transfer for all of them would be impossible, first because of the scope, and secondly because the available literature is inadequate for many cases. Thus the first task of this chapter is to reduce the conceptual picture of streams, valleys and floodplains to a simpler model – one in which process and morphology of streams, valleys and floodplains are seen as a general continuum (Figure 11.1).

At one end of the continuum are high-energy, low-resistance conditions with constant change, or inherently disequilibrium streams (Figure 11.1). Sediment flux and storage in such streams tend to be extremely episodic with the stream and valley materials being highly mobile and the morphology a relict of the last big event. Often, the channel may be poorly formed or even amorphous with little or no organic or 'genetic' (Graf, 1988) relation to the floodplain, which itself may be poorly formed and amorphous. Net storage gain can come

Sediment Cascades: An Integrated Approach Edited by Timothy Burt and Robert Allison

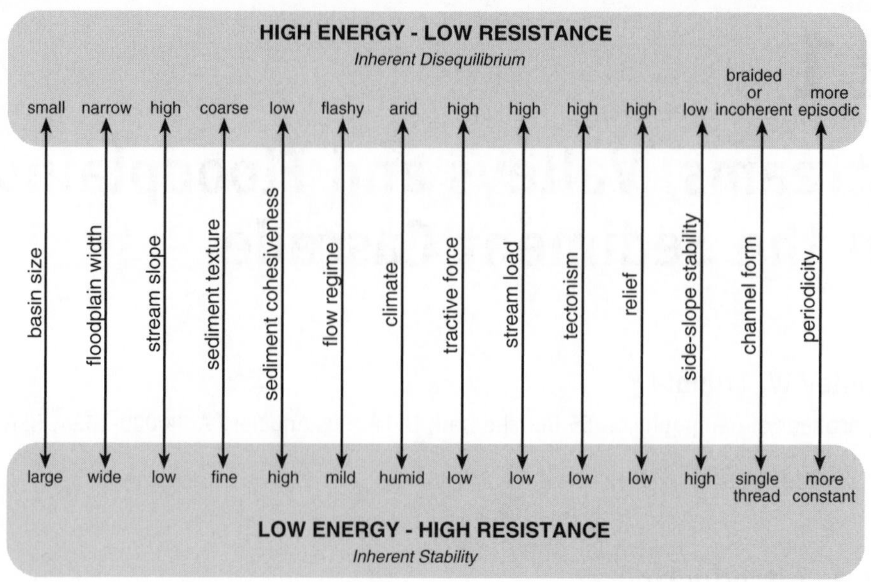

Figure 11.1 The continuum of stream types and some of the controlling variables. This is a schematic figure: variables are generalized and do not fit in every case (based in part on Ferguson, 1981, 1987; Richards, 1982; Graf, 1984; Nanson, 1986; Nanson and Croke, 1992). Most discussion in this chapter is related to the low-energy, high-resistance end of the continuum. The figure is not deterministic. For most streams, one or more, but probably not all, of the indicated variables would control stream behaviour

from vertical accretion or, in mountainous areas, from fans and mass movements from sometimes steep, unstable valley sides (e.g. Dietrich and Dunne, 1978; Kelsey, 1980; Lehre, 1981; Dietrich *et al.*, 1982; Pearce and Watson, 1986; Smith and Swanson, 1987; Benda, 1990; Slaymaker, 1993). While net accretion may come with smaller events, erosion of channels and floodplains often accompanies very large events which 'flush' sediment downstream (Schumm, 1977; Nanson, 1986; Nanson and Croke, 1992; Trimble, 1995; Moody and Meade, 2008). It appears that many arid streams also share these characteristics (Schick, 1977; Graf, 1983a; Boisson and Patton, 1985; Patton and Boison, 1986; Schick and Lekach, 1993). Streams at this end of the continuum will be given little in the way of further discussion in this chapter, in part because many of the processes that characterize the high-energy, low-resistance conditions are well covered elsewhere in this volume.

At the other end of the continuum (Figure 11.1) are low-energy, high-resistance conditions which normally see less change, and can be considered to be inherently in a condition of steady state. It is at or relatively near this end of the continuum that most studies of sediment transfer are found, so the focus here is on such streams. By far the best studied genre is the single-thread stream that transports mostly sandy or finer material, and often flows through a relatively wide valley with well-defined channel and banks with a vegetated, relatively fine-textured (fine sand or finer) floodplain which reflects the hydrological regime and sediment load of the stream. Steady-state streams, valleys and floodplains may experience both accretion and erosion, often simultaneously, and both approximately equal in terms of mass, in their constant transfer of sediment.

11.3 Role of Water and Sediment Flow Regime in Streams and Floodplains

11.3.1 Some Basic Processes

Expressed in simplest form, streams at a given range of discharges (flow regime) have the ability to transport a certain amount of sediment (capacity). If the catchment supplies more than the capacity, then the stream will deposit and sediment will accrete. But if the supply of sediment from the catchment is decreased, the stream will have excess capacity (sometimes termed 'hungry water') and will attempt to mobilize and transport sediment from its storage in bed and banks. If the material is not erodible, then the stream's efforts to mobilize the sediment are dissipated as heat and no change or work is done. Similarly, a decrease of peak flows with no reduction of sediment load, means that the stream will be beyond capacity and must deposit sediment. Conversely, an increase in flow without a commensurate increase in sediment load will result in downstream channel erosion, always assuming the downstream reach is erodible. These principles are shown schematically in the classic Lane scales (Figure 11.2). The following text explains the processes of sediment accretion and erosion.

Accretion. A long-held tenet (but now largely outdated) was that floodplains were mostly built by *lateral accretion* (Figure 11.3). Leopold, Wolman and Miller (1964) suggested that lateral accretion accounted for 0–80% of deposition. In this scenario, streams migrated laterally, eroding away a cut bank but at the same time depositing an approximately equal volume of sediment on the opposite bank (the point bar or slip-off slope). Thus, as Fenneman (1906) expressed it, floodplains could be 'built without floods'. Such a condition,

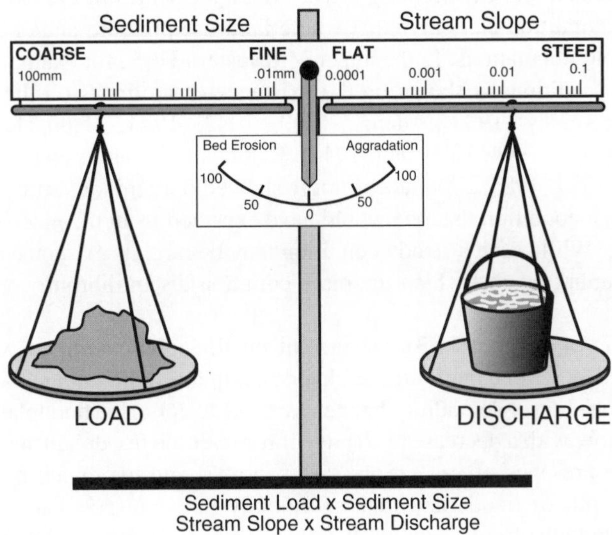

Figure 11.2 A schematic representation of stream behaviour as a function of discharge and slope on the one side and sediment load and size on the other (redrawn from Lane, 1955).

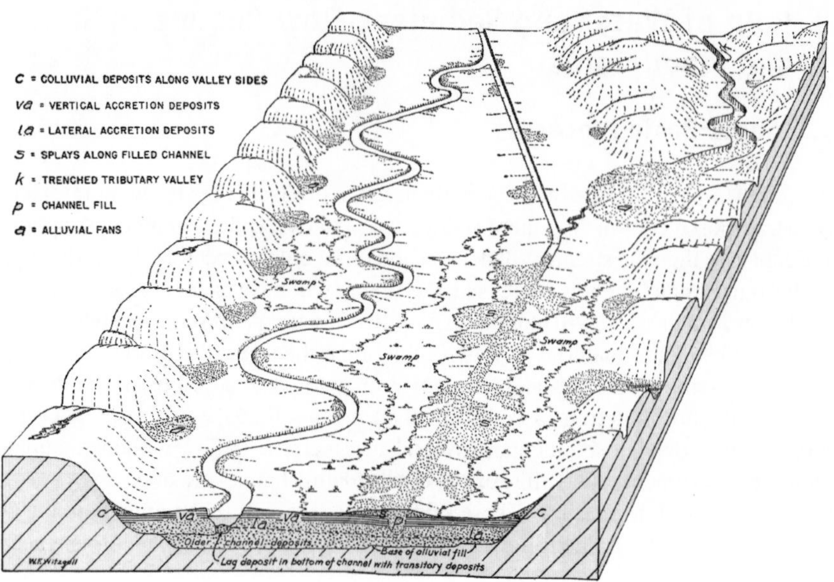

Figure 11.3 Schematic showing typical relations of the various types of stream, valley and floodplain deposits under conditions of accelerated sedimentation (from Happ, Rittenhouse and Dobson, 1940). View upstream.

where over a short period the deposition just matches the erosion, can be termed 'graded' (Mackin, 1948) but is more commonly now termed as 'steady state' or 'equilibrium'. Note that while any reach might experience gains or losses, the net result over short periods is no change. Such a condition can only exist where land-use pressure and/or climate are not subject to extreme fluctuations. In the upper Midwestern USA, for example, the evidence for such stable conditions is shown by the existence of mollisols on the pre-settlement floodplains (McKelvey, 1939; Adams, 1940, 1942, 1944; Happ, Rittenhouse and Dobson, 1940; Happ, 1944; Kunsman, 1944). Radiocarbon dates determined much later by Knox (1972, 1977, 1987, 2006) accordingly showed only minor vertical accretion on a millennial scale, a condition that one would have expected from the much earlier work of Happ and others. While such a steady condition may be rare, its existence and description provide a benchmark against which the more common disequilibrium conditions can be compared.

Vertical accretion (Figure 11.3) can prevail on floodplains with or without channel aggradation. And as will be demonstrated later, floodplain sediment may accrete simultaneously with an eroding or degrading channel. Vertical accretion on floodplains is facilitated by dispersal of flow with a decrease of tractive force over the floodplain, a situation greatly increased by the presence of vegetation (Piegay, 1997) and by a wide floodplain as this decreases the depth of floods and thus of velocity and transport capacity. While fine sediment (clay and silt) from overbank flow may be spread more or less evenly across a floodplain to create a fairly level surface, sand is more likely to be deposited near the stream bank as the hydraulic friction from the bank and bank vegetation slows the water, a process which often creates natural levees (Figure 11.3). While channels may aggrade as floodplains

Figure 11.4 Sand splays on the flood plain of Coon Creek WS after an extreme flood, August 2007. The splays are highly discontinuous and are from a few mm to over 30 cm in thickness (photograph by Edyta Zygmunt).

are accreting, that condition is not necessary for floodplain aggradation. During severe floods and/or where a natural levee breaks and permits deep flow to move across the floodplain, vertical accretion may be in the form of discontinuous and scattered splays of highly variable thickness, especially when sand and coarser material are available (Figures 11.3 and 11.4). Examples of this process during the flood of 1993 on the Mississippi and Missouri Rivers are described by Jacobson and Oberg (1997), Schalk and Jacobson (1997), and Magilligan *et al.* (1998). Extremely large events have been known to leave vertical accretion even on terrace surfaces (Moody and Meade, 2008).

Yet another form of vertical accretion on floodplains can be alluvial fans from tributaries (Figure 11.3). The fans can be quite thick, especially when composed of sand or coarser material. Fans from small ephemeral tributaries may cover only small areas but fans of large tributaries may extend across one side of the floodplain to the stream and sometimes can actually displace the stream laterally for considerable distances (Happ, Rittenhouse and Dobson, 1940; De Moor and Verstraeten, 2008; see also Latocha, 2009; Panin, Fuzeina and Belyaev, 2009; Schulte *et al.*, 2009; Zygmunt, 2009). Inasmuch as the entrance valley of a larger tributary creates a wide place in the floodplain and because the fan from the sediment load of the tributary is deposited there where the gradient decreases and the floodplain widens, the average thickness of sediment is greater. Not recognizing the often subtle fans, some investigators questionably ascribed the greater average floodplain thickness to the hydraulic effect of the wideness of the floodplain rather than to the fan itself (Magilligan, 1985).

The variability of depth and spatial discontinuity of vertical accretion due to sediment splays and fans, especially from large floods as described above, makes difficult and tenuous the attribution of sediment volumes to a particular flood. Individual point measurements based on cores demonstrate high variance and should be used to represent floodplain accretion only with a large sample and error analysis (Rommens *et al.*, 2006; Trimble, 2009). Long trenches (e.g. Brackenridge, 1984; Verstraeten *et al.*, 2009) and precise ground surveys should give better representation.

Erosion. At the more stable end of the stream–floodplain continuum (Figure 11.1), the remobilization of most sediment is by vertical erosion of stream channels (degradation) or by lateral erosion of stream banks. As discussed earlier with reference to the steady-state model, the lateral erosion of a cut bank tends to be offset fully by lateral accretion on the opposite side of the stream and the mean bottom elevation does not change. Thus, the area of the channel cross-section does not change. However, the channel may be displaced laterally. This steady state can be upset by two conditions, (i) a decrease in sediment load or sediment size, (ii) a decrease in storm flow, or both.

A decrease in sediment supply to a stream gives it additional energy to erode its bed and banks. If the channel bottom is erodible (mobile), the initial process is usually degradation or vertical cutting (Figure 11.5, '1969'). In some cases such as the Driftless Area of the north-central USA, channels in the upper parts of catchments are floored with cobbles and erode very little vertically (Knox, 1972, 1977). If the stream is flowing in mobile material, it may (for example) create a longitudinal profile that is adjusted to the decreased sediment supply and also engage in other adjustments such as net lateral erosion of the banks. Since the channel is enlarging, the channel- forming flows will be at a lower elevation, and lower floodplains will form as benches inset within the existing channel.

These sediment and flow conditions can operate independently but are often linked, perhaps following improvement in land-use practises or the construction of reservoirs upstream, both of which might reduce the sediment supply and reduce flood-flow peaks. Again, the result would be the lateral erosion of a high bank on one side of the stream and the

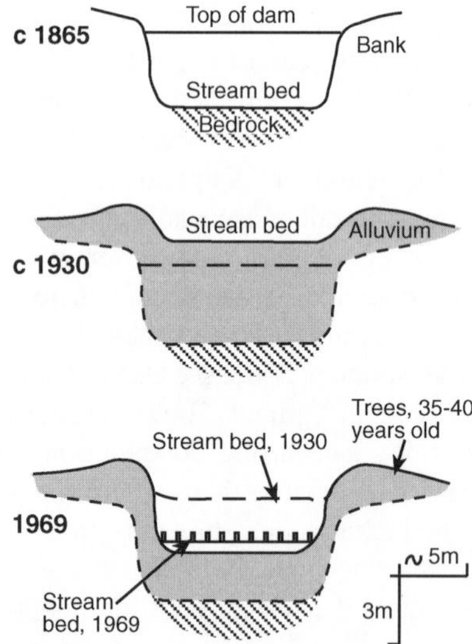

Figure 11.5 Aggradation and degradation as a wave of sediment passes through the Mulberry River at Mauldin Mill, Hall County, Georgia, USA, 1865–1969. Lateral erosion had not begun at that time. Refer to Figure 11.13
(from Trimble, 1969).

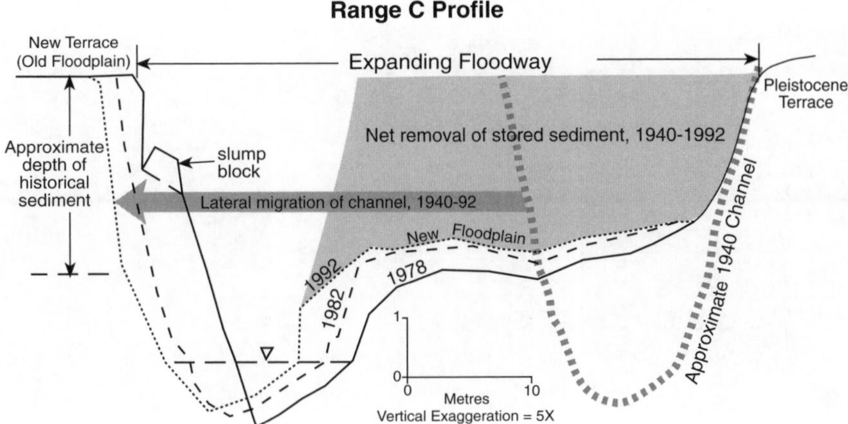

Figure 11.6 Sediment removal and channel transformation in upper main valley, Coon Creek, Wisconsin. The channel was 'flume-like' in about 1940 and flowed by the Pleistocene terrace at that time (approximate 1940 outline shown by ghost outline). Note that the historical floodplain had adjusted to the effective discharge of that time, had aggraded almost to the level of the Pleistocene terrace and many such terraces were buried. Since 1940, the channel has migrated (leftward) and as it continues to migrate, the lower new floodplain is built up by vertical accretion to the height of present-day effective discharge. Note that this 'meander belt' expansion has left the old floodplain as a relict floodplain or terrace. Net removal of material in this cross-section is about 100 m^2 and is shown in shadow. This storage loss since 1940 would be about 100 000 $m^3 km^{-1}$ or about 130 000 mg km^{-1} at a bulk density of 1.3. In this reach over recent years, the gain by vertical accretion on the new floodplains is almost as much as that lost from the high cut banks
(from Ward and Trimble, 2004).

construction of point bars on the other side and eventually, a lower floodplain adjusted to the lower channel-forming discharges of the milder hydrological regime (Figure 11.6). Like the formation of most floodplains, the lower floodplain would be initially formed by lateral point-bar-style accretion but it would eventually be capped by vertical overbank-style accretion. Severe channel and valley erosion can also be caused by increases in flood flows without commensurate increases of sediment supply and such channel erosion may be vertical or horizontal or both. With the work of Bryan (1925) in mind, Happ, Rittenhouse and Dobson (1940) studied this process in several regions in the USA and called it 'valley trenching'. In the early twentieth century, strong increases of runoff in the Driftless Area created gullies on hillsides and trenched stream valleys. Where a supply of coarse sediment was available, the combination caused single-thread streams to erode, widen, and sometimes even to braid (Figure 11.7A and B). But when floods were mitigated and the source of coarse sediment stabilized, the streams returned much to their original configuration (Figure 11.7C).

Gullying and valley trenching can play three especially significant roles. First, they effectively deliver the sediment from upland sources to downstream channel and floodplain reaches. Now referred to as 'coupling' (Caine and Swanson, 1989) or 'connectivity' (Burt and Pinay, 2005; Brierley, Fryirs and Vikrant, 2006; Fryirs *et al.*, 2007), this is really the old sediment-delivery problem (Walling, 1983, 1990; De Vente *et al.*, (2007) first explored systematically by Roehl (1962). Second, the gullies and trenched valleys themselves are

Figure 11.7 Evolution of tributary stream channel from changing flood regime and sediment supply, North Branch, Whitewater River, Minnesota, about 2 km north of Elba, Minnesota. For reference, note barn, bridge, and road intersection in all three frames. 1905: note narrow, apparently deep stream lines with trees and dense riparian vegetation. 1940: By lateral erosion, stream had widened two to five times in response to increased storm flow and the abundant supply of bedload from hillside gullies and the eroded stream itself. Sinuosity has increased greatly. 1975. Stream has again narrowed to similar dimensions as seen in 1905 with dense riparian vegetation (© Elsevier)

important sediment sources directly linked to the streams. Thirdly, the increased drainage density increases peak flow which can erode downstream reaches. The combined result is greatly enhanced sediment delivery and yield to streams. Conversely, a decrease in drainage density caused by the disuse of rills and gullies can radically decrease sediment delivery. For example, the disappearance of gullies in the Driftless Area resulting from agricultural improvements helped to reduce the sediment delivery ratio from 100% in 1934 to 8% in 1975 as more eroded material was deposited as colluvium (Figure 11.8). See also (Fraczek, 1987). An outstanding review of gullies and their significance is given by Poesen *et al.*, 2003; see also Dotterweich, 2005; Panin, Fuzeina and Belyaev, 2009). A particularly insightful paper on the effect of agricultural field boundaries on sediment delivery during 7 millennia in Germany is given by Houben (2008).

Perhaps the best known and most-studied examples of valley trenching are of arroyos or large gullies in the southwestern USA (Bryan, 1925; Happ, Rittenhouse and Dobson, 1940; Cooke and Reeves, 1977; Schumm, 1977). Whether caused by overgrazing or climate change, there was often an initial period of sediment accumulation on valley floors, many of which were previously unchannelled, followed by a period of severe cutting. In recent decades, these arroyos have been stabilized and even filled (Emmett, 1974; Leopold, 1976; Graf, 1983b, Schumm and Rea, 1994). Still uncertain is whether this stabilization is due to amelioration of the original problem (i.e. overgrazing? erosive rainfall?) or due to feedback effect from channel forms or vegetation (Schumm, 1977), all of which may vary by time and location. Incised channels and their responses to variable controls have received increasing attention in recent years (Simon, 1989; Prosser *et al.*, 1994; Prosser, 1996; Darby and Simon, 1999).

Urban streams often act in a similar manner but with an important difference. The initial urban construction may give a burst of sediment and downstream accretion (Wolman, 1967),

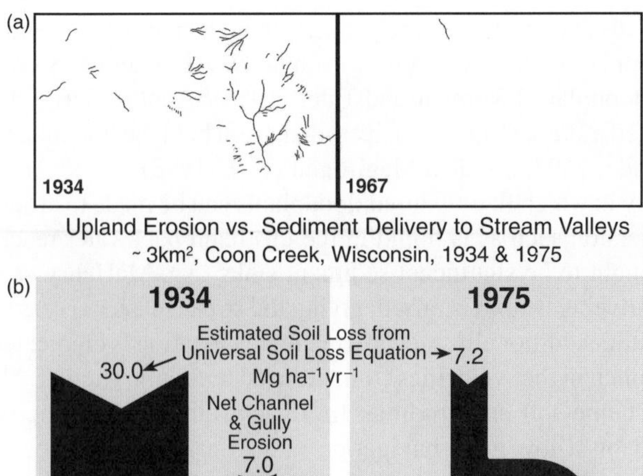

Figure 11.8 'Connectivity' of uplands and valleys with and without gullies, and the effects on sediment delivery in Coon Creek, Wisconsin. (a) Decrease of gullies in a 180 ha area from 1934 to 1967 as measured from aerial photographs. (b) Decrease of sediment delivery 1934 to 1975. In the 1930s, not only was most eroded soil from upland erosion being delivered to valleys, but it was also being augmented by erosion of the gullies (data from Trimble and Lund, 1982). By the 1970s, however, 'connectedness' was greatly reduced and most eroded material was consequently deposited as colluvium

but later waterproofing of the basin surfaces can cause severe channel erosion. Unlike the arroyos, urban channels continue to erode with the effect moving downstream because most urban areas become increasingly waterproof. Thus, urban channel erosion usually must be controlled by structures or channel hardening.

While 'vertical stripping' has been reported on steep, coarse-grained floodplains (Nanson, 1986), vertical erosion directly on the surface of fine-textured floodplains is rare for several reasons. First, the fine material is more resistant to erosion. Second, fine-textured floodplains, at least in humid regions, are usually vegetated, further increasing their resistance to erosion and decreasing velocity. Moreover, flood flows over floodplains are usually shallow and of low velocity, limiting the erosive force that can be can be exerted, even without vegetation. This is especially true of wide floodplains which spread the flow more effectively. The usual exception to these generalities is found in 'chutes' which are incipient channels usually running parallel to streams, or across meander necks (e.g. Gay *et al.*, 1998). When under cultivation and thus devoid of vegetation and especially after ploughing, they are especially susceptible to erosion. And as they erode and deepen, the erosive force becomes greater—an example of positive feedback.

Even extreme discharge events rarely exert enough tractive force to erode vegetated fine-textured floodplains. One catastrophic event in New England only eroded as much as it deposited on a floodplain (Wolman and Eiler, 1958). Another large flood, this time in Montana, deposited more sediment than it eroded, from both the floodplain and a low terrace (Moody and Meade, 2008; see also Magilligan *et al.*, 1998).

Yet another way in which fine-textured floodplains can be made to erode is by permanent but shallow immersion such as is found in the upstream backwater reaches of reservoirs, areas usually thought to be significant sediment sinks (e.g. McHenry *et al.*, 1984). Under these conditions, river elevation is raised, giving the stream easier access to flow across and down the floodplain. Additionally, most permanent vegetation is eradicated by the standing water, allowing much higher velocities from reduced hydraulic friction. Moreover, the bare sediment may become soft and erodible. In a backwater reach of the upper Mississippi River submerged by a low-head navigation dam, Carson (2003) found that there was much sediment flux with new channels being formed on the old floodplain. Indeed, newly eroded channels, or enhanced chutes, in the floodplain became so effective in conveying water, that stage was lowered for higher discharges. In this floodplain reach however, accretion was still occurring from all sources including lateral tributaries, roughly equalling erosion.

The foregoing discussion of sediment storage gain and loss underlines what seems to a basic principle in the sediment cascade for the type of streams considered here: vertical accretion on floodplains allows accumulation rates limited only by supply so that massive amounts of sediment can accumulate over a short period. But because sediment removal is mostly limited to channel erosion, it can take much longer, perhaps by orders of magnitude with larger basins where the floodplain is wide. Thus, a burst of sediment might fill a valley in decades or centuries but the removal of that sediment might require millennia. The general process might be expressed as 'fast in, slow out'. This principle will be more thoroughly demonstrated in the section on sediment budgets.

11.3.2 Role of Soil-Water Budgets in Streambank Stability

While the inherent strength of stream banks is intrinsically a function of sediment texture, the presence of pore water also plays an important role because it (i) increases pore pressure thus weakening the soil and (ii) greatly increases the bulk density thus increasing the force of gravity on the bank. Thus a very wet vertical cut bank can be very unstable, sometimes collapsing from its own weight (Hooke, 1980; Lawler, 1992). This significant decrease of resistance can be event-related but it also has a seasonal component related to the soil-water budget in any climate. Since streamflow also displays seasonal behaviour, the timing of bank erosion is partially a function of climate when weak banks (low resistance) and high flows (high force) coincide. In Mediterranean regions (Koeppen: Csa), for example, weak banks and high flows coincide in winter during the rainy season (Figure 11.9) so that most banks erode during a relatively short period of the year. In humid continental climates (Koeppen; Daf, Dbf), most rivers flood in summer when banks are liable to be dry and initially resistant to erosion. However, bank moisture and pore pressure are usually at a maximum during the spring thaw when high flows are not likely. Ironically, dry stream banks have been observed to resist high flows during the summer and then collapse of their own weight during extreme wetness during the next spring thaw. When a flood is prolonged, even in summer, and water

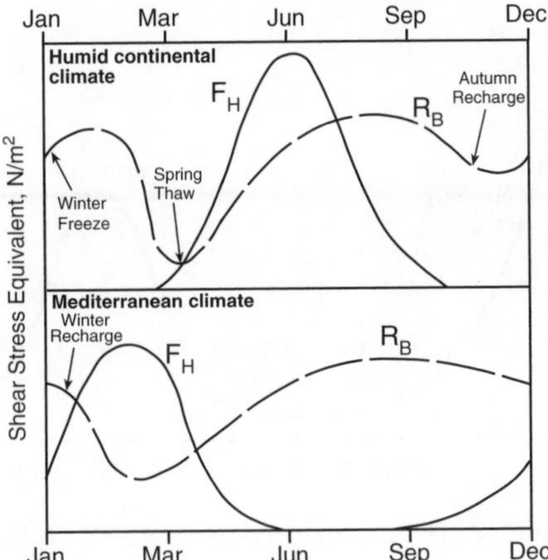

Figure 11.9 Comparison and contrast of hydraulic force and resistance on stream banks for two climate zones. Curve F_H is hydraulic force from streamflow and R_B is resistance to bank erosion. In the upper Midwest, higher force and lower resistance tend to occur at about the same time of the year. But in southern California, higher force and lower resistance are out of phase. See text for fuller discussion

has time to infiltrate deeply into the banks, the banks may collapse as the stream level falls after the flood. Especially graphic examples of this last phenomenon are the 'terras caidas' that collapse during the falling stages of the annual floods on the Amazon River (Sternberg, 1975) and which are washed downstream on the rising stages of the next year's flood (Meade, 2007).

11.3.3 Role of Vegetation in and Along Streams and on Floodplains

Fluvial geomorphologists agree that vegetation helps to stabilize stream channels and floodplains. Rather, the question is whether the better cover is formed by trees or by grass (Montgomery, 1997; Simon and Collison, 2002). Trees had traditionally been the preferred vegetation but that belief was first questioned in 1967 (Zimmerman, Goodlet and Comer, 1967) and by 1997, strong evidence had accumulated to show that grass often made for more stable channels (Davies-Colley, 1997; Trimble, 1997a). Trees generally have the advantage of strong, deep roots and, during the growing season, the ability to keep banks dryer and thus stronger. But the liability of trees is that they can induce turbulence and bank erosion during high flows, and if they collapse into the stream, turbulence in the cavity left by their root crown can destabilize a reach of bank. Large woody debris (LWD) from fallen trees has often been seen as a stabilizing influence because it can decrease *average* velocity but often forgotten is the fact that flow *accelerates* around the LWD. The stabilizing effect may be true at low to moderate flows and may be why LWD is often considered to be a stream stabilizer in climates with moderate rainfall intensities and flow regimes as found in west coast marine climates (Koeppen: Cbf) in western Europe, or the Northwest USA. However, in climates with extremely intense rain and high flow events, the accelerated flow around

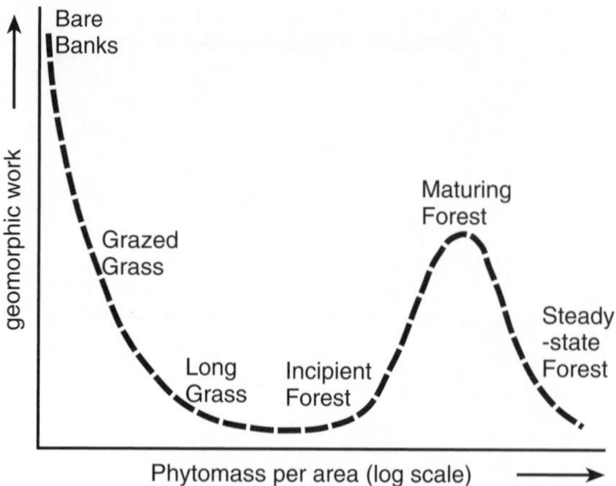

Figure 11.10 Schematic relation of unit riparian phytomass to net geomorphological work (erosion) accomplished along certain stream channels. Long grass and very young woody vegetation give the most stable channel but maturing forests result in the greatest erosion. Steady-state forest results in maintenance of larger channels (Trimble, 2004)

LWD can cause turbulence at higher flows that actually increases channel erosion (Trimble, 2004).

While grass has the disadvantage of being shallow rooted, it tends to lie flat on the surface of banks and floodplains and thus creates little turbulence while it protects the surface from erosion. Indeed, prone grass forms a 'thatch' that tends to trap any sediment and appears to make banks and floodplains accrete. A study by Trimble (1997a) showed that grassy stream reaches in Wisconsin were smaller in cross-section, storing 2100–8800 m^3 more sediment per kilometre than forested channels. Similar results were reported by Davies-Colley (1997) and Lyons, Trimble and Paine (2000). While geomorphologists generally relate stability to phytomass, the relation here to stream-channel stability is not so simple (Figure 11.10). However, it is important to emphasize that other studies have identified streams where trees induced more stability and there is still much to investigate about the many variables controlling these processes (Hey and Thorne, 1986; Stott, 1997; see Bennett and Simon, 2004).

Stabilization by vegetation has often been an important point in stream 'restoration' and some such projects have seen some success although this practise is normally combined with artificial structures. But sometimes there is a tendency to forget that young willows which hold stream banks so well, eventually grow to maturity, die and create LWD.

Although depending on natural vegetation as a stabilizer of stream channels may provide solutions in the short term, it can be problematic in the longer term, especially in meandering rivers. Many riparian trees depend upon a regular shifting of the channels, point bars, and proximal floodplains for their propagation, and for their orderly plant succession from opportunistic species to more mature forest species. The dynamic interaction of channel shifting and tree succession in the upper Amazon basin of Peru has been well described by Salo *et al.* (1986). In a reach of the Missouri River between two large reservoirs, the riparian cottonwood trees (*Populus deltoides*) are in a state of decline because the now heavily regulated river channel is no longer shifting and providing new substrates for regular propagation (Johnson, 1992, 1998).

Streams in arid regions normally have little vegetation. However, invasive exotic phreatophytes such as *Arundo donax* (Coffman, 2008a, 2008b) and *Tamarisk* (Graf, 1978; Dudley, 1998) have been accidentally introduced. These plants now cover long stream reaches in the southwestern USA, reducing streamflow by prolific transpiration from groundwater and their thick growth retards flow and traps sediment (Robinson, 1958, 1965; Graf, 1978). Their removal by herbicides and mechanical means has been very problematic.

11.3.4 Effects of Grazing Animals on Stream Channels and Banks

Grazing and browsing animals spend much time 'lounging' in and by streams and have a direct effect on bank erosion by trampling banks and even breaking them down (Trimble and Mendel, 1995). Another major destabilizing effect is indirect: by removing much of the vegetative cover and trampling the stream bed and banks, they lower the erosional resistance so that the force of a given stream discharge is much more effective. Moreover, when large animals enter and exit the stream, they can break down banks into 'ramps'" that alter the hydraulic characteristics and cause even more erosion (Trimble, 1994).

11.3.5 Effects of Hydraulic Structures in Streams

By far the most significant and widespread hydraulic structures found in streams are dams with their reservoirs. While reservoirs themselves are covered in another chapter of this book (Foster, Chapter 12), this chapter will discuss the effects on sediments in streams, valleys and floodplains, both upstream of the reservoir and downstream of the dam.

As a streamflows into the quiet, impounded water of the reservoir, its velocity is reduced so that some of the sediment it carries is deposited to form a delta. Because the slope of the topset delta is less than that of the stream, the stream is slowed and thus deposits sediment in order to increase the slope at the foreset face of the growing delta (Mackin, 1948). As the delta extends farther into the reservoir, maintaining the same stream slope requires the stream to extend a wedge of aggrading sediment farther upstream. The aggrading reach of stream may also aggrade the floodplain or not depending primarily on the conditions described earlier in the chapter.

Perhaps the most notable example of the aggrading wedge upstream of a reservoir is at the Elephant Butte reservoir, built in 1912 on the Rio Grande about 300 km downstream of Albuquerque, New Mexico (Eakin, 1936). By the 1930s, the Rio Grande had aggraded many kilometers upstream of the head of the lake. Because the sediment load was sand, most of the aggradation was in the channel with the creation of natural levées. Drainage from the distal land on to the floodplain was blocked from entering the stream by the natural levees and thus the floodplain became a wetland and was even ponded in many places. Near the reservoir, the stream level was superposed, or higher than the floodplain. By the 1930s, a local village, San Marcial, began to be flooded and was finally evacuated (Eakin, 1936). It was feared that the wedge of sediment would eventually extend all the way to Albuquerque but that never happened (Mackin, 1948). It would seem intuitive that a stream would assume a longitudinal profile parallel to the old one but that does not appear to happen (Leopold and Bull, 1979). Apparently, the stream makes adjustments to its cross-sectional profile and hydraulic roughness, which allows it to flow at a lower slope which intercepts the old slope some distance upstream.

The formation of such an upstream wedge is not yet completely understood nor can we yet predict how far upstream the wedge will reach. Sediment texture certainly seems to be a factor with coarse material appearing to favour the wedge. Flow regime could also be important and vegetation growth, especially in the delta, appears to promote the aggradation of the delta and thus its extension upstream.

Because the reservoir traps much of the sediment flowing into it and tends to moderate the downstream flow regime, very different conditions obtain downstream of the dam. The stream without this sediment load is energized and seeks to entrain more sediment (Figure 11.2). Thus, mobile sediment downstream of the dam is entrained and transported farther. Phillips (2001) reported that one river in Texas had restored its natural stream load only 16 km downstream of a dam. As much as 8 m of degradation has been recorded and this can leave water intakes and other stream-associated facilities stranded far above water level (Williams and Wolman, 1984). The sediment transported downstream can also have consequences.

The other main effect of reservoirs is to even out flow regimes. While it is difficult to totally control very high flows, most flood peaks are at least moderated and low flow is generally increased. This would decrease sediment transport and thus offset the sequestration effect to some degree. Perhaps the most notable effect of a moderated flood regime has been in Nebraska where the Platte River was changed from a wide braided stream to a relatively narrow single thread stream (Williams, 1978).

11.3.6 Stabilizing Unstable Streams with Structures

A review of the control of unstable streams by hydraulic structures is beyond the scope of this chapter but note that controlling the most severe channel erosion where storm flow has been greatly enhanced (e.g. in urban areas) calls for structures including hardening or paved channels. Indeed it might be argued that the only solution for stability in some urban areas is hard channels.

Most unstable channels that must be hardened (usually with concrete) also undergo an improvement of their hydraulic cross-section, generally enlargement, a reduction of width/depth ratio, and straightening to increase the slope. These changes together with the decreased hydraulic friction of the relatively smooth concrete cause the water in the new channel to move much faster with concomitantly higher stream power. Thus, the hardening becomes all the more necessary to prevent erosion. By design such a channel would overflow rarely, if ever, so that former floodplains are not used or affected. One of the most famous examples of such an urban stream structure is the Los Angeles River. Once a small discontinuous stream, the huge volume of storm water from the extensive impermeable surfaces of the city made the small original channel inadequate so that flooding increased and the channel eroded (Gumprecht, 2001). The end result is a concrete channel almost 80 km long.

Another example of eroding urban channels and the need for structures was found in Orange County, California, just south of Los Angeles. The filling of Newport Bay with sediment from San Diego Creek in the 1970s prompted local action to control upland erosion from poor agricultural practise but it turned out that urbanization of the San Diego

Creek was causing massive channel erosion with excellent connectivity to Newport Bay. Thus, about two-thirds of the sediment leaving the basin was actually coming from channel erosion (Trimble, 1997b). Most of these channels have been treated with hardened channels or small sediment detention reservoirs and the flow of sediment into the bay has apparently been reduced.

How useful are structures designed to promote stability and decrease sediment movement in rural or even wild streams? The general view in the USA is that such structures do not work and are a waste of money (Beschta and Platts, 1986; Reeves *et al.*, 1991). Results have been mixed and need more careful monitoring to determine the utility of these structures in different regions. Fish shelters, often built into cut banks, are one type of structure and have been built along thousands of kilometres of USA streams. With 9–19 years of precise survey data, Trimble (1997c) was able to measure relative stability of these fish shelters using the diagnostics of (i) change of channel size and (ii) lateral shift of the channel. The structures were effective but barely so. Grade-control structures concentrate energy at one place and thus are highly problematic unless well engineered. In a 6-year controlled experiment, Trimble (1994) demonstrated that a well-designed low dam decreased upstream bank erosion, even when the banks were grazed. In the UK, long-term sediment yield from rural streams has been significantly reduced by low dams (Brown, 1987) but, as with the effects of LWD discussed earlier, that reduction might be partially a function of the moderate flood regime.

11.4 Sediment Budgets

Having focused on sediment movement and storage, this chapter now turns to some well-studied examples of cascading sediment in streams, floodplains and valleys. While earlier investigators had spoken of sediment budget conditions (Gilbert, 1917; Happ, Rittenhouse and Dobson, 1940; Mackin, 1948; Happ, 1975), the first diagram showing sediment storage gain or loss over time was by Vita-Finzi (1969). However, sediment budget concepts, especially storage fluxes, became indispensable as investigators began to explore the erosional effects of human land use and other environmental drivers and discovered that sediment yield alone was not diagnostic. As the necessity for sediment budgets was expressed in *Science*:

> 'Studies that attempt to correlate basin erosion factors (such as vegetation, climate, and land use) with only sediment yields may be suspect. Under disturbed conditions (in most present basins), sediment yield may be more a direct function of sediment storage than of present upland erosional phenomena'
>
> (*Trimble, 1976, p. 871; see also Trimble, (1975)).*

Complete accounting of all components of a sediment budget is necessary in order to fully understand the response of hillslope and fluvial systems to catchment disturbances: depending on the (i) intensity, (ii) the spatial extent, (iii) the duration, and (iv) the type of perturbation, the various sinks and sources of a catchment's sediment budget will react differently (Verstraeten, Lang and Houben, 2009).

Sediment budgets have become very important in understanding the sediment cascade because the several components are isolated and analysed. On the basis of conservation of mass for example, Trimble and Crosson (2000) questioned the published high estimates of soil erosion for the USA. Sediment budgets are now seen to be 'an organizing framework' in fluvial geomorphology (Reid and Dunne, 2003; see also Dietrich and Dunne, 1978; Phillips, 1986; Reid and Dunne, 1996; Marston and Person, 2004; Brown *et al.*, 2009). Sediment budgets are also important as a management tool (Trimble, 1993; Phillips, 1986; Slaymaker, 2003; Walling and Collins, 2008; Houben, Wunderlich and Schrott, 2009). The chapter will first consider small to mid-size catchments at various timescales and then will turn to the largest river, the Amazon.

11.4.1 Sediment Budgets in Small and Mid-Sized Catchments

In small to mid-size streams, the seemingly most common form of sediment response happens when a burst or 'wave' of sediment, whether caused by humans or by natural forces, moves down the stream system of a catchment with the excess accumulating in the stream and valleys. Perhaps the most famous case and certainly the first studied, was the impacts of the hydraulic mining of gold in California where high-pressure streams of water were directed at gold-bearing Tertiary terraces in the foothills of the Sierra Nevada. The efflux of sediment, grading up to gravel size, was deliberately sluiced into valleys, filling them to depths as great as 25 m. The mining began *c.* 1850 and was abruptly stopped by court order in 1884 because so much farmland on floodplains downstream had been buried by that time with the sediment also filling the Sacramento River and flowing into San Francisco Bay. The Yuba River, a tributary of the Sacramento River, was intensively studied by Gilbert (1917) and again by Adler (1980). The lower Yuba River upstream of Yuba City was inundated by sediment, the overall rate peaking about 1890 (Figure 11.11).

Hydraulic mining ceased in 1884, and shortly after 1900, the entire basin was losing sediment from storage (Figure 11.11). This is a clear demonstration of the principle described earlier that a stream not loaded to capacity will seek supplementary load, thus degrading the bed and perhaps also eroding its banks. Such downstream migration of stored sediment can continue the 'wave' for long periods. A downstream cross-section of the Yuba River (Figure 11.12) shows that the wave of migrating sediment continued to aggrade the

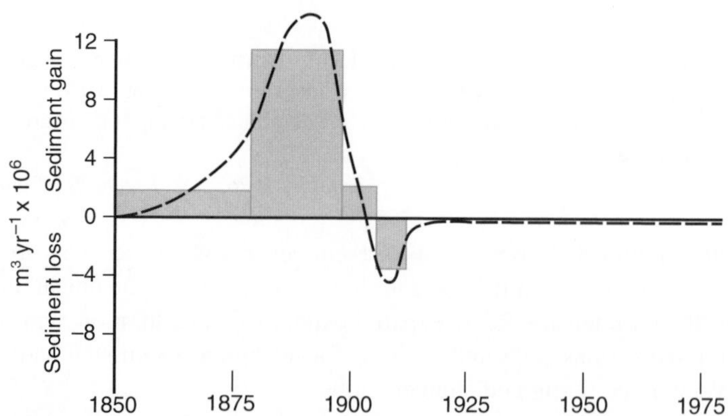

Figure 11.11 Sediment budget, lower Yuba River, California, 1850–1979 (modified from Adler, 1980).

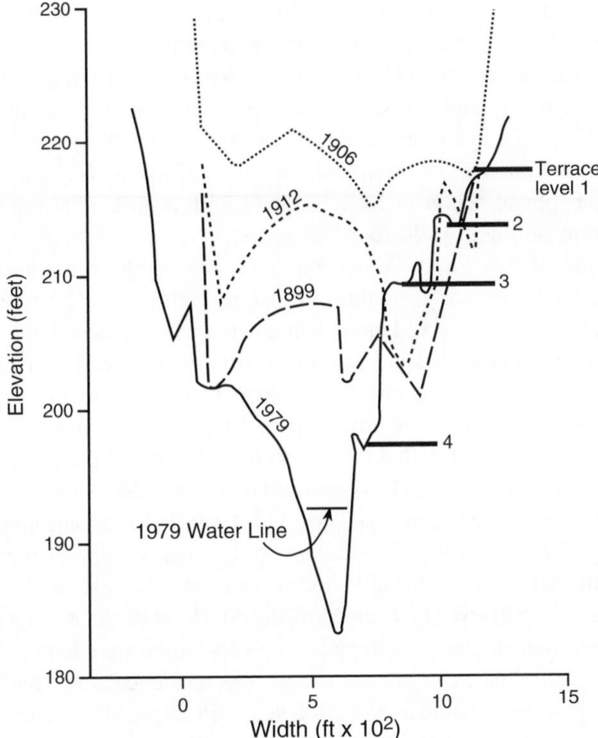

Figure 11.12 A 'wave' of sediment moving through the Yuba River, California. Note the peak in 1906 from migrating sediment, reached about three to four decades after the cessation of hydraulic mining (modified from Adler, 1980).

stream until 1906 but thereafter degraded. As the stream degraded, it paused long enough at different periods to erode laterally and thus create incipient, lower floodplains. And as the stream degraded further, it left small remnants of those relict floodplains as fluvial terraces. The two-decade lag described above was due not only to the travel time of sediment but also to its augmentation by upstream channel erosion. James (1989, 1999) found similar processes in nearby California streams as did Knighton (1989) in Tasmania.

Nevertheless, only a small portion of the total sediment storage from hydraulic mining in the Yuba River has been remobilized and while the stream erosion is apparently continuing, the rate of loss is asymptotic to the steady-state line. Expressed another way, the rate of sediment loss due to migration is asymmetrical, being skewed right much like a flood hydrograph. The primary reason is quite simple (see the 'fast in, slow out' principle expressed earlier): stream erosion can proceed here only by degradation and lateral erosion of the banks. Because the storage is spread over a floodplain several kilometres wide, it would take millennia for the stream to migrate back and forth across the floodplain and/or remove the stored material. But another important constraint on the Yuba River is that long reaches of the channel have been engineered not to erode.

When a stream does migrate across the floodplain to leave new, lower floodplains, it usually is removing not only the most recent storage, but also older material

(Duijsings, 1987). Set into geological time, this 'legacy sediment' with its 'wave' can be Holocene or older. For example, much of the sediment carried by streams in British Columbia comes from near-stream glacial deposits (Church and Ryder, 1972; Church and Slaymaker, 1989), stream banks in New York furnished about 60% of total stream loads (Nagle *et al.*, 2007), and a eye-opening review by Ashmore (1993) concludes that stream and valley sediment sources are dominant for all of Canada. Gordon (1979) has also suggested a similar phenomenon in New England with a millennial duration which has furnished a constant sediment yield for 8000 years.

Another example of this fill-and-erode model is the southern Piedmont of the USA (Figure 11.13A), There, severe agricultural erosion starting *c.* 1700 filled streams with sediment but the headwaters received more than downstream reaches (Figure 11.13B). With the advent of soil conservation and cropland reversion to forest and pasture, and the consequent reduction of sediment yield to the stream, headwater streams started degrading, sending sediment down to larger streams (Figure 11.13C). This process continues to the present day, and lateral erosion with the creation of new, lower floodplains has since been reported in the Maryland Piedmont (Jacobson and Coleman, 1986). Only a small portion of the historical sediment has been exported (Trimble, 1969, 1975; Meade and Trimble, 1974; Meade, 1982; see Figure 11.13C). One recent study suggests that at the present rate of evacuation it would take six to ten millennia to evacuate the historical sediment from a Piedmont catchment of 540 km^2 (Jackson *et al.*, 2005). However, that study is based largely on modelling rather than precise ground measurements which are uncommon in catchment studies. In a very active basin in New Zealand, the residence time 'half-life' of stored floodplain sediment was estimated to be >2000 years (Phillips, Marden and Gomez, 2006). In a much larger study, Holocene alluvium stored in the Rhine River catchment amounted to 0.5 t ha^{-1} per year over a 7500 year period (Houben *et al.*, 2009; Hoffman *et al.*, 2007). Kelsey (1987) and Malmon, Dunne and Reneau (2002, 2003) have shown that there can be long lags in the cascade of sediment downstream due to storage on floodplains and colluvial slopes. The foregoing are examples of the 'fast in, slow out' principle. Indeed, it is now accepted that much of the present sediment yield, rather than being the product of contemporary soil erosion, consists of material that was eroded anywhere from a day to several millennia earlier. The long residence time for sediment on floodplains is also important in that other matter including carbon and pollutants, sometimes adsorbed on to sediment particles, can be sequestered for long periods (e.g. Graf, 1990; Haycock and Burt, 1993; Smith *et al.*, 2001; Malmon, Dunne and Reneau, 2002; Hoffman, Glatze and Dikau, 2009).

11.4.2 Studies at Coon Creek

Both the hydraulic mining and the Piedmont demonstrate that different processes can obtain in different parts of the basin and that upstream morphology and processes can strongly affect downstream processes and morphology. This is part of what Schumm (1973) called the 'complex response', meaning that processes within a stream catchment can be out of phase. This complexity of the sediment cascade is clearly demonstrated by a 70-year study of erosion, sediment storage and movement in Coon Creek, a basin of about 320 km^2 in the Driftless Area of Wisconsin. Agriculture began in the 1850s and by the 1920s, erosion was so severe that streams and floodplains were aggrading at the rate of about 15 cm per year.

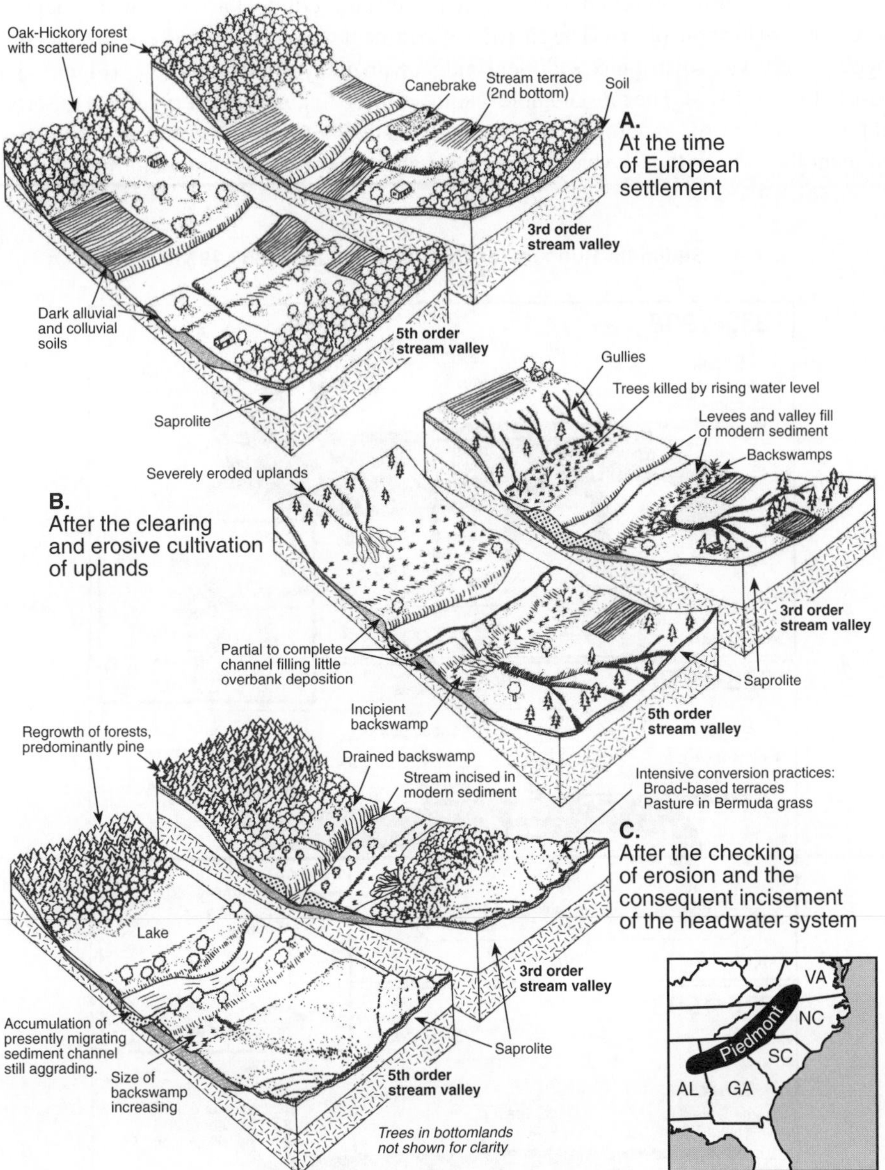

Figure 11.13 The historic evolution of the southern Piedmont riverine landscape 1700–1970, showing the migration of historic sediment and concomitant morphological–environmental changes. The stream in the upstream block might drain 25–60 km^2 while the downstream block might portray 125–250 km^2. Changes in 'connectivity' (or 'connectivity') are clearly demonstrated. Note especially the rills and gullies in 'B' which effectively conduct sediment to the floodplain. Refer to Figure 11.5 for a tributary cross-section
(from Trimble, 1974).

Roads, bridges, farms and even houses were buried and sediment accumulated to depths of more than 5 m in some places. The cascade of sediment was complex and one product of the ongoing study was a distributed sediment budget with many sources and sinks (Figure 11.14; see also Beach, 1994). The most notable feature of these fluxes is that very little of the eroded sediment reached the mouth of the stream no matter how great the erosion or upstream sediment flux was within the basin. Expressed another way, sediment yield from the basin

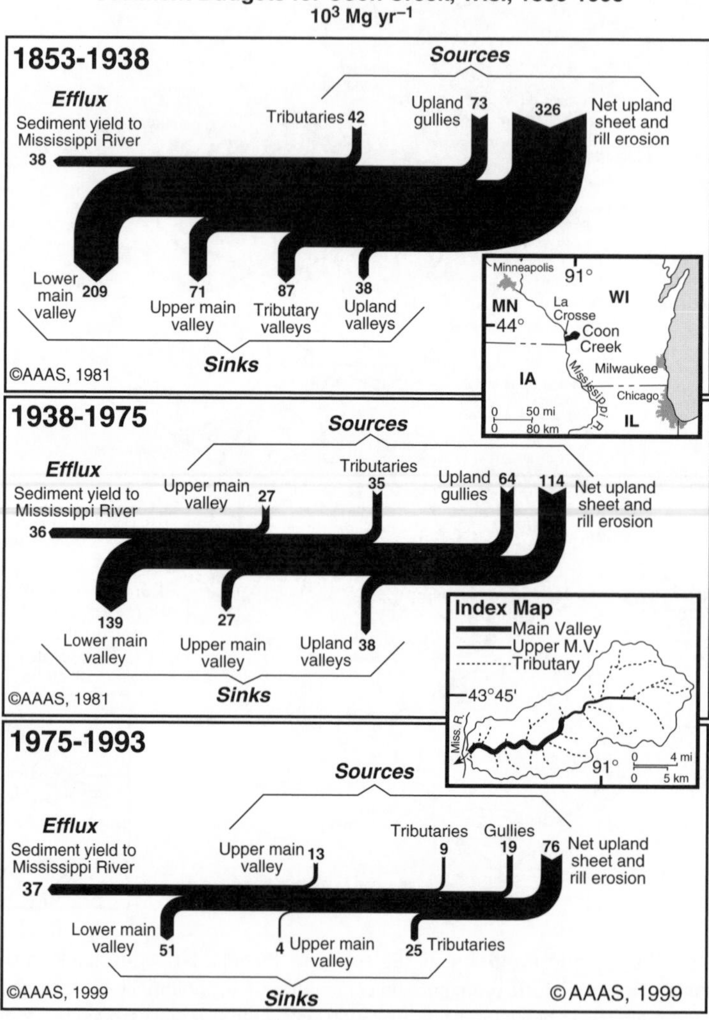

Figure 11.14 Sediment budgets for Coon Creek, Wisconsin, 1853–1993. This basin is about 25 km southeast of La Crosse, Wisconsin, and has a drainage area of 360 km^2. Numbers are annual averages for the periods in 10×3 Mg yr^{-1}. All values are direct measurements except net upland sheet and rill erosion, which is the sum of all sinks and the efflux minus the measured sources. The lower main valley and tributaries are net sediment sinks whereas the upper main valley is a net sediment source (Trimble, 1999).

was not sensitive to the extreme sediment fluxes within the basin. This suggests that we must not only think of stream capacity but also stream valley conveyance capacity. Because more sediment moves at high flows and it is the *floodplain* which conveys them, conveyance capacity would include such factors as floodplain width, morphology, and hydraulic friction as well as the existing flood regime. The Lane equilibrium concept was earlier presented to show the potential for sediment conveyance in the channel alone. Figure 11.15 is a simple conceptual model which attempts to combine Lane's concepts with the complexities of floodplains to explain sediment conveyance.

Coon Creek has a low conveyance capacity but there are few studies with which to compare it. Conventional wisdom has been that extreme floods might flush out stored sediment (Schumm, 1977) but a storm in Coon Creek, with a return period perhaps of >500 years, in 2007 apparently moved little sediment to the mouth. In the delta built into the Mississippi River, the accretion rate during the period 1992–2008, which includes the extreme event, was actually less than the rates for the period 1975–1992. A wide, non-erosive, hydraulically rough floodplain creates an excellent sediment trap. Given the sediment-trapping characteristics of the floodplain, losing the stored sediment from the Coon Creek catchment is akin to emptying a large container of a dense liquid through a pinhole.

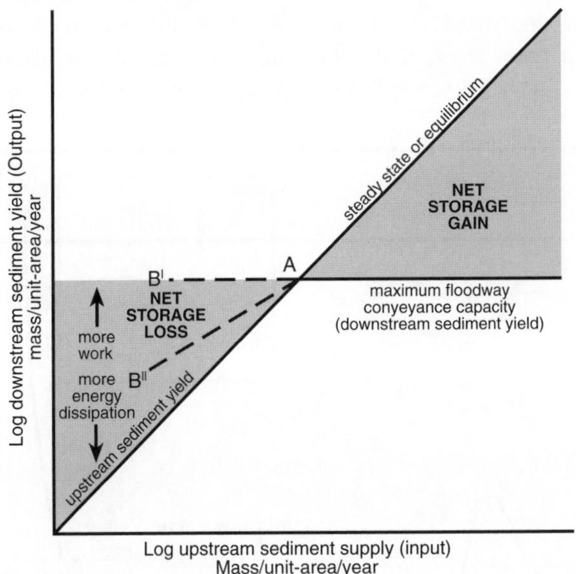

Figure 11.15 Schematic relation of downstream sediment yield and upstream sediment supply. Value A is the stream valley conveyance capacity. When upstream sediment supply exceeds A, the excess goes into storage, largely as vertical accretion on floodplain or by channel filling. When upstream sediment supply drops below A, the stream attempts to use its excess energy to obtain sediment load from the bed and banks. If sediment is easily obtainable and can be mobilized, it will be taken from storage, with the perfect case being line AB′, with 100% of the available energy used for geomorphological work. Generally, much of the available energy will be dissipated, so line AB″ is more likely
(redrawn from Trimble, 1983).

Accurate distributed sediment budgets are difficult to achieve because detailed ground surveys of many stream cross-sections over many years are required to show precisely the processes slowly evolving at different rates and times across different parts of the catchment. While the processes differ continuously in the basin, data limitations allowed only the capture of somewhat similar regions. Thus, Coon Creek is divided into three regions, tributaries, upper main valley, and lower main valley (Figure 11.16).

The *tributaries* were the first to receive historical sediment from agriculture (Figure 11.17). The vertical accretion on the floodplain increased bank height, meaning that the deeper flows exerted more stream power and channel erosion. But channel bottoms consisting largely of cobbles, usually did not erode significantly. Besides, flood peaks were increasing from poor land use, continuing to place sediment on the floodplain. By the 1920s and 1930s, the addition of coarse sediment, mostly sand to cobbles, widened the channels as seen in Figures 11.7 and 11.17. These are similar to 'meander plains' as described by Melton (1936). Channel erosion of tributaries sent much sediment downstream so that tributaries were simultaneously a sediment sink and a source. After the advent of modern soil conservation measures in the 1930s and 1940s, old, coarse meander plains had by the 1970s been covered by vertical accretion of fine material and were grassy. Indeed, they had become new floodplains adjusted to the more equable flood regime resulting from improved land use. Tributaries are becoming stable and are again a sediment sink. Note the role of both force and resistance in this scenario. Decreased force (*c.* 1950, Stage 3) allowed the deposition of fine material on the new floodplains which, in turn, allowed the development of a grass cover thus increasing resistance at the same time force was decreasing. This was a negative feedback allowing a more rapid stabilization of tributaries.

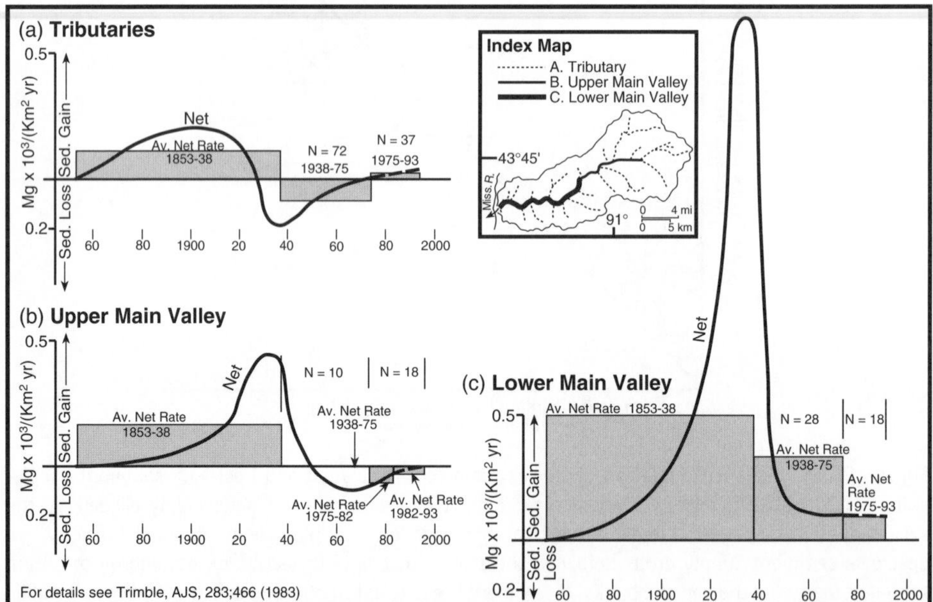

Figure 11.16 Differential stream and valley sediment budgets, Coon Creek, Wisconsin, 1853–1993: (a) tributaries; (b) upper main valley; (c) lower main valley (© Elsevier).

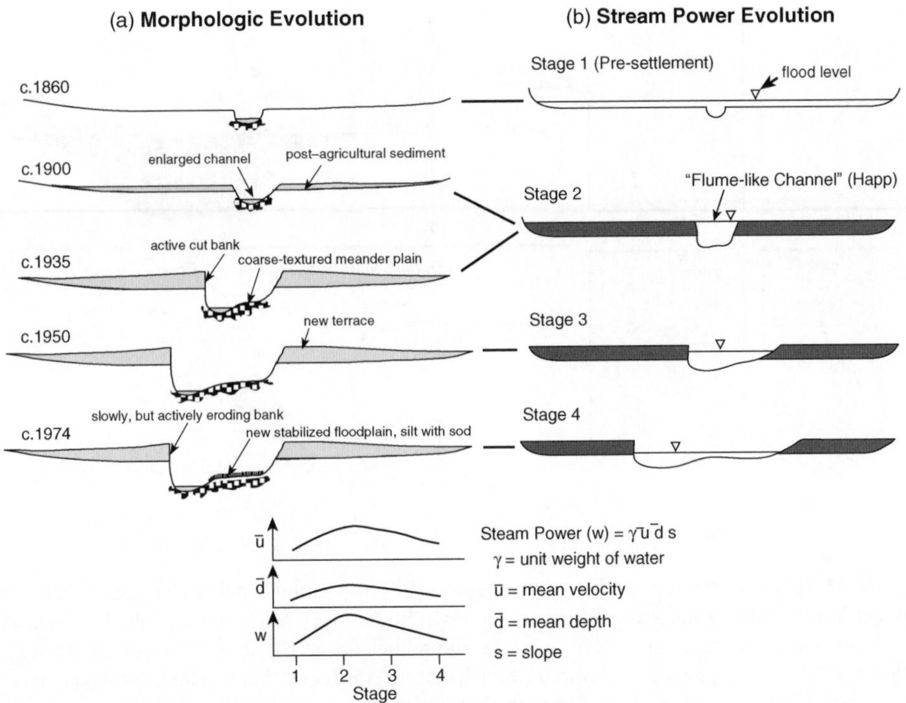

Figure 11.17 (a) Schematic model of changes of historic stream and valley morphology for Coon Creek and other Driftless Area tributaries, 1860 to 1974 (from p. 16 of a mimeographed pamphlet by S. W. Trimble for a field trip to the Driftless Area, April 1975, sponsored by the Association of American Geographers and led by George Dury, James C. Knox, W. C. Johnson and S. W. Trimble). (b) Changes of stream power and the transformation of stream and valley morphology. With the small stream channel of stage 1, floods spread out over the floodplain, keeping depth, velocity, and stream power low. With accretion of the floodplain and stream banks with historic sediment in stage 2 (*c.* 1900), greater flows were restricted to the channel, thus increasing depth, velocity, and stream power so that the channel erosion shown in the 1900 stage (left) must have been very rapid. In stages 3 and 4, the channel erodes laterally, so that floods are spread, with decreases of depth, velocity, and stream power (see Figure 11.7). By the latter stage, fine sediment covered the old gravel meander plains and new floodplains are formed as shown to the left (from presentation to the Association of American Geographers, San Diego, CA, 20 April 1992). Ron Shreve made important suggestions for preparation of this diagram in 1991. For a similar presentation of stream power, see Knox and Hudson, (1995) (from Ward and Trimble, 2004).

The old historical floodplain in the *upper main valley* continued to accrete until the early 1940s when streams flowed though deep, flume-like reaches. But, the curtailment of sediment inflow from upstream starting in the 1940s allowed channel erosion. As in the tributaries, the cobble channel bottoms allowed little vertical erosion but lateral bank erosion went apace. As high banks (typically 4 m) were eroded away, they were replaced on the other side of the channel by a lower bank, typically 2 m high, again adjusted to the moderated flood regime (Figure 11.6). From the evacuated cross-section one can gauge what a prolific source of sediment this reach was after about 1940. Predictions that this net loss would last for decades (Trimble, 1981) proved to be wrong, however. As the new floodplains

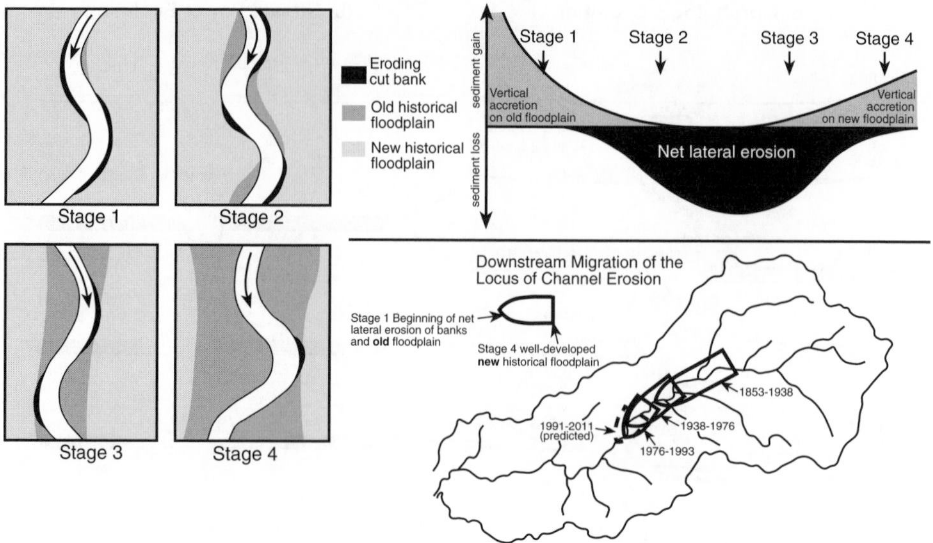

Figure 11.18 Transformation of the upper main valley and accompanying changes of sediment budget from sediment source to sediment sink (refer to Figure 11.6). Note the morphological feedback to process and, in turn, the process feedback to morphology, (a) as the stream moves from stage 1 to stage 4, it decreasingly has contact with the high banks, (b) the increasingly wider floodplain provides more opportunity for sediment storage with that eventually surpassing the losses from cut banks. Thus, the reach becomes a sediment sink
(redrawn from Trimble, 1994).

widened, they became much larger storage compartments and at the same time, the cut bank lost more contact with the stream. So over time, as with the tributaries this reach again became a sediment sink rather than a source (Figure 11.18), again demonstrating the complexities of the sediment cascade and emphasizes the potential role of floodplains to trap and store sediment. To be emphasized here is that only a fraction of the historical sediment in storage has been removed and, presently, both the tributaries and the upper main valley are gaining sediment (Figure 11.16a and b). What this portends for eventual removal is unknown but it underlines the 'slow out' part of the 'fast in, slow-out' principle. Knox (1989) contends that this region exhibits a 'slow in, fast out' behaviour but the mechanism for that is not apparent. Perhaps this may relate only to very small basins with narrow floodplains or to climatically induced cycles in the Holocene. However, the process mechanism is not clear, especially in the case of downstream reaches with wide floodplains. Based on contemporary analogues, severe trenching is upstream reaches would result in downstream accretion. Presumably, the complex response has always been operative.

The *lower main valley* has been a reach of continuous accretion during the historical period (Figure 11.19). Pre-historical accretion rates were very low. To the south of Coon Creek, Knox (1977, 1987, 2006) has measured pre-agricultural rates as being about 0.02 cm per year. But after decades of agriculture, accretion rates had increased to about 15 cm per year by the 1920s (Trimble and Lund, 1982). The present rates of vertical accretion, even with augmentation of sediment load coming from upstream channel erosion, are only about 0.5 cm per year (Trimble, 1999). However, the supply of sediment coming

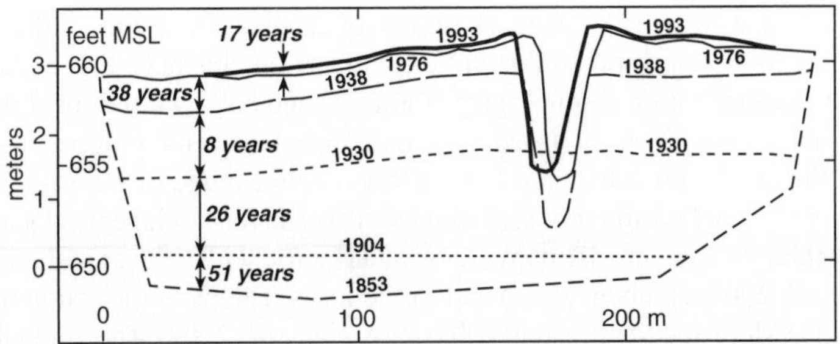

Figure 11.19 A cross-section in the lower main valley of Coon Creek, a perennial sediment sink. Succeeding higher stream and floodplain levels dating from 1853 to 1993 show vertical accretion but note the difference in rates
(Trimble, 1999).

from the erosion of often coarse upstream channels and banks is sorted, with at least some of the finer materials being transported out of the basin but with the sand tending to be retained in the channel and moved downstream, aggrading the channel. With high flows, the sand is carried overbank and quickly deposited along the stream as natural levees (Trimble, 2009). The result is that the stream aggrades more rapidly than the distal floodplain. This combined with the blockage of lateral drainage by the natural levees means that the distal bottoms are incipient back swamps, becoming ever wetter. Some have become perennial lakes. As seen earlier in the Piedmont (Figure 11.13C) and in the case of hydraulic mining, the transport of this legacy sediment downstream can have deleterious effects. In some parts of the Driftless Area, rising stream levels have increased flooding and led to higher water tables. In the city of Arcadia, Wisconsin, migrating sand from local tributaries has raised the normal surface elevation of the Trempeleau River by 1 m in recent years, leading to wet basements and increased risk of flood danger (Trimble, 1993, 2009).

11.4.3 Studies in Northwest European Loess Areas

In contrast to most catchments in the USA, including Coon Creek and the Piedmont region discussed above, the majority of river systems in the temperate regions of Europe have experienced human impact and associated sediment pulses over much longer time periods. Furthermore, for several catchments, the impact through intensive agricultural practises is still quite strong. Therefore, the typical 'sediment wave', as observed for instance in Coon Creek (Figure 11.16), is not common for many European rivers. This is in particular true for the fertile loess areas in northwest and central Europe. A good example comes from the Dijle catchment in central Belgium and one of its main tributaries, the Nethen (55 km^2) (Rommens *et al.*, 2006, 2007; Notebaert *et al.*, 2009; Verstraeten *et al.*, 2009). Hillslope erosion started in the neolithic period but increased from about 3000 years ago, as evidenced by major sediment storage in dry valley bottoms (Rommens *et al.*, 2007). However, the floodplains of the Nethen and Dijle only start to accrete at high rates from the mediaeval period onwards. The efficient coupling of hillslopes and valley systems since the Middle Ages was established through changing spatial patterns of land use, more intensive land use, but also through the introduction of floodplain management techniques that facilitated the delivery of sediment to the river network (drainage channels) and limited the conveyance capacity of the floodplain itself (mills and their diversion channels, fish ponds, and road obstructions). Sedimentation chronologies of alluvial and colluvial sediment deposits thus show a pronounced non-linearity in sediment dynamics between the slopes, the dry valley bottoms and the floodplain, which has been clearly demonstrated in the Nethen catchment (Verstraeten *et al.*, 2009). Similar observations were made for catchments at various spatial scales in the Rhine catchment, Germany (Lang and Hoenscheidt, 1999; Houben *et al.*, 2009; Hoffmann *et al.*, 2007; Houben, 2008) and in the loess regions in the southern tip of The Netherlands (De Moor and Verstraeten, 2008; De Moor *et al.*, 2008). The gradual movement of sediment down the whole cascade system from hillslopes to tributaries to main valley is apparent from these case studies. Even more than the examples from the 'New World', the principle of 'fast-in, slow out' is valid here. The majority of the sediment eroded during the past few thousand years in the Dijle catchment is still stored on the hillslopes (40%) or in the floodplains (42%) (Notebaert *et al.*, 2009). Moreover, current-day erosion processes still provide the fluvial system with 'new' inputs of sediment.

Overland flow and soil erosion are discussed further in Chapter 7.

11.4.4 Studies of Powder River, Montana

At a much shorter timescale than considered thus far, the cascade metaphor also may be applied to high-intensity low-frequency events, in which larger quantities of sediment are moved down-valley during shorter periods of time. The outcome of such a short-term down-valley cascade is shown in Figure 11.20, which represents the transfers of sediment within a 58-km reach of the valley of a river of moderate size during a single 8-day flood in Powder River, Montana, a semi-arid region. Of the six identifiable valley features – high terrace, middle terrace, low terrace, floodplain, point bar, and channel – all six were eroded, mostly by lateral attrition; and all but one (high terrace) were inundated and received new sediment deposits that accumulated vertically during the flood. The quantities of sediment transferred between features, or eroded from one feature and redeposited a few kilometres farther down-valley on the same feature, exceeded the mass of sediment transported out of the flood reach. The low terrace (Lightning) gained twice as much new sediment through aggradation as it lost by bank erosion. In fact, the mass of newly deposited sediment atop the Lightning terrace, taken by itself, was slightly greater than the mass transported down-river out of the reach during the flood. Any individual sediment particle that was transported into the reach from up-river during the flood was much more likely to have been deposited on one of the flooded surfaces within the reach than to have been transported downriver out of the reach.

In the years of more average runoff following the flood of 1978, this reach of Powder River has continued to convey and accumulate sediment, but at much less rapid rates, and only in the channels, point bars, and lower floodplains – that is, in the areas that usually are inundated at least every year or two. The 1978 flood cleared out segments of the pre-1978 floodplain by remobilizing their sediment for transfer onto higher surfaces such as the Lightning terrace, or for conveyance further downstream. This 8-day scouring process left many former floodplain sites scoured down to their coarse-gravel substrates. During the decades since 1978, Powder River has slowly brought in new sediment to replace the material that was scoured out by the flood, and new floodplains have regrown in the older sites (Moody, Pizzuto and Meade, 1999; Pizzuto, Moody and Meade, 2008). The cascading

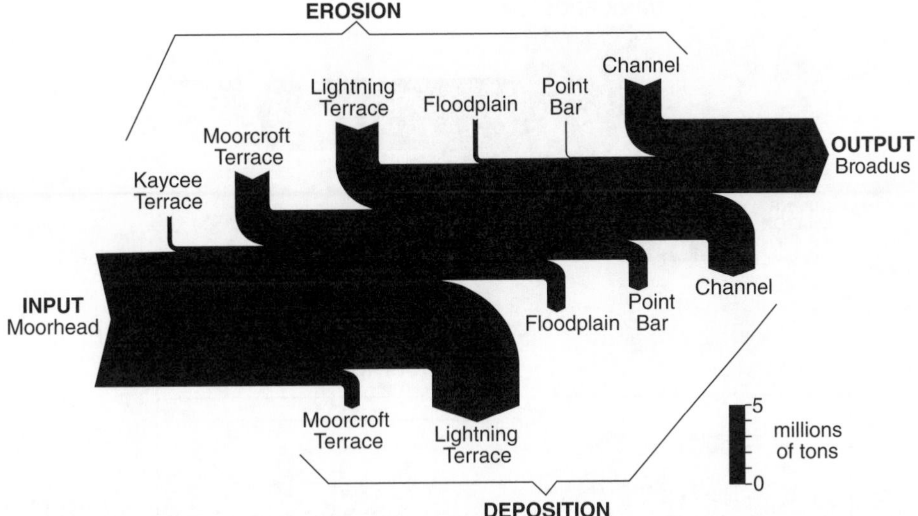

Figure 11.20 Sediment budget for a 58-km reach of Powder River valley (Montana, USA) for the 8-day duration of the flood of May 1978. Drainage area of Powder River is 20 800 km^2 at Moorhead gauging station. Shown are the input of sediment to the reach from up-river sources, the output from the down-river end of the reach, and the transfers of sediment during the flood to and from six identifiable segments of the river-valley system that were inundated or otherwise affected by the flood. These segments were, in order of decreasing elevation, three terraces (Kaycee, Moorcroft, Lightning), floodplains, point bars, and channels
(redrawn from Moody, J. A. and Meade, R. H. 'Terrace Aggradation During the 1978 Flood on Powder River, Montana, USA' Geomorphology, 99, 1–4, pg. 17,© 2008. Reprinted with permission from Elsevier).

processes continue at all stages of river discharge, at rates of sedimentation and remobilization that are connected exponentially to the rates of flow.

11.4.5 Studies in the Amazon Basin

For large rivers, an example of cascading floodplain sedimentation comes from the Amazon River of Brazil (Figure 11.21). The Brazilian reaches of the Amazon's main channel (which includes Rio Solimões, the segment of the Amazon main stem between the Peru–Brazil border and the Amazon–Negro confluence) flow through or alongside large tracts of floodplain, which are tens of kilometres wide (Mertes, Dunne and Martinelli, 1996; Mertes and Dunne, 2007). No engineering works (levees, bank revetments, control structures) inhibit the natural processes of channel migration, bank erosion, or overbank flooding. The pattern of annual flooding is regular and dependable, and, during each year (with very few exceptions), large masses of sediment are exchanged between channels and floodplains – so much so that in the1500-km reach of the Amazon main stem illustrated in Figure 11.21 (upper panel), the quantities of sediment exchanged between channels and floodplains significantly exceed the net downriver flux.

The cascading effect of these exchanges – the laying down of floodplain sediment and its subsequent remobilization by the laterally shifting (and bank-eroding) river – is better

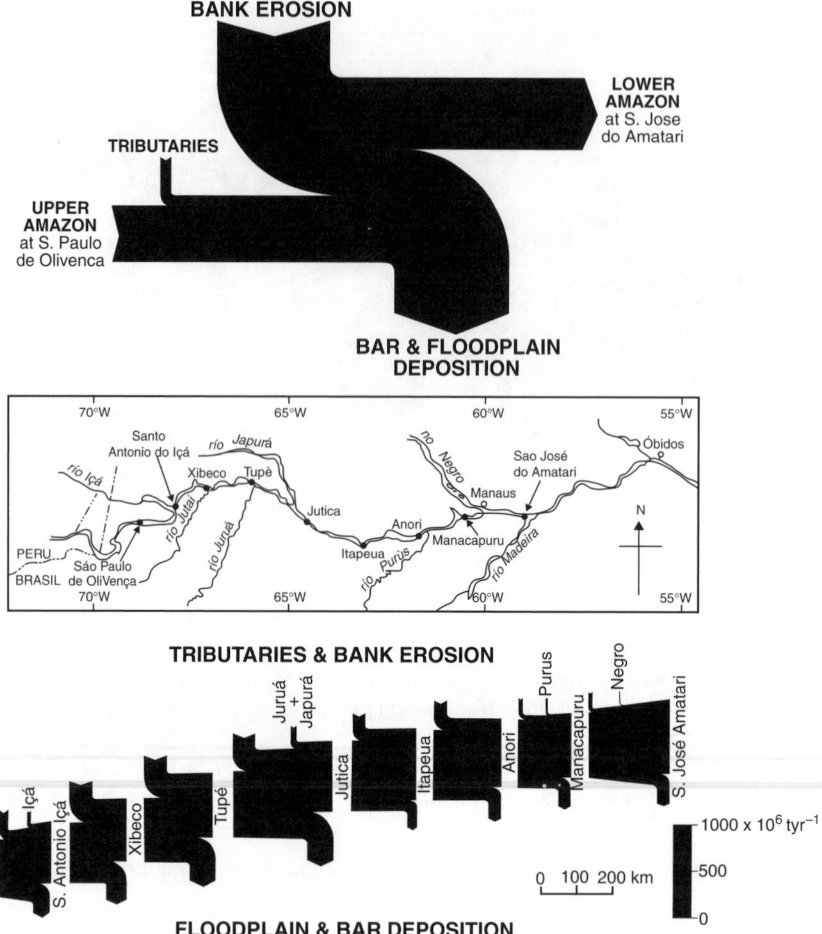

Figure 11.21 Budget diagrams showing cascading sediment in a 1500-km reach of the Amazon River in Brazil (from Meade, 2007). (Upper panel) Consolidated budget for entire 1500-km reach, showing that overall exchanges between channel and floodplain exceed the net down-river fluxes of sediment in the channel itself. Numbers represent averages for the 15-yr period, 1974–1989. (Lower panel) Exploded version of the upper panel, showing the individual sediment budgets for each of eight consecutive reaches of the main stem Amazon between São Paulo de Olivença and São José do Amatari. The upper left corners show the quantities of sediment contributed to the eight segments from bank erosion. The lower right corners show the quantities removed from the channel by deposition onto floodplains and point bars. Details of the calculations and compilations are given by Dunne *et al.*, (1998)
(redrawn from Meade, 2007).

visualized in the lower panel of Figure 11.21, which is an exploded version of the upper panel. As Meade ((2007), p. 56) well described it:

> 'By sweeping the eye more or less horizontally from left to right across the exploded diagram, the reader might gain the impression (or at least the supposition) that any given sediment

particle moving downstream past São Paulo de Olivença (especially during high-water season when much sediment-laden water is flowing on to the floodplain) is likely to be carried out of the channel and be deposited on the floodplain before the thread of the water flow reaches São José do Amatari. Conversely, by looking upstream from the section at São José do Amatari, the reader might just as easily visualize the likelihood that any sediment particle passing through this cross-section (especially during falling-water season when river-bank collapse is maximal for the year) had come out of one of those eroding banks, rather than having been swept this year off an Andean slope somewhere in Peru or Ecuador.,

(*Meade, 2007, p.506; see also Meade et al.*, 1985)

11.5 Summary and Conclusion

Streams, valleys, and floodplains are the 'zone of transfer' in the cascade of sediment from uplands to the sea. Only in the past few decades have we begun to learn how complex that movement is and there is still much to discover. Indeed, documenting the complex fluxes and storage of sediment through space and time along with establishing the interrelationships and drivers is a pressing need. Time is also a critical variable with all indications being that the cascade is slow to very slow in low-energy high-resistance catchments. Fluvial systems are responding to present drivers but also greatly constrained or even locked into their histories.

The cascade metaphor can be extended into the channels and floodplains of the world's largest rivers, although there are differences of scale that must be taken into account. Whereas the processes of bank erosion and point-bar formation and the associated lateral shifting of the cross-channel point-bar–river-bed–cutbank profile can be scaled up fairly smoothly from the smallest alluvial rivers to the very largest, the processes of floodplain formation by overbank deposition in very large alluvial river systems are markedly different from those in smaller rivers. Modern-day processes observed in the Amazon River of Brazil (Mertes, Dunne and Martinelli, 1996; Mertes and Dunne, 2007), and late Holocene floodplain deposits that have been cored and mapped in the lower Mississippi River valley of northern Louisiana (Aslan, Autin and Blum, 2005) strongly support the notion that the construction of floodplains by large rivers is more a matter of deposition 'by avulsion-related crevassing and lacustrine sedimentation rather than by overbank flooding' (Aslan and Autin, (1999), p. 800). But, regardless of the individual depositional processes, the basic spatial paradigms of *erosion* as a mostly *lateral* process, and of *deposition* as a mostly *vertical* process, seem to hold true for all but the very smallest rivers. Thus, the principle of 'fast in, slow out' as discussed in this chapter seems to hold at most scales.

Acknowledgements

I am indebted to Robert H. Meade, U.S. Geological Survey Emeritus, who gave the chapter two thorough critical readings and also contributed significant text. Likewise, Gert Verstraeten, Catholic University of Leuven, gave a critical reading and contributed text. Critical readings were also given by Claudio Vita-Finzi (Natural History Museum, London) and Greg Nagle (Cornell University). The Figures were done by Chase Langford, University of California Los Angeles and the photograph in Figure 11.3 was taken by Edyta Zygmunt, University of Silesia.

References

Adams, C. (1940) Modern sedimentation in the Galena River Valley, Illinois and Wisconsin. MS thesis, University of Iowa, Iowa City.

Adams, C. (1942) Accelerated sedimentation in the galena River basin, Illinois and Wisconsin. PhD dissertation, University of Iowa, Iowa City.

Adams, C. (1944) Mine waste as a source of Galena River bed sediment. *Journal of Geology*, **52**, 275–287.

Adler, L. (1980) Adjustment of the Yuba River to the influx of hydraulic mining debris, 1849–1979. MA thesis, University of California, Los Angeles.

Ashmore, P. (1993) Contemporary erosion of the Canadian Landscape. *Progress in Physical Geography*, **17**, 190–204.

Aslan, A. and Autin, W.J. (1999) Evolution of the Holocene Mississippi River floodplain, Ferriday, Louisiana: Insights on the origin of fine-grained floodplains. *Journal of Sedimentary Research*, **69** (4), 800–815.

Aslan, A., Autin, W.J. and Blum, M.D. (2005) Causes of river avulsion; Insights from the late Holocene avulsion history of the Mississippi River, USA. *Journal of Sedimentary Research*, **75**, 650–664.

Beach, T. (1994) The fate of eroded soil: Sediment sinks and sediment budgets of agrarian landscapes in southern Minnesota, 1851–1988. *Annals of the Association of American Geographers*, **84**, 5–28.

Benda, L. (1990) The influence of debris flows on channels and valley floors in the Oregon Coast Range, USA. *Earth Surface Processes and Landforms*, **15**, 457–466.

Bennett, S. J. and Simon, A. (eds) (2004) *Riparian Vegetation and Fluvial Geomorphology*, Water Science and Application 8, American Geophysical Union, Washington, DC.

Beschta, R. and Platts, H. (1986) Morphological features of small streams: significance and function. *Water Resources Bulletin*, **22**, 369–379.

Boisson, P. and Patton, P. (1985) Sediment storage and terrace formation in Coyote Gulch basin, south-central Utah. *Geology*, **13**, 31–34.

Brackenridge, G.R. (1984) Alluvial stratigraphy and radiocarbon dating along the Duck River, Tennessee: Implications regarding floodplain origins. *Geological Society American Bulletin*, **95**, 9–25.

Bridge, J.S. (2003) *Rivers and Floodplains: Forms, Processes and Sedimentary Record*, Blackwell Science, Oxford.

Brierley, G., Fryirs, K. and Vikrant, J. (2006) Landscape connectivity: the geographic basis of geomorphic applications. *Area*, **38**, 165–174.

Brown, A. (1987) Long-term sediment storage in the Severn and Wye catchments, in *Paleohydrology in Practice* (eds K. Gregory and J. Thornes), John Wiley and Sons, Chichester.

Brown, A., Cary, C., Erkens, G. *et al.* (2009) From sedimentary records to sediment budgets: Multiple approaches to catchment sediment flux. *Geomorphology*, **108** (1–2), 35–47.

Burt, T.P. and Pinay, G. (2005) Linking hydrology and biogeochemistry in complex landscapes. *Progress in Physical Geography*, **29** (3), 297–316.

Bryan, K. (1925) Date of channel trenching (arroyo cutting) in the arid Southwest. *Science*, **62**, 338–344.

Caine, N. and Swanson, F. (1989) Geomorphic coupling of hillslope and channel systems in two small mountain basins. *Zeitschrift fur Geomorphlogie*, **33**, 189–203.

Carson, J. (2003) Dynamic fluvial processes in upper Mississippi River backwaters. Unpublished MA thesis, Los Angeles, University of California.

Church, M. and Ryder, J.M. (1972) Paraglacial sedimentation: a consideration of fluvial processes conditioned by glaciation. *Geological Society of America Bulletin*, **83**, 3059–3072.

Church, M. and Slaymaker, O. (1989) Disequilibrium of Holocene sediment yield in glaciated British Columbia. *Nature*, **337**, 352–354.

Coffman, G. (2008a) Factors influencing invasion of Giant Reed (*Arundo donax*) in riparian ecosystems of Mediterranean-type climates. PhD dissertation. Los Angeles, University of California.

Coffman, G. (2008b) Giant reed (*Arundo donax*): effect on streams and water resources, in *Encyclopedia of Water Science* (ed. S. Trimble), CRC Press, Boca Raton, Florida.

Cooke, R. and Reeves, R. (1977) *Arroyos and Environmental Change in the American South-West*, Clarendon Press, Oxford.

Darby, S. and Simon, A. (1999) *Incised River Channels*, John Wiley and Sons, Chichester.

Davies-Colley, R. (1997) Stream channels are narrower in pasture than in forest. *New Zealand Journal of Marine and Freshwater Research*, **31**, 599–608.

De Moor, J.J.W. and Verstraeten, G. (2008) Alluvial and colluvial sediment storage in the Geul River catchment (The Netherlands) — Combining field and modelling data to construct a Late Holocene sediment budget. *Geomorphology*, **95**, 487–503.

De Moor, J.J.W., Kasse, C., Van Balen, R. *et al.* (2008) Human and climate impact on catchment development during the Holocene — Geul River, the Netherlands. *Geomorphology*, **98**, 316–339.

De Vente, J., Poesen, J., Arabkhedri, M. and Verstraeten, G. (2007) The sediment delivery problem revisited. *Progress in Physical Geography*, **31**, 155–178.

Dietrich, W. and Dunne, T. (1978) Sediment budget for a small catchment in mountainous terrain. *Zeitschrift fur Geomorphologie Supplement Band*, **29**, 191–206.

Dietrich, W., Dunne, T., Humphrey, N. and Reid, L. (1982) Construction of sediment budgets for drainage basins, in *Sediment Budgets and Routing in Forested Drainage Basins* (eds F. Swanson, R. Janda, T. Dunne and D Swanson), Forest Service General Technical Report PNW-141, U.S. Department of Agriculture, Washington, DC, pp. 5–23.

Dotterweich, M. (2005) High-resolution reconstruction of a 1300 year old gully system in northern Bavaria, Germany: a basis for modeling long-term human-induced landscape evolution. *The Holocene*, **15**, 994–1005.

Dudley, T. (1998) Exotic plant invasions in California riparian areas and wetlands. *Fremontia*, **26**, 24–29.

Duijsings, J. (1987) A sediment budget for a forested catchment in Luxembourg and its implications for channel development. *Earth Surface Processes and Landforms*, **12**, 173–184.

Dunne, T., Mertes, L.A.K., Meade, R.H. *et al.* (1998) Exchanges of sediment between the floodplain and channel of the Amazon River. *Geological Society of America Bulletin*, **110**, 450–467.

Eakin, H. (1936) Silting of Reservoirs, Technical Bulletin 524, U.S. Department of Agriculture, Washington, DC. Revised by C. Brown in 1939.

Emmett, W.W. (1974) Channel aggradation in western United States as indicated by observations at Vigil Network sites. *Zeitschrift fur Geopmorphologie*, **21**, 52–62.

Fenneman, N.M. (1906) Floodplains produced without floods. *American Geographical Society Bulletin*, **38**, 89–91.

Ferguson, R.I. (1981) Channel form and channel changes, in *British Rivers* (ed. J. Lewin), Allen and Unwin, London, pp. 9–125.

Ferguson, R.I. (1987) Hydraulic and sedimentary controls of channel pattern, in *Rivers–Environment and Process* (ed. K. Richards), Blackwell, Oxford, pp. 129–158.

Fraczek, W. (1987) Assessment of the effects of changes in agricultural practices on the magnitude of floods in Coon Creek Watershed using hydrograph analysis and air photo interpretation, Unpublished MS Thesis, University of Wisconsin, Madison, WI.

Fryirs, K., Brierley, G., Preston, N. and Kasai, M. (2007) Buffers, barriers and blankets: The (dis)connectivity of catchment-scale sediment cascades. *Catena*, **70**, 49–67.

Gay, G.R., Gay, H.H., Gay, W.H. *et al.* (1998) Evolution of cutoffs across meander necks, Powder River, Montana, USA. *Earth Surface Processes and Landforms*, **23**, 651–662.

Gilbert, G.K. (1917) Hydraulic mining debris in the Sierra Nevada. *U.S. Geological Survey Professional Paper*, 105.

Gordon, R. (1979) Denudation rate of central New England determined from estuarine sedimentation. *American Journal of Science*, **279**, 1997–2006.

Graf, W.L. (1978) Fluvial adjustments to the spread of tamarisk in the Colorado Plateau region. *Geological Society of America Bulletin*, **89**, 1491–1501.

Graf, W.L. (1983a) Variability of sediment removal in a semiarid watershed. *Water Resources Research*, **19**, 643–652.

Graf, W.L. (1983b) The arroyo problem:_paleohydrology and paleohydraulics in the short term, in *Background to Paleohydrology* (ed. K. Gregory), John Wiley and Sons, New York.

Graf, W.L. (1984) A probabilistic approach to the spatial assessment of river channel instability. *Water Resources Research*, **20**, 953–962.

Graf, W.L. (1988) *Fluvial Processes in Dryland Rivers*, Springer, Berlin.

Graf, W.L. (1990) Fluvial dynamics of thorium-230 in the Church Rock event, Puerco River, New Mexico. *Annals of the Association of American Geographers*, **80**, 327–342.

Gumprecht, B. (2001) *The Los Angeles River: Its Life, Death and Possible Rebirth*, Johns Hopkins Press, Baltimore.

Happ, S.C. (1944) Effects of sedimentation on floods in the Kickapoo valley, Wisconsin. *Journal of Geology*, **52**, 53–68.

Happ, S.C. (1975) Valley sedimentation as a factor in sediment-yield determinations, in *Present and Prospective Technology for Predicting Sediments Yields and Sources*, ARS Publication, U.S. Department of Agriculture, Washington, DC, pp. 5–40, 57–60.

Happ, S.C., Rittenhouse, G. and Dobson, G.C. (1940) *Some Principles of Accelerated Stream and Valley Sedimentation*. Technical Bulletin 695, U.S. Department of Agriculture, U.S. Government Printing Office, Washington, DC.

Haycock, N.E. and Burt, T.P. (1993) Role of floodplain sediments in reducing the nitrate concentration of subsurface runoff: a case study in the Cotswolds, UK. *Hydrological Processes*, **7**, 287–295.

Hey, R. and Thorne, C.R. (1986) Stable channels with mobile gravel beds. *Journal of Hydraulic Engineering*, **112**, 671–689.

Hooke, J.M. (1980) Magnitude and distribution of rates of river bank erosion. *Earth Surface Processes*, **5**, 143–157.

Hoffman, T., Glatze, S. and Dikau, R. (2009) A carbon storage perspective on alluvial sediment storage in the Rhine catchment. *Geomorphology*, **108** (1–2), 127–137.

Hoffman, T., Erkens, G., Cohen, K. *et al.* (2007) Holocene floodplain sediment storage and hillslope erosion within the Rhine catchment. *The Holocene*, **17**, 105–118.

Houben, P. (2008) Scale linkage and contingency effects of field-scale and hillslope-scale controls of long-term soil erosion: Anthropogenic sediment flux in agricultural loess watershed of southern Germany. *Geomorphology*, **101**, 172–191.

Houben, P., Wunderlich, J. and Schrott, L. (2009) Climate and long-term human impact on sediment fluxes in watershed systems. *Geomorphology*, **108** (1–2), 1–7.

Houben, P., Hoffman, T., Zimmerman, A. and Dikau, R. (2006) Land use and climatic effects on the Rhine system: Quantifying sediment fluxes and human impact with available data. *Catena*, **66**, 42–52.

Jackson, C. Martin, D., Leigh, D. and West, L. (2005) A southeastern Piedmont watershed sediment budget: evidence for a multi-millennial agricultural legacy. *Journal of Soil and Water Conservation*, **60**, 298–310.

Jacobson, R.B. and Coleman, D. (1986) Stratigraphy and recent evolution of Maryland Piedmont flood plains. *American Journal of Science*, **286**, 617–637.

Jacobson, R.B. and Oberg, K. (1997) Geomorphic change of the Mississippi River floodplain at Miller City, Illinois, a result of the flood of 1993. U S Geological Survey *Circular*, 1120-J.

James, L.A. (1989) Sustained storage and transport of hydraulic gold mining sediment in the Big Bear River, California. *Annals Association of American Geographers*, **37**, 570–592.

James, L.A. (1999) Time and the persistence of alluvium: river engineering, fluvial geomorphology, and mining sediment in California. *Geomorphology*, **31**, 265–290.

Johnson, W.C. (1992) Dams and riparian forest: case study from the upper Missouri River. *Rivers*, **3**, 229–242.

Johnson, W.C. (1998) Adjustments of riparian vegetation to river regulation in the Great Plains, USA. *Wetlands*, **18**, 608–618.

Kelsey, H.M. (1980) A sediment budget and an analysis of geomorphic process in the Van Duzen River basin, north coastal California, 1941–1975: summary. *Geological Society of America Bulletin*, **91**, 190–195.

Kelsey, H.M. (1987) Stochastic model for the long-term transport of stored sediment in a river channel. *Water Resources Research*, **23**, 1739–1750.

Knighton, A. (1989) River adjustment to changes in sediment load: the effects of tin mining on the Ringarooma River, Tasmania, 1875–1984. *Earth Surface Processes and Landforms*, **14**, 333–359.

Knox, J.C. (1972) Valley alluviation in southwestern Wisconsin. *Annals of the Association of American Geographers*, **62**, 401–410.

Knox, J.C. (1977) Human impacts on Wisconsin stream channels. *Annals of the Association of American Geographers*, **67**, 323–342.

Knox, J.C. (1987) Historical valley floor sedimentation in the upper Mississippi Valley. *Annals of the Association of American Geographers*, **77**, 224–244.

Knox, J.C. (1989) Long- and short-term episodic storage and removal of sediment in watersheds of southwestern Wisconsin and northwestern Illinois, Publication 184, International Association of Hydrological Sciences, Wallingford, pp. 157–164.

Knox, J.C. (2006) Agricultural influence on landscape sensitivity in the Upper Mississippi River Valley. *Catena*, **42**, 193–224.

Knox, J.C. and Hudson, J.C. (1995) Physical and cultural change in the Driftless Area of southeast Wisconsin, in *Geographical Excursions in the Chicago Area* (ed. M. Conzen), Association of American Geographers, Washington, DC, pp. 17–31.

Kunsman, H. (1944) Stream and valley sedimentation in Beaver Creek valley, Wisconsin. MS thesis. University of Wisconsin, Madison.

Lane, E.W. (1955) Design of stable channels. *Transactions of the American Society of Civil Engineers*, **120**, 1234–1260.

Lang, A. and Hoenscheidt, S. (1999) Age and source of colluvial sediments at Vaihingen-Enz, Germany. *Catena*, **38**, 89–107.

Latocha, A. (2009) Land-use change and longer-term human –environment interactions in a mountain region (Sudetes Mountains, Poland). *Geomorphology*, **108** (1–2), 48–57.

Lawler, D. (1992) Process dominance in bank erosion systems, in *Lowland Floodplain Rivers* (eds P. Carling and G. Petts), John Wiley and Sons, Chichester, pp. 117–143.

Lehre, A. (1981) Sediment budget of a small California coast Range drainage near San Francisco. IASH *Pub*. 132, 123–139.

Leopold, L.B. (1976) Reversal of erosion cycle and climatic change. *Quaternary Research*, **6**, 557–562.

Leopold, L.B. and Bull, W.B. (1979) Base level, aggradation and grade. *Proceedings of the American Philosophical Society*, **123**, 168–195.

Leopold, L.B., Wolman, M.G. and Miller, J. (1964) *Fluvial Processes in Geomorphology*, Freeman, San Francisco.

Lyons, J.S., Trimble, S.W. and Paine, L.K. (2000) Grass versus trees: managing riparian areas to benefit streams of central North America. *Journal of the American Water Resources Association*, **36**, 919–930.

Mackin, J.H. (1948) Concept of the graded river. *Bulletin of the Geological Society of America*, **59**, 463–512.

Magilligan, F.J. (1985) Historical floodplain sedimentation in the Galena River valley, Wisconsin and Illinois. *Annals of the Association of American Geographers*, **75**, 583–594.

Magilligan, F.J., Phillips, J., James, A. and Gomez, B. (1998) Geomorphic and sedimentological controls on the effectiveness of an extreme flood. *Journal of Geology*, **106**, 87–95.

Malmon, D.V., Dunne, T. and Reneau, S. (2002) Predicting the fate of sediment and pollutants in river floodplains. *Environmental Science and Technology*, **36**, **2926**, 32.

Malmon, D.V., Dunne, T. and Reneau, S. (2003) Stochastic theory of particle trajectory through alluvial valley floors. *Journal of Geology*, **111**, 525–542.

Marston, R. and Person, M. (2004) Sediment budget, in *Encyclopedia of Geomorphology* (ed. A. Goudie), Routledge, UK, pp. 927–930.

McHenry, J.R., Ritchie, J.C., Cooper, C. and Verdon, J. (1984) *Contaminants in the Upper Mississippi River* (eds J. Wiener, R. Anderson and D. McConville), Butterworth, Boston, pp. 99–117.

McKelvey, V.E. (1939) Stream and valley sedimentation in the Coon Creek drainage basin. Master's thesis, Department of Geology, University of Wisconsin, Madison.

Meade, R.H. (1982) Sources, sinks, and storage of river sediment in the Atlantic drainage of the United States. *Journal of Geology*, **90**, 235–252.

Meade, R.H. (2007) Transcontinental moving and storage: The Orinoco and Amazon Rivers transfer the Andes to the Atlantic, in *Large Rivers: Geomorphology and Management* (ed. A. Gupta), John Wiley Sons, Chichester, pp. 45–63.

Meade, R.H. and Trimble, S.W. (1974) Changes in sediment loads in rivers of the Atlantic drainage of the United States since 1900. Publication 113, International Association of Hydrological Sciences, Wallingford, pp. 99–104.

Meade, R.H., Dunne, T., Richey, J.E. *et al.* (1985) Storage and remobilization of suspended sediment in the lower Amazon River of Brazil. *Science*, **228**, 488–490.

Melton, F.A. (1936) An empirical classification of flood-plain streams. *The Geographical Review*, **26**, 593–609.

Mertes, L.A.K. and Dunne, T. (2007) Effects of tectonism, climate change, and sea-level change on the form and behaviour of the modern Amazon River and its floodplain, in *Large Rivers: Geomorphology and Management* (ed. A. Gupta), John Wiley and Sons, Chichester, pp. 115–144.

Mertes, L.A.K., Dunne, T. and Martinelli, L.A. (1996) Channel-floodplain geomorphology along the Solimões–Amazon River, Brazil. *Geological Society of America*, **108**, 1989–1107

Montgomery, D. (1997) What's best on banks? *Nature*, **188**, 328–329.

Moody, J.A. and Meade, R.H. (2008) Terrace aggradation during the 1978 flood on Powder River, Montana, USA. *Geomorphology*, **99**, 387–403.

Moody, J.A., Pizzuto, J.E. and Meade, R.H. (1999) Ontogeny of a flood plain. *Geological Society of America Bulletin*, **111**, 291–303.

Nagle, G.N., Fahey, T., Ritchie, J. and Woodbury, P. (2007) Variations in sediment sources and yields in the Finger Lakes and Catskills regions of New York. *Hydrological Processes*, **21**, 828–838.

Nanson, G. (1986) Episodes of vertical accretion and stripping: a model of disequilibrium flood-plain development. *Geological Society of America Bulletin*, **97**, 1467–1475.

Nanson, G. and Croke, J. (1992) A genetic classification of floodplains. *Geomorphology*, **4**, 459–486.

Notebaert, B., Verstraeten, G., Rommens, T. *et al.* (2009) Establishing a Holocene sediment budget for the River Dijle. *Catena*, **77**, 150–163.

Panin, A., Fuzeina, J. and Belyaev, V. (2009) Long-term development of Holocene and Pleistocene gullies in the Protva River basin, central Russia. *Geomorphology*, **108** (1–2), 71–91.

Patton, P. and Boison, P. (1986) Processes and rates of formation of Holocene alluvial terraces in Harris Wash, Escalante River basin, south-central Utah. *Geological Society of American Bulletin*, **97**, 369–378.

Pearce, A.J. and Watson, A. (1986) Effects of earthquake-induced landslides on sediment budget and transport over a 50 year period. *Geology*, **14**, 52–55.

Phillips, J.D. (1986) The utility of the sediment budget concept in sediment pollution control. *Professional Geographer*, **38**, 246–252.

Phillips, J.D. (2001) Sedimentation in bottomland hardwoods downstream of an east Texas dam. *Environmental Geology*, **40**, 860–868.

Phillips, J.D., Marden, M. and Gomez, B. (2006) Residence time of alluvium in an aggrading fluvial system. *Earth Surface Processes and Landforms*, **32**, 307–316.

Piegay, H. (1997) Interactions between floodplain forest and overbank flows: data from three piedmont rivers of southeastern France. *Global Ecology and Biogeography Letters*, **6**, 187–196.

Pizzuto, J.E., Moody, J.A. and Meade, R.H. (2008) Anatomy and dynamics of a floodplain, Powder River, Montana, USA. *Journal of Sedimentary Research*, **78**, 16–28.

Poesen, J. Nachtergaele, T., Verstraeten, G. and Valentin, C. (2003) Gully erosion and environmental change; importance and research needs. *Catena*, **50**, 91–133.

Prosser, I. (1996) Thresholds of channel initiation in historical and Holocene times, Southeastern Australia, in *Advances in Hillslope Processes* (eds M.G. Anderson and S.M. Brooks), John Wiley and Sons, Chichester, pp. 687–708.

Prosser, I., Chappel, J. and Gillespie, R. (1994) Holocene valley aggradation and gully erosion in headwater catchments, South-eastern highlands of Australia. *Earth Surface Processes and Landforms*, **19**, 465–480.

Reeves, G., Hall, J., Roelofs, T. *et al.* (1991) Rehabilitating and modifying stream habitats. *American Fisheries Society Special Publication*, **19**, 519–557.

Reid, L. and Dunne, T. (1996) *Rapid Evaluation of Sediment Budgets*, Catena Verlag, Reiskirchen, Germany.

Reid, L. and Dunne, T. (2003) Sediment budgets as an organizing framework in fluvial geomorphology, in *Tools in Fluvial Geomorphology* (eds G. Kondolf and H. Piegay), John Wiley and Sons, Chichester.

Richards, K.S. (1982) *Rivers: Form and Process in Alluvial Rivers*, Methuen, London.

Robinson, T. (1958) Phreatophytes. *U.S. Geological Survey Water Supply Paper*, 1423.

Robinson, T. (1965) Introduction, spread and areal extend of salt cedar (*Tamarix*) in the western states. *U.S. Geological Survey Professional Paper*, **491A**, 1–12.

Roehl, J. (1962) Sediment source areas, delivery ratios, and influencing morphological factors, Publication 59, International Association of Hydrological Sciences, Wallingford, pp. 202–213.

Rommens, T., Verstraeten, G., Bogman, P. *et al.* (2006) Holocene alluvial sediment storage in a small river catchment in the loess area of central Belgium. *Geomorphology*, **77**, 187–201.

Rommens, T., Verstraeten, G., Peeters, I. *et al.* (2007) Reconstruction of late-Holocene slope and dry valley sediment dynamics in a Belgian loess environment. *The Holocene*, **17**, 777–788.

Salo, J., Kalliola, R., Haekkenen, I. *et al.* (1986) River dynamics and the diversity of Amazon lowland forest. *Nature*, **322**, 254–258.

Schalk, G.K. and Jacobson, R.B. (1977) Scour, sedimentation, and sediment characteristics at six levee-break sites in Missouri from the 1993 Missouri River flood. *U.S. Geological Survey Water-Resources Investigations Report*, 97-4110.

Schick, A.P. (1977) A tentative sediment budget for an extremely arid watershed in the southern Negiv, in *Geomorphology in Arid Regions* (ed. D. Doering), State University of New York, Binghampton, Publications in Geomorphology, pp. 139–163.

Schick, A.P. and Lekach, J. (1993) An evaluation of two ten-year sediment budgets. Nahel Yael, Israel. *Physical Geography*, **14**, 225–238.

Schulte, L., Veit, H., Burjachs, F. and Jula, R. (2009) Lütschine fan delta response to climate variability and land use in the Bernese Alps during the last 2400 years. *Geomorphology*, **108** (1–2), 107–121.

Schumm, S.A. (1973) Geomorphic thresholds and complex response of drainage systems, in *Fluvial Geomorphology* (ed. M. Morisawa), New York State University Pubs., Binghampton, pp. 299–310.

Schumm, S.A. (1977) *The Fluvial System*, John Wiley and Sons, New York.

Schumm, S.A. and Rea, D. (1994) Sediment yield from disturbed earth systems. *Geology*, **23**, 391–394.

Simon, A. (1989) A model of response in disturbed alluvial channels. *Earth Surface Processes and Landforms*, **14**, 11–26.

Simon, A. and Collison, A. (2002) Quantifying the mechanical and hydrologic effects of riparian vegetation on streambank stability. *Earth Surface Processes and Landforms*, **27**, 527–546.

Slaymaker, O. (1993) The sediment budget of the Lillooet River Basin, British Columbia. *Physical Geography*, **14**, 305–320.

Slaymaker, O. (2003) The sediment budget as conceptual framework and management tool. *Hydrobiologia*, **494**, 71–82.

Smith, R. and Swanson, F. (1987) Sediment routing in a small drainage basin in the blast zone of Mount St. Helens, Washington, USA. *Geomorphology*, **1**, 1–13.

Smith, S.V., Renwick, W.H., Buddenmeier, R. and Crossland, C. (2001) Budgets of soil erosion and deposition for sediemtns and sedimentary organic carbon across the contierminous United States. *Global Biogeochemical Cycles*, **15**, 697–707.

Sternberg, H.O. (1975) The Amazon River of Brazil. *Geographischie Zeitschift Beihefte*, **40**, 74.

Stott, T. (1997) A comparison of stream bank erosion processes on forested and moorland streams in the Balquhidder catchments, central Scotland. *Earth Surface Processes and Landforms*, **22**, 383–399.

Trimble, S. W. (1969) Culturally accelerated sedimentation on the middle Georgia piedmont, MA Thesis, University of Georgia, Athens. Reprinted and distributed in 1970 by USDA Soil Conservation Service, Fort Worth, Texas.

Trimble, S.W. (1974, 2008) *Man-Induced Soil Erosion on the Southern Piedmont, 1700–1970*, Soil and Water Conservation Society of America, Ankeny, Iowa, Enhanced edition, 2008.

Trimble, S.W. (1975) Denudation studies: can we assume stream steady state. *Science*, **188**, 1207–1208.

Trimble, S.W. (1976) Unsteady state denudation. *Science*, **191**, 871.

Trimble, S.W. (1981) Changes in sediment storage in the Coon Creek basin, Driftless Area, Wisconsin, 1853 to 1975. *Science*, **214**, 181–183.

Trimble, S.W. (1983) A sediment budget for Coon Creek basin in the Driftless Area, Wisconsin, 1853-1977. *American Journal of Science*, **283**, 454–474.

Trimble, S.W. (1993) The distributed sediment budget model and watershed management in the Paleozoic Plateau of the upper Midwestern United States. *Physical Geography*, **14**, 285–303.

Trimble, S.W. (1994) Erosional effects of cattle on streambanks in Tennessee, USA. *Earth Surface Processes and Landforms*, **19**, 451–464.

Trimble, S.W. (1995) Catchment sediment budgets and change, in *Changing River Channels* (eds A. Gurnell and G. Petts), John Wiley and Sons, Chichester.

Trimble, S.W. (1997a) Stream channel erosion and change resulting from riparian forests. *Geology*, **25**, 467–469.

Trimble, S.W. (1997b) Streambank fish-shelter structures help stabilize tributary streams in Wisconsin. *Environmental Geology*, **32**, 230–234.

Trimble, S.W. (1997c) Contribution of Stream channel erosion to sediment yield form an urbanizing watershed. *Science*, **279**, 1442–1444.

Trimble, S.W. (1999) Decreased rates of alluvial sediment storage in the Coon Creek basin, Wisconsin, 1975–93. *Science*, **285**, 1244–1246.

Trimble, S.W. (2004) Effects of riparian vegetation on stream channel stability and sediment budgets, in *Riparian Vegetation and Fluvial Geomorphology* (eds S.J. Bennett and A. Simon), Water Science and Application 8, American Geophysical Union, Washington, DC, pp. 153–169.

Trimble, S.W. (2009) Fluvial processes, morphology, and sediment budgets in the Coon creek basin, WI, USA, 1975–1993. *Geomorphology*, **108** (1–2), 8–23.

Trimble, S.W. and Crosson, P. (2000) U.S. soil erosion rates – myth and reality. *Science*, **289**, 248–250.

Trimble, S.W. and Lund, S.W. (1982) Soil conservation and the reduction of erosion and sedimentation in the Coon Creek basin, Wisconsin. *U.S. Geological Survey Professional Paper*, **1234**.

Trimble, S.W. and Mendel, A.C. (1995) The cow as geomorphic agent – a critical review. *Geomorphology*, **13**, 233–253.

Verstraeten, G., Lang, A. and Houben, P. (2009) Human impact on sediment dynamics-quantification and timing. *Catena*, **77**, 77–80.

Verstraeten, G., Rommens, T., Peeters, I. *et al.* (2009) A temporarily changing Holocene sediment budget for a loess-covered catchment (central Belgium). *Geomorphology*, **108** (1–2), 24–34.

Vita-Finzi, C. (1969) *The Mediterranean Valleys*, University Press, Cambridge.

Walling, D.E. (1983) The sediment delivery problem. *Journal of Hydrology*, **65**, 209–237.

Walling, D.E. (1990) Linking the field to the river, in *Soil Erosion on Agricultural Land* (eds J. Boardman, I. Foster and J. Dearing), John Wiley and Sons, Chichester, pp. 129–152.

Walling, D.E. and Collins, A. (2008) The catchment sediment budget as a management tool. *Environmental Science and Policy*, **11**, 136–143.

Ward, A.D. and Trimble, S.W. (2004) *Environmental Hydrology*, Lewis Publishers, Boca Raton, Florida.

Williams, G.P. (1978) The case of the shrinking channels – the North Platte and Platte Rivers in Nebraska. *U.S. Geological Survey Circular*, 781.

Williams, G.P. and Wolman, M.G (1984) Downstream effects of dams on alluvial rivers. *U.S. Geological Survey Professional Paper*, **1286**.

Wolman, M.G. (1967) A cycle of sedimentation in urban river channels. *Geografiska Annaler*, **49A**, 385–395.

Wolman, M.G. and Eiler, J. (1958) Reconnaissance study of erosion and deposition produced by the flood of August, 1955, in Connecticut. *American Geophysical Union Transactions*, **39**, 1–14.

Wolman, M.G. and Leopold, L.B (1957) River floodplains: some observations on their formation. *U.S. Geological Survey Professional Paper*, **282C**, 87–107.

Zimmerman, R., Goodlet, J. and Comer, G. (1967) The influence of vegetation on channel form of small streams. Publication 75, International Association of Hydrological Sciences, Wallingford, pp. 255–275.

Zygmunt, E. (2009) Alluvial fans as an effect of long-term man–landscape interactions and moist climatic conditions: A case study from the Glubczyce Plateau, SW Poland. *Geomorphology*, **108** (1–2), 58–70.

12

Lakes and Reservoirs in the Sediment Cascade

Ian D.L. Foster[1,2]

[1]Westminster Water Science, Department of Molecular and Applied Biosciences, University of Westminster, Cavendish Campus, 115 New Cavendish Street, London W1W 6UW, UK

[2]Geography Department, Rhodes University, Grahamstown 6140, Eastern Cape, South Africa

12.1 Introduction

This chapter examines the origins and history of lake and reservoir basins. It considers their development and longevity, and their possible impact on the transfer of sediment from river catchments to the oceans. It will also evaluate the impact of reservoirs on the fluvial system both upstream and downstream of the impoundment and will discuss the value of sedimentary archives contained within these basins for providing information on geomorphological processes operating in river catchments. It will also briefly consider the possible role that palaeoenvironmental data can play in the establishment of environmental quality objectives.

12.2 Lake and Reservoir Basins: Origins and History

Natural or artificially constructed freshwater sedimentary basins exist in almost all parts of the world. They are transitory features at the Earth's surface, having a birth, life and death related to geological, geomorphological and/or biological processes. In some cases natural basins have been in existence for several tens of thousands or even millions of years (e.g. Lakes Baikal, Siberia and Biwa, Japan). In middle to high latitudes, however, most natural lake basins were created during the wasting of ice sheets and glaciers at the end of the Pleistocene cold period and their sedimentary records therefore span only the past *c.* 8000–10 000 years. Many lakes at low latitudes have existed and disappeared

Sediment Cascades: An Integrated Approach Edited by Timothy Burt and Robert Allison

more than once over Quaternary timescales, such as those of the Mojave drainage system of southern California, where a number of temporary lakes formed and drained between the San Bernardino Mountains and Death Valley over the past *c.* 30 000 years (Meek, 1999; Reheis and Redwine, 2008). Changes in the distribution and extent of lake basins over Quaternary timescales at low latitudes in Africa, Asia, the Americas and Australia have also been identified from an analysis of sedimentary deposits and the extent of former shorelines (see Anderson, Goudie and Parker (2007) for a review). These currently dry basins contain stores of fine sediment that may be reworked for periods of thousands or even tens of thousands of years. Other temporary lakes have been produced, for example, as a result of volcanism or rapid uplift at plate margins. These events often induce landslides that can block drainage systems and can lead to the development and destruction of natural lake basins over a range of timescales (e.g. Korup, 2002; Dunning *et al.*, 2006; see also Chapter 4, this volume).

Reservoirs have been constructed for water storage, for flood regulation and for the generation of hydroelectric power (HEP). The earliest dated reservoir for river regulation is from *c.* 3000 BC in Egypt (Smith, 1971) but dams became widespread in the Mediterranean area by Roman times. By the late Middle Ages, small dams were being constructed throughout Europe (Beckinsale, 1972). The nineteenth and twentieth centuries saw the most rapid expansion of dam construction, and most of the world's major rivers are now regulated through reservoirs (see Smith, 1971; Beaumont, 1978; Petts, 1984; McCully, 1996; WCD, 2000; Kalff, 2002; Walling, 2008).

Not only do lakes and reservoirs exist in almost all climatic zones, they are located in a variety of positions within the drainage basin. Figure 12.1 provides a simplified schematic illustration of the possible location of natural lakes, ponds and reservoirs in a hypothetical

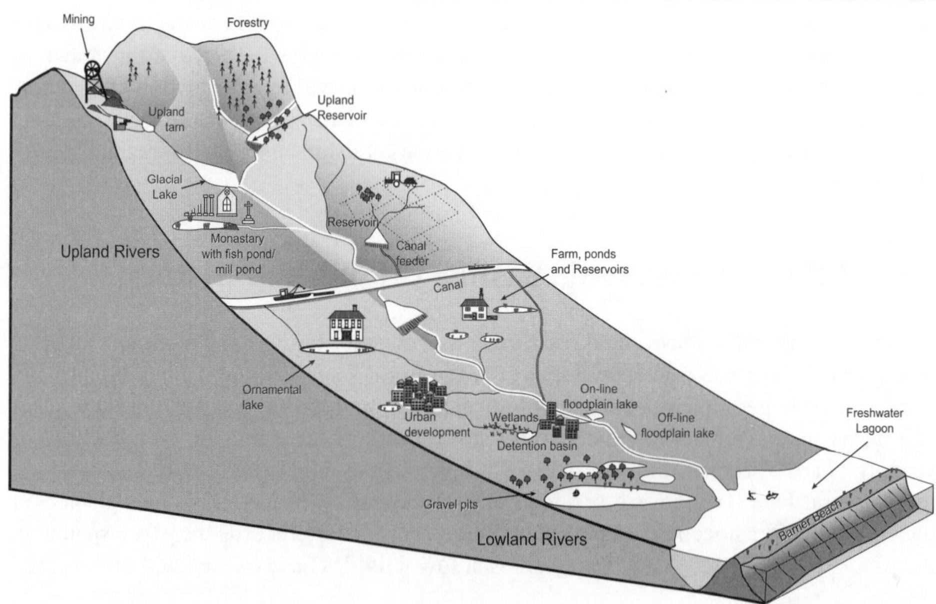

Figure 12.1 Lakes and reservoirs in the catchment sediment cascade; a hypothetical view of a temperate mid-latitude drainage basin

mid-latitude drainage basin. The scheme includes small tarns and glacial lake basins, upland water supply, river regulating and HEP generating reservoirs, canal feeder reservoirs, ornamental lakes, urban detention basins, fish and mill ponds, farm ponds and reservoirs, and freshwater or brackish lagoons created in coastal locations behind barrier beaches or wind-blown sand deposits. On river floodplains, small meander cut-off lakes are common and were defined by Foster (2006) as *off-line* if they were only connected to the river at periods of high flow (e.g. above bankfull) and *on-line* if they were in direct contact with the river at flow stages below bankfull. In both cases, a reduction in flow velocity will allow sediment to settle on the floodplain lake bed. The sediments contained in these small depressions will preserve at least a partial history of sediment and sediment-associated contaminant transport and will also provide opportunities for palaeoflood reconstruction (e.g. Foster and Charlesworth, 1996; Winter *et al.*, 2001; Paine, Rowan and Werritty, 2002; Wolfe *et al.*, 2006).

The hydrology and sedimentology of reservoirs and natural lakes differ significantly in several respects (Morris and Fan, 1997; Kalff, 2002; Shotbolt, Thomas and Hutchinson, 2005). In comparison with reservoirs, lakes are older, tend to fill natural depressions rather than river valleys, are dendritic and not regular in shape, and have a maximum depth near the centre rather than close to the outfall. Natural lakes have wind rather than flow-driven longitudinal gradients, have a shallower outfall depth, a longer water retention time and experience smaller water-level fluctuations. The latter may have significant implications for accelerated shoreline erosion and the resuspension of sediment in the shallow-water zone at maximum drawdown, both of which will have a significant impact on the reconstruction of sediment yields and sources (see below). Natural lake sediments are also more likely to contain a greater allochthonous contribution than reservoirs.

12.3 Impact of Reservoirs on Sediment Transport and Channel Adjustment

In all of the situations shown in Figure 12.1, the episodic movement of fine and/or coarse sediment from uplands to lowlands is interrupted. In Europe, Spain, the UK, France and Italy have each constructed over 400 major dams, with Spain having expanded dam construction particularly since the 1950s to a total of over 750 (EEA, 1999). In a recent review of the sediment yields of the world's major rivers, Walling (2008) identified dam construction as a key driver of reductions in global sediment delivery to the oceans. For example, construction of the Aswan High Dam has reduced the sediment load of the Nile from *c.* 100 Mt yr^{-1} to effectively zero. The Colorado and Rio Grande Rivers of the southwestern USA have seen their sediment loads decrease by 100% and 96% respectively as a result of dam construction (Vörösmarty *et al.*, 2003), while the annual runoff and sediment yield of the Indus has declined to *c.* 20% of its 1940s levels as a result of the development of irrigation schemes, including dam construction (Walling, 2007). In Europe, the sediment load of the Danube has decreased by *c.* 30% of its 1950 levels, but in this case there has been no significant reduction in annual runoff (Walling, 2006).

There is some uncertainty regarding the total amount of sediment trapped in reservoirs and of its global significance. Syvitski *et al.* (2005) suggested that global sediment yields have decreased by 3.6 Gt yr^{-1}, effectively reducing the contemporary gross flux of the world's rivers due to reservoir trapping by *c.* 16%. Walling (2008) disputed this calculation and suggested that reservoir sediment trapping accounted for *c.* 24 Gt yr^{-1}, effectively reducing the contemporary gross sediment flux of the world's rivers by *c.* 66%.

Many storage reservoirs constructed in regions of high sediment yield over the past 50 years have lost a large proportion of their initial storage capacity because of sedimentation and, in extreme cases, rates of sedimentation approaching *c.* 1 m yr^{-1} have been reported (e.g. Lahlou, 1996; Woodward and Foster, 1997). In the countries of North Africa, water reserves in reservoirs have reduced at a rate of 2–5% per year due to erosion and sedimentation (Fox *et al.*, 1997). These rapid rates of sedimentation suggest that storage reservoirs will have a relatively short-lived impact on sediment export from drainage basins in regions of the world where sediment yields are high.

Similar behaviour is reported in small reservoirs. Foster, Boardman and Keay-Bright (2007) and Foster, Boardman and Gates (2008), for example, report rapid sedimentation in farm reservoirs located in the Eastern Cape of South Africa. Earth dams constructed over the past *c.* 100–150 years are now full or almost full of sediment. Many have breached and remain unrepaired and will likely contribute to sustained high sediment yields until the reservoir basins are completely evacuated of sediment. Small dams are widespread in semi-arid areas of Europe, Africa, Asia and Australia and have often been used to estimate sediment yields or the magnitude and frequency of event-based sediment transport from their contributing small catchments (e.g. Laronne, 1990; Verstraeten and Poesen, 2002; Xiang-Zhou, 2004; Haregeweyn *et al.*, 2006; Tamene *et al.*, 2006; Boix-Fayos *et al.*, 2007; Xu and Milliman, 2009). The latter study calculated that, before the closure of the Three Gorges Dam (TGD) in China, sediment yields had already decreased by *c.* 60% and the hysteresis of seasonal rating curves in the upper reaches at Yichang station had shifted from clockwise to anticlockwise as a result of the construction of *c.* 50 000 small dams and the extensive reafforestation of the Loess Plateau. During the four years (2003–2006) after TGD impoundment, *c.* 60% of sediment entering the Three Gorges Reservoir was trapped, primarily during the high-discharge months (June–September).

In-channel structures, such as check-dams, are also widespread globally and create segmented longitudinal river profiles. They alter suspended and bedload sediment dynamics, the stability of river beds and banks, and may also change the characteristics of channel flood waves (Wohl, 2006). Check-dams have significant morphological and sedimentological effects on the riverbed (Boix-Fayos *et al.*, 2007) that can also change the hydraulic behaviour of the flow in extreme events. Many studies have shown that they interrupt sediment transport and decrease sediment yields (e.g. Simon and Darby, 2002; Martin-Rosales *et al.*, 2003; Surian and Rinaldi, 2003 but there are also suggestions that check-dams induce localized erosion (Martin-Rosales *et al.*, 2003; Castillo *et al.*, 2007), thereby affecting the sediment budget at the reach scale. In-channel debris dams may also cause similar effects to those brought about by the construction of check dams although they are usually more transient features. Woody debris is often delivered directly to the stream by wind-dominated processes that dislodge twigs and branches and occasionally lead to tree fall. Occasionally, woody debris will produce a watertight structure that will change the reach long profile and induce upstream sedimentation and/or downstream scour. In North America, the construction activities of beavers may lead to more permanent structures that also alter the long profile, the flow dynamics and sediment budgets of river reaches (Gurnell, Gregory and Petts, 1995; Gurnell, 1998; May and Gresswell, 2003; see also Chapter 2, this volume).

Grant, Schmidt and Lewis (2003) recently noted that, despite decades of research and numerous case studies on the downstream effects of dams on rivers, we have few general models predicting how a river is likely to adjust after impoundment. They point out that dams alter two critical elements in the downstream geomorphological system; the ability of the river to transport sediment and the amount of sediment available for transport. If transport capacity exceeds supply, a sediment deficit exists and the channel is predicted to evacuate sediment from its bed and/or banks. If the sediment transport capacity is lower than the sediment supply, the channel can be expected to accumulate sediment. These adjustments will include channel cross-section, bed material, planform and gradient but the exact response of a channel to sediment deficit or surplus varies (Brandt, 2000; Shields, Simon and Steffen, 2000). Typical downstream responses can include one or a combination of channel bed degradation or incision, textural changes such as coarsening or fining of bed sediment particle-size distributions, and lateral adjustments that are likely to involve both the expansion and contraction of channel width. Grant, Schmidt and Lewis (2003) also proposed a new methodology to predict the likely trajectory of rivers downstream of dams in response to specific changes in particle size and/or flow conditions and argued that this methodology offered river managers a more comprehensive way of evaluating trade-offs between management goals.

Upstream effects of reservoirs on sedimentation are less well documented.

Geomorphological theory would suggest that upstream rivers and streams would adjust to a new base level and would change their mainstream and tributary gradients as a result of sediment deposition in their lower reaches (Xu, 2001). This process increases the elevation of the riverbed and is likely to induce flooding upstream. Evidence for flooding and sedimentation upstream of a lake has been reported at Slapton Ley, Devon, England where the artificial raising of the overspill weir of Slapton Ley appears to have induced rapid sedimentation for at least 1 km upstream of the lake inflow in the Start valley and which has also led to periodic flooding of local roads (Owens, 1990; Foster, Owens and Walling, 1996; Owens *et al.*, 1997). Analysis of floodplain sediment cores in the latter two studies showed that ^{137}Cs was detectable to depths of over 70 cm in floodplain sediments (see sediment budgets below). A more extreme example is given by Eakin and Brown (1939: cited by Trimble and Wilson, in press). The town of San Marcial was located upstream of Elephant Butte Reservoir, which was completed in 1916 on the Rio Grande River in New Mexico. Despite the fact that the town was located 2 km upstream of the reservoir, and at several metres higher elevation, it was partially flooded and eventually evacuated by the 1930s as a result of sedimentation and increased flooding.

12.4 Reconstructing Environmental Change

While the origin of palaeolimnology is strongly rooted in the biological sciences, Oldfield (1977) argued that lake sediments preserved records of inputs from contributing catchments and could provide an uninterrupted record of catchment change since the time of lake formation or reservoir construction. Research since the publication of this key paper has seen a major growth in the use of palaeoenvironmental reconstruction for a number of geomophological purposes that include the reconstruction of sedimentation rates and sediment yields to elucidate the relative significance of climate change and human activity. Other research has reconstructed the history of river and coastal flooding (storm surges and tsunamis), atmospheric dust accumulation, the frequency of volcanic tephra production, earthquake-induced landslide activity, gullying, blanket peat erosion and the reworking of sediment by the rapid rise and fall in reservoir water levels (see Foster *et al.*, 1991; Hyatt and Gilbert, 2000; Zolitschka *et al.*, 1998; Banerjee, Murray and Foster, 2001; Edwards and Whittington, 2001; Nesje *et al.*, 2001; Lamoureux, 2002; Yeloff *et al.*, 2005; Foster *et al.*, 2006; Wagner *et al.*, 2007; Bertrand, Castiaux and Juvigné, 2008; Chiverrell *et al.*, 2008; Coombes, Chiverrell and Barber, 2008; Couch and Eyles, 2008; Holliday, Warburton and Higgitt, 2008; Koi *et al.*, 2008; Rowntree *et al.*, 2008; Chu *et al.*, 2009). Beyond direct applications in geomorphology, reconstructing environmental change using palaeoecological remains and chemical characteristics of deposited sediment provides a powerful methodology for understanding and quantifying the recent pollution histories of rivers and lakes (see Smol, 2008).

The development of a methodology to reconstruct sediment yield and sediment-associated contaminant histories has enabled geomorphologists to quantify the magnitude of the impacts of climate change and human disturbance on catchment systems over the past century (Foster, 1995; Foster and Lees, 1999a; Foster *et al.*, 2003; Foster, 2006; Foster, Boardman and Gates, 2008) and over much longer periods of time throughout the Holocene and late Quaternary (Likens and Davis, 1975; Oldfield, Worsley and Appleby, 1985; Dearing, 1992; O'Hara, Alayne Street-Perrot and Burt, 1993; Page and Trustram, 1997; Zolitschka, 1998; Edwards and Whittington, 2001; Foster *et al.*, 2008; Pittam, Foster and Mighall, 2009). Interpretations of catchment disturbance associated with changes in sediment sources have been refined by examining the physical, geochemical, mineral magnetic and radionuclide signatures of deposited sediments in comparison with potential catchment sources; a methodology described as 'forensic geomorphology' by Foster (2000). This approach has also benefited from the development of a range of methods for qualitative and quantitative sediment source fingerprinting (Walling, Woodward and Nicholas, 1993; Walden, Slattery and Burt, 1997; Foster *et al.*, 1998, 2002; Heppell *et al.*, 2002; Small, Rowan and Franks, 2002; Foster, 2006; Rowntree *et al.*, 2008). Furthermore, by linking lake-sediment-based studies to other studies of erosion and sediment transport in drainage basins, sediment budgets can be reconstructed over similar timescales (see below).

Sediment yield reconstruction requires the estimation of the total mass of sediment accumulating in a lake or reservoir basin. The following section reviews methods that have been developed in order to calculate a time-averaged yield for the life of the basin and/or to subdivide this record into shorter time intervals.

12.4.1 Time-Averaged Sediment Yields

Time-averaged sediment yields for reservoirs can be obtained by either of the following two approaches.

1. Estimating the water storage loss between the date of construction and subsequent capacity estimates based on bathymetric resurvey and measuring or estimating sediment density in order to calculate sediment mass (Rowan, Goodwill and Greco, 1995; Lahlou, 1996; White, Labadz and Butcher, 1996, 1997; Zorzou, 2004; Ferrari, 2006).

2. Ground surveys of sediment volume and density during natural or deliberate draw-down (see Duck and McManus, 1987; Verstraeten and Poesen, 2002; Haregeweyn *et al.*, 2006; Tamene *et al.*, 2006; Foster, Boardman and Keay-Bright, 2007).

These two approaches provide rapid reconnaissance methods to obtain local or regional estimates of sediment yield. The use of Global Positioning Systems directly linked to high-resolution echo sounding has significantly improved the accuracy of these resurvey techniques (Ferrari, 2006). However, they suffer four major disadvantages in that first, the original storage capacity estimates may be questionable or unknown; second, sediment yields are averaged over variable time periods depending on the date of reservoir construction and date(s) of resurvey; third, it may not be possible to link temporal patterns in sediment yield to changes in climate or catchment management; and fourth, estimates of sediment density are still required in order to calculate sediment mass. The major advantage of these methods is that they integrate estimates of sediment yield over several years, thereby eliminating problems of annual variability that frequently appear in river data (Reid and Ogdentrend, 2006).

Of major importance in reconstructing temporal changes in the sediment yield record has been the development of reliable methods for dating lake and reservoir sediments.

12.4.2 Dating Sedimentary Sequences

Several methods have been used for dating lake sediment sequences. Some minerals, especially quartz and feldspar, trap electrons as a result of radioactive decay of surrounding cosmogenic or primordial nuclides. These traps are emptied on exposure to sunlight, which resets the store to zero. Analytically, these traps are emptied by heating or exposure to light. This is known as luminescence. The amount of luminescence emitted is proportional to the accumulated dose since the minerals were last exposed to sunlight during sediment transport. At present, the technique is largely applicable to sediments in the sand size fraction. The decay of several radionuclides, including ^{14}C, ^{32}Si and ^{210}Pb, has also been used to date sedimentary deposits over timescales ranging from decades to thousands of years, while other radionuclides provide chronological markers relating to atmospheric fallout (e.g. ^{137}Cs). Incremental methods of dating include the use of annually laminated sediments (varve chronologies), tephra layers, biostratigraphic markers (e.g. pollen and diatoms), spherical carbonaceous particles (SCPs) and other pollution histories (e.g. heavy metals from atmospheric pollution), and palaeomagnetism. These have added to the number of available techniques for developing independent chronologies or for confirming the chronology provided by radiometric dating methods over similar timescales.

Laminated sediments are generally caused by the alternate deposition of organic- and minerogenic-rich layers in the form of annual couplets. The thickness of the minerogenic layer is used to estimate the annual influx of sediment from a catchment. In environments where laminations are shown to be annual, they provide precise estimates of sediment age (see Håkanson and Jansson, 1983; Saarnisto, 1986; Gilbert and Desloges, 1987; Boygle, 1993; Desloges, 1994; Zolitschka, 1998; Smith, Bradley and Abbot, 2004; Bogen, 2008).

In most environments, laminations are either not annual or are not preserved. Here, a number of radiometric methods have been used to date sediment sequences including ^{14}C dating of Holocene sedimentary records, luminescence dating, where particles of appropriate size and mineralogy are preserved in the sediments, and the use of cosmogenic ^{32}Si dating (half life *c.* 178 years) in sediments rich in biogenic silica (e.g. Likens and Davis, 1975; Nijampurkar *et al.*, 1998; Banerjee, Murray and Foster, 2001; Reimer *et al.*, 2004; Pittam, Foster and Mighall, 2009). Despite the appeal of ^{32}Si to bridge the gap between dates provided by ^{14}C and ^{210}Pb dating, the lengthy preparation procedure, long count times and the high cost of analysis has precluded its widespread application (Morgenstern, personal communication, 2004). Incremental methods for dating lake sediments include the use of tephra chronologies, pollen markers, SCPs and palaeomagnetism (e.g. Rose *et al.*, 1995; Vukic and Appleby, 2005; Walker, 2005; Smol, 2008; Pittam, Foster and Mighall, 2009).

The main focus of this chapter is on the past *c.* 100 years of environmental change where ^{137}Cs and ^{210}Pb dating methods have been used routinely for dating sedimentary sequences. The application of these dating methods has been reviewed recently by Foster (2006) and only a brief overview with examples will be presented here.

Caesium-137 has been used for dating sedimentary sequences since the 1970s (Pennington, Cambray and Fisher, 1973; Ritchie, McHenry and Gill, 1973). It is a radionuclide that was first released into the environment as a result of atmospheric thermonuclear weapons testing in the early 1950s and has a half-life of *c.* 30 years. The pattern of atmospheric fallout, peaking in 1963 (prior to the atmospheric test-ban treaty), and in some areas in 1986 (from fallout after the Chernobyl nuclear accident) often provide clear marker horizons in sedimentary sequences (Figure 12.2). The majority of examples presented in Figure 12.2 show clearly identifiable peaks in activity, but rapid increases in sedimentation, and the consequent dilution of ^{137}Cs by an influx of subsoil bearing no ^{137}Cs, may make dating using the ^{137}Cs record alone problematic (Lees *et al.*, 1997; Foster and Lees, 1999a; Foster, Boardman and Keay-Bright, 2007). Furthermore, the ^{137}Cs profile for Widdicombe Ley (Figure 12.2b), a shallow freshwater coastal lagoon in South Devon, UK, shows some evidence of upward migration of ^{137}Cs in the lake sediments and does not produce distinctive dateable horizons despite the fact that the total inventory for the lake sediment core is not significantly different from the regional atmospheric fallout inventory. It has recently been reported that the upward migration of ^{137}Cs in this and other shallow

Figure 12.2 Caesium-137 in: (a) atmospheric fallout, Oxford, UK, 1954–1992 (original measured fallout and decayed fallout to 1992) (reprinted from Foster, I. D. L. 'Lake and Reservoir Bottom Sediments' in Foster, I. D. L, Gurnell, A. M. & Webb, B. W., Sediment and Water Quality in River Catchments 265–283 © 1995 with permission from John Wiley & Sons Ltd.). (b) Widdicombe Ley, a natural freshwater coastal barrier lagoon, South Devon, UK. (c) The Old Mill water supply reservoir, South Devon, UK, constructed 1942 (after Foster, I. D. L. and Walling, D. E. 'Using Reservoir Deposits to Reconstruct Changing Sediment Yields and Sources' Hydrological Sciences Journal 39 347–368 © 1994 Reproduced with permission of IAHS Press). (d) Seeswood Pool, canal feeder reservoir, north Warwickshire, UK constructed 1765 (after Foster, I. D. L. *et al.* 'Quantifying Soil Erosion and Sediment Transport in Drainage Basins' in Variability in Stream Erosion and Sediment Transport, Proc. Canberra Symposium, 244, 55–64 © 1994. Reproduced with permission of IAHS Press). (e) Fillingham Lake, Lincolnshire, UK, a 'Capability Brown' ornamental lake, constructed 1790, and (f) Silsden water supply reservoir, North Yorkshire, UK, constructed 1867 (after Foster, I. D. L. and Lees, J. A. 'Changing Headwater Suspended Sediment Yields' Hydrological Processes, 13, 1137–1153, © 1999. Reproduced with permission from John Wiley & Sons Ltd.). (g) River Pool, River Avon valley (Brandon Marshes, Warwickshire, UK), an on-line floodplain lake created by mining subsidence in 1952

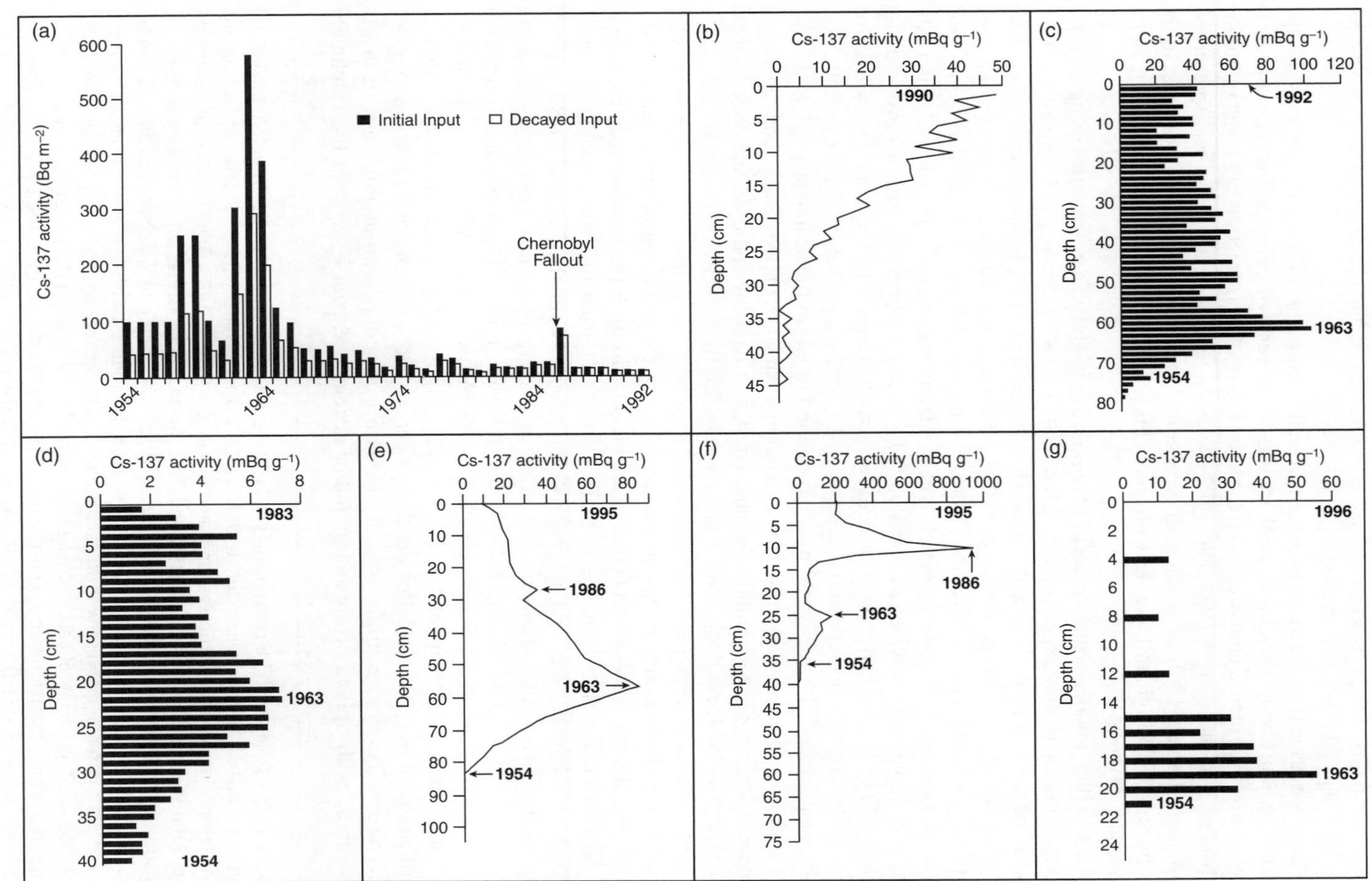
(a)
Cs-137 activity (Bq m−2)
Initial Input
Decayed Input
Chernobyl Fallout
1954
1964
1974
1984
1992
(b)
Cs-137 activity (mBq g−1)
Depth (cm)
1990
(c)
Cs-137 activity (mBq g−1)
Depth (cm)
1992
1963
1954
(d)
Cs-137 activity (mBq g−1)
Depth (cm)
1983
1963
1954
(e)
Cs-137 activity (mBq g−1)
Depth (cm)
1995
1986
1963
1954
(f)
Cs-137 activity (mBq g−1)
Depth (cm)
1995
1986
1963
1954
(g)
Cs-137 activity (mBq g−1)
Depth (cm)
1996
1963
1954

coastal lagoons in Southwest England is probably driven by sea water incursion through the basal sediments of the Ley as a result of tidal fluctuations and pressure differences between sea level and lake level (Foster *et al.*, 2006).

Despite seasonal fluctuations in water levels, the sediments of River Pool (Figure 12.2g), a small *on-line* floodplain lake on the Warwickshire River Avon, central England, appears to preserve a ^{137}Cs record consistent with the known date of its formation as a result of mining subsidence in the early 1950s. The 1963 weapons testing peak is also well preserved, although Chernobyl fallout is not well represented in Midland England in contrast to the sediments of Fillingham Lake, Lincolnshire (Figure 12.2e) and Silsden Reservoir, North Yorkshire (Figure 12.2f).

Lead-210 is produced from the decay of radioactive ^{238}U. It can be measured using the same instrumentation (gamma spectrometry; Murray *et al.*, 1987; Gilmore and Hemingway, 1995) and at the same time as ^{137}Cs and can also be measured using alpha spectrometry. Goldberg (1963) first suggested the possible use of ^{210}Pb for dating but it did not come into common use until about a decade later. ^{210}Pb derives from the escape of radon gas from the Earth's crust and its subsequent decay through a series of short-lived radionuclides to ^{210}Pb. As ^{210}Pb has a half-life of *c.* 22 years, it provides an opportunity for dating sedimentary deposits over the past *c.* 100–120 years (5–6 half-lives) depending on activities within the sediment. The calculation of sediment age requires separating the supported ^{210}Pb (in equilibrium with the parent ^{226}Ra) and unsupported ^{210}Pb (delivered by atmospheric fallout) in the sediments.

Depth–age curves are constructed most commonly using one of two models (constant rate of supply, 'CRS'; or constant initial concentration, 'CIC'; Appleby and Oldfield, 1978; Oldfield and Appleby, 1984; Appleby *et al.*, 1986; Appleby, 2001; Appleby *et al.*, 2003).

While the measurement and calculation of sediment age using ^{210}Pb is more problematic than the identification of dateable horizons using ^{137}Cs, it often allows sediments to be dated over the entire twentieth century and, on occasions, to the late-nineteenth century; thereby extending the chronology obtained from ^{137}Cs dating. The examples given in Figure 12.3 demonstrate; first, potential differences between the CIC and CRS modelled depth–age curves and sedimentation rates in two cores from Merevale Lake, north Warwickshire (Figure 12.3a); second, the consistency in ^{210}Pb depth–age correlation between two cores from Seeswood Pool, north Warwickshire, which is supported by independent mineral magnetic and ^{137}Cs data (Figure 12.3b); and, third, the potential application of the CRS modelled depth–age curves and sedimentation rates in a range of UK reservoirs. For example, recent increases in sedimentation rates are probably associated with upland afforestation and/or accelerated peat erosion in the catchment

Figure 12.3 Example ^{210}Pb chronologies in (a) Merevale Lake, North Warwickshire, UK, ornamental lake, constructed 1880 (after Foster, I. D. L. *et al.* 'Lake Catchment Based Studies of Erosion and Denudation', Earth Surface Processes and Landforms, 10, 45–68 © 1985. Reproduced with permission from John Wiley & Sons Ltd.). (b) Seeswood Pool, north Warwickshire, UK, canal feeder reservoir, constructed 1765 (after Foster, I. D. L. *et al* 'Historical Trends in Catchment Sediment Yields', Hydrological Sciences Journal, 31, 427–443, © 1986, Reproduced with permission from IAHS Press). (c) Fontburn water supply reservoir, Northumberland, UK, constructed 1905, (d) Elleron Lake, North York Moors, UK, ornamental parkland lake, constructed 1919, and (e) Barnes Loch, army camp water supply reservoir, Scottish Borders, constructed 1915 (after Foster, I. D. L. and Lees, J. A. 'Changing Headwater Suspended Sediment Yields' Hyrdological Processes, 13, 1137–1153, © 1999. Reproduced with permission from John Wiley & Sons Ltd.).

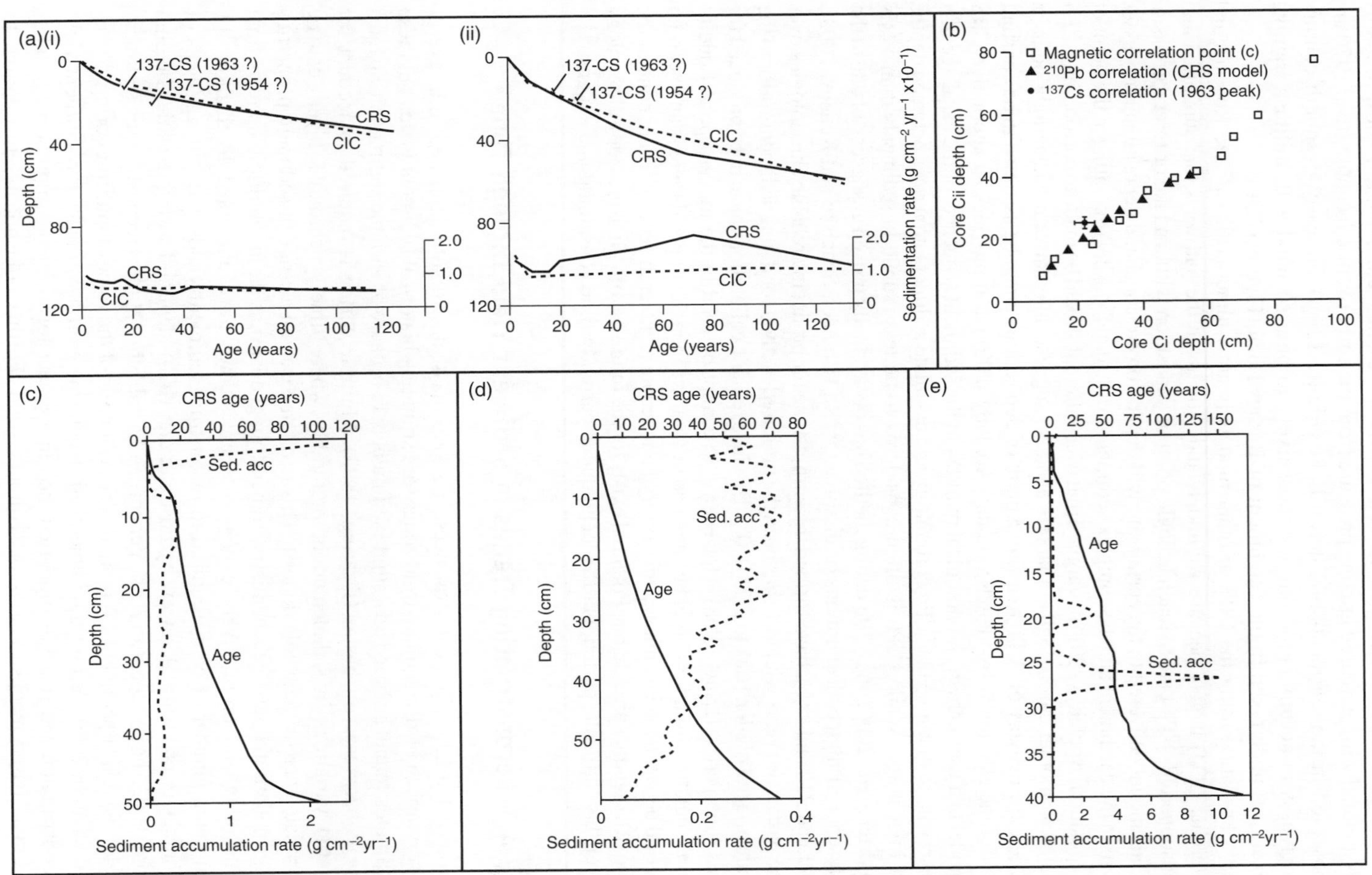
(a)(i)
137-CS (1963 ?)
137-CS (1954 ?)
CRS
CIC
Depth (cm)
Age (years)
(ii)
137-CS (1963 ?)
137-CS (1954 ?)
CIC
CRS
Age (years)
Sedimentation rate (g cm−2 yr−1 x10−1)
(b)
Magnetic correlation point (c)
210Pb correlation (CRS model)
137Cs correlation (1963 peak)
Core Cii depth (cm)
Core Ci depth (cm)
(c)
CRS age (years)
Sed. acc
Age
Depth (cm)
Sediment accumulation rate (g cm−2yr−1)
(d)
CRS age (years)
Sed. acc
Age
Depth (cm)
Sediment accumulation rate (g cm−2yr−1)
(e)
CRS age (years)
Age
Sed. acc
Depth (cm)
Sediment accumulation rate (g cm−2yr−1)

of Fontburn Reservoir (Figure 12.3c). Slowly increasing sedimentation rates, probably associated with increased grazing pressure from the early twentieth century to *c.* 1970 are observed in the sediments of Elleron Lake (Figure 12.3d) and early twentieth century increases in accumulation rate are most likely to be associated with military training exercises undertaken in the catchment of Barnes Loch (Figure 12.3e).

^{210}Pb dating using the CRS or other models is not without criticism (see Robbins and Herche, 1993) largely because we understand little about the pathways that supported and unsupported ^{210}Pb take to reach a single point of deposition within a lake or reservoir basin. The amount delivered to the coring site will depend on erosional and depositional processes within the catchment and the complex sedimentology (e.g. focusing) operating at the point of deposition. Appleby (2001) has argued that every attempt should be made to corroborate ^{210}Pb dates with other chronological markers and also suggests alternative models might be used where the delivery of ^{210}Pb to a single point of deposition cannot be assumed to be constant with time as implied by the CRS model. Appleby (2001) also provides an alternative CRS model that uses known chronological markers in the sediment sequence, or the basal age of a reservoir, to fine-tune the ^{210}Pb chronology and calculate sedimentation rates between depths of known age. While ^{210}Pb is often considered to be less mobile in sediments than ^{137}Cs (Foster *et al.*, 2006), the ^{210}Pb dating method has occasionally proved unsuccessful due to the presence of high concentrations of dissolved ^{226}Ra (Brenner, Schelske and Kenney, 2004).

Lead-210 and ^{137}Cs have formed the cornerstone for producing reliable chronologies over the past *c.* 100 years while ^{14}C has been the preferred method for dating sediments prior to the industrial revolution back to *c.* 25 000 yr BP since reliable calibrations have been produced for this time period (Lang, 2008). However, ^{14}C cannot be used in the modern period due to contamination from fossil fuel burning and nuclear weapons testing and there is an uncertain period between the late nineteenth century and the late eighteenth century where none of the above techniques are reliable. Potential exists for the future use of ^{32}Si to provide a chronology over this 'difficult' period, but the technique is currently time consuming and expensive.

12.4.3 Reconstructing Trends in Sediment Yield Through Time

From dated lake-sediment sequences, the total mass of sediment deposited in a lake or reservoir can be divided into time zones for estimating historical sediment yields but these estimates cannot be based on a single sediment core because of variable sedimentation rates across lake/reservoir basins (Dearing, 1992). Multiple coring is required to account for spatial variations in sedimentation, preferably where time-synchronous layers can be correlated across the whole deposit. This is often based on a range of sediment properties (e.g. pollen, diatoms, geochemistry, mineral magnetism) as dating multiple cores is time consuming and prohibitively expensive. Few studies have attempted to identify the minimum number of cores required to reconstruct sediment yields with known levels of accuracy or precision. A review by Foster *et al.* (1990) identified coring densities ranging from 0.0035 ha per core to 12.1 ha per core in 15 published studies reconstructing sediment yields using lake sediments. A more recent evaluation of the value of multiple coring (Evans and Church, 2000) demonstrated the need to determine error estimates for sediment yield reconstruction and strongly supported the requirement for multicore approaches to lake-sediment-based reconstructions of sediment yield. Further adjustments are required to account for dry bulk density, the area of sedimentation on the lake bed, contributions from biological productivity and shoreline erosion, and losses resulting from variable lake-trap efficiencies (see Foster *et al.*, 1985, 1990; Foster, Dearing and Appleby, 1986; Dearing, 1986; Dearing and Foster, 1993; Foster and Walling, 1994; Foster and Lees, 1999a; Verstraeten and Poesen, 2000; Foster, Boardman and Gates, 2008).

In much of the published literature, trap-efficiency curves developed by Brown (1944), Churchill (1948; cited in Trimble and Bube, 1990), Brune (1953), Chen (1975) or Heinemann (1981) have been used to estimate sediment retention. The Brune (1953) curve, which predicts trap efficiency from the annual capacity inflow ratio, has even been used in small dam sedimentation surveys (e.g. Neil and Mazari, 1993) despite its theoretical and practical limitations. Various attempts have been made to improve trap-efficiency estimates for large lakes and reservoirs (e.g. Trimble and Bube, 1990; Rowan *et al.*, 2001) although the development of these models requires good data on sediment influx and outflow for calibration that may not be available for most lakes and reservoirs in which sediment yield reconstructions are undertaken. Verstraeten and Poesen (2001b) have developed methods for estimating sediment-trap efficiency for small ponds since they argued that existing prediction methods for large reservoirs are unsuitable for these sites. A physically based numerical model was developed to simulate sediment deposition and to calculate sediment-trap efficiency. The model was used to simulate the long-term (33 year) trap efficiency of small ponds in Belgium based on 10-minute rainfall data. However, the model required detailed data on runoff, sediment inflow, outflow and deposition that are often lacking at such sites. A number of alternative models have been developed for predicting trap efficiency that have been reviewed in detail recently by Trimble and Wilson (in press).

12.4.4 Sediment Yields, Sediment Delivery and Sediment Budgets

Reconstructed sediment yields provide valuable information about the response of catchments to disturbance over time periods that are considerably longer than those for which directly monitored data are available. Examples are provided from natural lake basins in southern Sweden (Dearing *et al.*, 1990) and western Germany (Zolitschka, 1998) over the past 14 000 years in Figure 12.4a. The former was dated using pollen zones, where the timescale was derived from correlation with the regional ^{14}C dated history of vegetation. The latter was dated absolutely in calendar years by varve counting. Interpretation of these Holocene records largely reflects two mechanisms: an early postglacial adaptation to a still unstable palaeoclimate, with high rates of erosion on poorly developed soils, and a variety of human impacts associated with clearance and cultivation. In the Holzmaar catchment, high sediment yields dated at *c.* 1900 yr BP correlate with the time of Roman occupation. This feature is missing from the Swedish lake sediment record, probably due to the fact that Roman occupation did not extend into this area. The increased sediment yields in both catchments after *c.* 1200 yr BP probably reflect medieval and modern forest clearance and cultivation.

For more recent timescales (decades to centuries), sediment yield histories have been reconstructed from lake and reservoir bottom-sediments in a number of UK environments (see Foster (2006) for a review). Examples are presented in Figure 12.4 for undisturbed (forested) and lowland farmland catchments in north Warwickshire, (Figure 12.4b), lowland farmland catchments in Devon (Figure 12.4c) and Lincolnshire (Figure 12.4d), an upland farmland catchment in North Yorkshire (Figure 12.4e) and in a plantation forest in the North York Moors (Figure 12.4f). From these records spanning the past 50–130 years it is evident that the most dramatic increases in soil erosion and sediment transport occurred in cultivated

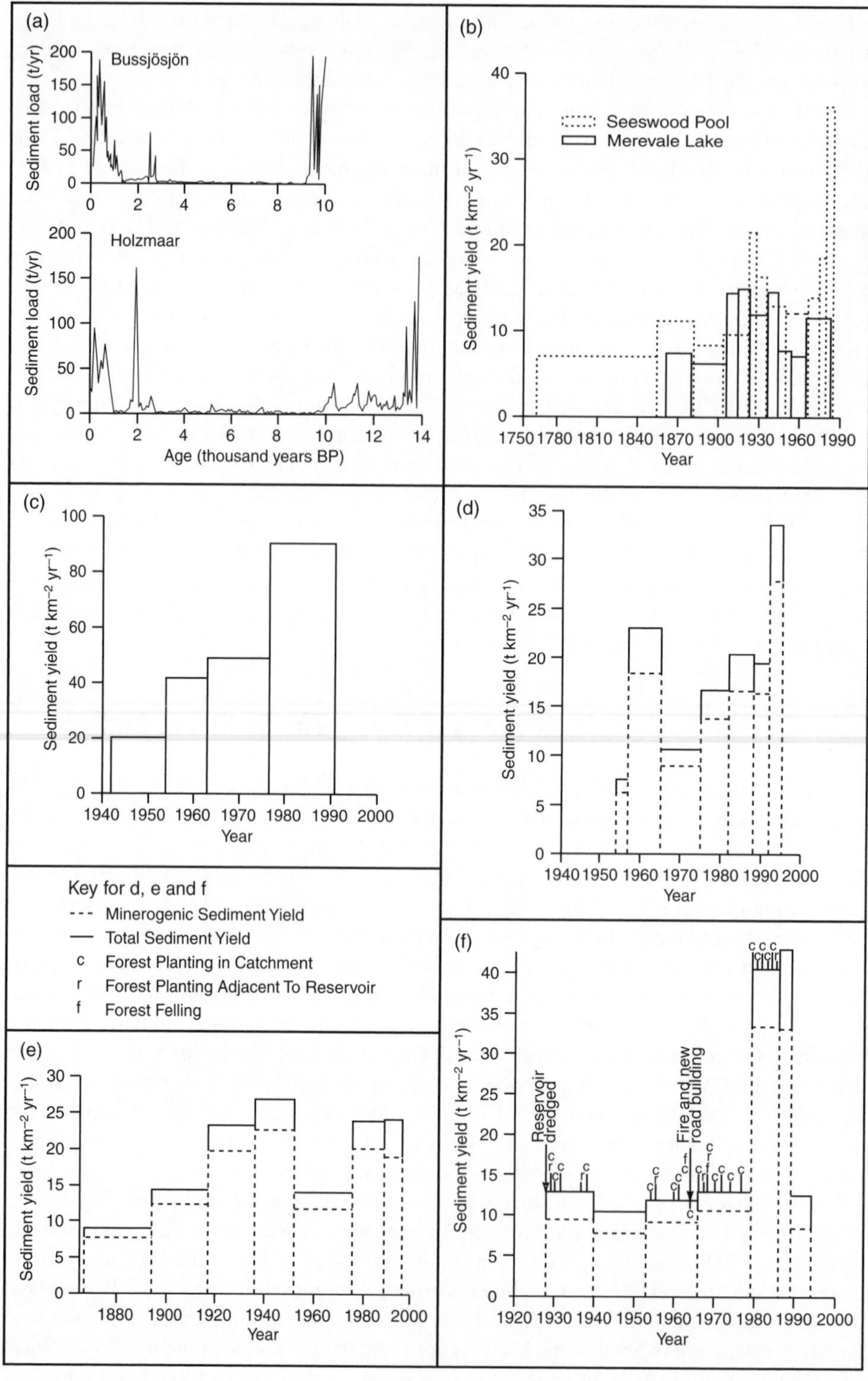
(a)
Bussjösjön
Sediment load (t/yr)
Holzmaar
Sediment load (t/yr)
Age (thousand years BP)
(b)
Seeswood Pool
Merevale Lake
Sediment yield (t km−2 yr−1)
Year
(c)
Sediment yield (t km−2 yr−1)
Year
(d)
Sediment yield (t km−2 yr−1)
Year
Key for d, e and f
Minerogenic Sediment Yield
Total Sediment Yield
c Forest Planting in Catchment
r Forest Planting Adjacent To Reservoir
f Forest Felling
(e)
Sediment yield (t km−2 yr−1)
Year
(f)
Reservoir dredged
Fire and new road building
Sediment yield (t km−2 yr−1)
Year

catchments from the early 1960s onwards, although short-term increases in erosion are also associated with forest planting in the Boltby reservoir catchment in the 1980s. Since 1990, however, this catchment has largely recovered from the initial disturbance. Lowland undisturbed catchments record suspended sediment yields of *c.* $10\,t\,km^{-2}\,yr^{-1}$, and are little different from those of lowland agricultural catchments prior to the Second World War. From the 1960s onwards, sediment yields appear to have increased by a factor of about four in the catchments of Seeswood Pool, the Old Mill reservoir and Fillingham lake – all of which are located in intensively cultivated landscapes in different parts of the UK.

Sediment yields, and their changes through time, reflect two major external controls:

1. human activities in drainage basins;

2. weather, climate and climate change.

However, the internal regulation of sediment transfer from hillslopes to catchment outlets (sediment delivery) should also be considered in any attempt to interpret sediment yields and their changes through time.

Human activities and/or climate change may increase or decrease rates of soil erosion either by changing the amount of sediment available for transport or by modifying the pathways, magnitude and timing of runoff and sediment transport. Increased runoff may also result in significant changes to the flow regime of rivers and may result in additional fine sediment inputs through processes such as river-channel adjustment and channel-bank erosion (e.g. Ashbridge, 1995).

The most common explanation for increased erosion and sediment yield has been the impact of forest clearance and cultivation. The long-term sediment yield histories of Figure 12.4a, for example, were largely interpreted as a response to human settlement and cultivation history (Dearing *et al.*, 1990; Zolitschka, 1998). However, Foster (1995) showed that while there was no detectable trend in annual rainfall in Midland England over the 130 years of record, there had been a significant increase in the number of intense daily rainfalls (>25 mm) per year and in the proportion of the total annual rainfall that was delivered by these events since the mid-1960s. Since high rainfall intensities provide the energy for soil detachment and transport, it was argued that the observed post-1960 increases in sediment yield in the Seeswood Pool

Figure 12.4 Reconstructed sediment yields. (a) Lakes Bussjösjön, southern Sweden (after Dearing, J. A. et al 'Recent and Long Term Records of Soil Erosion' in Boardman, J., Foster, I. D. L. and Dearing, J. A. (eds) Soil Erosion on Agricultural Land, © 1990. Reproduced with permission from John Wiley & Sons Ltd) and Holzmaar, Eiffel Mountains, western Germany (after Zolitschka, 'A 14,000 Year Sediment Yield', Geomorphology, 22, 1–17 © 1998. Reproduced with permission from Elseiver). (b) Seeswood Pool and Merevale Lake, north Warwickshire, UK (after Foster, I. D. L et al 'Magnitude and Frequency of Sediment Transport' 153–171 in Boardman, J., Foster, I. D. L. and Dearing, J. A. (eds) Soil Erosion on Agricultural Land, © 1990. Reproduced with permission from John Wiley & Sons Ltd.). (c) The Old Mill Reservoir, South Devon, UK (after Foster, I. D. L. and Walling, D. E. 'Using Reservoir Deposits to Reconstruct Changing Sediment Yields and Sources' Hydrological Sciences Journal 39 347–368 © 1994 Reproduced with permission of IAHS Press). (d) Fillingham Lake, Lincolnshire, UK, (e) Silsden reservoir, North Yorkshire, UK, and (f) Boltby Reservoir, North York Moors, UK (after Foster, I. D. L. and Lees, J. A. 'Changing Headwater Suspended Sediment Yields' Hyrdological Processes, 13, 1137–1153, © 1999. Reproduced with permission from John Wiley & Sons Ltd.).

catchment (Figure 12.4b) could in part be interpreted as a response to changing weather patterns. In the review presented by Foster (2006), six separate factors were identified as having an impact on sediment yield as reconstructed from sedimentary records. These were forest management, changes in land management (increases/decreases in the proportion of arable to pasture land), changes in stocking density, installation of land drains, changes in weather and climate and changes in sediment delivery.

While the amount of sediment transported by rivers may reflect the impact of both climate change and human activity, sediment yields are not necessarily directly analogous to rates of soil erosion. This is apparent at the global scale since suspended sediment yields are less than 30% of estimated global soil erosion rates (Walling and Webb, 1996). Many catchment studies have demonstrated that not all of the sediment eroded from hillslopes and field systems reaches the river and that a proportion of the amount initially mobilized may enter storage behind hedgerows, in river channels, floodplains and lakes (Walling, 1990). This factor is usually expressed as the sediment delivery ratio (SDR: the ratio between the amount of sediment transported by rivers and the amount eroded from hillslopes). Poor understanding of the sediment delivery process may lead to erroneous interpretation of the relative impact of catchment disturbance on erosion and sediment transport and may confound attempts to quantitatively separate the impacts of human activity and climate change on sediment yield (Boardman, Dearing and Foster, 1990). There is further discussion of sediment delivery in Chapter 6 (this volume).

Figure 12.5a illustrates potential relationships between soil erosion, sediment yield, valley storage (in fields and floodplains) and the SDR and suggests that the interpretation of changes in the sediment yield record must take account of the internal regulation of sediment delivery. In (i) all factors remain constant with respect to time. However, in (ii) it is suggested that sediment yields could remain constant with respect to time, despite increases in soil erosion, because opportunities may exist for an increase in valley sediment storage that will reduce the SDR through time. In catchment studies, the SDR generally decreases with increasing catchment size (Walling, 1990; de Vente *et al.*, 2007). In a review of lake sediment studies, Foster *et al.* (1990) plotted sediment yield against the ratio of catchment to lake area in order to account for the lake basin sediment delivery ratio (LBSDR; Figure 12.5b). In common with catchment data, the relationship is negative as yields generally decrease with an increase in the LBSDR.

Figure 12.5 The sediment delivery ratio (SDR). (a) Hypothetical relationships between soil erosion (SE) and sediment yield (SY) as regulated by sediment storage (simplified to Valley Storage; VS) and the sediment delivery ratio (after Foster, I. D. L. et al. 'Magnitude and Frequency of Sediment Transport in Agricultural Catchments' in Boardman, J., Foster, I. D. L. and Dearing, J. A. (eds) Soil Erosion on Agricultural Land, © 1990. Reproduced with permission from John Wiley & Sons Ltd). (b) Relationships between sediment yield and the catchment to lake area ratio (after Foster, I. D. L., Dearing, J. A., Grew, R. and Orend, K. 'The Sedimentary Data Base: An Appraisal of Lake and Reservoir Sediment Based Studies of Sediment Yield' 19–43 in Erosion, Transport and Deposition Processes. IAHS Publication No. 189 © 1990. Reproduced with permission from IAHS Press); (c) Sediment budget for the Yiwanshui catchment, Changshou Province, China (based on data in Lu, 1998); (d) sediment budgets for the Old Mill Reservoir and Start Valley Catchments, South Devon, UK (based on data in Foster and Walling, 1994; Foster, Owens and Walling, 1996 and Owens et al., 1997); (e) sediment budgets for the Seeswood Pool Catchment, North Warwickshire, UK (after Foster, Grew and Dearing, 1990b).

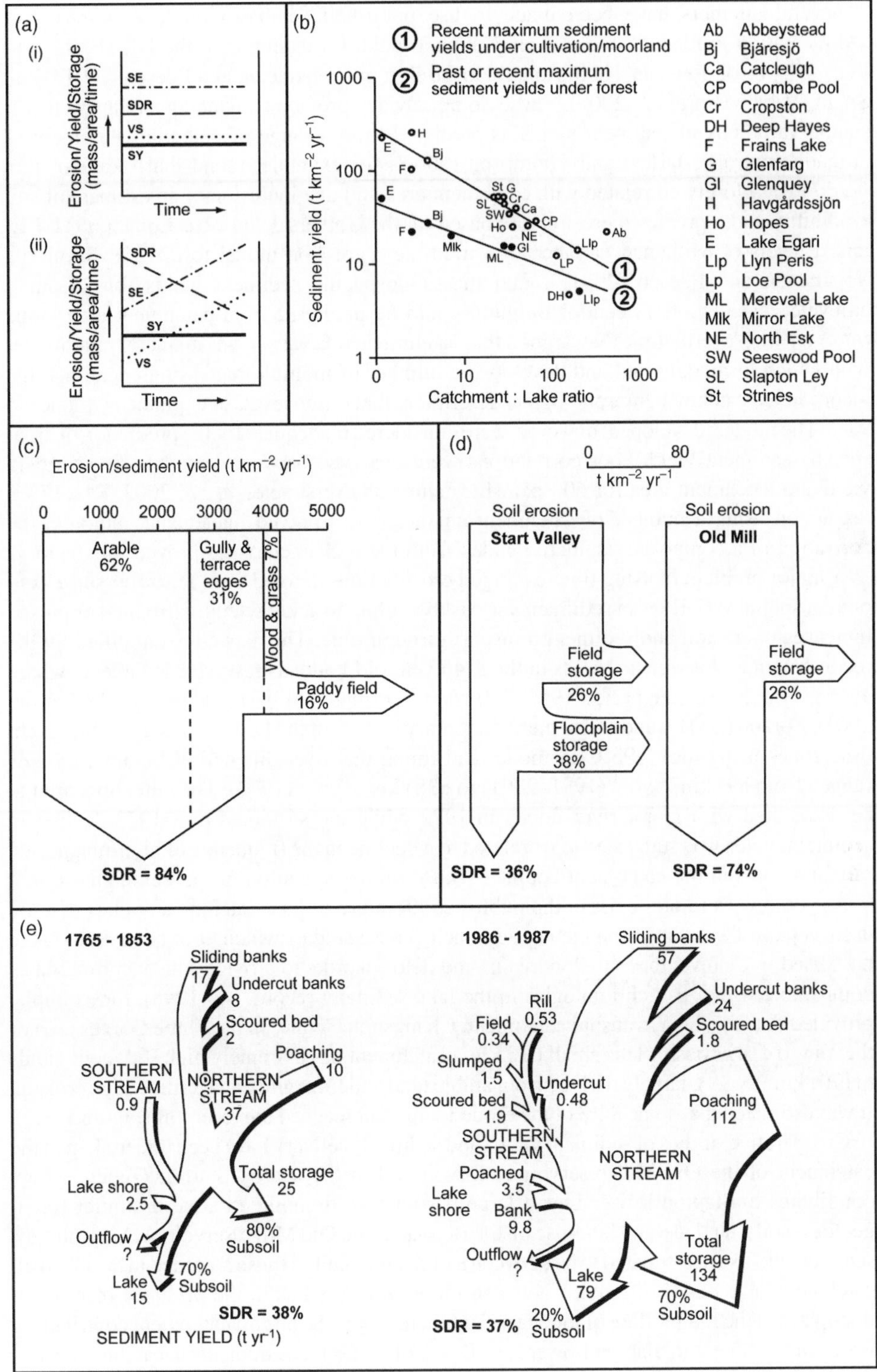

(a)
(i)
SE
SDR
VS
SY
Erosion/Yield/Storage (mass/area/time)
Time
(ii)
SDR
SE
SY
VS
Erosion/Yield/Storage (mass/area/time)
Time
(b)
① Recent maximum sediment yields under cultivation/moorland
② Past or recent maximum sediment yields under forest
Sediment yield (t km⁻² yr⁻¹)
Catchment : Lake ratio
Ab Abbeystead
Bj Bjäresjö
Ca Catcleugh
CP Coombe Pool
Cr Cropston
DH Deep Hayes
F Frains Lake
G Glenfarg
Gl Glenquey
H Havgårdssjön
Ho Hopes
E Lake Egari
Llp Llyn Peris
Lp Loe Pool
ML Merevale Lake
Mlk Mirror Lake
NE North Esk
SW Seeswood Pool
SL Slapton Ley
St Strines
(c)
Erosion/sediment yield (t km⁻² yr⁻¹)
0 1000 2000 3000 4000 5000
Arable 62%
Gully & terrace edges 31%
Wood & grass 7%
Paddy field 16%
SDR = 84%
(d)
0 80 t km⁻² yr⁻¹
Soil erosion
Start Valley
Field storage 26%
Floodplain storage 38%
SDR = 36%
Soil erosion
Old Mill
Field storage 26%
SDR = 74%
(e)
1765 - 1853
Sliding banks 17
Undercut banks 8
Scoured bed 2
Poaching 10
SOUTHERN STREAM 0.9
NORTHERN STREAM 37
Total storage 25
Lake shore 2.5
80% Subsoil
Outflow ?
Lake 15
70% Subsoil
SDR = 38%
SEDIMENT YIELD (t yr⁻¹)
1986 - 1987
Sliding banks 57
Undercut banks 24
Scoured bed 1.8
Rill 0.53
Field 0.34
Slumped 1.5
Undercut 0.48
Scoured bed 1.9
Poaching 112
SOUTHERN STREAM
NORTHERN STREAM
Poached 3.5
Lake shore
Bank 9.8
Outflow ?
Lake 79
Total storage 134
70% Subsoil
20% Subsoil
SDR = 37%

Several attempts have been made to use published databases on lake or reservoir sedimentation to identify controls on sediment yield, for example, in the UK (Foster and Lees, 1999a; Barlow and Thompson, 2000), Belgium (Verstraeten and Poesen, 2001a) and Spain (Verstraeten *et al.*, 2003). These do not always provide convincing explanations for spatial variations in sediment yields as predicted from a range of potential catchment or climatic controls. Barlow and Thompson (2000), for example, found that sediment flux ($t\,yr^{-1}$) was closely correlated with catchment area and that inclusion of an estimate of soil erodibility and a predicted erosion rate based on the Universal soil Loss Equation (USLE) raised levels of explained variance in a multiple regression model to *c.* 75%. Similarly, Verstraeten and Poesen (2001) found that area-specific sediment yields for 26 small cultivated catchments in central Belgium could be predicted from catchment area alone ($R^2 = 64\%$). Nevertheless, they argued that catchment area was an unsatisfactory parameter with which to model yield and developed a number of multiple regression equations that incorporated catchment morphological descriptors that raised levels of explained variance to 92%. The models developed, however, were considered inadequate for the prediction of area-specific sediment yield. Poor correlations were observed between area-specific sediment yield and catchment area for 60 Spanish catchments (Verstraeten *et al.*, 2003; $R^2 = 17\%$). Prediction using a number of catchment characteristics proved inadequate, although the derivation of a composite qualitative index from these characteristics proved satisfactory.

A major problem in using time-averaged erosion rates derived from reservoir surveys to predict spatial variations in sediment yield is the failure to acknowledge sustained or pulsed increases in erosion and sediment transport through time. This is well exemplified by the reconstruction of sediment yields in the *c.* 45 year old Ladonas reservoir in Greece, where average yields over the period 1955–2001 were estimated to be $163\,t\,km^{-2}\,yr^{-1}$ (Zorzou, 2004). Zorzou (2004) subdivided the sedimentary history of the Ladonas reservoir into eight time zones using the ^{210}Pb CRS model and found that over this period, sediment yields ranged from $89\,t\,km^{-2}\,yr^{-1}$ (1991–2001) to $388\,t\,km^{-2}\,yr^{-1}$ (1972). The latter appeared to be associated with major river floods in 1972 while the sustained post-1972 decline in sediment yields was suggested to be related to a decline in the frequency of high-magnitude rainfall events in the catchment coupled with rural depopulation of the Peloponnese.

As yet, few examples exist in the published literature of lake-catchment studies that are used as part of a sediment budgeting approach. Those studies which have been undertaken have used ^{137}Cs inventories in floodplains and fields in order to provide data over timescales comparable with ^{137}Cs chronologies in the lake sediment record. Lu (1998), for example, provided data for the Yiwanshui catchment, Changshou County in the Three Gorges area of the Yangtze river basin. This small ($0.7\,km^2$) catchment has extremely high sediment yields ($4170\,t\,km^{-2}\,yr^{-1}$), largely derived from arable land and from gully and terrace edge erosion. It was estimated that some 84% of the eroded sediment reached the reservoir (Figure 12.5c).

Comparative studies of sediment yields and sediment delivery have been undertaken in the catchment of the Old Mill reservoir in Devon and in the nearby Start catchment which contributes to Slapton Lower Ley, a large barrier lake draining an area of similar relief, geology, soils, land use and land-use history to that of the Old Mill reservoir. At Slapton Ley, sediment yields are estimated to be between a third and a half of those recorded in the Old Mill catchment. The major difference between these two sites lies in the presence of a major floodplain in the main valley of the Start catchment that is the largest catchment contributing water and sediment to Slapton Lower Ley. Based on ^{137}Cs measurements in catchment soils, floodplain and lake sediments, a sediment budget has been reconstructed (Foster, Owens and Walling, 1996) and refined (Owens *et al.*, 1997) at this site averaged over the past *c.* 50 years (i.e. since ^{137}Cs first entered the environment). The sediment budget of Figure 12.5d summarizes the main findings of the research and demonstrates that field erosion rates are high by UK standards (*c.* 80 t km^{-2} yr^{-1}). Sediment yields entering the main valley (*c.* 60 t km^{-2} yr^{-1}) are similar to those from the catchment of the Old Mill reservoir over the same time period. However, once sediment enters the main river, a large proportion is lost to floodplain storage in the Start valley and less than 30 t km^{-2} yr^{-1} (36% of that eroded from fields) over the past 40 years actually reached the lake. By contrast, it is tentatively estimated that sediment delivery in the Old Mill Reservoir is *c.* 74%. Figure 12.5e presents a catchment sediment budget for the Seeswood Pool catchment, north Warwickshire (Foster, Grew and Dearing, 1990) based on catchment monitoring and on lake-sediment-based reconstruction. In the earliest period of reconstruction (1765–1853) some 15 t of sediment were delivered annually to the lake, representing a sediment delivery ratio of *c.* 38%. While sediment yields and sources changed significantly between 1765–1783 and the monitored period of 1986–1987, the sediment delivery ratio remained essentially the same (37%). The similarity in sediment delivery ratios suggests that trends in sediment yield, while underestimating soil erosion, would provide a consistent record of changes in erosion through time at this site.

Interpreting palaeoenvironmental records is often problematic. In some cases, sedimentary records can be directly linked to instrumental records or to documented climate change in the historical period while others have shown that lake sediment records may sometimes be insensitive to documented climate forcing (e.g. Hasholt, Walling and Owens, 2000; Battarbee *et al.*, 2002; Cameron *et al.*, 2002; Tiljander *et al.*, 2003; Smith, Bradley and Abbot, 2004). In part this may reflect the scale of the lake or of its contributing catchment and/or the delay or insensitivity of the response in a large catchment to an environmental disturbance in headwater tributaries. For example, Walling *et al.* (2003) noted that despite major changes in sediment yields of headwater tributaries of the Yorkshire Ouse and Tweed catchments over the twentieth century, as reported by Foster and Lees (1999a), there was no apparent change in downstream sediment yields or in overbank sedimentation over similar timescales in the larger catchment. There also remains the traditional problem of deciphering the relative impacts of climate and human activity in triggering increases or decreases in erosion and sediment transport (Dotterweich, 2008).

12.4.5 Sediment Source Tracing

Information about sediment sources and possible changes in sediment sources through time can be obtained from a physical, biological, mineral magnetic and/or chemical analysis of the deposited sediments (e.g. Oldfield *et al.*, 1979; Dearing and Foster, 1987; Walling and Woodward, 1992; Charlesworth and Foster, 1993; Walling, Woodward and Nicholas, 1993; Foster and Walling, 1994; He and Owens, 1995; Loughran and Campbell, 1995; Dalgleish and Foster, 1996; Sanders, 1997; Foster *et al.*, 1998, 2002; Foster and Lees, 1999b; Foster, 2000; Heppell *et al.*, 2002; Small, Rowan and Franks, 2002; Foster, Boardman and Keay-Bright, 2007; Rowntree *et al.*, 2008; Pittam, Foster and Mighall, 2009).

Irrespective of which fingerprinting methodology is used, there are a number of important assumptions that are made concerning the ability of a tracer to distinguish between appropriate source types and of its conservative behaviour during transport or after deposition. The latter issue is particularly relevant to studies using lake and reservoir sediments and several studies have shown that post-depositional processes may limit the validity of some tracers. In hypereutrophic lakes, for example, magnetite dissolution may take place within strongly reduced organic sediments (Anderson and Rippey, 1988; Foster *et al.*, 1998), while the formation of authigenic ferrimagnetic iron sulphides, such as greigite, appears to occur in freshwater mud in contact with sulphur-rich marine sediments (Snowball and Thompson, 1988). There may also be situations in which the magnetic properties of lake sediment cores may include contributions from bacterial magnetite (Kirschvink, Jones and McFadden, 1985; Lovely *et al.*, 1987; Kim, Kodama and Moeller, 2005; Oldfield, 2007; Chiverrell *et al.*, 2008). Further detailed discussion of these issues, and of their relevance to the interpretation of the mineral magnetic signatures is given by Oldfield (1991, 1994, 2007), Lees (1997, 1999), van der Post *et al.* (1997), Foster *et al.* (1998, 2002, 2008), Foster (2000); Heppell *et al.* (2002) and Gruszowski *et al.* (2003).

12.5 Problems and Prospects

Lakes, reservoirs and ponds exist in almost all parts of the drainage basin and preserve records over varying timescales of the movement of fine particulate material through the sediment cascade. Several developments have taken place over the past three decades that have produced improved methods for the collection, dating, analysis and interpretation of the records contained within these sedimentary basins. One of the most significant developments has been in the use of high-resolution bathymetric surveys in order to calculate changes in reservoir storage volume. Although the resolution of these surveys probably remains too low to be of major use in most slowly accumulating UK lakes and reservoirs, the methods have great potential for providing good quality sediment yield data in many parts of the world as long as sediment density can be measured or modelled accurately in order to calculate sediment mass.

Coupled with other sources of geomorphological, archaeological and historical data (see Trimble (2008) for a review) the information provided by palaeoenvironmental reconstruction based on lake and reservoir sediments provides important insights into the factors that accelerate or reduce erosion rates. Reconstruction of sediment yield histories also allows estimation of the magnitude of the change, the response and relaxation time of the geomorphological system, and the time taken for a return to the former equilibrium or the establishment of new equilibrium conditions. While many detailed scientific questions concerning the interpretation of these data remain unanswered, and in some cases are unanswerable, the value of sedimentary records in integrated catchment research on sediment dynamics should not be underestimated. In isolation, long-term records of sediment yield produced from lakes and reservoirs provide information in regions where data from catchment monitoring is sparse or absent. Even where river monitoring has been used to estimate sediment yield, bedload is a more difficult parameter to measure in comparison with the suspended sediment load. As trap efficiency estimates show that lakes and reservoirs are *c.* 100% trap efficient for sand and coarser particle sizes, all transported sediment will remain in the reservoir basin. These coarse sediments are usually trapped in deltas that develop at the lake or reservoir inflow(s) and these could be surveyed separately from deep-water fine sediments to provide functional estimates of bedload transport.

The value of information derived from these sediment sinks increases as the analysis of lake sediments is incorporated into broader studies of erosion and sediment transport at the catchment scale as part of an overall strategy aimed at placing contemporary process studies in a historical context and in exploring controls on the sediment delivery process at different timescales. While process studies can help elucidate many of the mechanisms responsible for increasing erosion rates and sediment yields, history is often required in order to evaluate the response to, and recovery from, catchment disturbance over timescales of decades to centuries (Deevey, 1969).

Despite the potential for an analysis of lake sediments to provide quantitative data on sediment yields at a range of temporal and spatial scales within the drainage basin, two significant conceptual problems and several practical problems remain to be resolved. The first conceptual problem in interpreting changes in the sediment yield record relates to the need to separate 'association' from 'causation'. Erosion is a complex phenomenon and the link between erosion and sediment yield is still poorly understood at the catchment scale. Despite the complexity of the erosion process, single causal factors, especially 'human disturbance' through a variety of specific mechanisms, is the most frequently cited explanation for increases in sediment yield from reconstructed lake-sediment records. Climate change, and especially changes in weather patterns which include such factors as the timing, magnitude and intensity of storm events, are rarely analysed, even over timescales where such data exist. Secondly, few lake-catchment studies have attempted to quantify the sediment delivery ratio over different time periods. It is often implicitly assumed that the sediment yield record is a reliable surrogate for trends in soil erosion through time. Where there is no practical demonstration that sediment delivery has remained constant through time, the explanation for changes in the sediment yield record cannot dismiss changes in sediment delivery as a potential controlling factor.

Practical problems are associated first with the use of tracers to determine the provenance of sediments deposited in lake basins. Assumptions concerning an exclusively catchment-derived detrital record are invalidated in some of the research reviewed here. Second, inadequate trap efficiency estimates will potentially lead to large errors in sediment yield estimates. This is especially true for small reservoirs and methods for estimating trap efficiency require significant improvement in order to provide more parsimonious solutions to the problem. Third, few studies have measured or estimated the contribution derived from bank erosion or sediment resuspension during the rise and fall of reservoir levels although sediment fingerprinting could provide information about shoreline erosion provided that an appropriate signature unique to this source can be found. Finally, few attempts have been made to test the validity of predicted sediment yields by comparing modelled estimates with independent data derived from palaeoenvironmental reconstruction.

Despite these problems, analysis of material deposited in sedimentary basins throughout a catchment remains the only practical source of information in the absence of long-term monitoring records.

While this chapter has focused on the contribution that lakes and reservoirs make in changing the dynamics of the sediment cascade and in providing a wealth of palaeoenvironmental information that enables practitioners to understand catchment dynamics, few studies have discussed the potential for these reconstructions to contribute to the broad field of policy and governance. Environmental quality standards are at the forefront of attempts by policy makers to limit the impact of human activities on the environment and establish achievable goals, especially in the European Union following the enactment of the European Water Framework Directive (Crane, 2003; Collins and Anthony, 2008). To date, with the exception of an annual average suspended sediment concentration of $25\,mg\,l^{-1}$ established as part of the European Freshwater Fisheries Directive, targets for sediment and sediment-associated contaminants are largely absent (Collins and Anthony, 2008). The

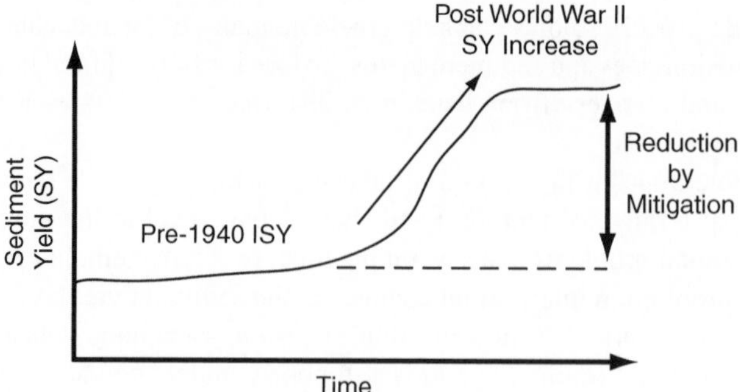

Figure 12.6 The intrinsic sediment yield concept (see text for explanation)

absence of environmental quality standards for sediment is also apparent in most recently published catchment consultation plans in the UK. For example, the Thames consultation plan (Environment Agency, 2008) mentions sediment only seven times but identifies sediment and sediment-associated contaminants as significant indicators of environmental quality. However, the plan provides no standards against which mitigation methods could be assessed. Furthermore, the most recent instructions provided by the UK Department for the Environment and Rural Affairs (Defra) to the UK Environment Agency (Defra, 2008) again largely ignores sediment. In part this reflects the difficulty in choosing the most appropriate sediment compartment to analyse (Crane, 2003), but palaeoenvironmental reconstruction, using the methods outlined in this review, could provide a key data source in order to establish environmental quality standards, including nutrient and contaminant concentrations and sediment yields. Figure 12.6 illustrates how palaeoenvironmental reconstruction could be used to establish a sediment yield target, here termed the intrinsic sediment yield (ISY). It requires identification of a time in the sediment yield record that pre-dates the most significant impact of land use on the environment. The sediment yield and sediment-associated phosphorus data reviewed by Foster (2006) suggest that the most dramatic impact of lowland England agriculture on the sediment yield record, for example, occurred after the Second World War and that the intensity of pre-war agricultural systems produced yields and phosphorus concentrations almost indistinguishable from those of records obtained from largely 'natural' catchments. While the early twentieth century might provide intrinsic sediment yield and sediment-associated phosphorus standards, other time periods may be more suitable for establishing targets in catchments polluted, for example, by the impacts of urbanization and/or industrial development. Nevertheless, the concept is one that should enable regional targets to be established in order for mitigation methods to be evaluated and implemented. Failure to establish such standards means that we are as yet unable to evaluate the success of erosion mitigation strategies on sediment yields.

12.6 Acknowledgements

Much of the primary research undertaken by the author, some of which has been reviewed in this chapter, has been supported by a number of funding agencies including Coventry

University, the British Society for Geomorphology (formerly the British Geomorphological Research Group), The Environment Agency, the UK Natural Environment Research Council and British Academy, the South African National Research Foundation and Rhodes University. I would like to acknowledge the contribution of many of my co-workers on several research projects including Peter Appleby, University of Liverpool; John Boardman, Oxford University; Sue Charlesworth, Jason Jordan and Joan Lees, Coventry University; John Dearing, University of Southampton; Tim Mighall, Aberdeen University; Phil Owens, University of Northern British Columbia; Graham Leeks, Centre for Ecology and Hydrology, Wallingford; Kate Rowntree and Leanne Du Preez, Rhodes University and Des Walling, University of Exeter. I also thank a number of individuals for field and laboratory assistance including Richard Bold, Tony Chapman, Heather Dalgleish, Libby Foster, Christopher Fry, Rob Grew, Ron Hannas, Bob Hollyoak, John Patterson, Andy Small and Liz Turner. The diagrams were drawn by Stuart Gill of Coventry University. I am especially indebted to Stan Trimble (University of California Los Angeles) for his constructive comments and suggestions on an earlier draft of this contribution.

References

Anderson, D.E., Goudie, A.S. and Parker, A.G. (2007) *Global Environments through the Quaternary*, Oxford University Press, Oxford.

Anderson, N.J. and Rippey, B. (1988) Diagenesis of magnetic minerals in recent sediments of a eutrophic lake. *Limnology and Oceanography*, **33**, 1476–1492.

Appleby, P.G. (2001) Chronostratigraphic techniques in recent sediments, in *Tracking Environmental Change using Lake sediments. Volume 1: Basin Analysis, Coring and Chronological Techniques* (eds W.M. Last and J.P. Smol), Kluwer, Dordrecht.

Appleby, P.G. and Oldfield, F. (1978) The calculation of Pb-210 dates assuming a constant rate of supply of unsupported Pb-210 to the sediment. *Catena*, **5**, 1–8.

Appleby, P.G., Nolan, P., Gifford, D.W. *et al.* (1986) Pb-210 dating by low background gamma counting. *Hydrobiologia*, **141**, 21–27.

Appleby, P.G., Haworth, E.Y., Michel, H. *et al.* (2003) The transport and mass balance of fallout radionuclides in Blelham Tarn, Cumbria (UK). *Journal of Paleolimnology*, **29**, 459–473.

Ashbridge, D. (1995) Processes of river bank erosion and their contribution to the suspended sediment load of the River Culm, Devon, in*Sediment and Water Quality in River Catchments* (eds I.D.L. Foster,A.M. Gurnell and B.W. Webb), John Wiley and Sons, Chichester, pp. 229–245.

Banerjee, D., Murray, A.S. and Foster, I.D.L. (2001) Scilly Isles, UK: optical dating of a possible tsunami deposit from the 1755 Lisbon Earthquake. *Quaternary Science Reviews*, **20**, 715–718.

Barlow, D.N. and Thompson, R. (2000) Holocene sediment erosion in Britain as calculated from lake-basin studies, in *Tracers in Geomorphology* (ed. I.D.L. Foster), John Wiley and Sons, Chichester, pp. 455–472.

Battarbee, R.W., Grytnes, J.-A., Thompson, R. *et al.* (2002) Comparing palaeolimnological and instrumental evidence of climate change for remote mountain lakes over the last 200 years. *Journal of Paleolimnology*, **28**, 161–179.

Beaumont, P. (1978) Man's impact on River Systems: a world-wide view. *Area*, **10**, 38–41.

Beckinsale, R.P. (1972) Clearwater erosion: its geographical and international significance. *International Geography*, **2**, 1244–1246.

Bertrand, S., Castiaux, J. and Juvigné, E. (2008) Tephrostratigraphy of the late glacial and Holocene sediments of Puyehue Lake (Southern Volcanic Zone, Chile, 40°S). *Quaternary Research*, **70**, 343–357.

Boardman, J., Dearing, J.A. and Foster, I.D.L. (1990) Soil erosion studies: some assessments, in *Soil erosion on agricultural land* (eds J. Boardman, I.D.L. Foster and J.A. Dearing), John Wiley and Sons, Chichester, pp. 659–672.

Boix-Fayos, C., Barberá, G.G., López-Bermúdez, F. and Castillo, V.M. (2007) Effects of check-dams, reforestation and land-use changes on river channel morphology: Case study of the Rogativa catchment (Murcia, Spain). *Geomorphology*, **91**, 103–123.

Bogen, J. (2008) The impact of climate change on glacial sediment delivery to rivers, Publication 325, International Association of Hydrological Sciences, Wallingford, pp. 432–439.

Boygle, J. (1993) The Swedish varve chronology – a review. *Progress in Physical Geography*, **17**, 1–19.

Brandt, S.A. (2000) Classification of geomorphological effects downstream of dams. *Catena*, **40**, 375–401.

Brenner, M., Schelske, C.L. and Kenney, W.F. (2004) Inputs of dissolved and particulate ^{226}Ra to lakes and implications for ^{210}Pb dating recent sediments. *Journal of Paleolimnology*, **32**, 53–66.

Brown, C.B. (1944) Discussion of "Sedimentation in Reservoirs" by B.J. Witzig. *Transactions of the American Society of Civil Engineers.*, **109**, 1080–1086.

Brune, G.M. (1953) Trap efficiency of reservoirs. *Transactions – American Geophysical Union*, **34**, 407–418.

Cameron, N.G., Schnell, Ø.A., Rautio, M.L. *et al.* (2002) High-resolution analyses of recent sediments from a Norwegian mountain lake and comparison with instrumental records of climate. *Journal of Paleolimnology*, **28**, 79–93.

Castillo, V.M., Mosch, W.M., Conesa García, C. *et al.* (2007) Effectiveness and geomorphological impacts of check dams for soil erosion control for in a semiarid Mediterranean catchment: El Cárcavo (Murcia, Spain). *Catena*, **70**, 416–427.

Charlesworth, S.M. and Foster, I.D.L. (1993) Effects of urbanization on lake sedimentation: the history of two lakes in Coventry, UK- preliminary results, in *Geomorphology and Sedimentology of Lakes and Reservoirs* (eds J. McManus and R.W. Duck), John Wiley and Sons, Chichester, pp. 15–29.

Chen, C.N. (1975) Design of sedimentation basins.Proceedings of the National Symposium on Urban Hydrol and Sediment Control, Kentucky, Lexington, USA, pp. 58–68.

Chiverrell, R.C., Oldfield, F., Appleby, P.G. *et al.* (2008) Evidence for changes in Holocene sediment flux in Semer Water and Raydale, North Yorkshire, UK. *Geomorphology*, **100**, 70–82.

Chu, G., Sun, Q., Zhaoyan, G. *et al.* (2009) Dust records from varved lacustrine sediments of two neighboring lakes in northeastern China over the last 1400 years. *Quaternary International*, **194** (1–2), 108–118.

Churchill, M.A. (1948) Discussion of 'Analysis and use of reservoir sedimentation data', Proceedings of the Federal. Inter Agency Sedimentation Conference, Denver 1947 (ed. L.C. Gottschalk). Bureau of Reclamation, U.S. Department of the Interior, Washington, DC, pp. 139–140.

Collins, A.L. and Anthony, S.G. (2008) Assessing the likelihood of catchments across England and Wales meeting 'good ecological status' due to sediment contributions from agricultural sources. *Environmental Science and Policy*, **11**, 163–170.

Coombes, P.M.V., Chiverrell, R.C. and Barber, K.E. (2008) A high-resolution pollen and geochemical analysis of late Holocene human impact and vegetation history in southern Cumbria, England. *Journal of Quaternary Science*, **24**, 224–236. doi:10.1002/jqs.1219.

Couch, A.G. and Eyles, N. (2008) Sedimentary record of glacial Lake Mackenzie, Northwest Territories, Canada: Implications for Arctic freshwater forcing. *Palaeogeography, Palaeoclimatology, Palaeoecology*, **268**, 26–38.

Crane, M. (2003) Proposed development of Sediment Quality Guidelines under the European Water Framework Directive: a critique. *Toxicology Letters*, **142**, 195–206.

Dalgleish, H.Y. and Foster, I.D.L. (1996) ^{137}Cs losses from a loamy surface water gleyed soil (*Inceptisol*); a laboratory simulation experiment. *Catena*, **26**, 227–245.

Dearing, J.A. (1986) Core correlation and total sediment influx, in *Handbook of Holocene Palaeoecology and Palaeohydrology* (ed. B. Berglund), John Wiley and Sons, Chichester, pp. 247–270.

Dearing, J.A. (1992) Sediment yields and sources in a Welsh upland lake catchment during the past 800 years. *Earth Surface Processes and Landforms*, **17**, 1–22.

Dearing, J.A. and Foster, I.D.L. (1986) Limnic sediments used to reconstruct sediment yields and sources in the English Midlands since 1765, in*International Geomorphology* (ed.V. Gardiner), John Wiley and Sons, Chichester, pp. 853–868.

Dearing, J.A. and Foster, I.D.L. (1993) Lake sediments and geomorphological processes: some thoughts, in *Geomorphology and Sedimentology of Lakes and Reservoirs* (eds J. McManus and R. W. Duck), John Wiley and Sons, Chichester, pp. 5–14.

Dearing, J.A., Alström, K., Bergman, A. *et al.* (1990) Recent and long-term records of soil erosion from Southern Sweden, in *Soil Erosion on Agricultural Land* (eds J. Boardman, I.D.L. Foster, J.A. Dearing), John Wiley and Sons, Chichester, pp. 173–196.

Defra (2008) *Water Framework Directive. Directions to the Environment Agency on Classification of Water Bodies*, Defra, London.

Deevey, E.S. (1969) Coaxing history to conduct experiments. *BioScience*, **19**, 40–43.

Desloges, J.R. (1994) Varve deposition and sediment yield record at three small lakes of the southern Canadian Cordillera. *Arctic and Alpine Research*, **26**, 130–140.

De Vente, J., Poesen, J., Arabkhedri, M. and Verstraeten, G. (2007) The sediment delivery problem revisited. *Progress in Physical Geography*, **31**, 155–178.

Dotterweich, M. (2008) The history of soil erosion and fluvial deposits in small catchments of central Europe: Deciphering the long-term interaction between humans and the environment – A review. *Geomorphology*, **101**, 192–208.

Duck, R.W. and McManus, J. (1987) Sediment yields in lowland Scotland derived from reservoir surveys. *Transactions of the Royal Society of Edinburgh, Earth Science*, **78**, 369–377.

Dunning, S.A., Rosser, N.J., Petley, D.N. and Massey, C.I. (2006) The formation and failure of the Tsatichhu landslide dam, Bhutan Himalaya. *Landslides*, **3**, 107–113.

Eakin, H. and Brown, C.B. (1939) *Silting of Reservoirs*. Technical Bulletin 524, U.S. Department of Agriculture, Washington, DC.

Edwards, K.J. and Whittington, G. (2001) Lake sediments, erosion and landscape change during the Holocene in Britain and Ireland. *Catena*, **42**, 143–173.

EEA (1999) Lakes and Reservoirs in the European Environmental Agency Area. Project Report 1. European Environment Agency, Copenhagen.

Environment Agency (2008) 'Water for life. A consultation on the draft river basin management plan Thames river basin district'. UK Environment Agency, Bristol. December 2008.

Evans, M. and Church, M. (2000) A method for error analysis of sediment yields derived from estimates of lacustrine sediment accumulation. *Earth Surface Processes and Landforms*, **25**, 1257–1267.

Ferrari, R.L. (2006) 'Reconnaisance techniques for reservoir survey'. U.S. Department of the Interior, Bureau of Reclamation, Technical Service Center, Sedimentation and River Hydraulics Group Denver, Colorado, April,. p.139.

Foster, I.D.L. (1995) Lake and reservoir bottom sediments as a source of soil erosion and sediment transport data in the UK, in *Sediment and Water Quality in River Catchments* (eds I.D.L. Foster, A. M. Gurnell and B.W. Webb), John Wiley and Sons, Chichester, pp. 265–283.

Foster, I.D.L. (ed.) (2000) *Tracers in Geomorphology*, John Wiley and Sons, Chichester.

Foster, I.D.L. (2006) Lakes and Reservoirs in the Sediment Delivery System – Reconstructing Sediment Yields, in *Soil Erosion and Sediment Redistribution in River Catchments* (eds P.N. Owens and A. Collins), CAB International, Wallingford, pp. 128–142.

Foster, I.D.L. and Charlesworth, S.M. (1996) Heavy metals in the hydrological cycle: trends and explanation. *Hydrological Processes*, **10**, 227–261.

Foster, I.D.L. and Lees, J.A. (1999a) Changing headwater suspended sediment yields in the LOIS catchments over the last century: a palaeolimnological approach. *Hydrological Processes*, **13**, 1137–1153.

Foster, I.D.L. and Lees, J.A. (1999b) Physical and geochemical properties of suspended sediments delivered to the headwaters of LOIS River Basins over the last 100 years: an analysis of lake and reservoir bottom-sediments. *Hydrological Processes*, **13**, 1067–1086.

Foster, I.D.L. and Walling, D.E. (1994) Using reservoir deposits to reconstruct changing sediment yields and sources in the catchment of the Old Mill reservoir, South Devon, UK over the past 50 years. *Hydrological Sciences Journal*, **39**, 347–368.

Foster, I.D.L., Boardman, J. and Gates, J.B. (2008) *Reconstructing historical sediment yields from the infilling of farm reservoirs, Eastern Cape, South Africa*, Publication 325, International Association of Hydrological Sciences, Wallingford, pp. 440–447.

Foster, I.D.L., Boardman, J. and Keay-Bright, J. (2007) The contribution of sediment tracing to an investigation of the environmental history of two small catchments in the uplands of the Karoo, South Africa. *Geomorphology*, **90**, 126–143.

Foster, I.D.L., Dearing, J.A. and Appleby, P.G. (1986) Historical trends in catchment sediment yields: a case study in reconstruction from lake sediment records in Warwickshire, UK. *Hydrological Sciences Journal*, **31**, 427–443.

Foster, I.D.L., Grew, R. and Dearing, J.A. (1990) Magnitude and frequency of sediment transport in agricultural catchments.: a paired lake catchment study in Midland England, in *Soil Erosion on Agricultural Land* (eds J. Boardman, I.D.L. Foster and J.A. Dearing), John Wiley and Sons, Chichester, pp. 153–171.

Foster, I.D.L., Owens, P.N. and Walling, D.E. (1996) Sediment yields and sediment delivery in the catchments of Slapton Lower Ley, South Devon, UK. *Field Studies*, **8**, 629–661.

Foster, I.D.L., Dearing, J.A., Simpson, A. *et al.* (1985) Lake catchment based studies of erosion and denudation in the Merevale Catchment, Warwickshire, UK. *Earth Surface Processes and Landforms*, **10**, 45–68.

Foster, I.D.L., Dearing, J.A., Grew, R.G. and Orend, K. (1990) The sedimentary data base: An appraisal of lake and reservoir sediment based studies of sediment yield, Publication 189, International Association of Hydrological Sciences, Wallingford, pp. 15–43.

Foster, I.D.L., Albon, A.J., Bardell, K.M. *et al.* (1991) High energy coastal sedimentary deposits: an evaluation of depositional processes in Southwest England. *Earth Surface Processes and Landforms*, **16**, 341–356.

Foster, I.D.L., Dalgleish, H., Dearing, J.A. and Jones, E.D. (1994) Quantifying soil erosion and sediment transport in drainage basins; some observations on the use of Cs-137, Publication 224, International Association of Hydrological Sciences, Wallingford, pp. 55–64.

Foster, I.D.L., Lees, J.A., Owens, P.N. and Walling, D.E. (1998) Mineral magnetic characterisation of sediment sources from an analysis of lake and floodplain sediments in the catchments of the Old Mill reservoir and Slapton Ley, South Devon, UK. *Earth Surface Processes and Landforms*, **23**, 685–703.

Foster, I.D.L., Lees, J.A., Jones, A.R. *et al.* (2002) The possible role of agricultural land drains in sediment delivery to a small reservoir, Worcestershire, UK: a multiparameter fingerprint study, Publication 276, International Association of Hydrological Sciences, Wallingford, pp. 433–442.

Foster, I.D.L., Chapman, A.S., Hodgkinson, R.M. *et al.* (2003) Changing suspended sediment and particulate loads and pathways in underdrained lowland agricultural catchments, Herefordshire and Worcestershire, UK. *Hydrobiologia*, **494**, 119–126.

Foster, I.D.L., Mighall, T.M., Proffitt, H. *et al.* (2006) Post-depositional ^{137}Cs mobility in the sediments of three shallow coastal lagoons, SW England. *Journal of Paleolimnology*, **35**, 881–895.

Foster, I.D.L., Oldfield, F., Flower, R.J. and Keatings, K. (2008) Trends in mineral magnetic signatures in a long core from Lake Qarun, Middle Egypt. *Journal of Paleolimnology*, **40**, 835–849.

Fox, H.R., Moore, H.M., Newell Price, J.P. and El Kasri, M. (1997) *Soil erosion and reservoir sedimentation in the High Atlas Mountains, Southern Morocco*, Publication 245, International Association of Hydrological Sciences, Wallingford, pp. 233–240.

Gilbert, R. and Desloges, J.R. (1987) Sediments of ice-dammed, self draining Ape Lake, British Columbia. *Canadian Journal of Earth Science*, **24**, 1735–1747.

Gilmore, G. and Hemingway, J. (1995) *Practical Gamma Spectrometry*, John Wiley and Sons, Chichester.

Goldberg, E.D. (1963) Geochronology with Pb^{210}. Radioactive Dating (Proceedings of the Athens Symposium) International Atomic Energy Authority, Vienna, pp. 121–131.

Grant, G.E., Schmidt, J.C. and Lewis, S.L. (2003) A geological framework for interpreting downstream effects of dams on rivers. *Water Science and Application*, **7**, 209–225.

Gruszowski, K.E., Foster, I.D.L., Lees, J.A. and Charlesworth, S.M. (2003) Sediment sources and transport pathways in a rural catchment, Herefordshire, UK. *Hydrological Processes*, **17**, 2665–2681.

Gurnell, A.M. (1998) The hydrogeomorphological effects of beaver dam-building activity. *Progress in Physical Geography*, **22**, 167–189.

Gurnell, A.M., Gregory, K.J. and Petts, G.E. (1995) The role of coarse woody debris in forest aquatic habitats: implications for management. *Aquatic Conservation: Marine and Freshwater Ecosystems*, **5**, 1–24.

Håkanson, L. and Jansson, M. (1983) *Principles of Lake Sedimentology*, Springer-Verlag, Berlin.

Haregeweyn, N., Poesen, J., Nyssen, J. *et al.* (2006) Reservoirs in Tigray (Northern Ethiopia): characteristics and sediment deposition problems. *Land Degradation and Development*, **17**, 211–230.

Hasholt, B., Walling, D.E. and Owens, P.N. (2000) Sedimentation in arctic proglacial lakes: Mittivakkat Glacier, south-east Greenland. *Hydrological Processes*, **14**, 679–699.

He, Q. and Owens, P.N. (1995) Determination of suspended sediment provenance using caesium-^{137}, unsupported Pb-210 and radium-226: a numerical mixing model, in *Sediment and Water Quality in River Catchments* (eds I.D.L. Foster, A.M. Gurnell and B.W. Webb), John Wiley and Sons, Chichester, pp. 207–227.

Heinemann, H.G. (1981) A new reservoir trap efficiency curve for small reservoirs. *Water Resources Bulletin*, **17**, 825–830.

Heppell, C.M., Burt, T.P., Walden, J. and Foster, I.D.L. (2002) Investigating contemporary and historical sediment inputs to Slapton Higher Ley: an analysis of the robustness of source ascription methods when applied to lake sediment data. *Hydrological Processes*, **16**, 3467–3486.

Holliday, V.J., Warburton, J. and Higgitt, D.L. (2008) Historic and contemporary sediment transfer in an upland Pennine catchment, UK. *Earth Surface Processes and Landforms*, **33** (14),2139–2155. doi:10.1002/esp.1660.

Hyatt, J.A. and Gilbert, R. (2000) Lacustrine sedimentary record of human-induced gully erosion and land-use change at Providence Canyon, southwest Georgia, USA. *Journal of Paleolimnology*, **23**, 421–438.

Kalff, J. (2002) *Limnology: inland water ecosystems*, Prentice Hall, Inc., Upper Saddle River, NJ, USA.

Kirschvink, J.L., Jones, D.S. and McFadden, B.J. (eds) (1985) *Magnetite Biomineralization and Magnetoreception in Organisms*, Plenum, New York.

Kim, B.Y., Kodama, K.P. and Moeller, R.E. (2005) Bacterial magnetite produced in water column dominates lake sediment mineral magnetism: Lake Ely, USA. *Geophysics Journal International*, **163**, 26–37.

Koi, T., Hotta, N., Ishigaki, I. *et al.* (2008) Prolonged impact of earthquake-induced landslides on sediment yield in a mountain watershed: The Tanzawa region, Japan. *Geomorphology*, **101**, 692–702.

Korup, O. (2002) Recent research on landslide dams – a literature review with special attention to New Zealand. *Progress in Physical Geography*, **26**, 206–235.

Lahlou, A. (1996) Environmental and socio-economic impacts of erosion and sedimentation in North Africa, Publication 236, International Association of Hydrological Sciences, Wallingford, pp. 491–500.

Lang, A. (2008) Recent advances in dating and source tracing of fluvial deposits, Publication 325, International Association of Hydrological Sciences, Wallingford, pp. 3–12.

Laronne, J.B. (1990) Probability Distribution of event sediment yields in the northern Negev, Israel, in *Soil Erosion on Agricultural Land* (eds J. Boardman, I.D.L. Foster and J.A., Dearing), John Wiley and Sons, Chichester, pp. 481–492.

Lamoureux, S. (2002) Temporal patterns of suspended sediment yield following moderate to extreme hydrological events recorded in varved lacustrine sediments. *Earth Surface Processes and Landforms*, **27**, 1107–1124.

Lees, J.A. (1997) Mineral magnetic properties of mixtures of environmental and synthetic materials: linear additivity and interaction effects. *Geophysical Journal International*, **131**, 335–346.

Lees, J.A. (1999) Evaluating magnetic parameters for use in source identification, classification and modelling of natural environmental materials, in *Environmental Magnetism: A Practical Guide* (eds J. Walden, F. Oldfield and J. Smith), QRA Technical Guide 6, Quaternary Research Association, London, pp. 113–138.

Lees, J.A., Foster, I.D.L., Jones, E.D. *et al.* (1997) Sediment yields in a changing environment: a historical reconstruction using reservoir bottom sediments in three contrasting small catchments, North York Moors, UK, Publication 245, International Association of Hydrological Sciences, Wallingford, pp. 169–179.

Likens, G.E. and Davis, M.B. (1975) Postglacial history of Mirror lake, New Hampshire, USA: an initial report. *Verhandlungen – Internationale Vereinigung fuer Theoretische und Angewandte Limnologie*, **19**, 982–993.

Loughran, R.J. and Campbell, B.L. (1995) The identification of catchment sediment sources, in *Sediment and Water Quality in River Catchments* (eds I.D.L. Foster, A.M. Gurnell and B.W. Webb), John Wiley and Sons, Chichester, pp. 189–205.

Lovely, D.R., Stolz, J.F., Nord, G.L. and Phillips, E.J.P. (1987) Anaerobic production of magnetite by a dissimilatory iron-reducing microorganism. *Nature*, **330**, 279–281.

Lu, X. (1998) Soil erosion and sediment yield in the Upper Yangtze, China. Unpublished PhD Thesis, University of Durham, Durham, UK.

Martin-Rosales, W., Pulido-Bosch, A., Gisbert, J. and Vallejos, A. (2003) Sediment yield estimation and check dams in a semiarid area (Sierra de Gador, southern Spain), Publication 279, International Association of Hydrological Sciences, Wallingford, pp. 51–58.

May, C.L. and Gresswell, R.E. (2003) Processes and rates of sediment and wood accumulation in headwater streams of the Oregon coast range, USA. *Earth Surface Process and Landforms*, **28**, 409–424.

McCully, P. (1996) *Silenced Rivers. The Ecology and Politics of Large Dams*, Zed Books, London.

Meek, N. (1999) New discoveries about the late Wisconsian history of the Mojave river system, in *Tracks Along the Mojave: A Field Guide from Cajun Pass to the Calico Mountains and Coyote Lake*, (eds R.E. Reynolds and J. Reynolds). *San Bernardino Museum Quarterly*, **46**, 113–117.

Morris, G.L. and Fan, J. (1997) *Reservoir Sedimentation Handbook: Design and Management of Dams, Reservoirs and Watersheds for Sustainable Use*, McGraw-Hill, New York.

Murray, A.S., Marten, R., Johnston, A. and Martin, P. (1987) Analysis for naturally occurring radionuclides at environmental concentration by gamma spectrometry. *Journal of Radioanalytical and Nuclear Chemistry-Articles*, **115**, 263–288.

Neil, D.T. and Mazari, R.K. (1993) Sediment yield mapping using small dam sedimentation surveys, Southern Tablelands, New South Wales. *Catena*, **20**, 13–25.

Nesje, A., Dahl, S.O., Matthews, J.A. and Berrisford, M.A. (2001) A *c.* 4500-yr record of river floods obtained from a sediment core in Lake Atnsjøen, eastern Norway. *Journal of Paleolimnology*, **25**, 329–342.

Nijampurkar, V.N., Rao, D.K., Oldfield, F. and Renberg, I. (1998) The half-life of ^{32}Si: a new estimate based on varved lake sediments. *Earth and Planetary Science Letters*, **163**, 191–196.

O'Hara, S., Alayne Street-Perrot, F. and Burt, T.P. (1993) Accelerated soil erosion around a Mexican highland lake caused by prehispanic agriculture. *Nature*, **362**, 48–51.

Oldfield, F. (1977) Lakes and their drainage basins as units of sediment based ecological study. *Progress in Physical Geography*, **1**, 460–504.

Oldfield, F. (1991) Environmental magnetism - a personal perspective. *Quaternary Science Reviews*, **10**, 73–85.

Oldfield, F. (1994) Toward the discrimination of fine-grained ferrimagnets by magnetic measurements in lake and near-shore marine sediments. *Journal of Geophysical Research*, **99**, 9045–9050.

Oldfield, F. (2007) Sources of fine-grained magnetic minerals in sediments: a problem revisited. *The Holocene*, **17**, 1265–1271.

Oldfield, F. and Appleby, P.G. (1984) Empirical testing of ^{210}Pb dating models for lake sediments, in *Lake sediments and environmental history* (eds E.Y. Hawarth and J.W.G. Lund), Leicester University Press, Leicester, pp. 93–124.

Oldfield, F., Worsley, A.T. and Appleby, P.G. (1985) Evidence from lake sediments for recent erosion rates in the highlands of Papua New Guinea, in *Environmental change and Tropical Geomorphology* (eds I. Douglas and T. Spencer), Allen and Unwin, London, pp. 185–196.

Oldfield, F., Rummery, T.A., Thompson, R. and Walling, D.E. (1979) Identification of suspended sediment sources by means of magnetic measurements: some preliminary results. *Water Resources Research*, **15**, 211–218.

Owens, P.N. (1990) Valley sedimentation at Slapton, South Devon and its implications for the estimation of lake-sediment based erosion rates, in *Soil Erosion on Agricultural Land* (eds J. Boardman, I.D.L. Foster and J.A. Dearing), John Wiley and Sons, Chichester, pp. 193–200.

Owens, P., Walling, D.E., He, Q. *et al.* (1997) The use of caesium-137 measurements to establish a sediment budget for the Start catchment, Devon, UK. *Hydrological Sciences Journal*, **42**, 405–423.

Page, M.J. and Trustram, N.A. (1997) A late Holocene lake sediment record of the erosion response to land use change in a steepland catchment, New Zealand. *Zeitschrift für Geomorphologie NF*, **41**, 369–392.

Paine, J.L., Rowan, J.S. and Werritty, A. (2002) Reconstructing historic floods using sediments from embanked floodplains: a case study of the River Tay in Scotland, Publication 276, International Association of Hydrological Sciences, Wallingford, pp. 211–218.

Pennington, W., Cambray, R.S. and Fisher, E.M. (1973) Observations on lake sediments using fallout ^{137}Cs as a tracer. *Nature*, **242**, 324–326.

Petts, G.E. (1984) *Impounded Rivers: Perspectives for Ecological Management*, John Wiley and Sons, Chichester.

Pittam, N., Foster, I.D.L. and Mighall, T.M. (2009) An integrated lake-catchment approach for determining sediment source changes at Aqualate Mere, Central England. *Journal of Paleolimnology*. doi:10.1007/s10933-008-9272-9

Reheis, M.C. and Redwine, J.L. (2008) Lake manix shorelines and afton canyon terraces: implications for incision of afton canyon, in *Late Cenozoic Drainage History of the Southwestern Great Basin and Lower Colorado River Region: Geologic and Biotic Perspectives*, (eds M.C. Reheis, R. Herschler and D.M. Miller). *Geological Society of America Special Paper*, **439**, 227–259.

Reid, M.A. and Ogdentrend, R.W. (2006) Variability or extreme event? the importance of long-term perspectives in river ecology. *River Research and Applications*, **22**, 167.

Reimer, P.J., Baillie, M.G.L., Bard, E. *et al.* (2004) INTCAL04 terrestrial radiocarbon age calibration, 0–26 Cal Kyr BP. *Radiocarbon*, **46**, 1029–1058.

Ritchie, J.C., McHenry, J.R. and Gill, A.C. (1973) Dating recent reservoir sediments. *Limnology and Oceanography*, **18**, 254–263.

Robbins, J.A. and Herche, L.R. (1993) Models and uncertainty in ^{210}Pb dating of sediments. *Verhandlugen-Internationale Vereinigung fur Theoretische undLimnologie*, **25**, 217–222.

Rose, N.L., Harlock, S., Appleby, P.G. and Battarbee, R.W. (1995) Dating of recent sediments in the United-Kingdom and Ireland using spheroidal carbonaceous particle (S.C.P.) concentration profiles. *The Holocene*, **5**, 328–335.

Rowan, J.S., Goodwill, P. and Greco, M. (1995) Temporal variability in catchment sediment yield determined from repeated bathymetric surveys: Abbeystead Reservoir, UK. *Physics and Chemistry of the Earth*, **20**, 199–206.

Rowan, J.W., Proce, L.E., Fawcett, C.P. and Young, P.C. (2001) Reconstructing historic sedimentation rates using data based mechanistic modelling. *Physics and Chemistry of the Earth, Parts B*, **26**, 77–82.

Rowntree, K.M., Foster, I.D.L., Mighall, T. *et al.* (2008) Post-European settlement impacts on erosion and land degradation; a case study using farm reservoir sedimentation in the Eastern Cape, South Africa, Publication 325, International Association of Hydrological Sciences, Wallingford, pp. 139–142.

Saarnisto, M. (1986) Annually laminated lake sediments, in *Handbook of Holocene Palaeoecology and Palaeohydrology* (ed. B. Berglund), John Wiley and Sons, Chichester, pp. 343–370.

Sanders, R.M. (1997) The characterisation of drainflow sediments from agricultural soils using magnetics, radionuclides and geochemical techniques. Unpublished MSc Thesis, Coventry University.

Shields, F.D. Jr, Simon, A. and Steffen, A.J. (2000) Reservoir effects on downstream river channel migration. *Environmental Conservation*, **27**, 54–66.

Shotbolt, L.A., Thomas, A.D. and Hutchinson, S.M. (2005) The use of reservoir sediments as environmental archives of catchment inputs and atmospheric pollution. *Progress in Physical Geography*, **29**, 337–361.

Simon, A. and Darby, S.E. (2002) Effectiveness of grade-control structures in reducing erosion along incised river channels: The case of Hotophia Creek, Mississippi. *Geomorphology*, **42**, 229–254.

Small, I., Rowan, J.S. and Franks, S.W. (2002) Quantitative sediment fingerprinting using a Bayesian uncertainty estimation framework, Publication 276, International Association of Hydrological Sciences, Wallingford, pp. 443–450.

Smith, N. (1971) *A History of Dams*, Peter Davies, London.

Smith, S.V. Jr, Bradley, R.S. and Abbot, M.B. (2004) A 300 year record of environmental change from Lake Tuborg, Ellesmere Island, Nunavut, Canada. *Journal of Paleolimnology*, **32**, 137–148.

Smol, J.P. (2008) *Pollution of Lakes and Rivers: A Palaeoenvironmental Perspective*, 2nd edn, Arnold, London.

Snowball, I. and Thompson, R. (1988) The occurrence of greigite in sediments from Loch Lomond. *Journal of Quaternary Science*, **3**, 121–125.

Surian, N. and Rinaldi, M. (2003) Morphological response to river engineering and management in alluvial channels in Italy. *Geomorphology*, **50**, 307–326.

Syvitski, J.P.M., Vörösmarty, C.J., Kettner, A.J. and Green, P. (2005) Impact of humans on the flux of of terrestrial sediment to the global coastal ocean. *Science*, **308**, 376–380.

Tamene, L., Park, S.J., Dikau, R. and Vlek, P.L.G. (2006) Reservoir siltation in the semi-arid highlands of northern Ethiopia: sediment yield–catchment area relationship and a semi quantitative approach for predicting sediment yield. *Earth Surface Processes and Landforms*, **31**, 1364–1383.

Tiljander, M., Saarnisto, M., Ojala, A.E.K. and Saarinen, T. (2003) A 3000-year palaeoenvironmental record from annually laminated sediment of Lake Korttajärvi, central Finland. *Boreas*, **32**, 566–577.

Trimble, S.W. (2008) The use of historical data and artifacts in geomorphology. *Progress in Physical Geography*, **32**, 3–29.

Trimble, S.W. and Bube, K.W. (1990) Improved Reservoir Trap Efficiency Prediction. *The Environmental Professional*, **12**, 255–272.

Trimble, S.W. and Wilson, B.N. (In press) Reservoir and lake trap efficiency, in *Encyclopedia Of Lakes, Reservoirs, and Paleolimnology* (eds R. Fairbridge and R. Herschey), Springer, New York.

Van der Post, K.D., Oldfield, F., Haworth, E.Y. *et al.* (1997) A record of accelerated erosion in the recent sediments of Blelham Tarn in the English Lake District. *Journal of Palaeolimnology*, **18**, 103–120.

Verstraeten, G. and Poesen, J. (2000) Estimating trap efficiency of small reservoirs and ponds: methods and implications for the assessment of sediment yield. *Progress in Physical Geography*, **24**, 219–251.

Verstraeten, G. and Poesen, J. (2001a) Factors controlling sediment yield from small intensively cultivated catchments in a temperate humid climate. *Geomorphology*, **40**, 123–144.

Verstraeten, G. and Poesen, J. (2001b) Modelling the long-term trap efficiency of small ponds. *Hydrological Processes*, **15**, 2797–2819.

Verstraeten, G. and Poesen, J. (2002) Using sediment deposits in small ponds to quantify sediment yield from small catchments: possibilities and limitations. *Earth Surface Processes and Landforms*, **27**, 1425–1439.

Verstraeten, G., Poesen, J., de Vente, J. and Koninckx, X. (2003) Sediment yield variability in Spain: a quantitative and semiqualitative analysis using reservoir sedimentation rates. *Geomorphology*, **50**, 327–348.

Vörösmarty, C.J., Meybeck, M., Fekete, B. *et al.* (2003) Anthropogenic sediment retention: major global impact from registered river impoundments. *Global and Planetary Change*, **39**, 169–190.

Vukic, J. and Appleby, P.G. (2005) Spheroidal carbonaceous particle record in sediments of a small reservoir. *Hydrobiologia*, **504**, 315–325.

Wagner, B., Bennike, O., Klug, M. and Cremer, H. (2007) First indication of Storegga tsunami deposits from East Greenland. *Journal of Quaternary Science*, **22**, 321–325.

Walker, M.J. (2005) *Quaternary Dating Methods*, John Wiley and Sons, Chichester.

Walden, J., Slattery, M.C. and Burt, T.P. (1997) Use of mineral magnetic measurements to fingerprint suspended sediment sources: approaches and techniques for data analysis. *Journal of Hydrology*, **202**, 353–372.

Walling, D.E. (1990) Linking the field to the river, in *Soil Erosion on Agricultural Land* (eds J. Boardman, I.D.L. Foster and J.A. Dearing), John Wiley and Sons, Chichester, pp. 129–152.

Walling, D.E. (2006) Human impact on land-ocean sediment transfer by the world's rivers. *Geomorphology*, **79**, 192–216.

Walling, D.E. (2007) Global change and the sediment loads of the world's rivers. inProceedings of the 10th International Symposium on River Sedimentation (Moscow, August 2007), Vol. 1, pp. 112–130.

Walling, D.E. (2008) The changing sediment load of the world's rivers, Publication 325, International Association of Hydrological Sciences, Wallingford, pp. 323–338.

Walling, D.E. and Webb, B.W. (1996) Erosion and sediment yield: a global overview, Publication 236, International Association of Hydrological Sciences, PL Wallingford, pp. 3–19.

Walling, D.E. and Woodward, J.C. (1992) Use of radiometric fingerprints to derive information on suspended sediment sources, Publication 210, International Association of Hydrological Sciences, Wallingford, pp. 153–164.

Walling, D.E., Woodward, J.C. and Nicholas, A.P. (1993) A multi-parameter approach to fingerprinting suspended-sediment sources, Publication 215, International Association of Hydrological Sciences, Wallingford, pp. 329–338.

Walling, D.E., Owens, P.N., Foster, I.D.L. and Lees, J.A. (2003) Changes in the fine sediment dynamics of the Ouse and Tweed basins in the UK over the last 100–150 years. *Hydrological Processes*, **17**, 3245–3269.

WCD (2000) Dams and development: a new framework for decision making. *Report of the World Commission on Dams, An overview*, pp1–40, http://www.dams.org.

White, P., Labadz, J.C. and Butcher, D.P. (1996) Sediment yield estimates from reservoir studies: an appraisal of variability in the Southern Pennines of the UK, Publication 236, International Association of Hydrological Sciences, Wallingford, pp. 491–500.

White, P., Labadz, J.C. and Butcher, D.P. (1997) Reservoir sedimentation and catchment erosion in the Strines catchment, UK. *Physics and Chemistry of the Earth*, **22**, 321–328.

Winter, L.T., Foster, I.D.L., Charlesworth, S.M. and Lees, J.A. (2001) Floodplain lakes as sinks for sediment-associated contaminants - a new source of proxy hydrological data? *Science of the Total Environment*, **266**, 187–194.

Wohl, E. (2006) Human impacts to mountain streams. *Geomorphology*, **79**, 217–248.

Wolfe, B.B., Hall, R.I., Last, W.M. *et al.* (2006) Reconstruction of multi-century flood histories from oxbow lake sediments, Peace-Athabasca Delta, Canada. *Hydrological Processes*, **20**, 4131–4153.

Woodward, J.C. and Foster, I.D.L. (1997) Erosion and suspended sediment transfer in river catchments. *Geography*, **82**, 353–376.

Xiang-Zhou, X. (2004) Development of check-dam systems in gullies on the Loess Plateau, China. *Environmental Science & Policy*, **7**, 79–86.

Xu, J. (2001) Adjustment of mainstream - Tributary relation upstream from a reservoir: An example from the Laohahe River, China. *Zeitschrift für Geomorphologie, NF Hauptbände*, **45**, 359–372.

Xu, K. and Milliman, J.D. (2009) Seasonal variations of sediment discharge from the Yangtze River before and after impoundment of the Three Gorges Dam. *Geomorphology*, **104** (3–4), 276–283.

Yeloff, D.E., Labadz, J.C., Hunt, C.O. *et al.* (2005) Blanket peat erosion and sediment yield in an upland reservoir catchment in the southern Pennines, UK. *Earth Surface Processes and Landforms*, **30**, 717–733.

Zolitschka, B. (1998) A 14,000 year sediment yield record from western Germany based on annually laminated lake sediments. *Geomorphology*, **22**, 1–17.

Zorzou, M. (2004) Reservoir sedimentation and sediment source tracing in a mountainous Mediterranean environment: the Ladonas catchment, Greece. Unpublished PhD Thesis, University of Leeds, UK.

13 Continental-Scale River Basins

David L. Higgitt
Department of Geography, National University of Singapore, Singapore

13.1 Introduction

The role of large rivers in conveying sediments from continents to oceans has long fascinated geoscientists. This allure has been motivated by four broad sets of questions. First, there is the quest to determine the global flux of continent-to-ocean sediment transfer and, by extension, contemporary denudation rates. The extent to which global variations in denudation can be explained by climatic, topographic, human disturbance and geological factors has attracted a number of investigations and employment of numerous techniques for quantifying the various characteristics of earth surface systems within defined drainage basins. Second, concern about land degradation has been scaled up from field studies to ask questions about the global extent of soil erosion and the implications of the downstream transfer of eroded sediment. The lifespan of major water resource projects (especially megadam projects such as Aswan or the Three Gorges) has come under the spotlight and, more generally, the question of the significance of human impact on sediment cascades in the world's largest rivers. Third, with increasing emphasis on global environmental change, attention has switched to the role of sediment in biogeochemical cycles. Large rivers act as the conduits for transferring nutrients from terrestrial sources to the coastal shelf. Human-induced changes in sediment cascades, particularly those associated with the retention of sediment behind impoundments, have implications for large-scale nutrient dynamics and, in turn, for the functioning of aquatic ecosystems. There is potential for dramatic impacts on the productivity of nearshore fish stocks. Fourth, large rivers have provided inspiration for geologists understanding fluvial stratigraphy. The present is the key to the past, enabling large rivers to be studied as analogues for interpreting strata in sedimentary facies. The alluvial stratigraphy provides an archive from which to examine the development of large river basins and their response to environmental change from scales of decades to millions of years.

It is not without coincidence that ancient civilizations arose independently in the lower reaches of continental-scale river basins – the Nile, Tigris–Euphrates, Indus and Yellow

Sediment Cascades: An Integrated Approach Edited by Timothy Burt and Robert Allison

Rivers – their culture in part defined by an ability to contain and utilize the waters and sediment-associated nutrients. What ancient societies began to understand intrinsically remains a challenge for contemporary science. In this chapter current understanding of sediment cascades in large river basins is explored, beginning with assessment of the global pattern of sediment delivery to the oceans. After consideration of the available data and their shortcomings, the challenge of constructing sediment budgets within large river basins is discussed. As the global environmental change agenda has influenced river science, the next sections consider how the impact of disturbance and environmental change has been approached, before summarizing the challenges ahead.

13.2 Global Sediment Yield

Attempting to estimate the global sediment load delivered to oceans has been a popular and persistent theme among geoscientists for more than 50 years. The task has taken on higher levels of sophistication as digital elevation models, geodatabases and computer modelling routines have enabled better representation of river basin characteristics. Concurrently, the increased availability of sediment load data forming the empirical base upon which model output can be tested has both advanced and complicated understanding of the controls on global sediment yield. One of the earliest and perhaps the most celebrated attempts to explain global variations in sediment yield was prepared by Langbein and Schumm (1958). Their analysis of the relationship between mean annual sediment yield and mean annual precipitation was based on 265 drainage basins of varying sizes across the USA, but predominantly from areas of agricultural production in the eastern part of the country. Acknowledging the inherent limitations of the available data, Langbein and Schumm (1958) produced the iconic curve of sediment yield against precipitation, identifying a peak at approximately 300 mm annual precipitation, equating to semi-arid environments. The curve reflects the interplay between erosivity and vegetation cover. In conditions wetter than 300 mm, the enhancement of vegetation cover is assumed to protect the ground from erosion processes, whereas below 300 mm it is assumed to be insufficient to inhibit erosion effectively. Of course, it is apparent that the relation between climate and both vegetation cover and erosion processes depends on many climatic variables beyond mean annual precipitation, most notably rainfall intensity, seasonality and temperature. Consequently, the location of peak sediment yield should be expected to vary between climatic zones. Nevertheless the Langbein–Schumm rule has become a key conceptual model in geomorphology and its potential to explain the impact of past climate change on sediment delivery systems has long been recognized (Scott, Ritter and Knott, 1968; Knox, 1972).

The empirical relationship between fluvial sediment yields and environmental parameters represents one of the ways that a conceptual model can be extended to estimate sediment flux to the global oceans. Thus, Langbein and Schumm (1958) estimated an annual global flux of 10.8 billion tonnes ($10^9\,\mathrm{t\,yr^{-1}}$). Other published estimates vary considerably, ranging from $< 10 \times 10^9\,\mathrm{t\,yr^{-1}}$ (Ahnert, 1970) to $> 50 \times 10^9\,\mathrm{t\,yr^{-1}}$ (Fournier, 1960; Ohmori, 1983). An alternative approach has been based on compilation and extrapolation of available sediment yield data (e.g. Holeman, 1968; Milliman and Meade, 1983; Milliman and Syvitski, 1992). By the mid-1990s, Walling and Webb (1996) were able to synthesize previous studies and note the convergence of estimates around $20 \times 10^9\,\mathrm{t\,yr^{-1}}$, albeit with considerable uncertainty and some geographical bias remaining. The continued improvement of computer models to simulate flowpaths of rivers on the global land mass, coupled with models of water balance and sediment transport, has enabled estimates of global sediment flux to be refined. Syvitski *et al.* (2005), in a paper destined to become the Langbein and Schumm of its time, have computed flux of terrestrial sediment to the oceans under contemporary and pre-human conditions. The methodology is discussed further in the following section. Their calculations predict a global suspended sediment yield of $14.0 \times 10^9\,\mathrm{t\,yr^{-1}}$ under pre-human conditions, reducing to $12.6 \times 10^9\,\mathrm{t\,yr^{-1}}$ for the modern (Anthropocene) era. If bedload is included in the calculations, the pre-human and Anthropocene sediment fluxes are 15.5 and $14.1 \times 10^9\,\mathrm{t\,yr^{-1}}$, respectively. Furthermore, Syvitski *et al.* (2005) estimate that if the amount of sediment trapped by reservoirs were added to the calculations, the global annual flux would be 17.8×10^9 t. This amount is slightly less than the Milliman and Syvitski (1992) estimate of $20 \times 10^9\,\mathrm{t\,yr^{-1}}$ which was based on data from 280 river basins but did not include an upscaling analysis on a river by river basis (Table 13.1).

Following Langbein and Schumm (1958), numerous attempts have been made to explain the global patterns of denudation in terms of climate and/or vegetation cover (Douglas, 1967; Wilson, 1973; Jansen and Painter, 1974) or topography and uplift (Ahnert, 1970; Milliman and Syvitski, 1992; Summerfield and Hulton, 1994). Summarized histories of this work can give the impression of a contentious debate between those in favour of climate control and those espousing the significance of tectonics, a line of argument reflected in the title of a recent paper – 'geology, geography and humans battle for the dominance over the delivery of sediment to the coastal ocean' (Syvitski and Milliman, 2007). In fact many of the early papers specifically identified the significance of multiple controls on sediment yields and the limitations of considering a dominant factor. For example, Lee Wilson's (1973) paper on 'mean annual sediment yield as a function of mean annual precipitation', despite its partisan title, goes to considerable lengths to identify non-climatic factors and concludes that 'there is no simple relationship between mean annual precipitation and mean annual sediment yield, in part because there are too many additional factors that determine sediment yield variations' (Wilson, 1973, p. 347). In particular, Wilson was concerned that land-use factors should be considered appropriately. A similar conclusion was reached by Douglas (1967) whose compilation of data from tropical regions bemoaned the lack of data from undisturbed basins and hence the difficulty of extrapolating 'natural rates' of erosion from present-day disturbed environments. Wilson (1973) found support for the Langbein–Schumm rule if data were subdivided by climate regime. Extending this approach, Jansson (1988) used the Köppen climate classification as the basis for subdivision and found that Af climates (humid tropical without distinct dry season) had the highest sediment yields. Warm temperate climates with a dry season exhibited high sediment yields while boreal climates had low sediment yields, except for those basins in China. Jansson found no consistent correlation between runoff and sediment yield and therefore no evidence that a Langbein–Schumm rule could be imported from one region to another.

To some extent, studies which have attempted to identify the significance of a particular variable have been subject to the idosyncracies of geography, where an apparent pattern may be attributed to other factors, such as enhanced human disturbance in tropical river basins or coincidence between low relief and regions with boreal climates. Multiple regression approaches have therefore been popular. Fournier (1960) conducted a study using data from 96 basins and demonstrated a significant relationship between specific sediment yield ($\mathrm{t\,km^{-2}\,yr^{-1}}$) and a precipitation peakedness index, mean drainage basin

Table 13.1 Notable estimates of global sediment flux from continents to oceans, arranged in decreasing order of estimate. More recent work has converged towards estimates of around 12.5 billion tonnes for contemporary suspended sediment flux, 14 billion tonnes for pre-human impact and 17 billion tonnes for human-induced without trapping behind dams (Syvitksi *et al.*, 2005)

Author	Estimated global sediment yield ($\times 10^9\,t\,yr^{-1}$)	Key observation
Fournier (1960)	64.0	Multiple regression: precipitation peakedness (p^2/P) as key variable
Ohmori (1983)	56.6	Emphasis on precipitation and vegetation cover as controls on erosion
Jansen and Painter (1974)	26.7	Relative effect of controlling variables differs between climatic regions
Milliman and Syvitski (1992)	20.0	Significance of smaller mountainous basins on tectonically active margins
Walling and Webb (1996)	20.0	Review of previous work: Milliman and Syvitksi as best estimate to date
Wilson (1973)	19.3	Qualifies Langbein–Schumm rule on basis of improved dataset
Holeman (1968)	18.3	First attempt to assemble comprehensive global dataset for empirical analysis
Pinet and Souriau (1988)	16.2	Strong relationship between mechanical denudation and relief
Ludwig and Probst (1998)	16.0	Environmental characteristics extracted from geodatabases to construct hydro-geomorphological variables
Dedkov and Gusarov (2006)	15.5	Updated estimate based on large database
Syvitski *et al.* (2005)	14.0	Modelled pre-human load based on digital representation of >4000 basins
Milliman and Meade (1983)	13.5	Reappraisal and update of empirical database
Syvitski *et al.* (2005)	12.6	Modelled 'modern' load based on digital representation of >4000 basins
McLennan (1993)	12.6	Developed Chemical Alteration Index (CIA) relating sediment yield and weathering
Douglas (1967)	11.5	Human disturbance complicates any attempt to infer 'natural' rates
Langbein and Schumm (1958)	10.8	'Rule' that sediment yield peaks at mean annual precipitation of 300 mm
Ahnert (1970)	9.3	Attention to topography, lithology and geological history as determinants of sediment yield

elevation and slope. The precipitation peakedness index (p^2/P) combines the highest monthly rainfall with the mean annual precipitation and hence is a measure of seasonality. More recent studies have focused more towards the influence of topography on global sediment yield. Milliman and Syvitski (1992) identified a log-linear relationship between sediment yield ($t\,yr^{-1}$) and basin area and maximum elevation, drawing particular attention to the significance of relatively short, steep rivers in tectonically active areas (such as Taiwan and Indonesia) as major suppliers of global sediment flux to the oceans. Hovius (1998) has extended this analysis by undertaking a stepwise regression which combines five variables as the most efficient predictor of specific sediment yield. These variables are: basin area (km^2), maximum elevation (m), mean annual temperature (°C), annual temperature range (°C) and runoff ($mm\ yr^{-1}$). The equation explains just under 50 per cent of the variation in specific sediment yield based on data from 86 of the world's largest rivers, using data that were intended to reflect the natural (i.e. pre-human) sediment dynamics. While achieving almost 50 per cent explanation is a good performance, the implication is that just over 50 per cent of the variation remains unexplained by the estimator variables. Hovius (1998) considered this problem and identified that it may be due to errors and quality issues in the database including the problem of screening out human impact, to lag effects of contemporary sediment yields reflecting past environmental controls (such as the legacy of past glaciations), or to omission of variables that influence all aspects of the sediment cascade. A systematic trend in the standardized residuals of the multiple regression equation indicated the influence of a control variable that was not included in the analysis: the rate of uplift. By segregating the larger database of Milliman and Syvitski (1992) according to tectonic setting, Hovius (1998) was able to demonstrate that river basins located in areas of active contractional strain have high specific sediment yields (100–10 000 $t\,km^{-2}\,yr^{-1}$) while basins draining cratons rarely exceeded 100 $t\,km^{-2}\,yr^{-1}$.

Continental sediment cascades can therefore be regarded as being a function of six interrelated variables: basin size, climate, geology (and soil), topography (including tectonic setting), vegetation characteristics, and disturbance, both natural and imposed. Natural disturbances include fire, vegetation disease, slope failure, and volcanic or seismic activity, whereas imposed disturbances include construction, agriculture, urbanization, and many other land-use and water-use practices. These anthropogenic influences complicate the identification of controlling factors on sediment load.

13.3 Basin Settings, Data Availability and Uncertainty

Summary water and sediment discharge data for the world's largest river basins are provided in Table 13.2. There are 53 basins with a surface drainage area larger than 200 000 km^2. Geologically, the majority of contemporary large river basins have their headwaters in zones of active or relict uplift and flow through extensive cratonic areas. Tandon and Sinha (2007) demonstrate the architecture of basin-forming mechanisms in relation to plate tectonic processes. Continental collision belts are the most important locations to generate large river basins, the Himalaya providing the most spectacular example, spawning seven of the world's largest rivers (Indus, Ganges, Brahamaputra, Irrawaddy, Salween, Mekong, Yangtze). The last three run through deep gorges, spectacularly bending around the eastern margin of the Himalayan syntaxis, resulting from a combination of shear zone development and clockwise rotation as the Tibetan Plateau and Southeast Asia provinces interacted (Brookfield, 1998) and a complex history of river capture (Clark *et al.*, 2004). While the Ganges drainage basin has evolved to flow axial to the Himalaya mountain front, the Indus flow is transverse to the source area and is believed to have occupied broadly the same location – the axis of the Suliaman foredeep – since the onset of collision. As it has evacuated sediments from the collision zone and aggraded the alluvial plain, the channel course has shifted to east and west in response to subsidence and localized upwarping

Table 13.2 The size, water discharge and sediment yield data for the world's largest rivers (53 rivers > 200,000 km^2) arranged in rank order of basin size (locations in Figure 13.1). Data used for subsequent analysis are derived from Hovius (1998) with following qualifications: (i) the Hovius correction for 'natural' suspended sediment load is used for Nile, Colorado and Mississippi; (ii) revised figures from recent analysis for Irrawaddy and Salween (Robinson *et al.*, 2007); (iii) use of measured load for Yellow River (Milliman and Meade, 1983) rather than presumed pre-human load. The large range in reported values for basin size and sediment load are derived from multiple sources (Milliman and Meade, 1983; Milliman and Syvitksi, 1992; Summerfield and Hulton, 1994; Koronkevich, 2003; Wohl, 2007; Zhang *et al.*, 2008).

	River	Basin area (10^6 km^2)	Indicative range for basin area	Mean annual water discharge (km^3 yr^{-1})	Suspended sediment load (10^6 t yr^{-1})	Indicative range for sediment load	Specific sediment yield (t km^{-2} yr^{-1})	Notes
1	Amazon	6.15	5.90–6.92	6307	1150	900–1320	187	
2	Zaire (Congo)	3.70	3.34–3.75	1290	33	33–51	9	
3	Mississippi	3.34	3.22–3.34	580	400	210–605	120	Large difference between low contemporary and high historic sediment load
4	Nile	2.72	2.72–3.35	10	125	100–125	46	Post-Aswan sediment load effectively zero
5	Parana (La Plata)	2.60		1101	112	87–112	43	See also number 51 Uruguay, which also flows into La Plata estuary
6	Yenisei	2.58	2.58–2.62	561	13		5	
7	Ob	2.50	2.50–2.99	385	16		6	
8	Lena	2.43	2.43–2.49	511	12		5	
9	Chiang Jiang (Yangtze)	1.94	1.90–1.94	899	480	451–480	247	
10	Amur	1.86	1.85–2.05	325	52	25–57	28	
11	Mackenzie	1.45	1.45–2.00	310	125	100–125	86	
12	Zambezi	1.40	1.32–1.40	223	15	15–48	11	
13	Volga	1.35		265	26		19	Drains into Caspian Sea
14	St Lawrence	1.19		451	4	2–4	3	
15	Niger	1.11	1.11–2.27	190	32	32–40	29	
16	Shatt al Arab	1.05		46	103	46–50	98	Tigris and Euphrates

17	Orange	1.02		91	91	17–91	89	
18	Ganges	0.98	0.95–1.06	366	524	524–680	535	Many studies combine Ganges and Brahmaputra
19	Huang He (Yellow)	0.98	0.75–0.98	49	1080	100–1100	1102	Historic sediment yield. Assumed ‘natural’ load = 100
20	Indus	0.96	0.96–1.17	240	250	100–300	260	
21	Orinoco	0.95	0.96–1.10	1101	150	150–210	158	
22	Murray	0.91	0.91–1.07	22	30	30–33	33	Often referred to as Murray–Darling system
23	Chari	0.88		42	4	3–4	5	Drains into Lake Chad
24	Yukon	0.85		195	60	60–79	71	
25	Danube	0.82		210	70	67–74	85	
26	Mekong	0.81		470	160	160–176	198	
27	Rio Grande	0.67		3	30		45	
28	Columbia	0.67		250	15	8–15	22	
29	Kolmya	0.65		71	6		9	Majority of basin within Arctic Circle
30	Colorado	0.64		1	150	120–167	23	Contemporary water discharge and sediment load effectively zero
31	Sao Francisco	0.64		97	6		9	
32	Brahmaputra	0.61	0.58–0.63	609	520	520–1157	852	Many studies combine Ganges and Brahmaputra
33	Dnepr	0.50		52	2.1		4	
34	Zhu Jiang (Pearl)	0.46		300	80	54–80	174	Zhu Jiang is collective name for three systems (Xi Jiang, Bei Jiang and Dong Jiang) which drain into the delta
35	Limpopo	0.44		5	33		75	
36	Senegal	0.44		24	1.9		4	
37	Don	0.42		27	6		14	

(continued)

Table 13.2 (*Continued*)

	River	Basin area (10^6 km^2)	Indicative range for basin area	Mean annual water discharge (km^3 yr^{-1})	Suspended sediment load (10^6 t yr^{-1})	Indicative range for sediment load	Specific sediment yield (t km^{-2} yr^{-1})	Notes
38	Irrawaddy	0.41		429	364	260–364	888	Sediment load estimate quoted in many studies is from a nineteenth century study (Gordon, 1885)
39	Volta	0.39		40	19		49	
40	Indigirka	0.36		55	14		39	Majority of basin within Arctic Circle
41	Sevemaya Dvina	0.35		106	4.5		13	
42	Salween	0.33	0.27–0.33	300	110	100–164	333	Perhaps least modified river in Asia, though dam construction is planned
43	Pechora	0.32		106	6.1		19	Basin partly within Arctic Circle
44	Amu Darya	0.31	0.31–0.47	46	94		303	Larger basin size estimate may include Syrdar'ya River. Drains into Aral Sea
45	Godavari	0.29		92	170		586	
46	Ural	0.28		9	3		11	Drains into Caspian Sea
47	Magdalena	0.26		220	220	150–220	846	
48	Krishna	0.26		51	65		250	
49	Rhine	0.26		71	0.7		3	
50	Jana	0.24		29	3		13	Majority of basin within Arctic Circle
51	Uruguay	0.24		158	11		46	
52	Fraser	0.22		112	20		91	
53	Syr Darya	0.22		18	12		55	Drains into Aral Sea

(Jorgensen *et al.*, 1993). A second tectonic setting for large river basins to form is the commencement of rifting (Tandon and Sinha, 2007) which generates new relief. Examples include the Congo River, which drains away from the shoulder of the East African Rift, while the Blue Nile rises within the Ethiopian Rift system. The third geological category for the formation of large rivers is a cratonic setting. Doming associated with mantle plumes may provide the topographic highs from which drainage flows. Examples include the Orange and Zambezi systems in Africa and lower parts of the Paraná system in South America (Tandon and Sinha, 2007). Several large basins have the majority of their flow length in cratonic settings but headwaters in former orogens that are no longer active.

The consequence of the tectonic influences on large river basin formation is reflected in the geography of the continental-scale drainage (Figure 13.1). The Amazon is by far the largest river basin, at 6.15 million km^2 equivalent to 5.8% of all non-ice-covered terrestrial surfaces. It discharges 16.4% of global freshwater to the oceans and 8.4% of the global sediment yield, essentially representing the transport of Andean material across a craton into the Atlantic Ocean, as Potter (1978) remarked: 'the translocation of sediment from the leading edge to the passive margin of a plate'. In North America, the Mississippi, at about half the size of the Amazon basin, nevertheless drains one sixth of the total land mass of North America. A remarkable array of huge catchments is found in the territory of the

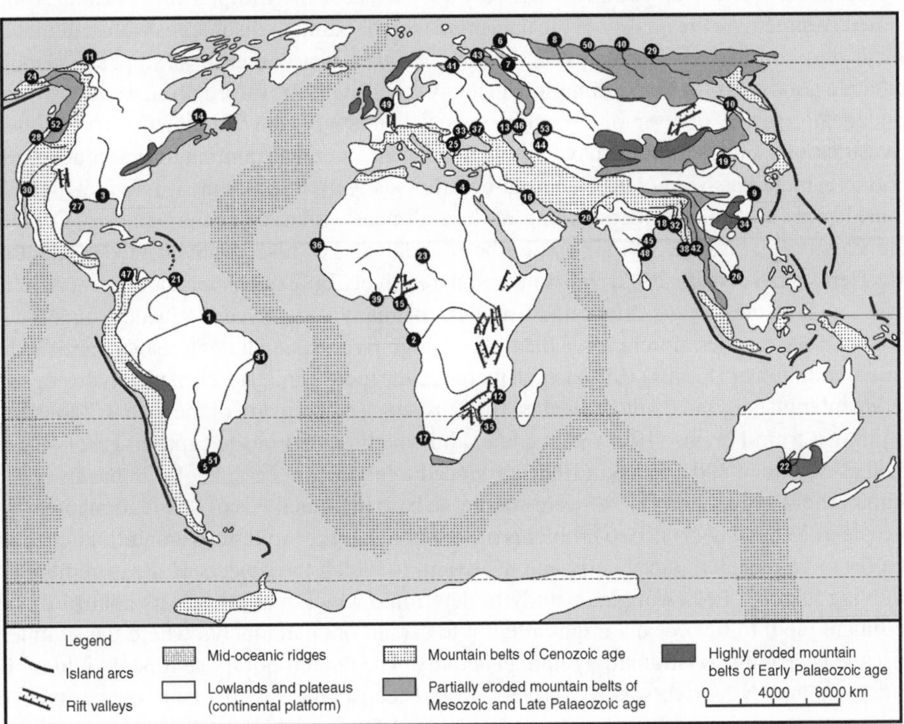

Figure 13.1 Location of large river basins (> 200,000 km^2) superimposed on major morphological features of the Earth (modified from Summerfield, 1991). The importance of a large shield area adjacent to a contemporary or relict uplifted zone is apparent. Key to river names in Table 13.2. Presentation based on an idea by Tandon and Sinha (2007)

former USSR, particularly the great Siberian rivers flowing northwards to the Arctic Ocean. The largest northward flowing rivers are, from west to east, the Sevemaya Dvina, Pecora, Ob, Yenesi, Lena, Jana, Indigirka and Kolmya, of which all but the first discharge north of the Arctic Circle. The major southerly flowing basins in Ukraine and western Russia are the Dnepr, Don, Volga and Ural. The first two drain into the Black Sea (the Don via the embayment of the Sea of Azov) and the latter two into the Caspian Sea. Internal drainage is also encountered in two great Central Asian rivers – the Syr Darya and Amu Darya which drain northwest from the uplifted Tien Shan and Pamir mountain ranges through the territories of Kazakhstan and Turkmenistan/Uzbekistan, respectively, into the Aral Sea. Of these two, the Amu Darya has a higher estimated annual specific sediment yield of $303\,t\,km^{-2}$, but the rivers are better known for the catastrophic impact of Soviet-era irrigation abstraction on the water balance of the Aral Sea and its consequent impact on the ecosystem (Nilsson and Berggren, 2000). A fifth large river draining to an internal water body is encountered in Africa where the Chari terminates in Lake Chad, although both water and sediment discharge is very low.

A key element in the development of estimates of global sediment yield, described in the previous section, has been the gradual build up of reliable hydrological data. Knowledge about the sediment dynamics of large rivers, or for that matter about the detailed geomorphological evolution of continental-scale river basins, remains patchy. In presenting new data on worldwide sediment delivery to the oceans, Milliman and Meade (1983) passed comment on the quality of the data upon which sediment discharges were calculated. While they considered Chinese rivers (Yellow, Yangtze, Pearl), the Mississippi and Danube to have a good basis for load estimation, in many other rivers the information was considered inadequate. In the quarter of a century since their study, the combined effort of field investigations and modelling have allowed some of the uncertainties to be reduced. For example, there has been concerted field-based investigation in South America, which has considerably advanced understanding of the sediment dynamics of the Amazon, Orinoco and Paraná (e.g. Meade, 2007), while the Magdalena has also received recent attention (Restrepo and Syvitski, 2006). In comparison, relatively little is known about great African rivers such as the Congo, Niger and Zambesi, or many Asian rivers outside China.

One of the most comprehensive datasets of large river basin characteristics published is the compilation of Hovius (1998) which recorded the morphometric, climatic, hydrological, sediment transport and denudation data for 97 rivers. The majority of the data in Table 13.2 are drawn from Hovius (1998) with some adjustments noted in the caption. Later studies such as Milliman and Syvitski (1992) produced a larger sample size of 225 basins, while Summerfield and Hulton (1994) were content to base their analysis of denudation rates on a sample of 33 large basins. Two problems of sediment load compilations are the tendency for figures to be recycled uncritically and the extent to which the objects of measurement are 'moving targets'. For example, a study to determine which natural factors control fluvial sediment yield to the oceans might attempt to screen out catchments where the sediment cascade is heavily constrained by human activity. The contemporary sediment yield of the Colorado and Nile rivers is effectively zero because the downstream conveyance of sediments from source area has been intercepted by impoundments. Pre-impoundment values of $150\,Mt\,yr^{-1}$ and $125\,Mt\,yr^{-1}$ have been used in Table 13.2. On the other hand, the investigator might wish to include the impacts of disturbance in the analysis and utilize the best available empirical data. The Huang He (Yellow River) provides an illustration of this point. Hovius (1998) suggested a sediment load of $120\,Mt\,yr^{-1}$, reflecting the perceived natural rate of denudation prior to human-induced soil erosion in the Loess Plateau. By contrast, Milliman and Meade (1983) provided a 'new' estimate of 1080 Mt yr^{-1}, but one considerably smaller than the 1890 Mt yr^{-1} used by Holeman (1968). The combined effect of the erodible terrain of the Loess Plateau, high-intensity convective storms, an actively expanding gully network, high rates of soil erosion from arable and grazing activities, and several millennia of human occupation have made suspended sediment concentrations in the Huang He by far the highest on the planet, where monthly average suspended concentrations could be as much as 70 g L^{-1} (Milliman and Meade, 1983). Milliman and Meade (1983) noted that the earlier estimates of sediment load used a gauging station on the main river close to where it exits from the Loess Plateau, whereas a record of more than three decades was available from a station at Lijin, more than 800 km further downstream. Between these two stations, the Huang He was observed to deposit between a third and half of its Loess Plateau load. Undoubtedly, the sediment load leaving the Loess Plateau has continued to drop as the combined influences of a proactive soil conservation campaign, reduced rainfall and increased water abstraction have reduced source input while the construction of the Sanmenxia Dam has reduced conveyance (Walling and Fang, 2003). Similarly, the current sediment load of the Mississippi (210 Mt yr^{-1}) is substantially less than historic values (Knox, 2007). The 'natural' sediment load probably lies between the two since the earlier part of the measured record includes the impact of European settlement and land-use change (see Chapter 11).

The reliability of data in Table 13.2 is also called into question by some discrepancies. Alongside the 'best estimate' Table 13.2 indicates ranges for catchment area and sediment load reported in the literature. Even catchment area estimates vary as much as 20% for the Amazon, Nile, Ob, Niger, Yellow, Indus, Salween and Amu Darya. Sediment load data compilations are derived from scientific papers and reports prepared by national or international organizations. Frequently, these data are recycled from one compilation to another. An interesting example is the reported sediment load for the Irrawaddy. Ranked thirty-eighth in terms of catchment area, but tenth in terms of water discharge and seventh in terms of sediment load, the Irrawaddy, together with its near neighbour the Salween, is a significant contributor of sediment to the global ocean but has received little attention, not least because of the difficulty of gaining access to Myanmar. In fact, the sediment load figures for the Irrawaddy are derived from a nineteenth-century study by Gordon (1885), a civil engineer in charge of river works for the Irrawaddy Flotilla Company. His measurements of discharge were undertaken at Seiktha in order to calculate the required dimensions of flood embankments. Gordon's report to the Royal Geographical Society presented monthly discharge data between 1869 and 1879 and a single year of sediment load data (1877–1878). An annual sediment load value of 261 Mt yr^{-1} was presented in the 1885 report and it is this figure which has been cited in subsequent studies (Stamp, 1940; Milliman and Syvitski, 1992; Hovius, 1998). Milliman and Meade (1983), while citing a slightly higher sediment load of 285 Mt yr^{-1}, note that the recycling of the nineteenth-century data in subsequent reports made it difficult to determine the validity of the original data, further commenting that the 1885 report did not discuss the methods used to calculate sediment load. In an unusual piece of detective work, Ruth Robinson (Robinson *et al.*, 2007) tracked down Gordon's original full report in the archives of the Royal Geographical Society. Running to 550 pages, this report contains the full daily discharge, rainfall and sediment concentration data sets together with the detailed description of sampling techniques not included in the summary 1885 report. The data are of remarkable quality, covering a full range of flows including monsoon peak discharge. During the year 1877–1878, sediment load was measured six days per week (excluding Sundays, holidays and adverse weather conditions) at three positions across the river and at three depths (about 1 m below surface, mid-depth and 1 m above river bed). Subsamples from each sampling depth were combined over the six days to produce 52 weekly samples. Robinson *et al.* (2007) have reanalysed Gordon's original data and compared them with new field-based measurements. The four sources of uncertainty in the original data concern the accuracy of the velocity measurements; whether the sampling protocol captured all the suspended sediment; the merging of stage–discharge relationships from two locations (Gordon used a longer stage record from a downstream station at Myanaung); and the relative contribution of suspended sediment near the channel bed to the daily depth-averaged concentrations. The re-evaluation identified some minor arithmetic errors in the original discharge calculations (accounting for a 4% decrease in discharge); a refit of the sediment–discharge rating curve (accounting for a 27% increase in sediment load); and a correction based on analysis of filtered sediment concentration data which suggest that Gordon underestimated the $> 0.45\ \mu m$ fraction by 18%. This suggests that the frequently cited sediment load of 261 Mt yr^{-1} may be a significant underestimate and the new estimate is 360 ± 60 Mt yr^{-1} (Robinson *et al.*, 2007). Hence, the combined Irrawaddy–Salween system plays a more significant role in Himalayan denudation and global continent to ocean sediment flux than hitherto realized. A further piece of detective work in Myanmar itself (Furuichi, Win and Wasson, 2009) has unearthed some suspended sediment measurements from mid-twentieth century, which appear to support the higher sediment load calculated by Robinson *et al.* (2007). Continuing field seasons in Myanmar are planned to refine these preliminary estimates of sediment and carbon flux.

Noting the limitations inherent in a compilation of sediment load data from the world's largest rivers, it is nevertheless possible to make some general comments. In terms of absolute sediment load the top ten largest contributors to the global oceans are the Amazon, Yellow, Ganges, Brahmaputra, Yangtze, Mississippi, Irrawaddy, Indus, Magdalena and Mekong. Specific sediment yield normalizes load to catchment area (t km^{-2} yr^{-1}). The top ten large rivers in terms of specific sediment yield are Yellow, Irrawaddy, Magdalena, Brahmaputra, Godavari, Ganges, Salween, Amu Darya, Indus and Krishna. These sediment yield data seldom represent natural levels. The two rivers draining the Indian Shield (Godavari, 586 t km^{-2} yr^{-1}; Krishna, 250 t km^{-2} yr^{-1}) reflect the impacts of human disturbance through deforestation (Subramanian, 1996; Kale, 2002). While an analysis of sediment load from 32 tributaries in the Magdalena catchment indicates that specific sediment yield can be explained partly in terms of natural variables, it is evident that anthropogenic changes, notably deforestation, agriculture, mining activities and urbanization, have contributed to the spectacular contribution of sediment to the Caribbean (Restrepo and Syvitski, 2006). Smaller river basins ($< 200\,000$ km^2) with very significant sediment loads include the Red River (Hong He) flowing through China and Vietnam (160 Mt yr^{-1}); Purari-Fly system in Papua New Guinea (110 Mt yr^{-1}); the Copper, draining the Alaska Range (70 Mt yr^{-1}); the Choshui, the largest river in Taiwan (66 Mt yr^{-1}); and the Liahe in northeastern China (41 Mt yr^{-1}) (Meade, 1996), while the combined contribution of smaller rivers draining Indonesia is also important to the global inventory (Milliman and Syvitski, 1992). Figure 13.2 plots the cumulative percentage of terrestrial land area, water discharge and sediment load (using data of Syvitski *et al.* (2005) for total land area, discharge and sediment load). Half the non-ice-covered Earth surface is drained by the

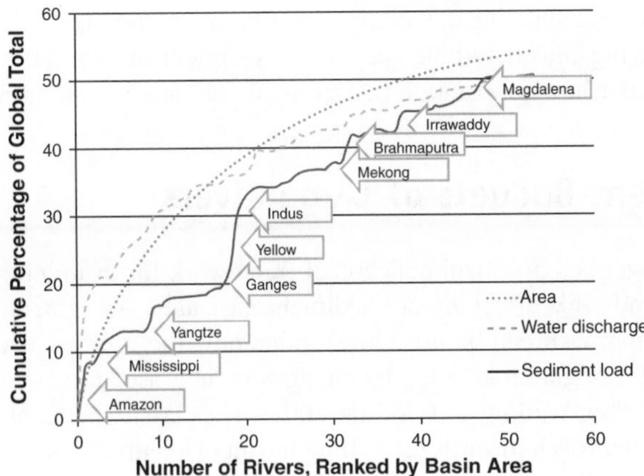

Figure 13.2 Cumulative proportion of global land surface, water and sediment discharge contributed by river basins $> 200{,}000\,km^2$

48 largest rivers. As specific sediment yield is known to be scale dependent (as opportunities for sediment storage increase with catchment area), it is no surprise that cumulative sediment load lags behind cumulative land area drained, though the contribution of individual rivers is apparent in the steps in the cumulative plot.

As noted above, the configuration of collisional zones and adjacent shields provides the necessary scale for continental-scale river basins to develop. An enormous quantity of water is drained into the Arctic Ocean from Eurasian and North American rivers but, with the exception of the Mackenzie, which has formed a sizeable delta in Canada's Northern Territory, the amount of sediment transferred to the Arctic Ocean is limited. Slaymaker (2008) cautions that sediment budget studies in cryospheric environments have been limited and must be improved to assess future impact of global warming in cold environments. By contrast, much of the sediment enters the oceans in tropical regions. Indeed, of the top ten sediment suppliers, only the Yellow (37°N), Yangtze (32°N), Mississippi (29°N) and Indus (24°N) debauch into oceans outside the Tropic *sensu strictu*.

Moving beyond the limitations of the empirical data base, Syvitski *et al.* (2005) have used a modelling approach to examine the structure of sediment load distribution. The approach uses the Simulated Topological Network 30 minute (STN-30p) model to calculate potential river flowpaths and hence to organize data into defined river basins. A total of 6292 river basins of area $>100\,km^2$ were defined of which 4464 were not covered by ice sheets in Antarctica, Greenland or northern Canada. Next, the authors used the University of New Hampshire water balance and transport model (WBM/WTM) to estimate discharge, followed by a drainage basin flux model (DBFM) to predict flux of suspended sediment for each of the major climatic regions. A version of this model, the Area Relief Temperature sediment delivery model (ART), was used to estimate pre-human sediment loads. For the Anthropocene period, Syvitski *et al.* (2005) merged empirical data from major rivers with another version of the DBFM model, the Discharge Relief Temperature model (QRT), which is able to account for human-induced changes in runoff due to abstractions and diversions and the trapping of sediments in reservoirs. As the data in Table 13.2 purport to represent the 'natural', or at least pre-Anthropocene sediment loads, the Syvitski *et al.* (2005) estimate of pre-human flux of $14\,030\,Mt\,yr^{-1}$, from $106\,M\,km^2$ of land area, is used as the basis for calculating the relative contribution of the world's largest rivers to the global oceans. Having considered the question of sediment supply to the oceans, it is now important to consider how to examine the sediment cascade *within* large river basins.

13.4 Sediment Budgets of Large Rivers

The sediment budget is a powerful conceptual framework for examining the relationship between source and sink in catchment sediment cascades. As a management tool, the sediment budget approach can be used to examine how the sources, sinks and pathways of sediment transfer might be affected by changes in land use, climate or extreme events (Reid and Dunne, 1996; Walling and Collins, 2008). The classic sediment budget is derived from a field-based approach towards estimating the inputs, outputs and storage of sediment and hence it has tended to be used for relatively small catchments. Continental-scale rivers go well beyond the manageable dimensions for field-based investigations and hence provide a special challenge to those compiling sediment budgets. However, the development of techniques in remote sensing (as well as increasing availability of satellite imagery) have made it more convenient to examine large river systems over considerable lengths (Mertes and Magadzire, 2007) and the improved resolution of geodatabases and digital elevation models have made it easier to derive spatially distributed representations of catchment characteristics.

Data on sediment transport by large rivers provide information on the sediment load at the terrestrial–marine interface but do not indicate the source area of the material or the history of delivery from source to sink. Questions about where the sediments are derived from, the locations and processes responsible for initializing erosion and the controls on how the sediment is delivered to the river network are issues that are more easily quantified in small basins. Sediment delivery, reflecting differences in the rate of sediment production and transport, varies spatially and temporally and is difficult to measure directly. Some progress can be made from examining sediment load at various gauging locations within the catchment, but a distinction between catchment and network sediment delivery (Richards, 1993) is necessary, the former concerning the proportion of sediment produced by gross erosion that is transferred into and through the river system and the latter the downstream conveyance of sediment load once delivered to the channel. Thus, while sediment budgets for large rivers may identify the key source areas as certain tributaries, the ability to specify the processes of sediment production and the rates of intermediate deposition in terms of landform element (e.g. alluvial floodplain, colluvium, channel bars) is often lacking. In tackling the challenge of constructing sediment budgets for large rivers, the topics of establishing suspended sediment archives and conducting fieldwork for rapid appraisal are discussed below. The discussion is framed around experience of working on sediment budgets for large Chinese rivers, in particular for the Upper Yangtze, during the construction phase of the Three Gorges Dam.

The availability and quality of suspended sediment data have already been identified as major constraints on determining the role of large rivers in transferring sediments from terrestrial to marine environments. Large volumes of secondary data on hydrological and water quality can be obtained from selected Asian rivers but much effort is required to put the records into a usable form. In China, data were published in Hydrological Year Books and released into the public domain until 1987. After 1987, data were not disclosed but have subsequently been made available for purchase, under certain restrictions. For the construction of sediment budgets for the Upper Yangtze, information from printed yearbooks was transcribed into a database over a period of several months. The database contained annual and monthly sediment discharge data (kg s^{-1}) from which a sediment flux could be calculated. A total of 255 gauging stations operated upstream of Yichang at various stages during the period 1956–1987, providing a database of 3820 station years from which 62 stations with long-term records were selected to explore spatial and temporal patterns of sediment delivery. There are several issues that impact on the analysis of suspended sediment records. First, scale is an important factor since sediment delivery ratios are strongly scale dependent (Walling, 1983). As basins increase in size, opportunities for sediment storage (accommodation space) increase disproportionally as mean gradient reduces. Second, sediment concentrations are strongly influenced by weather conditions. In the Yangtze, the vast majority of the annual sediment yield is transported in the summer monsoon period (June to September). Identifying underlying trends in sediment yield over decadal scales necessitates removing the influence of year to year hydrological variation. Third, the monthly sediment discharge data may not be representative of the mean value during the month in question. Milliman and Meade (1983) pointed out that an irregular sampling programme on monsoon-dominated rivers may considerably underestimate the maximum sediment concentrations associated with the flood peak, though this may be less of a problem than in smaller catchments where suspended sediment concentration is strongly controlled by individual storm events (cf. Meybeck *et al.*, 2003). Xu *et al.* (2005) were able to use tightly constrained rating curves from six stations along the Yangtze main stem to simulate sediment flux during the 1998 flood event (60-year recurrence). The analysis shows that the annual sediment yield from both the upper Yangzte (Three Gorges) and the last gauging station before the sea was close to twice the long-term average, while the amount of sediment deposited in the floodplains of the Middle Yangtze was more than six times the average level.

Alternative representations of network sediment delivery for selected large rivers (Orinoco, Ganges and Yangtze) are shown in Figure 13.3. Plotted alongside water discharge, these diagrams provide a simple yet effective way of visualizing the major sediment sources. Using multivariate techniques, it is possible to examine the degree to which sediment yield can be explained in terms of catchment characteristics. A variety of climatic, topographic, geological and land-use data can be extracted from geodatabases and manipulated in a GIS framework to determine variables for any defined subcatchment within the larger basin. More than 95% of the variation in sediment yields in the western part of the Upper Yangtze could be explained by the topographic variables: mean slope and mean elevation (Lu and Higgitt, 1999). However, in the eastern part of the basin, where there is greater human impact through agriculture, the multiple regression technique was less successful in explaining the spatial variability of sediment yields, even when land-use data and a sloping agricultural land index were introduced (Higgitt and Lu, 1999). Partly this reflects the limited (1 km) resolution of geodatabases available at the time of the analysis. As higher resolution data become available, there is much potential to explore spatial patterns of sediment yield. Similar multivariate approaches to examining spatial variability of sediment yield have been attempted in the Magdalena (Restrepo and Syvitski, 2006) and Upper Indus (Ali and de Boer, 2008).

Given the enormous dimensions of large rivers, it might appear that field-based investigations of sediment budgets would not be feasible. However, computer-based analysis of suspended sediment data are useful for identifying 'hotspots' of high sediment yield and/or

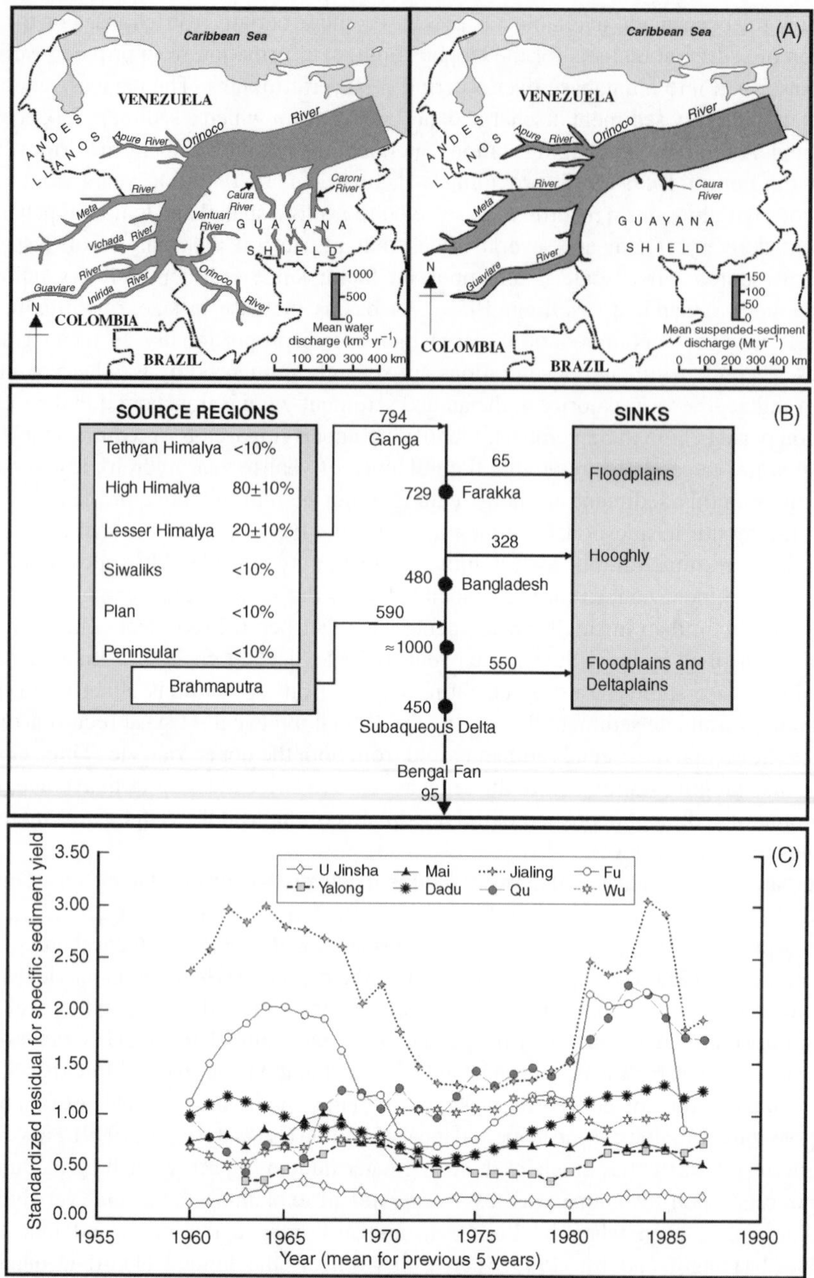

Figure 13.3 Representations of catchment sediment budgets for large rivers. (A) Average discharge of suspended sediment and water in the Orinoco River and its tributaries, indicating the essential structure of the sediment cascade as a conveyance of Andean sediment to the Atlantic (redrawn from Meade, 2007). (B) A tentative sediment budget for the Ganges–Brahmaputra basins (modified from Wasson, 2003). The contribution from the High Himalaya is based on Nd-Sr isotope signatures. (C) The changing contribution of tributaries to the sediment yield of the Upper Yangtze, 1956–1987

areas where sediment yield has increased over time. In the Upper Yangtze, analysis of time series of 62 subcatchments (Lu and Higgitt, 1998) identified the Dadu tributary in western Sichuan as an increasingly significant source area. In order to test the notion that the increase was due to the expansion of agricultural land, a short erosion audit expedition was conducted by traversing a distance of about 200 km along the Dadu valley, mapping the location of erosion features Higgitt (in press). The erosion features comprised mainly debris flows feeding into talus cones, rotational slides and gully complexes. Such a rapid appraisal method does not provide an opportunity to fully record the dimensions of every erosion feature, nor meet the detailed and meticulous standards of measurement that might be expected of a rigorous scientific investigation, but it can provide a concise picture of the key processes. In this case, the density of erosion features per 5 km reach and the extent of agricultural land on the valley slopes was recorded (Figure 13.4). The density of landslide scars approaches 5 km^{-1} (i.e. 25 features per 5 km reach) in the vicinity of Luding, where the valley sides had been extensively cleared for terrace field agriculture. South of Luding, the extent of agricultural conversion is slightly less and the density of erosion features decreases. At Shimian where the river turns east, the valley is very narrow and the forest cover remains intact. Where there has been limited disturbance of the land cover, very few erosion features are observed. The rapid appraisal therefore suggests that the conversion to agricultural land is the main driver of increased sediment production which is reflected in increased sediment yield. Many of the fields on the steep valley slopes have been abandoned as a major conservation initiative has sought to reafforest the valley, yet these abandoned fields represent a continuing sediment source. The field observations have been backed up by analysing false colour Landsat imagery from 1975, 1986 and 2001 which depict the progressive destruction of the vegetation cover along the river corridor. Some of the erosion features are large enough to be seen on the imagery.

Initiating monitoring programmes to measure suspended sediment directly in large rivers is an enormous logistical challenge and most of the investigations cited in the chapter have relied on secondary data from hydrographic stations operated by national authorities. However, in data-scarce environments, primary data collection may be the only viable means of investigating sediment dynamics. The field study in Myanmar described in the previous section has established gauging points on both the Irrawaddy and Salween rivers and a regular programme of water sampling from different depths, combined with acoustic Doppler current surveys (ADCP) surveys of velocity fields and turbidity. Preliminary results have been published (Robinson *et al.*, 2007; Bird *et al.*, 2008) but several years of data are required to ensure representativeness. On the Upper Niger River, Picouet, Hingray and Olivry (2001) determined the structure of hysteresis in annual floods based on a seven-year monitoring programme, while Coynel *et al.* (2005) maintained monthly monitoring of the Congo River and some of its major tributaries for periods of 2–6 years in order to evaluate the sediment and organic carbon yield. The Congo has very low sediment concentrations but high dissolved organic carbon. They estimate that the combined organic carbon input from the Amazon, Orinoco and Congo is 15–18% of the total global continent to ocean flux. In the Puruvian

Figure 13.3 *(Continued)* Contribution is expressed as a five year running mean of standardized residual from a sediment-yield–catchment-area regression equation. The plot reveals both a phase of lower sediment yields in the 1960s to 1970s and a change in the relative importance of particular tributaries. While the Jialing, draining the Sichuan Basin remains the dominant source, the Qu and Dadu supplied an increasing contribution during the time frame

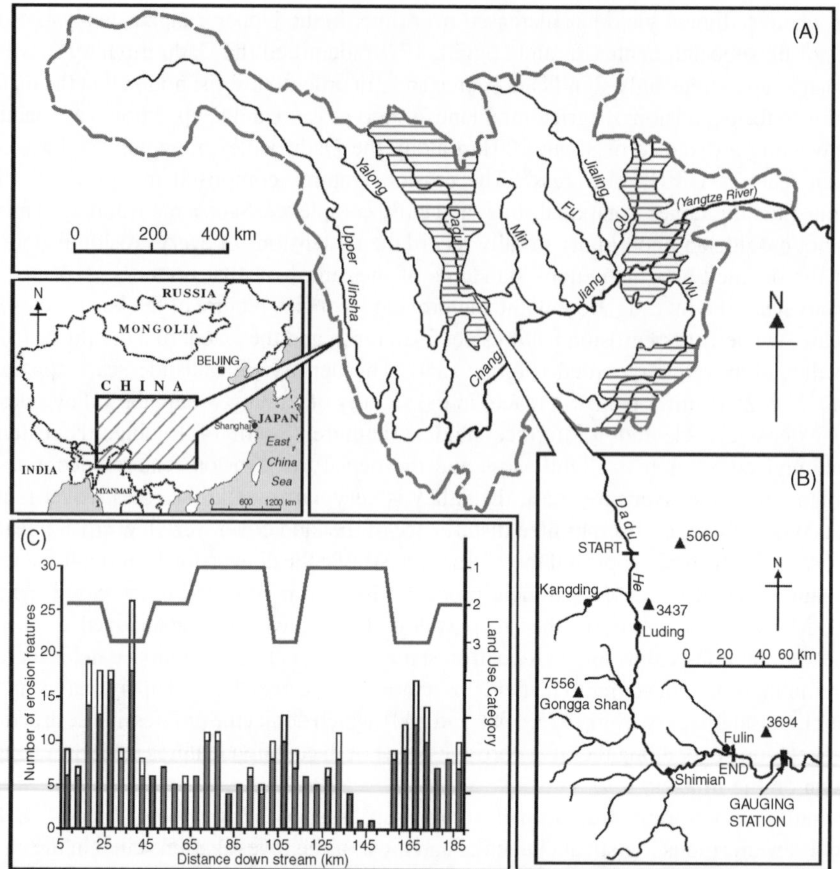

Figure 13.4 Rapid appraisal of erosion features in the Dadu River (tributary in the Yangtze Basin). (A) Parts of catchment within the Upper Yangzte basin exhibiting increased trend in suspended sediment load during the 1980s. (B) Location map of the Dadu Valley in the vicinity of Luding, Sichuan Province, indicating start and finish of erosion audit. (C) Distribution of erosion features along a 195 km reach survey of the Dadu Valley. Major features (e.g. large gully complexes) indicated as lighter shading. The valley-side land use categories refer to: (i) forest cover mainly intact; (ii) agriculture extending less than 50% up valley side; (iii) agriculture extending > 50% up valley side. The observed erosion features strongly coincide with land use

Andes, a major sediment source for the Amazon system, Townsend-Small *et al.* (2007) undertook a detailed year-long study of sediment and carbon dynamics and demonstrated that the majority of sediment is mobilized by individual storm events and by landslides.

13.5 Impact of Land Use, Climate Change and Human Activities on Large Rivers

As Syvitski *et al.* (2005) conclude, the combined effect of humans, increasing supply of sediment through soil erosion but decreasing downstream conveyance through retention in reservoirs, is a net reduction in the global sediment flux to the oceans. Meybeck (2003) considers the extent of human pressure on continental rivers to have usurped the control of Earth system processes. By defining the Anthropocene, he dramatically labels these riverine changes as 'syndromes': such as flow regulation, habitat fragmentation, chemical contamination, acidification and eutrophication, and he coins the term 'neo-arheism' to describe the global reduction of discharge due to abstraction, diversion and consumption of water resources. Vorosmarty *et al.* (2003) calculate that more than 40% of global discharge is now intercepted by large reservoirs (in excess of 0.5 km^3 storage capacity) and that sediment trapping increased rapidly from 1950 to the mid-1980s before stabilizing. The sensitivity of sediment fluxes to reservoir construction and to various forms of land disturbance has prompted many attempts to examine trends in sediment loads in large rivers. At the time when the construction of the Three Gorges Dam commenced on the Yangtze, the authorities were confident that the lack of a trend in the sediment yield at Yichang (the nearest gauging station) indicated that soil erosion was not contributing additional sediment to the river system. This apparent paradox, given evidence from provincial soil erosion audits that the extent and magnitude of soil erosion had increased dramatically in southern China since the 1950s, prompted an investigation of time series for different stations within the catchment (Lu and Higgitt, 1998). A complex spatial pattern emerged with some stations indicating increasing trajectories and others reducing trends, once again relating to land disturbance and reservoir impoundment, respectively. In the subsequent 20 years, some success in soil conservation measures and a considerable amount of additional sediment retention behind dams has seen the Yangtze sediment yield to the ocean reduce appreciably (Zhang *et al.*, 2009), a pattern that is repeated in many Asian rivers (Zhang *et al.*, 2007; Dai, Yang and Cai, 2008).

In an attempt to overview current trends in sediment loads, Walling and Fang (2003) undertook trend analysis of annual sediment load and runoff for 145 river basins exceeding 10 000 km^2, albeit with uneven geographical coverage. Just under half these rivers exhibited no significant trend. Of the remainder, 47% exhibited decreasing sediment yields and 5% increasing yields, leading the authors to concur that reservoir impoundment was the most likely factor. One problem of undertaking time-series analysis of annual sediment load data is segregating the response to year to year variability in runoff from any underlying influence on the delivery of sediment into or through the network. A double mass plot of cumulative runoff against cumulative sediment load is effective in visualizing trends, but more robust methods are needed to test the statistical significance of an apparent trend and to identify any departure points (i.e. abrupt changes) in the time series. In the Yangtze study, Lu and Higgitt (1998) used a procedure to detect trends in the residuals of a discharge–sediment-yield relationship (based on Helsel and Hirsch, 1992). In later work on the Pearl (Xu Jiang) River, a combination of Mann-Kendall and Pettit tests were employed (Zhang *et al.*, 2008). Examples of trend analysis for the Kolmya River in Siberia and the Pearl River in China are illustrated in Figure 13.5.

Beyond the instrumental record, the storage component in the sediment cascade provides its own archive of sediment yield response to environmental change. The floodplains of major rivers impose formidable logistical constraints on field-based investigations. Furthermore, impacts from discrete disturbance events are likely to be preserved close to source but blur as the scale increases downstream. In a remarkable study, Aalto *et al.* (2003) collected 276 cores of up to 1.6 m depth from floodplain transects in the Beni and Mamore Rivers, which drain 700 000 km^2 of the Bolivian Andes across the Andean–Amazon

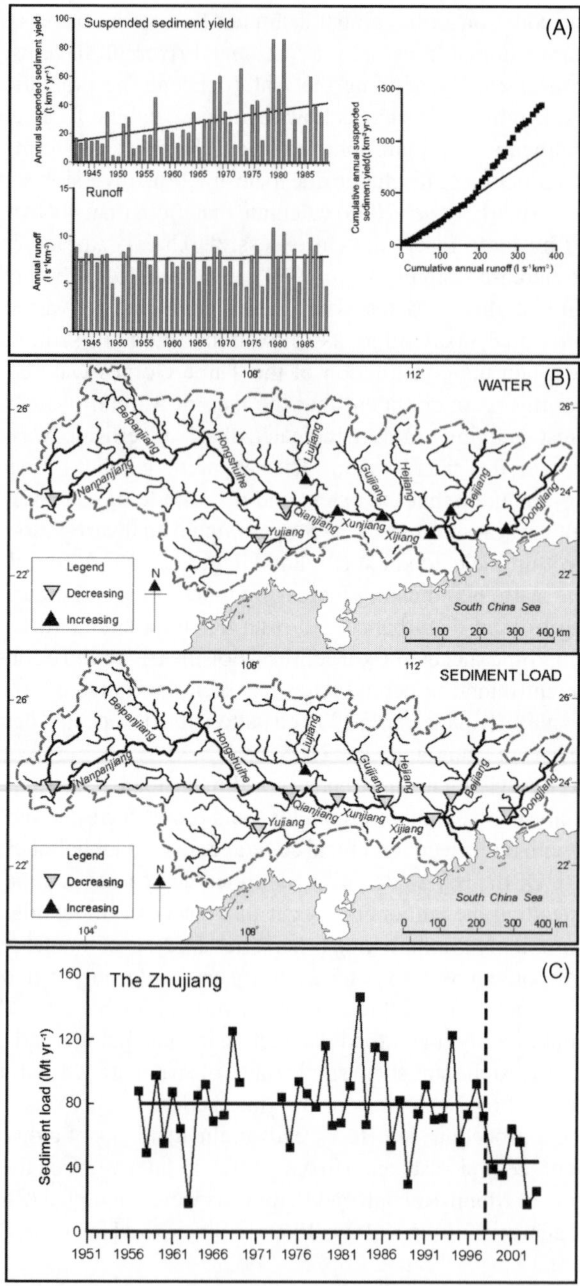

Figure 13.5 Representations of trend detection. (A) Changes in sediment load and annual runoff for the Kolmya River at Ust Srednekansk, eastern Siberia (1942–1989). The double mass plot of cumulative annual runoff against cumulative suspended sediment yield indicates a doubling of sediment load since the 1960s, associated with gold mining (redrawn from Walling and Fang, 2003). (B) Locations of gauging stations within the Zhujiang (Pearl River) exhibiting changes in water discharge (upper panel) and sediment discharge (lower panel). Only the station at Luijiang on the

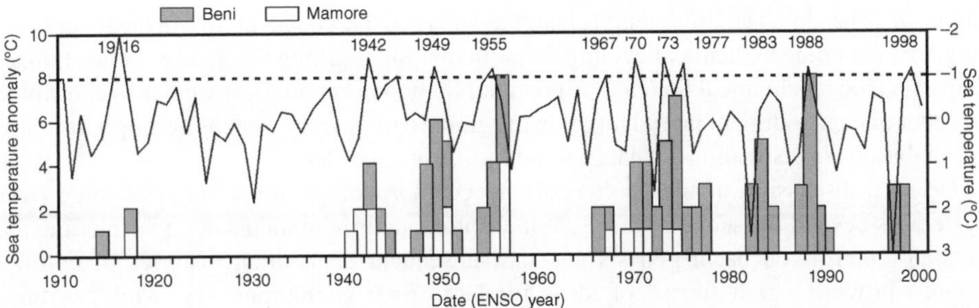

Figure 13.6 Floodplain accumulation events in the Beni and Mamore tributaries of the Amazon, Bolivia. The date of accumulation events is based on ^{210}Pb measurements using the CIRCAUS model. The average sea-surface temperature anomaly in the eastern equatorial Pacific for the early wet season in Bolivia (November to February) is superimposed. An anomaly $< 1°$C represents La Nina threshold and is coincident with many overbank events (from Aalto *et al.*, 2003).

foreland. Using clay-normalized ^{210}Pb activity, individual 'packages' of sedimentation were identified that are deposited as crevasse splays by large floods. The temporal distribution of floodplain accumulation events during the past 100 years closely corresponds to the occurrence of La Nina events (Figure 13.6). Such considerable interannual sediment load is not clear along the Amazon main stem, implying that this intermediate floodplain storage and exchange modulates sediment conveyance.

Alongside empirical investigations focused on past erosion–sediment-delivery dynamics, there is a challenge to provide predictions of sediment yield response to likely future scenarios. The signal from large rivers is obscured by a number of competing influences taking place in different parts of the basin. A simple simulation model (Higgitt and Lu, 2001), for example, demonstrated that the direction of land-cover change (whether deforestation proceeds from the lowest point in a tributary and proceeds upstream towards the headwaters or vice versa) can have a profound impact on the sediment yield time series at a catchment outlet. This is explained by the increased likelihood of sediment going into storage as the distance between source and catchment outlet increases. Applying physically based soil erosion models to large catchments is not feasible in terms of the input data and computer resource requirements, but conceptual models of sediment delivery processes may have some potential (Ferro and Minacapilli, 1995; Lu, Moran and Prosser, 2006).

13.6 Summary: Towards Defining Continental-Scale Sediment Cascades

This review has indicated that, while the role of continental scale rivers as suppliers of sediment to the global ocean has received considerable attention, many aspects of sediment cascades within individual large river basins, with a few notable exceptions, remain largely

Figure 13.5 *(Continued)* Luizhou River exhibits increasing sediment load despite increases in water discharge over the period. (C) Time series of the overall Zhujiang annual sediment load data (sum of load from Xijiang, Beijiang and Dongjiang Rivers). The Pettit test identifies a significant break (reduction) in the trend in 1996. For most upstream stations a similar abrupt decrease is detected in the 1990s

unknown. The global environmental change agenda has sharpened interest in the function of large rivers in biogeochemical cycling, in particular highlighting the need to define human impact on biogeochemical fluxes. The potential for geomorphological studies to contribute to better understanding of lateral fluxes in the global carbon cycles opens new opportunities for ambitious investigations of large-scale sediment cascades.

Detailed discussion of the geological and geomorphological evolution of large river systems is beyond the scope of the present chapter, but developments in understanding the tectonic controls on basin genesis and the subsequent denudation histories of uplifted regions provide a rich source of ideas for large river geomorphology, while, in turn, the elucidation of sediment cascades can provide the evidence to refine river basin histories. For example, river terrace sequences provide the archive for affirming the chronology of river capture in the eastern Himalayas (Clark *et al.*, 2004). During the Quaternary, the growth of Northern Hemisphere ice sheets and the related lowering of global sea levels caused profound changes in the geography of large rivers. The submarine fluvial geomorphology of continental shelves, drowned by episodes of sea-level rise, provide important clues to understanding sequence stratigraphy (Wescott, 1993), but have received scant attention compared to deeper oil-bearing structures (see Chapter 16). Many of the charismatic features of modern large rivers, such as the Middle Yangtze floodplain lakes or the seasonally reversing flow into Tonle Sap Lake in the lower Mekong, are no more than a few thousand years old. In higher latitudes, drainage was directly impacted by ice-sheet growth and during deglaciation phases spectacular floods of immense magnitude occurred, associated with the catastrophic collapse of ice-dammed lakes. The scabland channels in northwestern USA (within the present Columbia basin) have received the most detailed attention (Baker, 2009) but megafloods in Siberia and Central Asia appear to have been even larger. For a brief period around 18 000 years ago, a temporary river of some 8000 km length, exceeding the modern Amazon, formed as a series of spillways connecting ice-dammed lakes in the location of the present day Yenesei and Ob valleys across Kazakstan into the Aral Sea, overtopping into a southwestern spillway connecting it to the Caspian Sea (a freshwater lake more than twice its present size), spilling into the Don Valley and into the Black Sea (also a freshwater lake at the time) and eventually via another spillway into the palaeo-Mediterranean (Baker, 2007). Across sub-Saharan Africa, high level stands during humid phases in the early Holocene enabled lakes to overtop and reconfigure catchment boundaries. For a short period around 9000 yr BP, Megalake Chad overtopped and spilled water towards the Niger Delta, while the catchment area of the Nile was significantly enhanced as a sequence of closed lake basins in the Rift Valley overtopped and connected (Reid, 1993). The distinctive geochemistry of the sediments has enabled the drainage history of the area to be reconstructed.

In geological studies of large-scale weathering and denudation, the geochemical characteristics of fluvial sediment have become a key technique in inferring the erosional histories of basin areas and defining the bulk characteristics of crustal material. An assumption of steady-state erosion and sediment supply is inherent in these approaches but rarely tested (Gaillardet, Dupre and Allegre, 1995). As more detailed studies are undertaken, greater variability becomes apparent. For example, Viers *et al.* (2008) have investigated Nd–Sr isotopic compositions of Amazon tributaries and identified differences both between rivers and in seasonal variation within rivers which appears to reflect mechanisms of sediment supply. Wasson (2003) has used Nd–Sr signatures to identify the significance of the High Himalaya as a source area for the Ganges sediment budget (see Figure 13.3b), assisted by the clear distinction in isotope signal between the crystalline rocks of the High Himalaya and the metasediments of the Lesser Himalaya. The prospect that the geochemistry of the sediment might provide the means to infer source location and/or sediment production process has long intrigued geomorphologists. Known as 'fingerprinting', tracing techniques based on a variety of geochemical or physical characteristics of suspended sediment (e.g. particle size, colour, C/N ratio, trace chemistry, radionuclide signatures, magnetic properties) have been applied enthusiastically in small basins (Collins and Walling, 2004).The potential for fingerprinting in continental-scale basins is yet to be realized but the increasing emphasis on the global change agenda in determining biogeochemical fluxes suggests that further articulation of continental sediment cascades will be required.

Continental-scale rivers, as cradles of civilization, as sentinels of global environmental change and as the key pathway for terrestrial to marine sediment transfer, are an important focus for study. While the total flux from continents to oceans has been the object of many studies, our knowledge of the geomorphology of continental-scale sediment cascades is far from complete.

References

Aalto, R., Maurice-Bourgoin, L., Dunne, T. *et al.* (2003) Episodic sediment accumulation on Amazonian flood plains influenced by El Niño/Southern Oscillation. *Nature*, **425**, 493–497.

Ahnert, F. (1970) Functional relationships between denudation, relief and uplift in large mid-latitude drainage basins. *American Journal of Science*, **268**, 243–263.

Ali, K.F. and de Boer, D.H. (2008) Factors controlling specific sediment yield in the upper Indus River basin, northern Pakisatn. *Hydrological Processes*, **22**, 3102–3114.

Baker, V.R. (2007) Great floods and largest rivers, in *Large Rivers: Geomorphology and Management* (ed. A. Gupta), John Wiley and Sons, Chichester, pp. 65–74.

Baker, V.R. (2009) The Channeled Scabland: A Retrospective. *Annual Review of Earth and Planetary Sciences*, **37**, 393–411. doi: 10.1146/annurev.earth.061008.134726.

Bird, M.I., Robinson, R.A.J., Win Oo, N. *et al.* (2008) Preliminary estimate of organic carbon transport by the Ayerawady (Irrawaddy) and Thanlwin (Salween) Rivers of Myanmar. *Quaternary International*, **186**, 113–122.

Brookfield, M.E. (1998) The evolution of the great river systems of southern Asia during the Cenozoic India–Asia collision. *Geomorphology*, **22**, 285–312.

Clark, M.K., Schoenbohm, L.M., Royden, L.H. *et al.* (2004) Surface uplift, tectonics and erosion of eastern Tibet from large scale drainage patterns. *Tectonics*, **23**, TC1006. doi: 10.1029/2002TC001402.

Collins, A.L. and Walling, D.E. (2004) Documenting catchment suspended sediment sources: problems, approaches and prospects. *Progress in Physical Geography*, **28**, 159–196.

Coynel, A., Seyler, P., Etcheber, H. *et al.* (2005) Spatial and seasonal dynamics of total suspended sediment and organic carbon in the Congo River. *Global Biogeochemical Cycles*, **19**, GB4019. doi: 10.1029/2004GB002335.

Dai, S.B., Yang, S.L. and Cai, A.M. (2008) Impacts of dams on the sediment flux of the Pearl River, southern China. *Catena*, **76**, 36–43.

Dedkov, A.P. and Gusarov, A.V. (2006) Suspended sediment yield from continents into the World Ocean: spatial and temporal changeability, in *Sediment Dynamics and the Hydromorphology of Fluvial Systems*, IAHS Publication 306 (eds J.S. Rowan, R.W. Duck and A. Werrity), International Association of Hydrological Sciences, Wallingford, pp. 3–11.

Douglas, I. (1967) Man, vegetation, and the sediment yield of rivers. *Nature*, **215**, 925–928.

Ferro, V. and Minacapilli, M. (1995) Sediment delivery processes at basin scale. *Hydrological Sciences Journal*, **40**, 703–717.

Fournier, F. (1960) *Climat et erosion*, Presses Universitaires de France, Paris.

Furuichi, T., Win, Z. and Wasson, R.J. (2009) Suspended sediment transport in the Ayeyarwady River, Myanmar. *Hydrological Processes*, **23** (11), 1631–1641.

Gaillardet, J., Dupre, B. and Allegre, C.J. (1995) A global geochemical mass budget applied to the Congo Basin rivers – erosion rates and continental crust composition. *Geochemica et Cosmochimica Acta*, **59**, 3449–3485

Gordon, R. (1885) The Irawadi River. *Proceedings of the Royal Geographical Society*, **7**, 292–331.

Helsel, D.R. and Hirsch, R.M. (1992) *Statistical Methods in Water Resources*, Elsevier, Amsterdam.

Higgitt, D.L. (in press) Sediment delivery and environmental change in Asian rivers: Linking archives with field investigation. *Advances in Geosciences*, in press.

Higgitt, D.L. and Lu, X.X. (1999) Challenges in relating land use to sediment yield in the Upper Yangtze. *Hydrobiologia*, **410**, 269–277.

Higgitt, D.L. and Lu, X.X. (2001) Sediment delivery to the Three Gorges: 1 Catchment controls. *Geomorphology*, **41**, 143–156.

Holeman, J.N. (1968) The sediment yield of major rivers of the world. *Water Resources Research*, **4**, 737–747.

Hovius, N. (1998) Controls on sediment supply by large rivers, in *Relative Role of Eustasy, Climate and Tectonism in Continental Rocks,* SEPM Special Publication No. 59 (eds K.W. Shanley and P.J. McCabe), Society of Economic Paleontologists and Mineralogists, Tulsa, OK, pp. 3–16.

Jansen, J.M.L. and Painter, R.B. (1974) Predicting sediment yield from climate and topography. *Journal of Hydrology*, **21**, 371–380.

Jansson, M.B. (1988) A global survey of sediment yield. *Geografisker Annaler*, **70A**, 81–98.

Jorgensen, D.W., Harvey, M.D., Schumm, S.A. and Flam, L. (1993) Morphology and dynamics of the Indus River: Implications for the Mohen Jo Daro site, *Himalaya to the Sea* (ed. J.F. Shroder), Routledge, London, pp. 181–204.

Kale, V.S. (2002) Fluvial geomorphology of Indian rivers: an overview. *Progress in Physical Geography*, **26**, 400–433.

Knox, J.C. (1972) Valley alluviation in Southwestern Wisconsin. *Annals of the Association of American Geographers*, **62**, 401–410.

Knox, J.C. (2007) Historic valley floor sedimentation in the upper Mississippi valley. *Annals of the American Association of Geographers*, **77**, 224–244.

Koronkevich, N. (2003) Rivers, lakes, inland seas and wetlands, in *Physical Geography of Northern Eurasia* (ed. M. Shahgedanova), Oxford University Press, Oxford, pp. 122–148.

Langbein, W.B. and Schumm, S.A. (1958) Yield of sediment in relation to mean annual precipitation. *American Geophysical Union Transactions*, **39**, 1076–1084.

Lu, H., Moran, C.J. and Prosser, I.P. (2006) Modelling sediment delivery ration over the Murray Darling Basin. *Environmental Modelling and Software*, **21**, 1297–1308.

Lu, X.X. and Higgitt, D.L. (1998) Recent changes of sediment yield in the Upper Yangtze, China. *Environmental Management*, **22**, 697–709.

Lu, X.X. and Higgitt, D.L. (1999) Sediment yield variability in the Upper Yangtze, China. *Earth Surface Processes and Landforms*, **24**, 1077–1093.

Ludwig, W. and Probst, J.L. (1998) River sediment discharge to the oceans: present-day controls & global budgets. *American Journal of Science*, **298**, 265–295.

McLennan, S.M. (1993) Weathering and global denudation. *Journal of Geology*, **101**, 295–303.

Meade, R.H. (1996) River-sediment inputs to major deltas, in *Sea-Level Rise and Coastal Subsidence: Causes, Consequences and Strategies* (eds J.D. Milliman and B.U. Haq), Kluwer, Dordrecht, pp. 63–85.

Meade, R.H. (2007) Transcontinental moving and storing. The Orinoco and Amazon transfer the Andes to the Atlantic, in *Large Rivers: Geomorphology and Management* (ed. A. Gupta), John Wiley and Sons, Chichester, pp. 45–63.

Mertes, L.A.K. and Magadzire, T.T. (2007) Large rivers from space, in *Large Rivers: Geomorphology and Management* (ed. A. Gupta) John Wiley and Sons, Chichester, pp. 535–552.

Meybeck, M. (2003) Global analysis of river systems: from Earth system controls to Anthropocene syndromes. *Philosophical Transactions of the Royal Society B- Biological Sciences*, **358**, 1935–1955.

Meybeck, M., Laroche, L., Durr, H.H. and Syvitksi, J.P.M. (2003) Global variability of daily total suspended solids and their fluxes in rivers. *Global and Planetary Change*, **39**, 65–93.

Milliman, J.D. and Meade, R.H. (1983) World-wide delivery of river sediment to the ocean. *Journal of Geology*, **91**, 1–21.

Milliman, J.D. and Syvitski, J.P.M. (1992) Geomorphic/tectonic control of the sediment discharge to the ocean: The importance of small mountainous rivers. *Journal of Geology*, **100**, 525–544.

Nilsson, C. and Berggren, K. (2000) Alteration of riparian ecosystems caused by river regulation. *BioScience*, **50**, 783–792.

Ohmori, H. (1983) Erosion rates and their relation to vegetation from the viewpoint of world-wide distribution. *University of Tokyo, Department of Geography Bulletin*, **15**, 77–91.

Picouet, C., Hingray, B. and Olivry, J.C. (2001) Empirical and conceptual modelling of the suspended sediment dynamics in a large tropical river: the Upper Niger River basin. *Journal of Hydrology*, **250**, 19–39.

Pinet, P. and Souriau, M. (1988) Continental erosion and large scale relief. *Tectonics*, **7**, 563–582.

Potter, R.E. (1978) Significance and origin of big rivers. *Journal of Geology*, **86**, 13–33.

Reid, I. (1993) River landforms and sediments: Evidence of climatic change, in *Geomorphology of Desert Environments* (eds A.D. Abrahams and A.J. Parsons), Chapman & Hall, London, pp. 571–592.

Reid, L.M. and Dunne, T. (1996) *Rapid Evaluation of Sediment Budgets*, Catena Verlag, Reiskirchen.

Restrepo, J.D. and Syvitski, J.P.M. (2006) Assessing the effect of natural controls and land use change on sediment yield in a major Andean river: The Magdalena drainage basin, *Columbia. Ambio*, **35**, 65–74.

Richards, K.S. (1993) Sediment delivery and the drainage network, in *Channel Network Hydrology* (eds K.J. Beven and M.J. Kirkby) John Wiley and Sons, Chichester, pp. 221–254.

Robinson, R.A.J., Bird, M.I., Win Oo, N. *et al.* (2007) The Irrawaddy River sediment flux to the Indian Ocean: The original nineteenth-century data revisited. *Journal of Geology*, **115**, 629–640.

Scott, K.M., Ritter, J.R. and Knott, J.M. (1968) Sedimentation in Piru Creek watershed, southern California. *U.S. Geological Survey Water Supply Paper*, **1798-E**.

Slaymaker, O. (2008) Sediment budget and sediment flux studies under accelerating global change in cold environments. *Zeitschrift fur Geomorphologie*, **52** (Suppl 1), 123–148.

Stamp, L.D. (1940) The Irrawaddy River. *Geographical Journal*, **95**, 329–352.

Subramanian, V. (1996) The sediment load of Indian rivers, in *Erosion and Sediment Yield: Global and Regional Perspectives*, IAHS Publication 236 (eds D.E. Walling and B.W. Webb), International Association of Hydrological Sciences, Wallingford, pp. 183–189.

Summerfield, M.A. (1991) *Global Geomorphology*, Longman, London.

Summerfield, M.A. and Hulton, N.J. (1994) Natural controls of fluvial denudation rates in major world drainage basins. *Journal of Geophysical Research*, **99**, 13871–13883.

Syvitski, J.P.M. and Milliman, J.D. (2007) Geology, geography and humans battle for dominance over the delivery of fluvial sediment to the coastal ocean. *Journal of Geology*, **115**, 1–19.

Syvitski, J.P.M., Vorosmarty, C.J., Kettner, A.J. and Green, P. (2005) Impact of humans on the flux of terrestrial sediment to the global coastal ocean. *Science*, **308**, 376–380.

Tandon, S.K. and Sinha, R. (2007) Geology of large river systems, in *Large Rivers: Geomorphology and Management* (ed. A. Gupta), John Wiley and Sons, Chichester, pp. 7–28.

Townsend-Small, A., McClain, M.E., Hall, B. *et al.* (2007) Suspended sediments and organic matter in mountain headwaters of the Amazon River: Results from a 1-year time series study in the central Peruvian Andes. *Geochemica et Cosmochimica Acta*, **72** (3), 732–740. doi: 10.1016/j.gca.2007.11.020

Viers, J., Roddaz, M., Filizola, N. *et al.* (2008) Seasonal and provenance controls on Nd-Sr isotopic compositions of Amazaon rivers suspended sediments and implications for Nd and Sr fluxes exported to the Atlantic Ocean. *Earth and Planetary Science Letters*, **274**, 511–523.

Vorosmarty, C.J., Meybeck, M., Fekete, B. *et al.* (2003) Anthropogenic sediment retention: major global impact from registered river impoundements. *Global and Planetary Change*, **39**, 169–190.

Walling, D.E. (1983) The sediment delivery problem. *Journal of Hydrology*, **65**, 209–237.

Walling, D.E. and Collins, A.D. (2008) The catchment sediment budget as a management tool. *Environmental Science and Policy*, **11**, 136–143.

Walling, D.E. and Fang, D. (2003) Recent trends in the suspended sediment loads of the world's rivers. *Global and Planetary Change*, **39**, 111–126.

Walling, D.E. and Webb, B.W. (1996) Erosion and sediment yield, a global overview, in *Erosion and Sediment Yield: Global and Regional Perspectives*, IAHS Publication 236 (eds D.E. Walling and B.W. Webb), International Association of Hydrological Sciences, Wallingford, pp. 3–19.

Wasson, R.J. (2003) A sediment budget for the Ganga-Brahmaputra catchment. *Current Science*, **84**, 1041–1047.

Wescott, W.A. (1993) Geomorphic thresholds and complex response of fluvial systems–some implications for sequence stratigraphy. *American Association of Petroleum Geologists Bulletin*, **77** (7), 1208–1218. doi: 10.1306/BDFF8E3E-1718-11D7-8645000102C1865D

Wilson, L. (1973) Variations in mean annual sediment yield as a function of mean annual precipitation. *American Journal of Science*, **273**, 335–379.

Wohl, E. (2007) Hydrology and discharge, in *Large Rivers: Geomorphology and Management* (ed. A. Gupta), John Wiley and Sons, Chichester, pp. 29–44.

Xu, K.Q., Chen, Z.Y., Zhao, Y.W. *et al.* (2005) Simulated sediment flux during the 1998 big-flood of the Yangtze (Changjiang) River, China. *Journal of Hydrology*, **313**, 221–233.

Zhang, Q., Xu, C.Y., Singh, V.P. and Yang, T. (2009) Multiscale variability of sediment load and streamflow of the lower Yangtze River basin: Possible causes and implications. *Journal of Hydrology*, **368**, 96–104.

Zhang, S.R., Lu, X.X., Higgitt, D.L. *et al.* (2007) Water chemistry of the Zhujiang (Pearl River): natural processes and anthropogenic influences. *Journal of Geophysical Research*, **112**, 1–17.

Zhang, S.R., Lu, X.X., Higgitt, D.L. *et al.* (2008) Recent changes of water discharge and sediment load in the Zhujiang (Pearl River) Basin, China. *Global and Planetary Change*, **60**, 365–380.

14

Estuaries

Tom Spencer[1] and Denise J. Reed[2]
[1]Cambridge University, UK
[2]University of New Orleans, USA

14.1 Introduction: Background Contexts and Challenges

Estuaries, and the coastal wetlands that characterize their margins, form a major sediment sink at the end of the sediment cascade over millennial timescales. The degree of infill depends upon the relations between accommodation space provided by pre-existing drowned river-valley topographies, sea-level history and rates of sea-level change, and sediment supply (Figure 14.1). On shorter, centennial timescales, the changing volume and nature of these sinks reflects in part variations in the dynamic up-cascade processes of sediment transmission and temporary storage, both natural and human-affected; thus Syvitski *et al.* (2005) note that human activities have simultaneously increased the sediment transport by global rivers through soil erosion by $2.3 \pm 0.6\,\mathrm{Gt\,yr^{-1}}$, yet reduced the flux of sediment reaching the world's coasts by $1.4 \pm \mathrm{Gt\,yr^{-1}}$ because of retention within reservoirs. On contemporary timescales, an understanding of sediment dynamics is important because of its impact on rates of erosion and accretion and morphological change; but also because it provides pathways for pollutant transfer; and influences the optical properties of waters and thus primary productivity.

As they lie at the terrestrial–marine interface, estuarine environments are subject not only to gradient-driven, down-cascade riverine hydrodynamics but also to a bi-directional process regime driven by tides and waves. Dalrymple, Zaitlin and Boyd (1992) suggest a tripartite division of estuarine morphology on the basis of the changing axial balance of these processes, with an outer, marine-dominated zone, an inner, riverine-dominated zone and a central 'mixed' zone where both marine and fluvial processes are important (Figure 14.2). Wave-dominated estuaries are characterized by barriers breached by tidal inlets, flood-tide deltas and bayhead deltas (e.g. estuaries of New South Wales coast, Australia (Roy *et al.*, 1994)), whereas tide-dominated systems support longitudinal tidal sand bars at the mouth (Figure 14.3) and central zone channels surrounded by extensive intertidal mudflats and saltmarshes (e.g. northern Australia (Chappell, 1993); Humber estuary, UK (Hardisty *et al.*, 1998)).

Sediment Cascades: An Integrated Approach Edited by Timothy Burt and Robert Allison

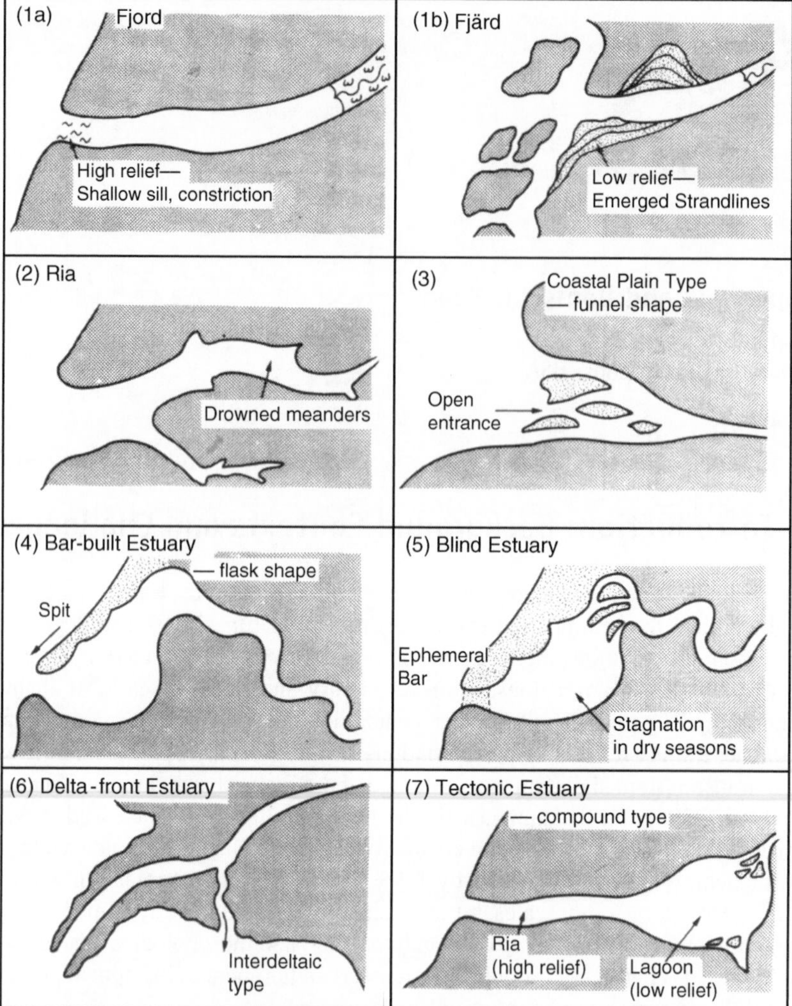

Figure 14.1 Basic estuarine physiographic types
(reproduced, with permission, from Nichols, M. M. and Biggs, R. B. 'Estuaries' in Davis, R. A. (ed.) 'Coastal Sedimentary Environments, 2nd Edition' Fig. 2–1 pg 79 © 1985 Springer-Verlag; after Fairbridge, 1980).
The EstSim Consortium (2007) has developed systems diagrams for these classes for application to UK estuaries.

Sediments are supplied to estuarine systems from the sediment cascade via fluvial processes, but also by local reworking of sea-floor deposits and regional longshore transport of sediment from areas of coastal erosion. The volume and nature of sediment discharged from an estuary may differ significantly from that initially supplied to it as a result of hydrodynamic, chemical and biological processes. And the interaction of tides, waves, river inflow and meteorological forces influences estuarine hydrodynamics. The influence of these processes on sediment fate is further complicated by seasonal cycles in, amongst

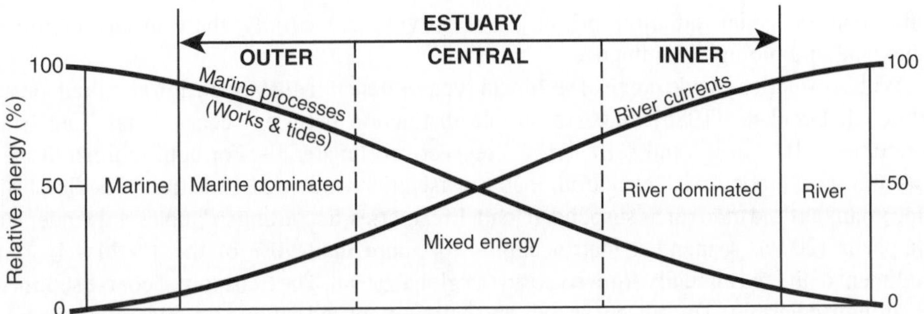

Figure 14.2 Schematic distribution of the physical processes operating within estuaries, and the resulting tripartite facies zonation
(reproduced, with permission, from Dalrymple, R. W., Zaitlin, B. A. and Boyd, R., 'Esturine Facies Models: Conceptual Basis and Stratigraphic Implications' *Journal of Sedimentary Petrology*, **62:6**, 131, Fig 1:B © 1992 SEPM (Society for Sedimentary Geology)).

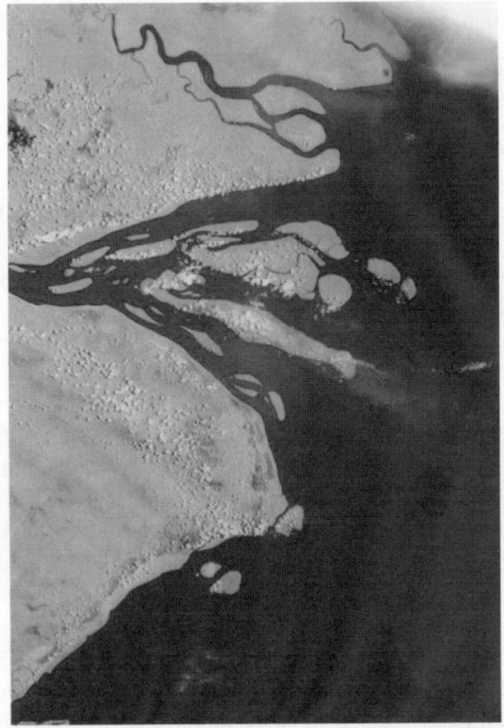

Figure 14.3 Landsat image of the tropical Fly River, Gulf of Papua, Papua New Guinea showing morphology typical of a tide dominated estuary (tidal range is 3.5 m at the entrance, increasing to 5 m at the apex (Wolanski, King and Galloway, 1997))
(courtesy of Earth Scan Laboratory, Coastal Studies Institute, Louisiana State University, USA (2009)).

other controls, solar radiation, primary productivity, and salinity; these in turn modulate chemical and biological influences.

Within estuaries a wide range of sediment types is generally in suspension simultaneously, although Dronkers (1986) makes a simple distinction between 'coarse' and 'fine' (fall velocities $\geq 10^{-1}$ m s^{-1} and $\leq 10^{-2}$ m s^{-1} respectively) materials. For both sediment types, two transport modes exist: as bottom load (coarse grains moving by traction in bedload and fine grains as fluid mud) or as suspended load. In estuaries, according to Uncles, Lavender and Stephens (2001), suspended load accounts for more than 90% of the 18–24 $\times$ 10^9 t of sediment delivered annually from rivers to the global ocean. The transport of coarse sediment is strongly determined by current speeds and entrainment and settling is rapid within the tidal cycle. By comparison, fine sediments only settle at very low current speeds and the settling time delay is significant. The result is a time lag in the sediment response to flow. Postma (1967) identifies both 'settling lag' and 'scour lag' as important controls on the fate of fine sediments in the intertidal zone noting that the two effects combined produce a tendency for fine sediment to accumulate at the upper limit of tidal flooding. However, a further complication is introduced by the fact that cohesive sediments tend to combine, either due to chemical flocculation as salinities increase or due to binding by biological matter, into larger aggregates which settle much more rapidly than their constituent particles.

Establishing sediment budgets (the balance in the quantity of sediment entering and leaving a defined area) in these circumstances relies on the quantification of the processes of erosion, sediment transport and deposition (e.g. see Postma (1967), Van Rijn (1993) and Soulsby (1997) for discussion of fundamental mechanisms). Estuaries can be characterized by point sources of fluvial inputs and broader sources and sinks from shelf seas. The relations between these end points are dynamic; thus, for example, river-flood pulses can under certain circumstances completely override tidal hydrodynamics and up-estuary sediment transport. In addition, line sources and sinks can be identified in terms of sediment transfers between the different morphodynamic units along the length of an estuarine channel. Changes in the volume bounded by surfaces in the control volume can be used to construct a budget. Such an approach works well in non-cohesive sediments where the volume of sediment held in the water column is low and the material types throughout the control volume are similar. In estuaries, however, the sediments are both non-cohesive and cohesive in character, large volumes of sediment are resident within the water column, and there is a high degree of variability in the sediment dynamics within the control volume. Thus, in the Humber estuary, UK on a spring tide *c.* 1.5 $\times$ 10^5 t tide^{-1} are moved, reducing to *c.* 0.8 $\times$ 10^5 t tide^{-1} on a neap tide. The residual fluxes are, however, of the range 7–3 $\times$ 10^3 t tide^{-1}, one to two orders of magnitude lower, yet it is these residual transports that play a key role in determining estuarine morphology (Hardisty, 2007). Under such conditions, it is usual to define the exchanges in mass to and from the water column to give a mass balance. This fits well with the modelling of suspended and bedload transport and field measurements of sediment fluxes and concentrations. However, there is often a need to integrate across volume and mass changes, which requires a knowledge of water density, bulk density and sediment density to achieve the correct comparisons. This can be problematic when the same sediment can have radically different properties under different conditions (e.g. the sediment density of mobile versus settled versus consolidated silt/clay (Townend and Whitehead, 2003)).

Interest in understanding the sediment dynamics and long-term morphological evolution of estuaries has intensified over the past decade (e.g. EMPHASYS Consortium, 2000;

EstProc Consortium, 2004). This interest is driven not only by the need to conform to national and international legislation on water quality and biological conservation, and to better anticipate the response of estuarine systems to near-future accelerated sea-level rise and changes in storminess, but also by the imperative to better predict the likely consequences of 'soft engineering' interventions – such as the managed realignment of shorelines from the setting-back of existing defence lines – in estuaries highly modified by centuries of land claim and other anthropogenic interventions (Townend and Pethick, 2002). The rapid development of relatively inexpensive numerical modelling of estuarine processes on a variety of computing platforms has now largely superseded the use of hydraulic scale models in simulating estuarine dynamics. However, model validation is still often restricted to the coarse structure of tidal dynamics (such as the semi-diurnal lunar M_2 constituent) and struggles to reproduce the complexity of mixing and sedimentation processes. This is due to the large range of sediment sizes present, the non-linear coupling between flow and sediment suspensions and the extreme spatial variability – vertical, axial and transverse – of sediment concentrations. Vertical and horizontal shear in tidal currents generates fine-scale turbulence which sets the overall rate of mixing over 'decay times' which are measured over hours, and are thus amenable to short-term modelling, but the flushing times for river inflows are measured over days and thus contain a time-bound 'memory' component (Prandle, 2009). These issues form part of a more fundamental difficulty in that estuaries provide a classic example of a morphodynamic problem, whereby there is a continual co-adjustment of processes and morphology. These interactions – such as the way in which changing bed roughness (both sediment type and morphology) affects sediment erosion and deposition – continually redefine system boundary conditions, effectively preventing the output from small-scale 'bottom up' hydrodynamic models to be extrapolated to long-term morphological evolution (EMPHASYS Consortium, 2000; Davies and Thorne, 2008). Finally, there is the overarching problem of the technical difficulty of within-estuary measurements and the paucity of comprehensive, high quality observational data to interpret sensitivity tests from modelling exercises. In these circumstances, it has been argued (e.g. Pethick, 1996) that a fruitful way forward is to adopt not 'bottom up' numerical models but rather 'top down', 'rule-based' approaches which aim to identify the steady-state response of an estuary to tidal exchanges and then examine the smaller scale, internal estuary dynamics within this framework. This viewpoint has considerable merit: thus, for example, tidal fluxes in the Mersey estuary, UK can reach 2×10^5 t on individual spring tides yet a century of bathymetric surveys has recorded remarkable morphological stability, with a net loss of just 0.1% of estuarine volume each year (Prandle, Murray and Johnson, 1990; Thomas, Spearman and Turnbull, 2002).

In this chapter, we begin by taking a large-scale view of estuarine morphodynamics and then focus on three estuarine environments: the subtidal reaches of estuaries; the low to middle intertidal zone sand and mudflats; and the upper intertidal saltmarshes. There are important differences in energy regime as one moves from central channels towards estuarine margins and important variations in sedimentary characteristics at different tidal levels, from predominantly sandy subtidal channels to mixed sand–mud surfaces in the mid-intertidal zone, to silt/clay substrates in upper intertidal environments (Klein, 1985; Figure 14.4). Furthermore – and this forms a key focus for this chapter – this approach ranges across a changing set of interactions between sedimentological and biological processes, from the subtidal zone where physical (and physio-chemical) processes are dominant, through a

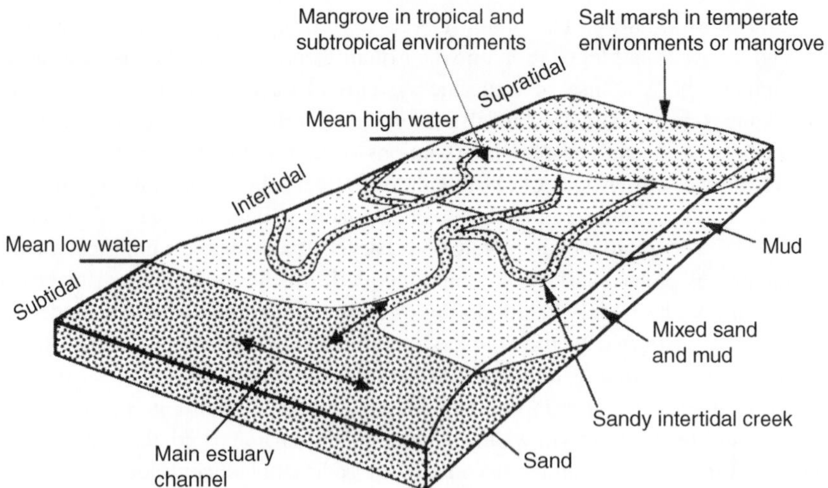

Figure 14.4 Morphology and sediment distribution across the intertidal margin of an estuarine channel
(from G. Masselink & M. G. Hughes, *Introduction to Coastal Processes and Geomorphology* © 2003, Hodder Arnold. Reproduced by permission of Edward Arnold (Publishers) Ltd).

complex set of interactions on tidal flats where the stability and erodibility and deposition of sediments is mediated by biological activity, to saltmarshes where substrate stability is strongly controlled by the presence of a permanent vegetation canopy.

14.2 The Estuary Scale: Tidal Dynamics and Estuarine Morphology

Many estuaries, including ria coastlines and fjords, represent topographies drowned by the postglacial rise is sea level and their planform is strongly controlled by geological structure (Figure 14.1). Where estuaries occur in alluvial systems, however, they characteristically show a funnel-shaped planform. Wright, Coleman and Thom (1973) have argued that this planform represents a stable morphological configuration. Furthermore, it is related to tidal range: mesotidal estuaries are relatively short and strongly-flared whereas macrotidal estuaries are typically longer and more funnel-shaped (Pethick, 1996). This observation supports the notion that stability results from the dissipation of tidal energy inputs at a level at, or below, that required to erode the subtidal channel bed and intertidal margins. With such a planform, tidal energy is forced into a channel of decreasing hydraulic efficiency and increasing frictional drag, ultimately reducing tidal energy to zero at the head of the estuary, thus encouraging the retention and deposition of fine sediments in landward reaches. Frictional dissipation also affects the form of the tidal wave. As the velocity of the tidal wave is determined by water depth, the trough progresses more slowly than the crest; this differential increases rapidly in water depths of < 5 m (Masselink and Hughes, 2003). This behaviour leads to a short-duration but high-velocity flood tide compared to the ebb tide and estuaries described as 'flood dominant'. Such systems import greater sediment volumes than they export. Deposition rates, often in the form of extensive intertidal mudflats and saltmarshes on estuarine margins (e.g. southeast China; Wang, Hu and Wu, 2002), or within-channel sand shoals (e.g. Ord River, Western Australia; Wright, Coleman and Thom, 1973) can be rapid under such circumstances. Intertidal deposition decreases the main channel depth. As the tide can be seen as a shallow water wave, deterministic wave theory predicts that a reduction in depth must not only affect velocity but also lead to a reduction in wavelength. When the tidal wavelength approaches a multiple of four times the estuarine length, tidal resonance, in the form of an estuarine standing wave, is set up. The standing wave has a nodal point of maximum velocities near the estuary mouth and an antinode, and zero velocities, near the estuary head (McDowell and O'Connor, 1977). This arrangement thus describes an energy gradient from mouth to tidal limit, a uniform tidal energy dissipation per unit bed area, and an associated morphological response with increasingly extensive intertidal mudflats and/or sand shoals up-estuary. In reality, frictional dampening can be opposed by channel narrowing and shoaling, or convergence, which increase tidal amplitude. Thus, several estuarine configurations are possible: synchronous estuaries, where friction and shoaling effects are balanced and tidal range is constant until close to the tidal limit; hyposynchronous estuaries, typical of short systems, where friction is dominant and tidal range decreases progressively landwards; and hypersynchronous estuaries, typical of long systems, where shoaling is dominant and tidal range increases with distance from the coast until close to the tidal limit (Dyer, 1997; Masselink and Hughes, 2003).

The standing-wave–tidal-resonance model also determines the timing of maximum and minimum velocities and thus the tidal stage where deposition takes place; under such conditions, maximum velocities are found at mid-tide and minimum velocities at high and low water (Le Hir *et al.*, 2000). Deposition is therefore concentrated above mid-tide level and the elevation of surfaces on estuarine margins rise in the tidal frame (see mathematical modelling by Roberts, Le Hir and Whitehouse, 2000); the resulting general form of the intertidal profile shows a strong change in slope at mid-tide level (e.g. Evans, 1965) and, in estuaries, a 'shoulder' between the flanking mudflats and the deep main channel. As Pethick (1996) points out, this leads to the apparent paradox of mean channel depth being less at high tide than at low tide. High stages are characterized by shallow water depths over often spatially extensive upper intertidal flats whereas low stages are confined to deep lower intertidal and subtidal channels. This state of affairs reverses tidal dynamics as the trough velocities are now relatively higher than the shallow-water crest velocities. Flood dominance is thus replaced by ebb dominance, and sediment import by sediment export, in the later stages of estuarine evolution. Mudflat and sand shoal sedimentation thus ceases, or is even reversed, with surface erosion of intertidal sediments and the flushing of estuarine sediments towards and from entrance channels (Friedrichs and Aubrey, 1988). These relationships work best for coarse clastic sediments because of the more direct relationship between bed shear stress, sediment entrainment and transport. For finer sediments additional factors complicate the simple prediction of sediment transport based on velocity asymmetry; we consider these complications further below.

14.3 The Subtidal Zone: Suspended Sediment Dynamics

Increasing concern regarding the transport of particulate material, and associated contaminants, from land to the coastal ocean has led to an increasing need to understand the processes controlling the fate of suspended sediment entering estuaries. Within the permanently flooded parts of the estuary, flows, and thus the fate of sediments, are influenced by density differences as salt and freshwater mix. Density gradients associated with this mixing can produce currents in addition to those dependent upon river and tidal flows. The magnitude of this gravitational circulation depends on the degree of mixing between river and tidal flow which in turn is controlled by the relative speeds and volumes of the two flows. Three types of mixing conditions have been recognized: salt-wedge, where isohalines generally parallel the bed; partially mixed, with a less pronounced halocline; and fully-mixed, with vertical isohalines (Figure 14.5). For sediment retention in estuaries, this mixing is important, not only for the current regime but also as it influences the flocculation of clay particles (generally occurring at salinities of approximately 2 ppt) which increases settling velocity. However, the size of 'flocs' is dependent on a number of factors, including shear, resulting in complex changes in particle settling velocity associated with tidal, seasonal and interannual variations in estuarine currents.

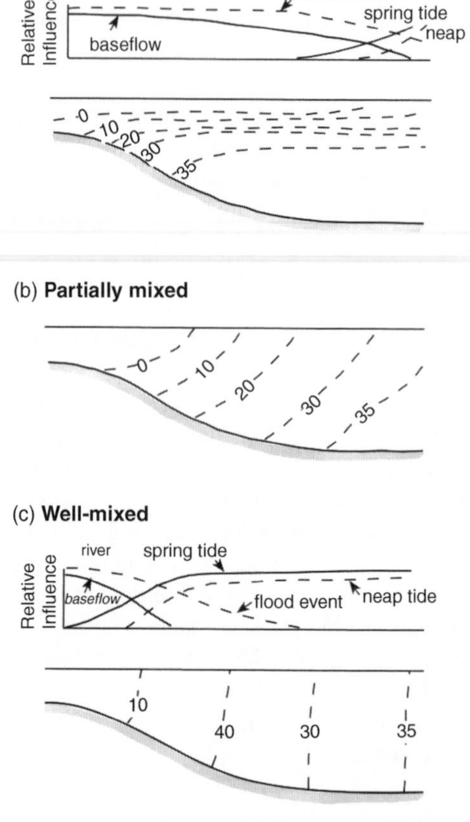

Figure 14.5 Circulation within estuaries for (a) salt-wedge; (b) partially mixed; and (c) well-mixed estuaries, showing time-variant relative influences of tides and river flow and salinity characteristics from typical pattern of isohalines (units = ppt)
(reproduced, with permission, from Woodroffe, C. D. 'Coasts: Form, Process and Evolution' © 2002, Cambridge University Press).

While the morphology of estuaries points to the long-term fate of sediments, the tidal dynamics of the subtidal areas can have important controls on the residence time of sediments, especially the role of the estuarine turbidity maxima (ETM) as an area of particle trapping and transformation. Dyer (1995) notes that concentrations of suspended sediment found in the ETM range from 100–200 mg L^{-1} in estuaries with low tidal range to of the order of 1,000–10,000 mg L^{-1} in high tidal range estuaries. Studies of sediment dynamics within estuaries have shown that in many systems estuarine turbidity maxima occur in association with an area of finer bed sediments which are resuspended and deposited by varying tidal flows (e.g. Schubel, 1968; Bartholdy and Madsen, 1985; Uncles and Stephens, 1993). This zone is termed the 'mud reach' by Wellershaus (1981) in his studies of the Wesser estuary. Correlation between observations of tidal velocities and suspended sediment concentrations (SSC) are found in many estuaries (e.g. Allen *et al.*, 1980; Wellershaus, 1981; Uncles, Barton and Stephens, 1994) although some systems show distinct time lags between maximum suspended sediment concentration and peak flow velocity (e.g. Barua, 1990). The role of tidal resuspension in ETM development is most apparent when there is both a direct spatial association between the mud reach and the ETM zone, and a correlation between flow speed and SSC. Most models of ETM development show cycles of resuspension and deposition of fine sediments (Lang *et al.*, 1989) which may vary on a spring–neap timescale (Avoine and Larsonneur, 1987; Grabemann *et al.*, 1997), in accordance with the energy available for resuspension.

In the Columbia River estuary, among others, while a well-defined ETM exists (Gelfenbaum, 1983; Reed and Donovan, 1994), bed sediments are dominated by medium to fine sand (Sherwood *et al.*, 1990; Hubbell and Glenn, 1973) and no mud reach is present. Gelfenbaum (1983) estimates that the amount of sediment present in the ETM would form a layer less than 2 mm in thickness if all the material was deposited on the bed. This presents the possibility that a transient bed deposit may exist which 'fuels' the ETM but which is too thin to be detected as a significant weight percent in grab samples of bed sediments. Alternatively, the ETM may result solely from internal circulation processes. Given that freshwater discharge from the rivers into the estuary is continuous, if variable in magnitude, there is a tidal-mean residual freshwater flow out of the estuary into the sea. This would have a tendency to wash out suspended material to the sea unless countercurrents exist which transport suspended sediment in a landward direction (Burchard and Baumert, 1998). Such near-bottom residual countercurrents have explicitly been measured by Lindsay, Balls and West (1996) in the Forth estuary in Scotland. The most commonly attributed causes of these currents are residual gravitational circulation (Postma and Kalle, 1955), tidal velocity asymmetry (Jay and Musiak,1994), and tidal mixing asymmetry (Geyer, 1993), all of which rely on some level of stratification (Figure 14.5) within the estuary.

The role of the ETM in determining particle fate in estuaries has recently been shown to be highly dependent on particle settling velocity. While fundamental properties of particles (i.e., size, density) will influence settling rates, several studies have noted the importance of enhanced settling within the ETM for the retention of particles. For example, Orton and Kineke (2001) noted that settling velocity was one of the most important factors limiting their ability to predict suspended sediment concentrations in the ETM of the Hudson River. In Chesapeake Bay, Sanford, Suttles and Halka (2001) found that seasonal differences in particle settling velocity (from January to October) had a strong influence on the retention of suspended sediment within the ETM. They suggest that the fate of sediments delivered to estuaries during large floods may depend on the season in which they occur. More detailed studies of ETM particles and their biology (e.g. Crump, Simenstad and Baross, 1998; Dyer and Manning, 1999; Hollibaugh and Wong, 1999) reveal the potential importance of organic substances in binding particles and increasing settling velocities rather than the physio-chemical process of flocculation. Studies in the Columbia River estuary (see Simenstad *et al.* (1994) for research approach and Reed and Donovan (1994) for more detail on sediment aspects of the ETM) have shown distinct changes in the character and composition of the ETM between flood and ebb, and spring and neap, tidal cycles. Figure 14.6 summarizes this 'cycle' of change and how aggregation of particles also changes. Crump and Baross (1996) noted differences in particle-associated bacterial carbon production over these same cycles illustrating the likely important effect of biological factors on particle settling and retention within this ETM. Studies of the dynamics of suspended sediment in estuarine turbidity maxima continue to develop as instrumentation for the finer scale identification of suspended sediment and velocity profiles, and *in situ* examination of particles, becomes available. However, the changing location of these turbidity maxima, as the salt wedge migrates back and forth within the estuary with each tide, and their seasonal

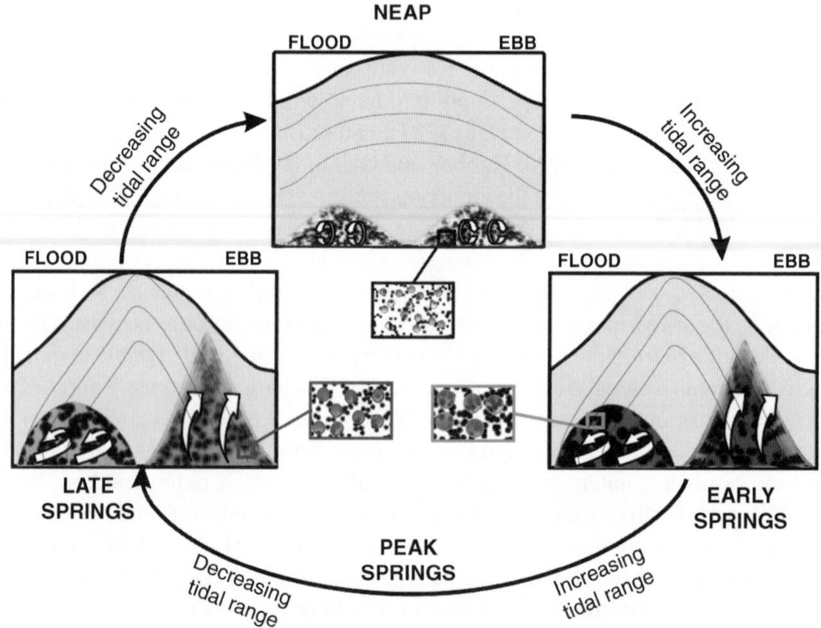

Figure 14.6 Idealized cycle of the form and composition of the Columbia River estuarine turbidity maximum (ETM) from Land Margin Ecosystem Research studies 1990–1999 based on Eulerian sampling at mid-channel location (Buoy 39). During neap tides (top panel) the water column is stratified on both flood and ebb (near horizontal isohalines - dashed lines). The ETM is of low intensity (low sediment concentration - greyscale shading), is composed of fine sediments and aggregation of particles is minimal. The ETM is likely to be a result of temporary resuspension from the bed. As tidal range increases (right panel) saline waters are washed out on ebb flows. Both flood and ebb ETM are more intense and are comprised of large aggregates. Sediment suspended on the ebb is likely to be washed out of the estuary. By late spring tides (left panel) the form of the ETM is similar to early springs but the intensity is less and the aggregates are smaller in size

variability, as the relative volume of tide and river flow change, will require intensive field observations of physical and biological characteristics, integrated with the modelling of biogeophysical processes. Only then will the accurate estimation of suspended sediment fluxes through estuaries, so important for global sediment budget calculations (Uncles, Stephens and Smith, 2002), be possible.

14.4 The Lower and Middle Intertidal Zone: Dynamics of Intertidal Flats

As the discussion above shows, estuary form and function have traditionally been defined with reference to tidal hydrodynamics. More recently, however, widespread evidence of intertidal habitat loss on estuarine margins (e.g. Allen, 2000; Van der Wal and Pye, 2004) has increased interest in the potential role of waves, and the interaction between continuous tidal current and intermittent wave processes (e.g. Green, Black and Amos, 1997) within estuaries in determining sediment dynamics.

It has been shown that shore-parallel currents dominate estuarine hydrodynamic forcing in subtidal and lowest intertidal channels. However, at the time of peak current flows in the main channel, water depths over the flanking mudflats are likely to be small and here it is shore-normal currents and/or wave activity that are likely to be dominant. Regime theory predicts that a large tidal volume will be associated with a large cross-sectional area (e.g. Spearman, Dearnaley and Dennis, 1998) and O'Brien's (1931) relationship between estuarine entrance cross-section and tidal prism has been shown to be widely applicable across a range of scales (e.g. Gao and Collins, 1994; Dronkers, 2005). Both records of the three-dimensional morphometry, in the form of hypsometric curves of estuaries subject to hyper- to microtidal ranges (Kirby, 2000), and mathematical modelling of intertidal profiles subject to shore-normal currents only (Roberts, Le Hir and Whitehouse, 2000), show that mudflat slopes increase with increasing tidal range. However, the cross-sectional profile is also a function of sediment supply; small boundary sediment concentrations lead to steep slopes whereas higher concentrations promote shallower slopes. At each point on the equilibrium mudflat there is a phase of erosion on the flood tide and a further phase on the ebb, bracketing a period of deposition around high water. The shape of the tidal flat must adjust so that the two erosive phases are balanced by the period of deposition. High sediment concentrations result in large settling fluxes during the high-water slack period; if the initial slope is steeper than the equilibrium slope, then deposition will take place and the slope will flatten (Roberts, Le Hir and Whitehouse, 2000).

In addition, middle to upper tidal flats can also be subjected to a band of high shear stress from waves moving across the tidal flat with the rising and falling tide. The rapidly oscillating flows under wind waves lead to much greater shear stresses than under the long wavelengths associated with the passage of the tidal wave. Waves disturb the mudflat surface and promote the suspension of sediments, which can then be redistributed by tidal currents (French, Burningham and Benson, 2008). Thus, wave activity can exert control on estuarine morphology. Dronkers (1986) was able to show that in the dominantly accretionary environment of the Wadden Sea, 50% of the mudflat area lies below mean tide level whereas the reverse is true of the dominantly erosive environment of the Eastern Scheldt basin (Figure 14.7).

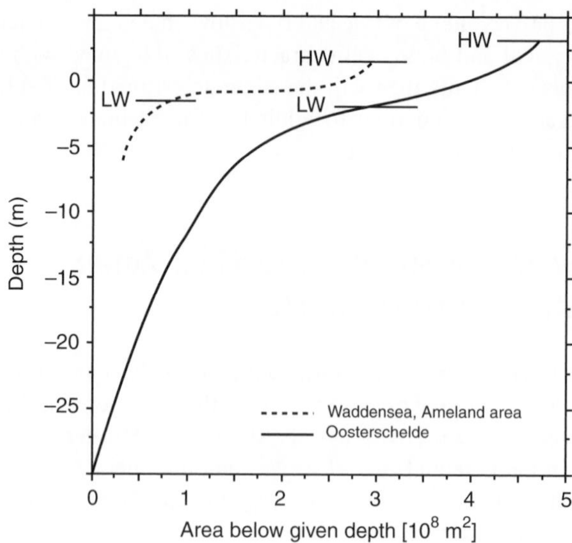

Figure 14.7 Hypsometric curves for an eroding estuary (Oosterschelde) and an accreting estuary (Waddensea, Ameland area). The eroding estuary mudflats lie higher in the tidal frame with a more pronounced convexity around low water (LW)
(reprinted from Pethick, J. S. in Nordstrom, K. F. and Roman C. T. (eds) Estuarine Shores © 1996 with permission from John Wiley & Sons Ltd; after Dronkers, 1986).

In some cases even small waves are able to erode large volumes of surface sediment (e.g. Anderson *et al.*, 1981; French *et al.*, 2000). Modelling suggests that increasing wave heights in the presence of cross-shore tidal currents steepen the intertidal profile; make the profile less convex/more concave; and move the position of mean water level towards the high water mark (Roberts, Le Hir and Whitehouse, 2000). In reality, the nature of wave control is complicated by both the intermittent nature of wave attack, being restricted to those times when mid- to upper intertidal surfaces are covered by the tide and, within this window, when wind speeds, durations and directions combine with the position and orientation of intertidal surfaces to create surface wave activity and by the depth-limited character of wave action. Thus Le Hir *et al.* (2000) demonstrate that there is a limiting wave height at a given water depth on the mudflat at Brouage, southwest France, as a result of wave energy dissipation by bottom friction.

In general, modelling considers tidal flats to be featureless planar slopes. In fact, many tidal flats support a diverse surface topography at a range of spatial scales (Figure 14.8). A survey of northwest European mudflats (Whitehouse *et al.*, 2000) identified four classes of feature: channels, creeks and gullies up to 1.5 m deep; ridge and runnel topography, with wavelengths of 1.5–3 m and amplitudes of 15–20 cm; ripples; and cliffs. These surface forms contribute to the general bed roughness (and hence bed friction regime) and, at finer scales, influence vertical flow structure, through form drag and the creation of wetting and drying pathways at times of tidal flat inundation and exposure (Whitehouse *et al.*, 2000). *In situ* measurements to establish erosion thresholds have found significant spatial and temporal variations in sediment properties and erodibility between the fluid muds in the runnels and the consolidated sediments of the ridges (Widdows, Brinsley and Elliott, 1998).

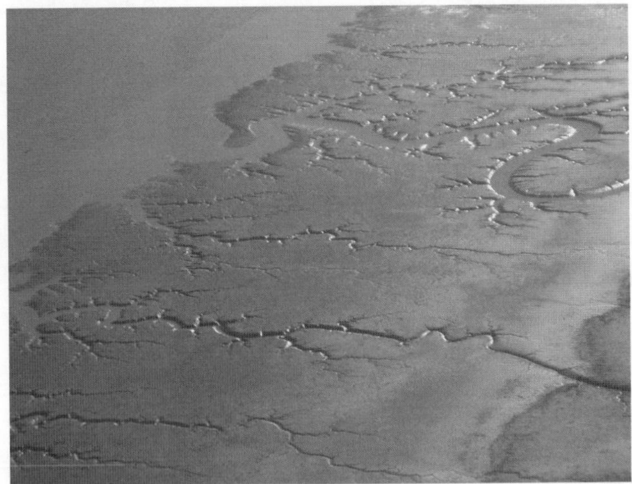

Figure 14.8 Exposed tidal flats in Cook Inlet, Alaska, USA showing variation in drainage density and channel patterns at a variety of scales
(photograph by D.J. Reed).

Drainage associated with these features can either be a surface channel flow or result from the lagged ebb-tide release of porewaters from perched water tables within low permeability clay substrates. Flow may be organized into either a dendritic network of channels or, most probably in conjunction with cross-shore tidal currents, a shore-normal topography of ridges and intervening runnels. Such a system on the Brouage mudflat was shown to be responsible for an offshore transport of 1000 kg m^{-1} per tide, in a period when a single surface erosion event removed 2000 kg m^{-1} (a 40 mm lowering over 1 km of tidal flat) and a residual onshore tidal flux of 20,000 kg m^{-1} (Bassoullet *et al.*, 2000).

Furthermore, the temporal dynamics of these tidal flat surfaces are complex (e.g. Christie and Dyer, 1998; Whitehouse and Mitchener, 1998). For example, a macrotidal mudflat at Peterstone Wentlooge, Severn estuary, UK showed long-term erosion; variations in level of ±10–25 cm (equivalent to a change in storage of 0.2 Mt) from accretion in spring–summer to erosion in autumn–winter; and development of height differences of ±1 cm over spring–neap and individual tidal cycles, offset by periodic surface lowering from wave-driven events (O'Brien, Whitehouse and Cramp, 2000). On the Groningen mudflat, Dutch Wadden Sea, accretion of sediments under calm, current-dominated conditions is replaced by erosion from combined tidal current and wave action in rough weather and by severe erosion under storm, wave-dominated episodes (Janssen-Stelder, 2000).

All these topographic controls and hydrodynamic processes are mediated by the vertical and lateral variation in sediment particle size distributions and textures present and in the differences in sediment erosion resistance properties. The latter are present at a number of scales and generally related to biological processes. Both significant shore-normal (e.g. Humber estuary, UK: Widdows, Brinsley and Elliott, 1998) and axial estuarine (e.g. Tamar estuary, UK: Bale, Stephens and Harris, 2007) gradients in sediment properties have been reported on intertidal flats. Below this scale, there is a 'patchiness', often related to the association of particular organisms with subtle variations in microtopography (e.g. Andersen *et al.*, 2002), and nested within this patterning are finer scale variations

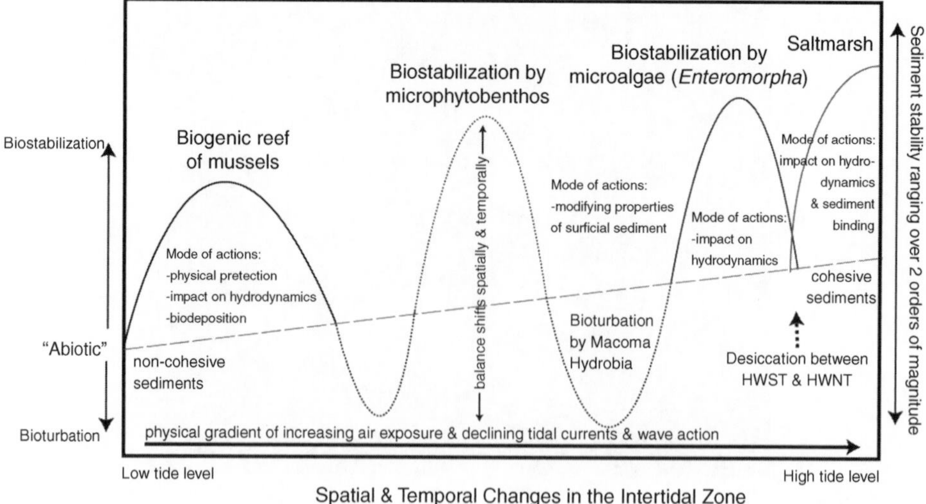

Figure 14.9 Conceptualization of the major biological and physical factors influencing sediment stability in the intertidal zone. Dotted line represents general shoreward increase in sediment stability with increasing exposure and decreasing tide/wave action. Solid line represents the long-term role of biota and the dashed oscillating line shows shorter terms switches from biostabilizers to destabilizers (reprinted from Widdows, J. and Brinsley, M. 'Impact of Biotic and Abiotic Processes on Sediment Dynamics and the Consequences to the Structure and Functioning of the Intertidal Zone *Journal of Sea Research* © 2002 with permission from Elsevier).

in erosion resistance. For example, Tolhurst *et al.* (2006) have shown subcentimetre-scale variations in erosion thresholds due to the presence of diatom biofilms; as the constituent organisms are sensitive to changes in irradiance these thresholds are temporally dynamic, with accompanying diurnal shifts in sediment stability (Friend, Lucas and Rossington, 2005). Biological controls can be broadly divided into biostabilization and biodestabilization, with sediment surface stability being ultimately dependent upon the balance between these two competing sets of processes (Figure 14.9). This balance varies spatially, both vertically within the tidal frame and horizontally along the estuarine salinity gradient, and temporally, on seasonal, interannual and perhaps longer timescales. These patterns have implications for estuarine morphology; strong biostabilization will lead to a flatter profile as flood and ebb tidal pulses will be less effective on tidal-flat surfaces. Conversely, destabilization will lead to lower critical shear stresses for erosion and thus result in steeper profiles. In general, biostabilization is associated with microorganisms and biodestabilization with a benthic macrofauna.

Study of these processes is not, however, straightforward. It is almost impossible to sample muddy sediments without disturbing sediment fabric, hydrological properties or fauna and this limits the utility of laboratory-derived data on sediment behaviour (e.g. Tolhurst, Reithmüller and Paterson, 2000). This fundamental difficulty has led to the development of a number of field-portable devices for the *in situ* measurement of erosion thresholds, although intercomparisons between devices remain problematic, as a result of both device design issues and varying operating protocols (e.g. Le Hir, Monbet and Orvain, 2007; Widdows *et al.*, 2007; Tolhurst, Black and Paterson, 2009).

Nevertheless, in general, biostabilization processes have been shown to include enhanced sediment cohesion from a microphytobenthos; binding by filamentous algae; a set of processes associated with bed armouring, biofiltration and biodeposition; and attenuation of near-bed flows and waves by vegetation. For example, experimental flume studies have shown that near-bed flows in the presence of the nuisance macroalga *Enteromorpha* spp. are reduced by 18–56% and sediment erodibility from two- to onefold, and sedimentation of fine muds increased by *c.* 50%, as surface coverage increases from 10 to 60% (Romano *et al.*, 2003). However, it is the widely distributed microbiota-like bacteria and microphytobenthos which play the major role in the cohesiveness and stability of estuarine sediments, achieved through the production of a biofilm that acts as a protective skin over surface sediments. The biofilm consists of microbial cells (diatoms, cyanobacteria and bacteria) embedded within mucus-like carbohydrates and/or EPS (extracellular polymeric substances); it is thought to raise the critical erosion threshold of the upper few hundred microns of the sediment–water interface 200- to 500-fold over abiotic control surfaces (Paterson, 1997). The difficulty, however, is that most measurements of biostabilization have not measured EPS directly but used levels of Chlorophyll *a* as a proxy indicator. This has produced poor correlations and conflicting results on different sediment mixtures and under different experimental settings (Le Hir, Monbet and Orvain, 2007). Macrofauna influence sediment erodibility by increasing bottom roughness elements (e.g. Nowell and Jumars, 1984); introducing particulate fluxes (from the water column to sediment through biodeposition from suspension feeders or, conversely, with bioresuspension from the production of faecal pellets (e.g. Graf and Rosenberg, 1997); and through bioturbation. Biodestabilization also includes bed corrasion by saltating shell material (Amos *et al.*, 2000).

While there are a wealth of both laboratory and field studies on the relations between erodibility and the presence or absence of particular individual organisms, there are few studies which extend these relationships to erosion rates and even fewer which extend any such analysis to the scale of the estuary cross-profile (Le Hir, Monbet and Orvain, 2007). However, for the Humber estuary, UK, Wood and Widdows (2002, 2003) have developed a simple, as yet unvalidated, cross-shore model incorporating biostabilization (in the form of Chlorophyll *a* content) and bioturbation (from the burrowing bivalve *Macoma balthica*). The model suggests that the erosion or deposition driven by natural fluctuations in biota densities are as large as the changes caused by variations in tidal range and currents over a spring – neap cycle or are equivalent to a doubling of the external sediment supply. Seasonal variations in the density of stabilizing diatoms can alter the magnitude of net deposition by a factor of two and interannual changes in *Macoma* density change deposition by a factor of 5. In a UK climate change scenario, milder winters will result in lower springtime recruitment of *Macoma*, leading to lower rates of bioturbation at mid-intertidal levels and lower sediment supply to the upper intertidal zone and its fringing saltmarshes (Wood and Widdows, 2003). Using the same two biotic groups, Paarlberg *et al.* (2005) extended Wood and Widdows (2002) approach, showing that changes in bioturbation and stabilization by microphytobenthos can potentially alter the mud content and elevation (by 5–10 cm) of shoal banks in the Westerschelde estuary, The Netherlands. In fact, field observations in the Westerschelde show that two stable sedimentological states are present at intermediate levels of bed shear stress, either a bare surface with low silt content or a high silt content supporting a high density of diatoms. This bimodal pattern results from the feedback links between the biota and silt content. Diatom growth rates are enhanced by the nutrients present in silt-rich sediments and diatom resuspension falls as diatom density increases, favouring the accumulation of silts. Loss of diatom cover sets this dynamic in reverse (van de Koppel *et al.*, 2001).

The development of benthic diatom cover is almost always seasonal so the effect over annual timescales may be small. Although the development of biofilms leads to surface protection, beneath this thin skin it is often accompanied by increased water contents and reductions in sediment density, with implications for enhanced erosion once the immediate surface layer is disrupted by episodic wave processes (Tolhurst, Consalvey and Paterson, 2008). By comparison, biodestabilization processes associated with a macrofauna are likely to vary on interannual scales. The same is not true of the macrofauna and here changes are likely to take place over interannual timescales. Rather than the controlled experimentation typical of laboratory studies, where the density of an individual species is related to a measure of sediment erodibility, the occurrence of a species in a particular location and its biogeomorphological role depends upon a complex set of interactions between: particle size, organic and microbial content of the sediments; hydrodynamic and chemical conditions; and biological interactions (which include its trophic level and predator–prey interactions). This kind of ecogeomorphological modelling, with assemblages of species, has yet to be tackled in a concerted manner for estuarine environments.

14.5 The Upper Intertidal Zone: The Role of Vegetation

The establishment of a year-round vegetation cover marks an important change in sedimentary processes on estuarine margins. A considerable literature has been expended on the sediment-trapping powers of a cover of saltmarsh halophytes, yet experimental testing of this role has been inconclusive. What is apparent is that the patterns of periodic surface elevation increase and loss seen on marsh-fronting mudflats become transformed into a pattern that is dominated by accretion, with the absence of surface lowering erosional events. Thus Neumeier and Ciavola's (2004) sediment-trap data show that sediment accumulation in a *Spartina* canopy is related more to erosion protection during storms than to enhanced sedimentation under non-storm conditions.

In general terms, marine macrophytes dampen currents, waves and turbulence in subtidal and intertidal environments (Koch *et al.*, 2009) and a considerable research effort has been focused on trying to establish vegetation friction factors. Early attempts were largely based on laboratory studies which transplanted single saltmarsh species into laboratory flumes and then reported time-averaged velocity profiles through these artificial stands, under unidirectional flow (e.g. Pethick, Leggett and Husain, 1990; Shi, Pethick and Pye, 1995; Coops *et al.*, 1996); these studies showed that vertical flow profiles are far from uniform and deviate from the logarithmic profile in areas lacking vegetation. The nature of this experimental set-up led to an emphasis on trying to explain modifications of flow fields through the effects of shoot size and spacing, drawing upon the body of work by Eckman and others (e.g. Eckman *et al.*, 1981; Rhoads and Boyer, 1982) concerned with the stabilizing and destabilizing roles of tube burrows. Although this may provide an acceptable model for flows around basal stems it is not as applicable to the upper, more complex canopy structures. More recently, detailed field measurements of flow hydrodynamics within saltmarsh canopies have been reported; these show a strong reduction in velocity within the canopy, although often with secondary velocity peaks in the lower canopy, and the

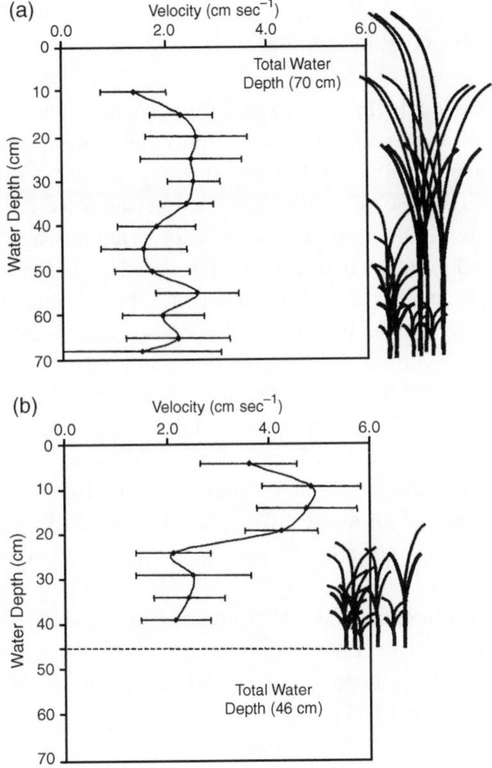

Figure 14.10 Velocity profiles (left) against general distribution of plant material in the water column (right) collected in moderately dense (150 stems m^{-2}) *Spartina alterniflora* canopies on mesotidal saltmarshes, southeastern North Carolina, USA. (a) Pre-manipulation vertical velocity profiles measured at 5 cm increments throughout the water column, giving depth-integrated mean velocity of 2.2 cm s^{-1}. (b) The canopy was then clipped to a uniform height of about 25 cm and flow velocity profiles remeasured. Profiles show within canopy depth-integrated mean velocity of 2.1 cm s^{-1}, with overlying region of skimming flow with a depth-integrated mean velocity of 4.5 cm s^{-1} (reprinted from Leonard, L. A. and Croft, A. L. 'The Effect of Standing Biomass on Flow Velocity and Turbulence in Spartina Alterniflora Canopies' Estuarine, Coastal and Shelf Science 69 3–4 pg 12 © 2006 with permission from Elsevier).

creation of skimming flow over dense vegetation covers (Leonard and Croft, 2006; Neumeier, 2007; Bouma *et al.*, 2007; Figure 14.10).

Field experimentation has shown that vegetation canopies significantly reduce tidal current flows between feeder creeks and marsh interiors (e.g. Leonard and Reed, 2002) and play a significant role in upper intertidal wave attenuation. Field studies have confirmed Brampton's (1992) earlier physical modelling that saltmarshes significantly increase the attenuation of incident waves compared to fronting mudflats and sand flats. Thus, for example, in a meso- to macrotidal open coast environment in North Norfolk, eastern England, measurements of wave energy dissipation rates over a 180 m wide saltmarsh were significantly higher ($x = 82\%$) than over a comparable width fronting sand flat ($x = 29\%$) (Moeller, Spencer and French, 1996; Möller *et al.*, 1999). The majority of the reduction in wave height has been shown to take place over the first 10–20 m of the vegetated marsh surface (Möller, Spencer and Rawson, 2002). How these patterns can be explained is a considerable challenge.

First, there is the need to deal effectively with the stochastic nature of the input conditions – on any given surface on any given day wave energy dissipation will vary in response to changing hydrodynamic conditions. Following Le Hir *et al.* (2000), one way to characterize this variability is through the use of relative wave height (the wave-height/water-depth ratio, H_{rms}/h). On the Essex mudflats, eastern England, the relative wave height is high (0.45) indicating that shoaling dominates over frictional losses. This high ratio is maintained over the transitional area between mudflat and marsh but once into the marsh the ratio declines. By the landward margin of the marsh – 100 m from the seaward edge – the ratio has been reduced to a value of 0.15 (Möller, 2006). The inference is that frictional losses become dominant over these surfaces; Möller *et al.* (1999) suggest a wave friction factor, f_w, of up to 0.4 for vegetated saltmarsh compared to $f_w = 0.01$ for an unvegetated, smooth bed sandflat under identical hydrodynamic conditions (and see also Teeter *et al.* (2001) who used f_w values of 0.08–0.9 from laboratory measurements (by Fonseca and Cahalan, 1992) over seagrasses). Such friction factors do not, however, allow the different contributors to wave energy dissipation to be separately isolated and quantified. The second challenge, therefore, follows from the saltmarsh environment being one of complex surfaces, including variable marsh/mudflat margin types (Möller and Spencer, 2002), bifurcating creek networks and marsh surface pans, where 'bed roughness' is due both to topographic variations in marsh surface geometry (on which almost no research has been focussed) and on vegetation community coverage and canopy characteristics (Figure 14.11).

The exact mechanisms by which ecological processes influence energy dissipation, the thresholds at which these dissipative controls are exceeded and the spatial and temporal scale at which these processes are significant remain largely ill-defined. Part of the

Figure 14.11 View south over the backbarrier marsh at Hut Marsh, Scolt Head Island, North Norfolk coast, UK. The marsh surface shows complex surface roughness features including salt pans, bifurcating creeks, and changes in vegetation character from creek margins to high marsh plain (photograph by T Spencer).

problem stems from the difficulties in recording appropriate measures of vegetation structure, although recent advances in digital imaging processing hold promise for generating metrics with hydrodynamic meaning (Neumeier, 2005; Möller, 2006). Möller and Spencer (2002) observed seasonal differences in wave energy dissipation along a transect from mudflat to upper saltmarsh in Essex, UK from the autumn months (September to December), through the winter months (December to February) into the spring and early summer months (March to July), particularly in the transition zone between saltmarsh and mudflat characterized by the seasonal growth of marsh edge pioneer species (Figure 14.12).

It is tempting to relate these changes to variations in vegetation characteristics – canopy height is at its greatest in the autumn but lower in the winter, a decline which continues into the spring – although the relative coverage of perennial versus annual species and, more subtly, the mix of species present in floristically diverse saltmarsh communities, and their

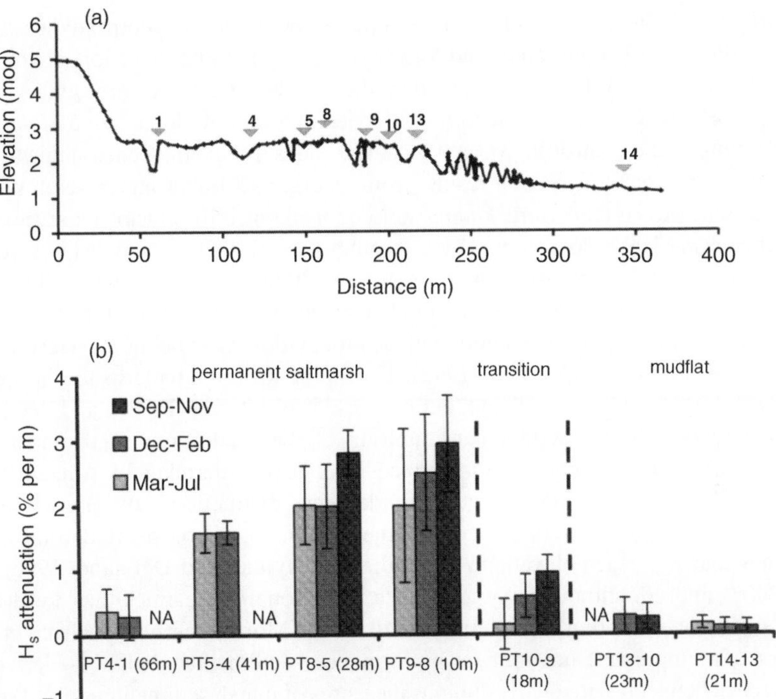

Figure 14.12 (a) Cross-shore profile (sea defence at +5 m OD Newlyn (where OD Newlyn approximates mean sea level) far left) and location of pressure transmitters for water level and wave burst measurements, Tillingham Marsh, Dengie Peninsula, Essex coast, UK. (b) Seasonal averages of significant wave height (H_s) attenuation over the period October 2000 to August 2001 (there were 19 occasions when all stations were inundated, with water depths > 30 cm at the seaward marsh edge). Little seasonal difference was observed over the mudflat or at the most landward saltmarsh sites but more pronounced differences were seen at the marsh edge. Consistent difference in attenuation was seen from autumn, through winter and into spring

(reprinted from Moller, I., Spencer, T. and Rawson, J. in Smith, J. M. (ed) Proceedings 28th ICCE 2002 pp. 651–663 © 2002 World Scientific).

differing growth performance over an annual growth cycle, may be of greater significance. Attempts to establish such relationships are also confounded by the co-variation of vegetation canopy development and varying seasonal hydrodynamic conditions. Under late growing season (September) conditions in Essex, when the maximum significant wave height was 19 cm, a very clear relationship was seen between relative wave height and wave energy dissipation (%H_s reduction) whereas in winter (December), when the maximum significant wave height was 32 cm and higher relative wave heights were present, this relationship began to break down. It can be suggested that once the relative wave height exceeds a value of $H_s/h = 0.55$, then the energy dissipating ability of the saltmarsh vegetation is overcome and must be taken up by other processes, including surface erosion (Möller, 2006). Such studies echo those at a larger scale which have tried to relate saltmarsh margin type to incident energy conditions and have started to refine where the process thresholds lie that control the 'switching' of these saltmarsh systems from one regime (e.g. accretional) to another (e.g. erosional) – something that has been frequently documented but not properly quantified.

A third set of challenges arises from the fact that whereas the short-term studies described above see the role of topography and vegetation to be passive, over longer timescales, because they influence rates of sedimentation, there are feedbacks between hydrodynamics and marsh surface. It is not possible to simply scale up from individual inundations to long-term behaviour as the controlling variables change. Thus, for example, at a single tide level, in systems where marsh accretion results from external sediment inputs, as in northwest Europe, distance from feeder creek appears to be the controlling factor on sedimentation rate (French *et al.*, 1995; Reed *et al.*, 1999; Temmerman *et al.*, 2003) and, in USA west coast marshes, not strongly controlled by vegetation (Culberson, Foin and Collins, 2004). However, at the annual scale, it is the number of inundations that is important and thus the control becomes surface elevation, with sedimentation rates being inversely related to surface height (French and Spencer, 1993). Thus quite different patterns to sedimentation result depending upon the spatial and temporal scales of interest (French and Reed, 2001). Numerical models of the sedimentary infilling of the tidal frame, and marsh surface evolution towards a quasi-equilibrium surface, have been undertaken by Allen (1990) and French (1993, 1994); these models make a fundamental distinction between autochthonous systems in which plant productivity and the accumulation of plant detrital material determines marsh surface elevation (e.g. Callaway, Nyman and DeLaune, 1996; Morris *et al.*, 2002) and allochthonous systems with accretionary dynamics that are driven by mineral sediment inputs (e.g. Temmerman *et al.*, 2003). More recently, French has criticized the notion of a simple measure of marsh performance – sedimentary balance – derived from the difference between statistically dubious measures of marsh sedimentation and sea-level rise and begun the 'exploratory modelling of the parameter space defined by marsh surface elevation, sedimentation, sea level rise, sediment supply and tidal range' (French, 2006, p. 121). One set of outcomes is the identification of the importance of the 18.6 year nodal tidal variation, particularly in macrotidal settings, and the role of storm events, particularly in micro- to mesotidal systems, in influencing marsh sedimentary responses and the dangers of short-term monitoring not identifying all the key controls.

Valuable as these modelling exercises are, they do not capture the lateral changes in marsh extent, and potential shifts in marsh position that may accompany surface responses to changes in controlling variables. The fourth challenge here then is of spatial scaling. For example, marsh performance may appear 'healthy' in a vertical sense yet may be subject to considerable lateral retreat and loss of area. Although it is common to link saltmarsh habitat loss to rising sea levels, databases on historical changes in intertidal mudflat and saltmarsh extent for UK estuaries (e.g. Pye and French, 1993; Cooper, Cooper and Burd, 2001; French and Burningham, 2003; Van der Wal and Pye, 2004) show more complex oscillations between periods of rapid mudflat accretion and saltmarsh progradation and periods of mudflat lowering and saltmarsh retreat, at odds with a single factor link to progressive sea-level rise. Various process links have been proposed at the level of broad correlation (including changes in water level, variations in storminess and the prevalence of different weather types (as typified, for example, by the North Atlantic Oscillation (French, 2006)), but none have been comprehensively tested through estuarine-scale modelling.

14.6 Conclusion

This chapter has shown that the sediment store that lies at the end of the sediment cascade in estuaries, over timescales relevant to human lives and livelihoods, is far from a passive accumulator. It is clear that estuarine dynamics are influenced by both changes in extrinsic forcing factors from both catchment and continental shelf directions but also by intrinsic feedbacks and thresholds within the systems themselves. It is apparent that change in any one area, be it natural or human-induced, can have impacts elsewhere within the system and that these impacts vary in temporal and spatial footprint. Whereas traditionally, estuarine dynamics was explained primarily by reference to tidal hydrodynamics, it is now becoming clear that wave processes, and the interaction between waves and tides (e.g. Grant and Madsen, 1979), are more important than thought previously. It is also now obvious that numerical models of estuarine sediment dynamics and morphology need to incorporate biological as well as physical processes in order to represent the natural estuarine environment. These biological impacts are not trivial: switches between biostable and bio-unstable states can alter erosion rates by over two orders of magnitude. In energetic subtidal estuarine channels physical processes remain dominant but their influence on the fate of sediments is strongly mediated by biological controls on sediment aggregation and settling velocity. On lower and middle intertidal zone mudflats and sandflats, this chapter has shown the potentially critical threshold in sediment stability between the protective role of thin biofilms associated with the microphytobenthos and their disruption by physical events or macrofauna. Such structures can give increasing erosion with depth, quite contrary to the usual, physically based expectation of decreasing erosion with depth with increasing sediment density (e.g. Parthenaides, 2007). On permanently vegetated upper intertidal zone margins, biological controls become pervasive, although even here there are clearly physical process thresholds between efficient energy dissipation and ecosystem degradation, retreat and areal loss.

Relations between the patterning of estuarine macrobenthic and littoral assemblages and hydrodynamic regime in this 'controlling biological overlay' (Tolhurst, Black and Paterson, 2009, p. 80) are typically made through the correlation of patterns of abundance with easily measurable but 'static' physical parameters, such as particle-size distribution or depth-integrated bulk density properties. In reality, of course, the occurrence of a plant or animal species in a particular location depends upon a complex set of interactions between: substrate particle size and texture; organic and microbial content of sediments; hydrodynamic, biogeochemical and perhaps meteorological conditions; and interspecific interactions (which for fauna may include predator–prey dynamics). Considerable progress has been made over the past decade in teasing out these interrelationships, initially in the laboratory (with the attendant problems of transferability to field situations) and more recently to field experimentation. Much of this experimentation has necessarily been at very small scales – in the form, for example, of Eulerian measurements of ETM character, point measurements of sediment properties on tidal flats or of the establishment of flow fields around individual stems in saltmarsh vegetation canopies – and there is a considerable challenge in progressing from this level to more estuarine scales. And in modelling terms, the horizontal resolution of the models is much greater than the patchiness of the physical properties and biota of interest and must rely on spatially averaged properties between widely spaced sampling points. There are thus considerable problems of upscaling (Murray, Meadows and Meadows, 2002).

The wider application of remote sensing, from airborne platforms calibrated with ground-based sampling and field spectral reflectance data, does, however, hold promise for providing input for improved modelling of the spatial distribution of intertidal sediment properties (e.g. Smith *et al.*, 2004; Murphy *et al.*, 2008) and estuarine suspended sediment concentrations. In saltmarsh environments, for example, there is a need for innovative scientific approaches to characterize the hydrodynamic 'roughness' of medium-scale areas (10s to 100s of metres) that consist of both the complex small-scale roughness elements that arise from the presence of a vegetation canopy of a saltmarsh and the topographic roughness elements which comprise tidal creeks, saltpans and marsh cliffs. The integration of multispectral and laser scanning airborne remote sensing now offers an opportunity to address this challenge. Hyperspectral imagery combined with field-collected reflectance spectra (e.g. Artigas and Yang, 2005 (and see also Li, Ustin and Lay (2005) for spectral mixture modelling)) can be used to map exposed vegetation (e.g. biomass, species) and the biophysical properties of surrounding mudflats and this information can be combined with digital elevation models derived from LiDAR (e.g. Yang, 2005). Ultimately, it would be of value to develop algorithms from this work which identify different scales of intertidal roughness elements and then use this information as input to shallow water flow and wave models.

Finally, there is a need to think about the consequences of global environmental change for estuaries. Prandle (2006) predicts changes in estuarine lengths of 0.5–1.0 km, with width changes of 5–250 m, with both increases and decreases of 25% in UK river flows. In addition, changes in sea level of +50 cm may lead to length changes of 1–2.5 km and width changes of 70–100 m. Furthermore, in systems where low-frequency, high-magnitude events may be of considerable significance in determining the pattern of suspended sediment dynamics (e.g. Ridderinkhof, van der Ham and van der Lee, 2000; Schoelhammer, 2002) the compound effect of rising sea level and potential changes in near-future storminess assumes significance. These changes could be considerable; for Immingham on the UK east coast, a storm surge level of 1.5 m is predicted to be exceeded every 120 years under the present climate. Under a UK Meteorological Office scenario of an increase in average global temperature of 3.5 °C by 2080, plus predicted changes in storminess, mean sea level and vertical land movements, this exceedance frequency reduces to the 1 in 7 year event (Lowe and Gregory, 2005). Finally, changes in air and water temperature will alter biological communities and their seasonal dynamics. Understanding how estuaries will respond to these and other associated changes, in terms of their morphology and role in sediment storage will require an integration of approaches across the tidal gradient, from the subtidal to the supratidal zone, and across the physical, chemical and biological drivers of estuarine sediment fates.

Acknowledgements

Figure 14.6 was drafted by Nina de Luca. Dr I. Möller, University of Cambridge, is thanked for valuable inputs into discussions on many of the issues raised in this chapter.

References

Allen, G.P., Salomon, J.C., Bassoullet, P. *et al.* (1980) Effects of tides on mixing and suspended sediment transport in macrotidal estuaries. *Sedimentary Geology*, **26**, 69–90.

Allen, J.R.L. (1990) Salt-marsh growth and stratification: A numerical model with special reference to the Severn Estuary, Southwest Britain. *Marine Geology*, **95**, 77–96.

Allen, J.R.L. (2000) Morphodynamics of Holocene saltmarshes: a review sketch from the Atlantic and Southern North Sea coasts of Europe. *Quaternary Science Reviews*, **19**, 1155–1231.

Amos, C.L., Sutherland, T.F., Cloutier, D. and Patterson, S. (2000) Corrosion of a remoulded cohesive bed by saltating littorinid shells. *Continental Shelf Research*, **20**, 1291–1315.

Andersen, T.J., Jensen, K.T., Lund-Hansen, L. *et al.* (2002) Enhanced erodability of fine-grained marine sediments by Hydrobia ulvae. *Journal of Sea Research*, **48**, 51–58.

Anderson, F.E., Black, L., Watling, L.E. *et al.* (1981) A temporal and spatial study of mudflat erosion and deposition. *Journal of Sedimentary Petrology*, **51**, 729–736.

Artigas, F.R. and Yang, J.S. (2005) Hyperspectral remote sensing of marsh species and plant vigour gradient in the New Jersey Meadowlands. *International Journal of Remote Sensing*, **26**, 5209–5220.

Avoine, J. and Larsonneur, C. (1987) Dynamics and behaviour of suspended sediment in macrotidal estuaries along the south coast of the English Channel. *Continental Shelf Research*, **7**, 1201–1306.

Bale, A.J., Stephens, J.A. and Harris, C.B. (2007) Critical erosion profiles in macro-tidal estuary sediments: implications for the stability of intertidal mud and the slope of mud banks. *Continental Shelf Research*, **27**, 2303–2312.

Bartholdy, J. and Madsen, P. (1985) Accumulation of fine grained materials in Danish tidal area. *Marine Geology*, **67**, 121–137.

Barua, D.K. (1990) Suspended sediment movement in the estuary of the Ganges-Brahmaputra-Meghna River system. *Marine Geology*, **91**, 243–253.

Bassoullet, P., Le Hir, P., Gouleau, D. and Robert, S. (2000) Sediment transport over an intertidal mudflat: field investigations and estimation of fluxes within the "Baie de Marenngres-Oleron" (France). *Continental Shelf Research*, **20**, 1635–1653.

Bouma, T.J., van Duren, L.A., Temmerman, S. *et al.* (2007) Spatial flow and sedimentation patterns within patches of epibenthic structures: Combining field, flume and modelling experiments. *Continental Shelf Research*, **27**, 1020–1045.

Brampton, A.H. (1992) Engineering significance of British saltmarshes, in *Saltmarshes: Morphodynamics, Conservation and Engineering Significance* (eds J.R.L. Allen and K. Pye), Cambridge University Press, Cambridge, pp. 115–122.

Burchard, H. and Baumert, H. (1998) The formation of estuarine turbidity maxima due to density effects in the salt wedge. A hydrodynamic process study. *Journal of Physical Oceanography*, **28**, 309–321.

Callaway, J.C., Nyman, J.A. and DeLaune, R.D. (1996) Sediment accretion in coastal wetlands: a review and a simulation model of processes. *Current Topics in Wetland Biogeochemistry*, **12**, 2–23.

Chappell, J. (1993) Contrasting Holocene sedimentary geologies of lower Daly River, northern Australia, and lower Sepik-Ramu, Papua New Guinea. *Sedimentary Geology*, **83**, 339–358.

Christie, M.C. and Dyer, K.R. (1998) Measurements of the turbid tidal edge over the Skeffling mudflats, in *Sedimentary Processes in the Intertidal Zone* (eds K.S. Black, D.M. Paterson and A. Cramp), (Special Publication 139, Geological Society of London), Geological Society Publishing House, Bath, UK, pp. 45–55.

Cooper, N.J., Cooper, T. and Burd, F. (2001) 25 years of saltmarsh erosion in Essex: Implications for coastal defence and nature conservation. *Journal of Coastal Conservation*, **9**, 31–40.

Coops, H., Geilen, N., Verheij, H.J. *et al.* (1996) Interactions between wave, bank erosion and emergent vegetation: an experimental study in a wave tank. *Aquatic Botany*, **53**, 187–198.

Crump, B.C. and Baross, J.A. (1996) Particle-attached bacteria and heterotrophic plankton associated with the Columbia River estuarine turbidity maxima. *Marine Ecology Progress Series*, **138**, 265–213.

Crump, B.C., Simenstad, C.A. and Baross, J.A. (1998) Particle-attached bacteria dominate the Columbia River estuary. *Aquatic Microbial Ecology*, **14**, 7–18.

Culberson, S.D., Foin, T.C. and Collins, J.N. (2004) The role of sedimentation in estuarine marsh development within the San Francisco Estuary, California, USA. *Journal of Coastal Research*, **20**, 970–979.

Dalrymple, R.W., Zaitlin, B.A. and Boyd, R. (1992) Estuarine facies models: conceptual basis and stratigraphic implications. *Journal of Sedimentary Petrology*, **62**, 1130–1146.

Davies, A.G. and Thorne, P.D. (2008) Advances in the study of moving sediments and evolving seabeds. *Surveys in Geophysics*, **29**, 1–36.

Dronkers, J. (1986) Tidal asymmetry and estuarine morphology. *Netherlands Journal of Sea Research*, **20**, 117–131.

Dronkers, J. (2005) *Dynamics of Coastal Systems*, vol. **25**, (Advanced Series in Ocean Engineering), World Scientific, Singapore.

Dyer, K.R. (1995) Sediment transport processes in estuaries, in *Geomorphology and Sedimentology of Estuaries*, vol. **53** (ed. G.M.E. Perillo), Elsevier, Amsterdam, pp. 423–449.

Dyer, K.R. (1997) *Estuaries: A Physical Introduction*, 2nd edn, John Wiley and Sons, Hoboken, NJ.

Dyer, K.R. and Manning, A.J. (1999) Observation of the size, settling velocity and effective density of flocs, and their fractal dimensions. *Journal of Sea Research*, **41**, 287–95.

Eckman, J.E., Duggins, D.O. and Sewell, A.T. (1981) Ecology of understory kelp environments. I. Effects of kelps on flow and particle transport near the bottom. *Journal of Experimental Marine Biology and Ecology*, **129**, 173–187.

EMPHASYS Consortium (2000) Modelling estuary morphology and processes. Final Report. Research by the EMPHASYS Consortium for MAFF Project FD1401. Report TR111. HR Wallingford: Crowmarsh Gifford.

EstProc Consortium (2004) Integrated research results on hydrobiosedimentary processes in estuaries. Final Report of the Estuary Process Research Project (EstProc). R & D Technical Report prepared by the Estuary Process Consortium for the DEFRA and Environment Agency Joint Flood and Coastal Processes Theme. Report No. FD1905/T2, Synthesis Report. HR Wallingford: Crowmarsh Gifford.

EstSim Consortium (2007) Development and demonstration of systems-based estuary simulators. R & D Technical Report, DEFRA and Environment Agency Joint Flood and Coastal Erosion Risk Management R & D Programme, Report No. FD2117/TR, DEFRA: London.

Evans, G. (1965) Intertidal flat sediments and their environment of deposition in The Wash. *Quarterly Journal of the Geological Society of London*, **121**, 209–245.

Fairbridge, R.W. (1980) The estuary: its definition and geodynamic cycle, in *Chemistry and Biogeochemistry of Estuaries* (eds E. Olausson and I. Cato), John Wiley and Sons, Chichester, pp. 1–36.

Fonseca, M.S. and Cahalan, J.A. (1992) A preliminary evaluation of wave attenuation by four species of seagrass. *Estuarine, Coastal and Shelf Science*, **35**, 565–576.

French, C.E., French, J.R., Clifford, N.J. and Watson, C.J. (2000) Sedimentation – erosion dynamics of abandoned reclamations: the role of waves and tides. *Continental Shelf Research*, **20**, 1711–1733.

French, J.R. (1993) Numerical simulation of vertical marsh growth and adjustment to accelerated sea-level rise, North Norfolk, UK. *Earth Surface Processes and Landforms*, **81**, 63–81.

French, J.R. (1994) Wetland response to accelerated sea-level rise: a European perspective. *Journal of Coastal Research*, **SI12**, 94–105.

French, J.R. (2006) Tidal marsh sedimentation and resilience to environmental change: Exploratory modelling of tidal, sea-level and sediment supply forcing in predominantly allochthonous systems. *Marine Geology*, **235**, 119–136.

French, J.R. and Burningham, H. (2003) Tidal marsh sedimentation versus sea-level rise: a southeast England estuarine perspective, in *Proceedings International Conference on Coastal Sediments 2003* (ed. R.A. Davis), World Scientific Publishing and East Meets West, Productions, Corpus Christi, pp. 1–14.

French, J.R. and Reed, D.J. (2001) Physical contexts for saltmarsh conservation, in *Habitat Conservation. Managing the Physical Environment* (eds A. Warren and J.R. French), John Wiley and Sons, Chichester, pp. 179–228.

French, J.R. and Spencer, T. (1993) Dynamics of sedimentation in a tide-dominated backbarrier salt marsh, Norfolk, UK. *Marine Geology*, **110**, 315–331.

French, J.R., Burningham, H. and Benson, T. (2008) Tidal and meteorological forcing of suspended sediment flux in a muddy mesotidal estuary. *Estuaries and Coasts*, **31**, 843–859.

French, J.R., Spencer, T., Murray, A.L. and Arnold, N.S. (1995) Geostatistical analysis of sediment deposition in two small tidal wetlands, Norfolk, UK. *Journal of Coastal Research*, **11**, 308–321.

Friedrichs, C.T. and Aubrey, D.G. (1988) Non-linear tidal distortion in shallow well-mixed estuaries: A synthesis. *Estuarine, Coastal and Shelf Science*, **27**, 521–545.

Friend, P.L., Lucas, C.H. and Rossington, S.K. (2005) Day-night variation of cohesive sediment stability. *Estuarine, Coastal and Shelf Science*, **64**, 407–418.

Gao, S. and Collins, M. (1994) Tidal inlet equilibrium in relation to cross-sectional area and sediment transport patterns. *Estuarine, Coastal and Shelf Science*, **38**, 157–172.

Gelfenbaum, G. (1983) Suspended sediment response to semi diurnal and fortnightly variations in a mesotidal estuary: Columbia River, USA. *Marine Geology*, **52**, 39–57.

Geyer, W.R. (1993) The importance of suppression of turbulence by stratification on the estuarine turbidity maximum. *Estuaries*, **16**, 113–125.

Grabemann, I., Uncles, R.J., Krause, G. and Stephens, J.A. (1997) Behavior of turbidity maxima in the Tamar (U.K.) and Weser (F.R.G.) estuaries. *Estuarine, Coastal and Shelf Science*, **45**, 235–246.

Graf, G. and Rosenberg, R. (1997) Bioresuspension and biodeposition: a review. *Journal of Marine Systems*, **11**, 269–278.

Grant, W.D. and Madsen, O.S. (1979) Combined wave and current interaction with a rough bottom. *Journal of Geophysical Research*, **84**, 1797–1808.

Green, M.O., Black, K.P. and Amos, C.L. (1997) Control of estuarine dynamics by interactions between currents and waves at several scales. *Marine Geology*, **144**, 97–116.

Hardisty, J. (2007) *Estuaries: Monitoring and Modelling the Physical System*, Blackwell Publishing, Oxford.

Hardisty, J., Middleton, R., Whyatt, D. and Rouse, H.L. (1998) Geomorphological and hydrodynamic results from digital terrain models of the Humber Estuary, in *Landform Monitoring, Modelling and Analysis* (eds S.N. Lane, K.S. Richards and J.H. Chandler), John Wiley and Sons, Chichester, pp. 422–433.

Hollibaugh, J.T. and Wong, P.S. (1999) Microbial processes in the San Francisco Bay estuarine turbidity maximum. *Estuaries*, **22**, 848–862.

Hubbell, D.W. and Glenn, J.C. (1973) Distribution of Radionuclides in Bottom Sediments of the Columbia River Estuary. United Sates Geological Survey: U.S. Government Printing, Office, Washington DC, Professional Paper: 433-L.

Janssen-Stelder, B. (2000) The effect of different hydrodynamic conditions on the morphodynamics of a tidal mudfalt in the Dutch Wadden Sea. *Continental Shelf Research*, **20**, 1461–1478.

Jay, D.A. and Musiak, J.D. (1994) Particle trapping in estuarine tidal flows. *Journal of Geophysical Research*, **9** (C10), 20445–20461.

Kirby, R. (2000) Practical implications of tidal flat shape. *Continental Shelf Research*, **20**, 1061–1077.

Koch, E.W., Barbier, E.B., Silliman, B.R. *et al.* (2009) Non-linearity in ecosystem services: temporal and spatial variability in coastal protection. *Frontiers in Ecology and the Environment*, **7**, 29–37.

Klein, G.deV. (1985) Intertidal flats and intertidal sand bodies, in *Coastal Sedimentary Environments*, 2nd edn (ed. R.A. DaviesJr), Springer-Verlag, New York, pp. 187–224.

Le Hir, P., Monbet, Y. and Orvain, F. (2007) Sediment erodability in sediment transport modelling: Can we account for biotic effects? *Continental Shelf Research*, **27**, 1116–1142.

Le Hir, P., Roberts, W., Cazaillet, O. *et al.* (2000) Characterization of intertidal flat hydrodynamics. *Continental Shelf Research*, **20**, 1433–1459.

Lang, G., Schubert, R., Markofsky, M. *et al.* (1989) Data interpretation and mumerical modelling of the Mud and Suspended Sediment Experiment 1985. *Journal of Geophysical Research*, **94**, 14381–14394.

Leonard, L.A. and Croft, A.L. (2006) The effect of standing biomass on flow velocity and turbulence in *Spartina alterniflora* canopies. *Estuarine, Coastal and Shelf Science*, **69**, 325–336.

Leonard, L.A. and Reed, D.J. (2002) Hydrodynamics and sediment transport through tidal marsh canopies. *Journal of Coastal Research*, **SI36**, 459–469.

Li, L., Ustin, S.L. and Lay, M. (2005) Application of multiple end member spectral mixing analysis (MESMA) to AVIRIS imagery for coastal saltmarsh mapping: a case study of China Camp, C.A., USA. *International Journal of Remote Sensing*, **26**, 5193–5207.

Lindsay, P., Balls, P.W. and West, J.R. (1996) Influence of tidal range and river discharge on suspended particulate matter fluxes in the Forth estuary. *Estuarine, Coastal and Shelf Science*, **42**, 63–82.

Lowe, J.A. and Gregory, J.M. (2005) The effects of climate change on storm surges around the United Kingdom. *Philosophical Transactions of the Royal Society of London, Series A*, **363**, 1313–1328.

Masselink, G. and Hughes, M.G. (2003) *Introduction to Coastal Processes & Geomorphology*, Hodder Headline, London.

McDowell, D.M. and O'Connor, B.A. (1977) *Hydraulic Behaviour of Estuaries*, Macmillian, London.

Möller, I. (2006) Quantifying saltmarsh vegetation and its effect on wave height dissipation: Results from a UK East Coast saltmarsh. *Estuarine, Coastal and Shelf Science*, **69**, 337–351.

Möller, I. and Spencer, T. (2002) Wave dissipation over macro-tidal saltmarshes: effects of marsh edge typology and vegetation change. *Journal of Coastal Research*, **SI36**, 506–521.

Möller, I., Spencer, T. and French, J.R. (1996) Wind wave attenuation over saltmarsh surfaces: preliminary results from Norfolk, England. *Journal of Coastal Research*, **12**, 1009–1016.

Möller, I., Spencer, T. and Rawson, J. (2002) Spatial and temporal variability of wave attenuation over a UK East-coast saltmarsh. Proceedings of the 28th International Conference on Coastal Engineering, Cardiff, July 2002, vol. 3.

Möller, I., Spencer, T., French, J.R. *et al.* (1999) Wave transformation over saltmarshes: a field and numerical modelling study from North Norfolk, England. *Estuarine, Coastal and Shelf Science*, **49**, 411–426.

Morris, J.T., Sundareshwar, P.V., Nietch, C.T. *et al.* (2002) Responses of coastal wetlands to rising sea-level. *Ecology*, **83**, 2869–2877.

Murphy, R.J., Tolhurst, T.J., Chapman, M.G. and Underwood, A.J. (2008) Spatial variation in chlorophyll on estuarine mudflats determined by field-based remote sensing. *Marine Ecology Progress Series*, **365**, 45–55.

Murray, J.M.H., Meadows, A. and Meadows, P.S. (2002) Biogeomorphological implications of microscale interactions between sediment geotechnics and marine benthos: a review. *Geomorphology*, **47**, 15–30.

Neumeier, U. (2005) Quantification of vertical density variations of saltmarsh vegetation. *Estuarine, Coastal and Shelf Science*, **63**, 489–496.

Neumeier, U. (2007) Velocity and turbulence variations at the edge of saltmarshes. *Continental Shelf Research*, **27**, 1046–1059.

Neumeier, U. and Ciavola, P. (2004) Flow resistance and associated sedimentary processes in a *Spartina maritima* salt-marsh. *Journal of Coastal Research*, **20**, 435–447.

Nicholls, M.M. and Biggs, R.B. (1985) Estuaries, in *Coastal Sedimentary Environments*, 2nd edn (ed. R.A. Davies Jr), Springer-Verlag, New York, pp. 77–186.

Nowell, A.R.M. and Jumars, P.A. (1984) Flow environments of aquatic benthos. *Annual Review of Ecology and Systematics*, **15**, 303–328.

O'Brien, D.J., Whitehouse, R.J.S. and Cramp, A. (2000) The cyclic development of a macrotidal mudflat on varying timescales. *Continental Shelf Research*, **20**, 1593–1619.

O'Brien, M.P. (1931) Estuary tidal prism related to entrance areas. *Civil Engineering*, **1**, 738–739.

Orton, P.M. and Kineke, G.C. (2001) Comparing calculated and observed vertical suspended-sediment distributions from a Hudson River estuary turbidity maximum. *Estuarine, Coastal and Shelf Science*, **52**, 401–410.

Paarlberg, A.J., Knaapen, M.A.F., de Vries, M.B. and de Wang, Z.B. (2005) Biological influences on morphology and bed composition of an intertidal flat. *Estuarine, Coastal and Shelf Science*, **64**, 577–590.

Parthenaides, E. (2007) *Engineering Properties and Hydraulic Behaviour of Cohesive Sediments*, CRC Press, Boca Raton.

Paterson, D.M. (1997) Biological mediation of sediment erodibility: ecology and physical dynamics, in *Cohesive Sediments* (eds N. Burt, R. Parker and J. Watts), Wiley Interscience, New York, pp. 215–230.

Pethick, J.S. (1996) The geomorphology of mudflats, in *Estuarine Shores: Evolution, Environments and Human Alterations* (eds K.F. Nordstrom and C.T. Roman), John Wiley and Sons, Chichester, pp. 185–211.

Pethick, J., Leggett, D. and Husain, L. (1990) Boundary layers under salt marsh vegetation developed in tidal currents, in *Vegetation and Erosion* (ed. J.B. Thornes), John Wiley and Sons, Chichester, pp. 113–123.

Postma, H. (1967) Sediment transport and sedimentation in the estuarine environment, in *Estuaries* (ed. G.H. Lauff), AAAS, Washington DC, pp. 158–180.

Postma, H. and Kalle, K. (1955) Die entstehung von trübungszonen im unterlauf der flüsse, speziell im hinblick auf die verhältnisse in der Unterelbe. *Deutsche Hydrographische Zeitschrift*, **8**, 137–144.

Prandle, D. (2006) Dynamical controls on estuarine bathymetries: assessment against UK database. *Estuarine, Coastal and Shelf Science*, **68**, 282–288.

Prandle, D. (2009) *Estuaries: Dynamics, Mixing, Sedimentation and Morphology*, Cambridge University Press, Cambridge.

Prandle, D., Murray, A. and Johnson, R. (1990) Analyses of flux measurements in the River Mersey, in *Residual Currents and Long Term Transport*, vol. **38** (ed. R.T. Cheng), (Coastal and Estuarine Studies), Springer-Verlag, Berlin, pp. 413–430.

Pye, K. and French, P. (1993) *Erosion and Accretion Processes in British Saltmarshes*, (Final Report to MAFF), vol. **5**, Cambridge Environmental Research Consultants Ltd, Cambridge.

Reed, D.J. and Donovan, J. (1994) The character and composition of the Columbia River estuarine turbidity maximum, in *Changes in Fluxes in Estuaries: Implications for Science and Management* (eds K.R. Dyer and B.A. Orth), ECSA/ERF Symposium, Plymouth, September 1992, Olsen and Olsen Press, Fredensborg, pp. 445–450.

Reed, D.J., Spencer, T., Murray, A.L. *et al.* (1999) Marsh surface sediment deposition and the role of tidal creeks: implications for created and managed coastal marshes. *Journal of Coastal Conservation*, **5**, 81–90.

Rhoads, D.C. and Boyer, L.F. (1982) Effects of marine benthos on physical properties of sediments. A successional perspective, in *Animal–Sediment Relations* (eds P.L. McCall and M.J.S. Tevesz), Plenum Press, New York, pp. 3–51.

Ridderinkhof, H., van der Ham, R. and van der Lee, W. (2000) Temporal variations in concentration and transport of suspended sediments in a channel–flat system in the Ems–Dollard estuary. *Continental Shelf Research*, **20**, 1479–1493.

Roberts, W., Le Hir, P. and Whitehouse, R.J.S. (2000) Investigation using simple mathematical models of the effect of tidal currents and waves on the profile shape of intertidal mudflats. *Continental Shelf Research*, **20**, 1079–1097.

Romano, C., Widdows, J., Brinsley, M. and Staff, F.J. (2003) Implications of *Enteromorpha intestinalis* mats on near-bed currents and sediment dynamics: flume studies. *Marine Ecology Progress Series*, **256**, 63–74.

Roy, P.S., Cowell, P.J., Ferland, M.A. and Thom, B.G. (1994) Wave-dominated coasts, in *Coastal Evolution: Late Quaternary Shoreline Morphodynamics* (eds R.W.G. Carter and C.D. Woodroffe), Cambridge University Press, Cambridge, pp. 1211–186.

Sanford, L.P., Suttles, S.E. and Halka, J.P. (2001) Reconsidering the physics of the Chesapeake Bay estuarine turbidity maximum. *Estuaries*, **24**, 655–669.

Schoelhammer, D.H. (2002) Variability of suspended sediment concentration at tidal to annual time scales in San Francisco Bay, USA. *Continental Shelf Research*, **22**, 1857–1866.

Schubel, J.R. (1968) Turbidity maximum of northern Chesapeake Bay. *Science*, **161**, 1013–1015.

Sherwood, C.R., Jay, D.A., Harvey, B. *et al.* (1990) Historical changes in the Columbia River estuary. *Progress in Oceanography*, **25**, 299–352.

Shi, Z., Pethick, J.S. and Pye, K. (1995) Flow structure in and above the various heights of a saltmarsh canopy: a laboratory flume study. *Journal of Coastal Research*, **11**, 1204–1209.

Simenstad, C.A., Reed, D.J., Jay, D.A. *et al.* (1994) Land-margin ecosystem research in the Columbia River estuary: investigations of the couplings between physical and ecological processes within estuarine turbidity maxima, in *Changes in Fluxes in Estuaries: Implications for Science and Management* (eds K.R. Dyer B.A. Orth), ECSA/ERF Symposium, Plymouth, September 1992, Olsen and Olsen Press, Fredensborg, pp. 437–444.

Smith, G.M., Thomson, A.G., Möller, I. and Kromkamp, J.C. (2004) Using hyperspectral imagery for the assessment of mudflat surface stability. *Journal of Coastal Research*, **20**, 1165–1175.

Soulsby, R.L. (1997) *Dynamics of Marine Sands: A Manual for Practical Applications*, Thomas Telford, London.

Spearman, J.S., Dearnaley, M.P. and Dennis, J.M. (1998) A simulation of estuary response to training wall construction using a regime approach. *Coastal Engineering*, **33**, 71–78.

Syvitski, J.P.M., Vörösmarty, C.J., Kettner, A.J. and Green, P. (2005) Impact of humans on the flux of sediment to the global coastal ocean. *Science*, **308**, 376–380.

Teeter, A.M., Johnson, B.H., Burger, C. *et al.* (2001) Hydrodynamic and sediment transport modelling with emphasis on shallow-water vegetated areas (lakes, reservoirs, estuaries and lagoons). *Hydrobiologia*, **444**, 1–24.

Temmerman, S., Govers, G., Wartel, S. and Meire, P. (2003) Spatial and temporal factors controlling short-term sedimentation in a salt and freshwater tidal marsh, Scheldt Estuary, Belgium, SW Netherlands. *Earth Surface Processes and Landforms*, **28**, 739–755.

Thomas, C.G., Spearman, J.R. and Turnbull, M.Y. (2002) Historical morphological change in the Mersey Estuary. *Continental Shelf Research*, **22**, 1775–1794.

Tolhurst, T.J., Black, K.S. and Paterson, D.M. (2009) Muddy sediment erosion: insights from field studies. *Journal of Hydraulic Engineering ASCE*, **135**, 73–87.

Tolhurst, T.J., Consalvey, M.C. and Paterson, D.M. (2008) Changes in cohesive sediment properties associated with the growth of a diatom biofilm. *Hydrobiologia*, **596**, 225–239.

Tolhurst, T.J., Reithmüller, R. and Paterson, D.M. (2000) *In situ* versus laboratory analysis of sediment stability from intertidal mudflats. *Continental Shelf Research*, **20**, 1317–1334.

Tolhurst, T.J., Defew, E.C., de Brouwer, J.F.C. *et al.* (2006) Small-scale temporal and spatial variability in the erosion threshold and properties of cohesive intertidal sediments. *Continental Shelf Research*, **26**, 351–362.

Townend, I. and Whitehead, P. (2003) A preliminary net sediment budget for the Humber estuary. *The Science of the Total Environment*, **314–316**, 755–767.

Townend, I. and Pethick, J. (2002) Estuarine flooding and managed retreat. *Philosophical Transactions of the Royal Society of London, Series A*, **260**, 1477–1495.

Uncles, R.J. and Stephens, J.A. (1993) Nature of turbidity maximum in the Tamar Estuary, U.K. *Estuarine, Coastal and Shelf Science*, **36**, 413–431.

Uncles, R.J., Barton, M.L. and Stephens, J.A. (1994) Seasonal variability of fine-sediment concentrations in the turbidity maximum regions of the Tamar Estuary. *Estuarine, Coastal and Shelf Science*, **38**, 19–39.

Uncles, R.J., Lavender, S.J. and Stephens, J.A. (2001) Remotely sensed observations of the turbidity maximum in the highly turbid Humber Estuary, UK. *Estuaries*, **24**, 745–755.

Uncles, R.J., Stephens, J.A. and Smith, R.E. (2002) The dependence of estuarine turbidity on tidal intrusion length, tidal range and residence times. *Continental Shelf Research*, **22**, 1835–1856.

Van de Koppel, J., Herman, P.M.J., Thoolen, P. and Heip, C.H.R. (2001) Do alternate stable states occur in natural ecosystems? Evidence from a tidal flat. *Ecology*, **82**, 3449–3461.

Van der Wal, D. and Pye, K. (2004) Patterns, rates and possible causes of saltmarsh erosion in the Greater Thames area (UK). *Geomorphology*, **61**, 373–391.

Van Rijn, L.C. (1993) *Principles of Sediment Transport in Rivers, Estuaries and Coastal Seas*, Aqua Publications, Amsterdam.

Wang, Y., Hu, D. and Wu, X. (2002) Tidal flats and associated muddy coast of China, in *Muddy Coasts of the World: Processes, Deposits and Function*, vol. **4** (eds T. Healy, Y. Wang and J-.A. Healy), (Proceedings in Marine Science), Elsevier, Amsterdam, pp. 319–345.

Wellershaus, S. (1981) Turbidity maximum and mud shoaling in the Weser estuary. *Archiv für Hydrobiologie*, **92**, 161–198.

Whitehouse, R.J.S. and Mitchener, H.J. (1998) Observations of the morphodynamic behaviour of an intertidal mudflat at different timescales, in *Sedimentary Processes in the Intertidal Zone* (eds K.S. Black, D.M. Paterson and A. Cramp), (Special Publication 139, Geological Society of London), Geological Society Publishing House, Bath, UK, pp. 255–271.

Whitehouse, R.J.S., Bassoullet, P., Dyer, K.R. *et al.* (2000) The influence of bedforms on flow and sediment transport over intertidal mudflats. *Continental Shelf Research*, **20**, 1099–1124.

Widdows, J. and Brinsley, M.D. (2002) Impact of biotic and abiotic processes on sediment dynamics and the consequences to the structure and functioning of the intertidal zone. *Journal of Sea Research*, **48**, 143–156.

Widdows, J., Brinsley, M. and Elliott, M. (1998) Use of *in situ* flume to quantify particle flux, in *Sedimentary Processes in the Intertidal Zone* (eds K.S. Black, D.M. Paterson and A. Cramp), Special Publication 139, Geological Society Publishing House, Bath, UK, pp. 85–97.

Widdows, J., Friend, P.L., Bale, A.J. *et al.* (2007) Inter-comparison between five devices for determining erodability of intertidal sediments. *Continental Shelf Research*, **27**, 1174–1189.

Wolanski, E.J., King, B.A. and Galloway, D. (1997) Salinity intrusion in the Fly River Estuary, Papua New Guinea. *Journal of Coastal Research*, **13**, 983–994.

Wood, R. and Widdows, J. (2002) A model of sediment transport over an intertidal transect, comparing the influences of biological and physical factors. *Limnology and Oceanography*, **47**, 848–855.

Wood, R. and Widdows, J. (2003) Modelling intertidal sediment transport for nutrient change and climate change scenarios. *The Science of the Total Environment*, **341–316**, 637–649.

Woodroffe, C.D. (2003) *Coasts: Form, Process and Evolution.* Cambridge University Press, Cambridge.

Wright, L.D., Coleman, J.M. and Thom, B.G. (1973) Processes of channel development in a high-tide range environment: Cambridge Gulf – Ord River Delta, W Australia. *Journal of Geology*, **81**, 15–41.

Yang, X. (2005) Use of LIDAR elevation data to construct a high-resolution digital terrain model for an estuarine marsh area. *International Journal of Remote Sensing*, **26**, 5163–5166.

15 The Continental Shelf and Continental Slope

David N. Petley
University of Durham, UK

15.1 Introduction

Globally, it is estimated that rivers and glaciers supply in the range of $15–20 \times 10^{12}\,kg\,yr^{-1}$ of sediment to the ocean (Walling and Webb, 1996). Almost all of this sediment either crosses or is retained by the continental shelf and continental slope system. The only exceptions are where tectonic processes have meant that there is no continental shelf present – the proportion of the world's rivers draining into such areas is very small. Thus, the continental shelf and slope system is an absolutely key component of the global sediment cascade. However, our understanding of the dynamics of this environment remains surprisingly poor, although it has improved considerably in the past two decades. In this chapter, a review is provided of the key components of the sediment cascade of the continental shelf and continental slope systems. The importance of these environments in the sedimentary system cannot be underestimated: the volume of sediments on the global continental shelf and slope system is variously estimated as $150 \times 10^6\,km^3$ (Emery and Milliman, 1970) or $163 \times 10^6\,km^3$ (Khain, Levin and Tuliani, 1971). This volume is probably higher than the total volume in the much larger deep ocean basins.

The surface area of the continental shelf and continental slope combined is about $82.5 \times 10^6\,km^2$, which is equivalent to about 55% of the Earth's subaerial land surface. Of this the continental shelf has an estimated surface area of $27.5 \times 10^6\,km^2$, while the continental slope has a surface area of about $55.0 \times 10^6\,km^2$. If we are to properly understand the global source to sink sediment cascade then a better understanding of shelf and slope systems must be a priority. It is also notable that the continental shelf represents a key environment for a range of human activities, most notably in terms of fisheries and hydrocarbon resources. Of perhaps greatest concern, generation of tsunami by submarine landslides represents a substantial threat to coastal populations (Masson *et al.*, 2006).

Sediment Cascades: An Integrated Approach Edited by Timothy Burt and Robert Allison

15.2 Definition and Classification of the Continental Shelf and the Continental Slope

Geologically, both the continental shelf and the continental slope are components of the continental landmass, being underlain by continental rather than oceanic crust. They differ from the remainder of the continental system only in that they lie beneath the ocean surface, but of course this means that they are subject to very different processes to their subaerial equivalents, and as a result they are usually covered in comparatively thick drapes of marine sediments.

In most cases the geometry of the shelf is that of a broad, sediment-draped surface (Figure 15.1). On the seaward side it is terminated by the shelf break, which represents the crown of the continental slope. Globally, the width of the continental shelf averages about 80 km, but this varies greatly. Thus, for example, in some tectonically active areas, especially close to subduction zones such as to the east of the Philippines, no continental shelf is present, whereas in some ancient continental areas (such as Siberia) the continental shelf can extend to 1500 km from the current coastline. The maximum depth of the continental shelf is controlled by the location of the continental break, which is almost always at about 140 m below current sea level. This value is remarkably constant globally, reflecting both the effects of previous global sea levels and isostatic processes. The continental shelf typically has a gentle (*c*. 0.5°) seaward gradient and very subdued surface topography due to the sediment drape, although in some cases it is incised by recently formed canyons.

Beyond the shelf break lies continental slope, which typically has a gradient of about 3–5° (although it can vary from *c*. 1° to > 12°). This typically extends about 100 km from the shelf break, to a depth of about 3000 m. Beyond this lies the continental rise, which is essentially the sediment accumulation zone for material that has cascaded down the continental slope, and then the abyssal plains (Figure 15.1). This chapter focuses only on the continental shelf and the continental slope.

Two key types of continental shelf are generally recognized (Nittrouer and Wright, 1994).

(1) **Epicontinental shelves** are semi-enclosed basins with limited connections to the deep ocean. Examples include the North Sea and the Adriatic Sea.

(2) **Pericontinental shelves** lie on the margin of continents and are open to the deep ocean.

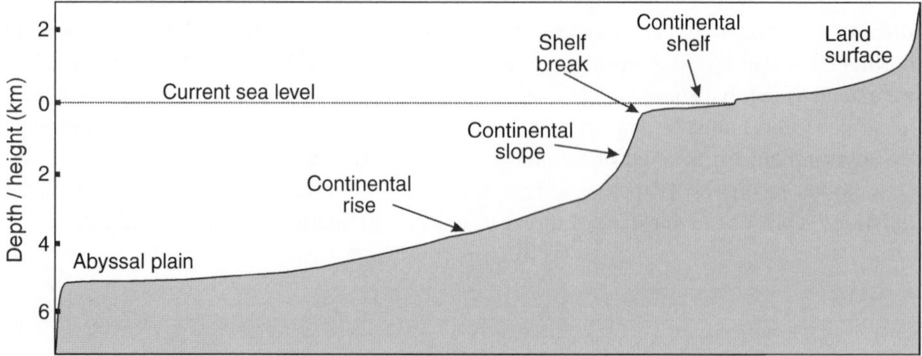

Figure 15.1 A summary of the main features of the continental shelf and continental slope system

Inevitably the nature of the sediment cascade varies greatly between these two environments, primarily because in epicontinental shelves sediment transportation to the deep ocean is a far less efficient process.

15.3 The Characteristics of the Continental Shelf

As the continental shelf is a transitional zone between the subaerial landmass and the deep oceans, it retains features of both. The bedrock geology is essentially that of the continent – the shelf represents no more than the flooded margin of the continent after all – but in terms of current environment it is clearly marine. However, over the course of the full-glacial–interglacial cycle a rather different pattern emerges. At the glacial maximum, when global sea level was typically about 135 m below the current mean level (Yokoyama *et al.*, 2000), large parts of the continental shelf would have been subject to subaerial conditions, and much of the rest would have been a very shallow marine environment. Even at average conditions for glacial conditions (*c.*120 m below current sea level), large parts of the continental shelf would have been exposed to subaerial conditions. In this environment rivers would have flowed across the shelf, incising to form valleys, while the rivers would have constructed deltas below what is now sea level. Marine sedimentation of the material brought down by the rivers would probably have occurred nearer to the edge of the continental shelf, and in some cases on to the continental slope.

The continental shelf is typically divided into the following three subzones based upon the physical processes that dominate sediment delivery and deposition.

15.3.1 The Inner Continental Shelf

The inner continental shelf is a high-energy zone extending from the shore to the zone in which the wind stress effects cease to be transmitted to the sea floor, which typically occurs at a depth of about 30 m, typically equating to distances of five to 15 km from the coastline. It is a high-energy environment that is contiguous with the coastal environment; as such it is dominated by coarse sediments. In this zone there is extensive sediment deposition of coarse-grained particles originating primarily from the terrestrial fluvial system.

15.3.2 The Mid-Shelf

The mid-shelf environment is a region in which the sea floor is largely not affected by wind stress but in which there is also no large-scale impact as a result of interactions with ocean currents or nearshore circulation. In this zone, deposition of finer grained sediment from the fluvial system occurs on quite a large scale. Comparatively small amounts of fine sediment reach this zone.

15.3.3 The Outer Shelf

The outer shelf is a zone that is dominated by processes that are affected by deep ocean circulation processes. This zone is dominated by relict sediment deposited when sea level was much lower than at present; as is shown below, comparatively little modern sediment is deposited in this region.

15.4 Current Sediment Environments on the Continental Shelf

While the aim of this chapter is not to review current sedimentary deposits on the continental shelf, it is important to briefly note the key features. Importantly, the vast majority of the sediments stored on the continental shelf are relict deposits that represent ancient shoreline and nearshore processes during periods when sea level was lower than at present (i.e. during the glacial periods). Thus, the outer-shelf deposits are typically coarse-grained sediments (sands and gravels) that are out of equilibrium with the current conditions. These were depoisted at and near to the mouth of the river systems flowing across the continental shelf. The inner-shelf deposits are typically similarly coarse-grained materials, in this case representing modern deposition of fluvial deposits. The mid-shelf deposits tend to be finer as the periods in which these areas have been at and close to the mouth of the river system are short due to the rate of sea-level change in the post-glacial period. However, in the modern environment sedimentation of finer grained particles occurs in this zone.

15.5 Contemporary Sources of Sediments on the Continental Shelf

Modern continental shelf environments receive sediment inputs from the following key sources, in addition to reworking of sediments already in place on the shelf:

- sediment delivered from river systems;
- sediment derived from coastal erosion, both in the form of removal of beach sediments and from the retreat of coastal cliffs;
- material transported through the aeolian processes, most notably dust from arid environments;
- material transported and dropped by ice;
- material brought laterally to the shelf from offshore, generally by ocean currents.

The amount of material delivered to the continental shelf from offshore is generally considered to be negligible (Johnson and Baldwin, 1986). Although it is difficult to derive reliable estimates for the mass of sediment transported into the oceans by rivers, the most reasonable estimate is about $18.5 \times 10^{12}\,\mathrm{kg\,yr^{-1}}$ (Lisitzin, 1996). The majority of this is transported across, and deposited upon, the continental shelf. However, the spatial pattern is undoubtedly highly variable, with the highest sediment yields being derived from the Pacific and Indian Ocean margins, where the combination of high relative relief, high precipitation rates and the presence of erodible materials facilitate high rates of erosion. It is unlikely that the other three sources of sediment input are really significant on a global scale compared with the input from river systems. For example, the total oceanic input of aeolian-derived sediment is estimated to be $1.6 \times 10^{12}\,\mathrm{kg\,yr^{-1}}$, but the vast majority ($1.3 \times 10^{12}\,\mathrm{kg\,yr^{-1}}$) of this represents an input directly to the deep ocean rather than routing via the continental shelf (Lisitzin, 1996). Similarly, coastal erosion is considered to deliver about $0.5 \times 10^{12}\,\mathrm{kg\,yr^{-1}}$ and glacial transportation about $1.5 \times 10^{12}\,\mathrm{kg\,yr^{-1}}$ (Lisitzin, 1996). The total terrigenous sediment delivery to the deep oceans is considered to be about $1.7 \times 10^{12}\,\mathrm{kg\,yr^{-1}}$ (Lisitzin, 1996), meaning that about $19.1 \times 10^{12}\,\mathrm{kg\,yr^{-1}}$ of terrigenous sediment is currently being deposited on the continental shelf and slope system.

15.6 Continental Shelf Sediment Cascades

Figure 15.2 presents a conceptual diagram of the main routes of sediment movement in the continental shelf and slope system. As shown above, the majority of sediment input on to the coastal shelf occurs as a result of fluvial input, with approximately 90% of this occurring as suspended sediment. Under normal (low flow) conditions, there is a rapid change in flow rate, and thus in carrying capacity, as the river enters the marine environment. This often causes rapid deposition of the coarser grained sediment, which has allowed the development of vast deltaic complexes on the continental shelf at the mouth of many of the world's largest river systems; for example, the submarine debris cone associated with the Ganges and Brahmaputra rivers has a volume of approximately $5 \times 10^6\,\mathrm{km^3}$ (Curray and Moore, 1974). Very large debris cones can also be found for the Nile, the Congo, the Orinoco, the Yangtze, the Yellow and many other major river systems. In many cases, these huge debris cones have inundated the continental slope to extend on to the abyssal plain.

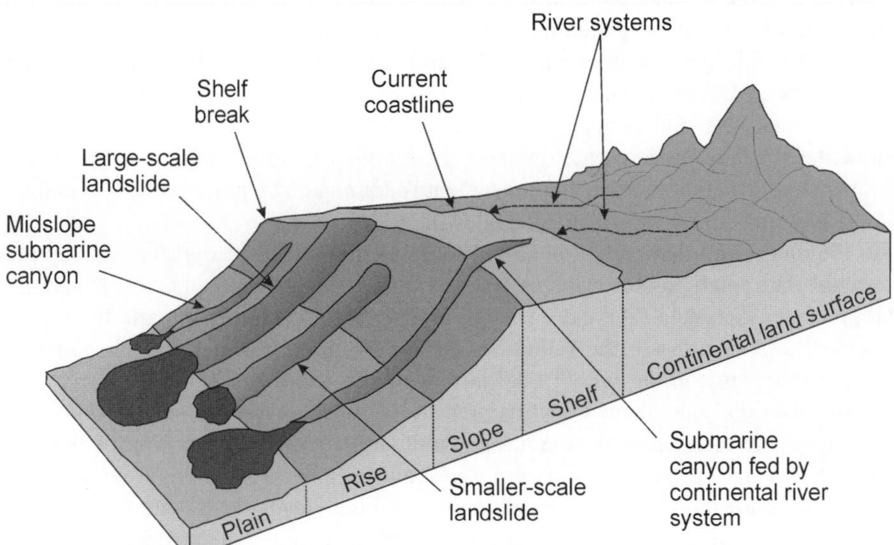

Figure 15.2 Key landforms on the continental shelf and slope associated with sediment mobility

Occasionally, if the density of the incoming flow is higher than that of the oceanic water into which it is flowing, then a hyperpycnal plume (a form of density current) may form (Imran and Syvitski, 2000). For a while it was thought that conditions for the formation of such flows into the ocean were comparatively rare as sea water has a specific gravity of 1.026, whereas the specific gravity of river water is close to unity. Thus, it was thought that the requisite high density of the incoming water could only be met when sediment concentrations were very high (typically $> 35\ \text{kg mL}^{-1}$, that is when the river is in spate, and even then only a comparatively small number of rivers were considered capable of achieving this high sediment load (Mulder and Syvitski, 1995). Perhaps surprisingly, these rivers tend not to be the very largest systems but instead to drain small mountainous watersheds (and thus to have an annual average discharge of $< 400\ \text{m}^3\ \text{s}^{-1}$) and to be formed from materials that can be easily eroded (Mulder and Syvitski, 1995). Such systems probably predominantly occur in highly tectonically active environments – for example there is strong evidence from the submarine geomorphology that such flows occur comparatively frequently on the rivers draining the eastern side of Taiwan (Ramsey *et al.*, 2006). Thus, hyperpycnal flows are a key aspect of oceanic sediment delivery in tectonically active areas.

However, recent research (McCool and Parsons, 2004; Wright and Friedrichs, 2006) has suggested that sediment-laden plumes may show negative buoyancy at surprisingly low concentrations ($< 1\ \text{kg m}^{-3}$), suggesting that these flows may occur more frequently than had been previously understood. Indeed, it is now considered by some that the buoyant plume seen emerging from a river mouth in flood may in fact be in effect a remnant feature, with the key marine transportation occurring in a hyperpycnal plume that forms a bottom current (e.g. McCool and Parsons, 2004; Parsons *et al.*, 2007). Nonetheless, Wright and Friedrichs 2006 noted that outflows from the largest rivers, such as the Amazon and the Yangtze, clearly show hypopycnal (not hyperpycnal) conditions. They found that in most cases hyperpycnal currents travel short distances from the river mouth, distributing sediment across the inner shelf. This sediment is then remobilized by water currents driven by waves and the wind. Long run-out gravity currents only normally occur when the hyperpycnal current intercepts an incised gully on the shelf.

Thus, although hyperpyncal currents do represent a mechanism that allows sediment entering the shelf environment to be transported directly to the continental slope and thus to the continental rise and beyond, most incoming sediment is deposited on the inner continental shelf. In modern environments the majority of this material is deposited on the inner shelf, where its subsequent movement is dominated by processes associated with tides, waves and storms (Johnson and Baldwin, 1986).

On the inner shelf there are a range of processes that affect the dynamics of sediment movement. In a small proportion of shelves (about 17% of the total global inner shelf area) tidal processes dominate. The inner shelves tend to occur primarily in partially enclosed basins such as the North Sea, the Yellow Sea and the South China Sea. Here, tidal ranges are greater than *c.* 3 m and the rate of tidal flow regularly exceeds 600 mm s^{-1}. In modern environments tidally dominated environments are almost always accommodation- rather than supply-controlled, allowing the formation of large areas of sand deposition that often displays a tabular or ridged structure. Rates of sediment movement in this environment are low, usually corresponding to slow migration of sand ridges. For example, Swift and Field (1981) measured migration rates of 2 m yr^{-1} for sand ridges on the Maryland sector of the North Atlantic continental shelf.

Sediment movement in most inner-shelf environments is dominated by storms and waves. Although these are clearly differentiated, the mechanics of these two movement types are generally considered together since they represent sections of the same spectrum of hydraulic processes. Two key types of water movement dominate.

(1) **Oscillatory and wave-drift currents:** wave action leads to oscillatory particle movements that vary in effect according to the depth of the water. At water depths of less than about 10 m even fair weather (low magnitude) waves can affect particle movement, but at greater depths movements are limited to storm waves. At greater depths the movements are primarily back and forth oscillations in which the shoreward component is generally greater than the seaward component. The net effect is to create a current that tends to move sediment entering the mid-shelf region back towards the shore. This is the so-called 'littoral energy fence' that generally reduces the movement of fine-grained coastal sediment on to the outer shelf and into the deeper ocean.

(2) **Wind-driven currents:** wind stress also induces water movement near to the surface, modified of course by the Ekman spiral. Quite high velocities have been recorded (*c.* 1 m s^{-1}), but generally only for short durations.

Exceptional storms, such as the movement onshore of a typhoon or a hurricane, may also induce seaward flowing bottom currents of quite high velocity. Here, the very strong winds drive surface water onshore, often creating damaging storm surges. These powerful surface currents may induce deep (sea floor) counter currents – flow velocities of 2 m s^{-1} have been estimated for very large hurricanes (Morton, 1981). These counter currents may be strong enough to transport substantial amounts of sediment to the outer shelf and even through to the continental break and thus down the continental slope. So, for example, Teague *et al.* (2007) reported 100×10^6 m^3 of bottom scour at an average depth of 60 m from an area of 525 km^2 in the northeast Gulf of Mexico during the passage of Hurricane Ivan in 2004. In this case the sediment was not thought to have descended the continental slope, but rather to have been deposited on the outer shelf close to the continental break, increasing the chances of a large-scale mass movement on the continental slope.

15.7 Sediment Cascades in Supply-Dominated Shelf Environments

Supply-controlled shelf environments, which are usually associated with tidally dominated deltaic systems, generally have a substantial input of fine-grained sediments. Most of this sediment originates in the form of suspended sediment from the fluvial system. Unlike the coarse-grained sediment above, this material follows a highly complex path across the continental shelf. On the shelf itself two key movement mechanisms are generally observed.

(1) **Sediment plumes:** here the material moves as a plume of sediment, with deposition along track. The trajectory of the plume is heavily affected by the local tidal and ocean currents.

(2) **Nepheloid layers:** these are layers of mobile sediment located in the boundary layer of the water body.

A substantial proportion of the fine-grained sediment released by the fluvial system is deposited on the shelf, but a substantial portion may also flow into the deeper ocean. As the input of sediment from the fluvial system is seasonal, and seasonal variations in weather patterns greatly influence the patterns of currents on the shelf, it is unsurprising that there is a strong seasonality in the deposition of fine-grained sediments. However, a very substantial amount of sediment is able to travel beyond the continental break – for example, at the edge of the northeast South American shelf at the Gulf of Paria it is estimated that 100×10^6 m^3 of fine-grained sediment enters the deep ocean each year (Johnson and Baldwin, 1986). Similarly, Kao *et al.* (2008) estimated that 85% of the *c.* 42×10^6 m^3 of fine-grained sediments that enter the Taiwan Strait flow into the deep ocean, despite the comparatively sheltered geometry of the shelf between Taiwan and China.

15.8 Sediment Cascades on Ocean-Dominated Shelf Environments

Due to the impact of the littoral energy fence, the outer shelf environment receives comparatively little coarse sediment. There is some input of finer grained sediment from plumes and nepheloid layers; although it used to be considered that these volumes were comparatively small, recent detailed sediment budgets have suggested that they may be substantial. Thus, for example, Sommerfield and Wheatcroft (2007) found that 20% of terrigenous sediment delivered to the ocean from the Eel River in California accumulates on the outer shelf, although note that this is a narrow shelf environment (*c.* 12–17 km). A small number of outer-shelf environments are affected by the effects of ocean currents; examples include the outer Saharan shelf, the Middle Atlantic Bight and the southeastern African shelf. In most cases the effect of the ocean currents is comparatively weak, requiring augmentation from other currents to allow sediment transport. Thus, the rates of sediment mobility in these environments are very low. However, in the case of the southeastern African Shelf, the Agulhas Current flows at a rate of up to 2.5 m s^{-1}, and thus is sufficiently powerful to strip an estimated 100×10^6 m^3 of sediment from the shelf per year.

15.9 Sediment Movements within the Continental Shelf Environment

Due to the effect of the distance from the dominant sediment sources (river mouths and the coast), the occurrence of strong currents that flow parallel to the coastline and the impact of the littoral energy fence, a large proportion of sediment movement on the continental shelf occurs as a result of remobilization of material already situated on the shelf. Indeed, in most environments the magnitude of along-shelf (shoreline parallel) movements of sediment is probably greater than the across-shelf component (Nittrouer and Wright, 1994). Movement primarily occurs in the form of suspended particles, especially where local currents change the flow of material emerging from river mouths and where wave action is able to mobilize basal material, primarily in the inner-shelf environment, plus some mobilization at low rates of bedload material. Thus, for example, for the Fly River in

Papua New Guinea approximately 50% of the sediment load of the river at the mouth is transported along the shelf (Harris *et al.*, 1993), whereas for the Yangtze River more than half the sediment load is transported southwards along the continental shelf (DeMaster *et al.*, 1985).

15.10 The Characteristics of the Continental Slope

Textbook depictions of the continental slope often give the impression that it is a comparatively short and steep slope with a planar morphology. This is highly misleading. First, the slope is generally rather long – the average distance between the continental break and the toe of the continental rise is about 75 km. In most cases the gradient of the continental slope is rather low – typically having a gradient in the range 1–6°, with an average of about 4°; this is a very low gradient compared with many slopes on the land surface. In many cases the slope gradient is at its maximum just below the shelf break, progressively reducing downslope, and with no abrupt break in slope marking the start of the continental rise. However, there remains a distinct difference between the continental slope and the continental rise in that the latter is founded on oceanic crust whereas the former is continental. In many locations the morphology differs from a long, gentle slope as the form of the continental slope system is highly dependent upon the local tectonic processes. Thus, for example, in some cases the continental slope consists of a much steeper, shorter cliff; this is the case offshore Chile for example where subduction has removed the continental rise and the sediments on the continental slope. On the other hand, offshore from Southern California the submarine topography mirrors the basin-and-range topography found onshore, with steep upper scarps and long approximately planar, sediment-mantled slopes below.

The surface of the continental slope is generally far from planar in form despite the very thick volumes of sediment located upon it. In most cases it is deeply incised with canyons and gullies, and it is often pockmarked with the scars of submarine landslides (Figure 15.3). The slope is formed from thick accumulations of primarily terrigenous deposits. Thus, on geological timescales the margin of the continental shelf cannot be considered to be a static entity; instead it progrades as sediment builds up upon the continental slope. The continental break may also migrate towards the shore if a large-scale landslide occurs within the sediment at the top of the slope. This appears to be an important process during large-scale earthquakes for example.

There are three fundamental mechanisms that allow the transport of sediment on to and down the continental slope. First, the continental slope receives a constant, low-intensity rain of fine-grained suspended sediment, primarily derived from the fluvial system. The volumes are small compared with that associated with turbidity currents and submarine landslides, but constant deposition over large areas means that this source of sediment cannot be ignored (Figure 15.2). Second, the continental slope is subject to density currents (turbidity currents and hyperpycnal plumes); that is, large-scale, high-velocity flows of sediment-laden water down the slope (Figure 15.3). Finally, the slope is subjected to submarine landslides of various types and on a range of scales. Whilst both of these processes have been the subject of considerable research in recent years, the difficulties associated with monitoring and measuring them means that they remain comparatively poorly understood. Their importance, however, is shown by the morphology of the

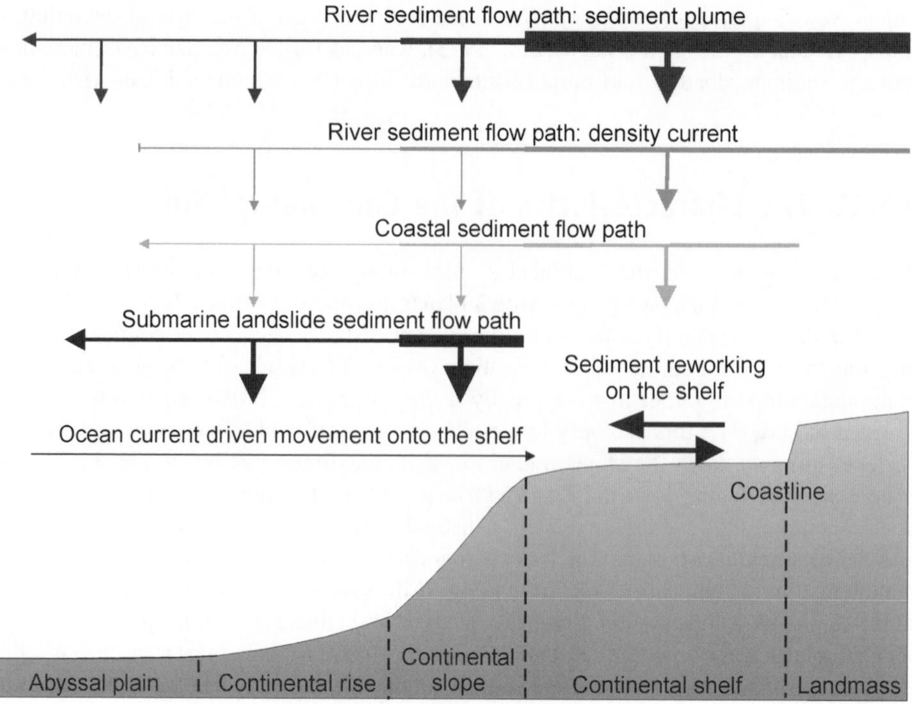

Figure 15.3 A conceptual summary of the sediment budget of the continental shelf and slope system

continental slope surface; the extensive canyons that generally occur on the slope mark the location of density current paths – sometimes the canyons extend to the top of the slope and can be traced to the mouth of terrestrial river systems. In other cases the canyons start midslope; here the scarp is often found to be a relict landslide initiation zone. Smaller scale gullies may also mark the location of lower magnitude failures, although they may also have formed as a result of spatial variations in depositional processes.

15.11 Mass Movements on the Continental Slope

The most comprehensive review of submarine mass movements is that of Huhnerbach and Masson (2004), who used a database of 260 submarine landslides from around the North Atlantic basin to review their characteristics. Although temporally quite rare, the scale of these events means that they are significant in terms of sediment movement down the continental slope. For example, for the 18 landslides analysed in the eastern North Atlantic, the median failure volume was 30 km^3, whilst the largest had a volume of 5580 km^3. The Storegga landslide in the North Sea, which is believed to have initiated in about one hour, involved about 3000 km^3 of sediment (Masson *et al.*, 2006), with a runout distance of about 800 km (Haflidason *et al.*, 2004). Interestingly, Huhnerbach and Masson (2004) found that the median water depth of the crown was at a depth in the range of 1000–1300 m below the surface, suggesting that the key role of submarine landslides is the transportation of sediment that has been deposited on the continental slope itself, rather than material on the margin of the continental shelf. Nonetheless, some submarine landslides do extend to the continental shelf edge, or even beyond; for example, the sea floor around the Canary Islands is littered with submarine landslide deposits, many of which may have originated as subaerial landslides (Parsons *et al.*, 2007).

Some of the very largest submarine landslides transition to form very long runout debris flows (e.g. Masson, 1994; Wynn *et al.*, 2002). It appears that progressive breakdown of blocks within the landslide mass allows the formation of a very dense, sediment-laden flow. It has long been recognized that these deposits can travel hundreds of kilometres. However, Talling *et al.* (2007) demonstrated that the capacity for these systems to transport very large volumes of sediment over huge distances is remarkable. Analysis of sediment cores from the Agadir basin, the Seine abyssal plain and the Madeira abyssal plain, all of which are located offshore from northwest Africa, demonstrated that very large submarine landslides can generate a turbidity current on the low-gradient slopes at the foot of the continental slope. These turbidity flows are able to travel across very low gradient slopes (in this case, *c.* 0.05°) without undergoing any deposition; indeed, in this phase the flow appears to have been erosive. After about 100 km a minor change in gradient (from 0.05° to 0.01°) induced deposition of a 2–3 m thick debris flow deposit, underlain by a coarser turbidite deposit. Thus, the landslide transported a volume of about 22.5×10^{13} kg of sediment (equivalent to 10 years of river sediment input to the global oceans) over a maximum distance of 1500 km (Talling *et al.*, 2007).

The implications of the Talling *et al.* (2007) study are profound. Clearly, such large-volume and very long runout systems could have a key role in the continental shelf and slope sediment cascade. However, at present our understanding of these systems is poor. Until we understand the likely number of such failures, and their magnitude–frequency relationship, it will be impossible to determine their true role. To a lesser extent this problem extends to submarine landslides in general – at present our limited knowledge means that it is difficult to estimate their relative importance in transporting material down the continental slope. Further work is needed to map in detail submarine landslide volumes and run-out distances, and to gain an age constraint upon them. Thus, the sediment cascade for the continental slope system remains remarkably poorly understood.

15.12 Density Currents on the Continental Slope

The term 'density current', also sometimes called a sediment gravity flow, represents a complex movement mechanism in which sediment moves as a result of its density relative to the surrounding fluid. A component of these flows are hyperpycnal plumes, described above, which occur when a fluvial system discharges sediment-rich water that is able to flow to the continental break and then down the continental slope, usually following a submarine canyon. In reality, density currents are a component of a complex spectrum of mass movements that occur on the submarine slope, ranging from submarine slides through to turbidity currents. Their origin is complex and poorly understood, but two key mechanisms have been identified in addition to the hyperpycnal plume mechanism described above.

(1) **Failure-induced formation:** a significant number of density currents originate as a result of a submarine landslide. Here, even a small failure is able to entrain material into the mobile flow, which bulks up and dilutes to form a turbidity current. Potential mechanisms for the generation of the initial failure include sediment accumulation at the continental break or on the continental slope; seismic activity; pore pressure generation as a result of hydrate release; and storm (especially tropical cyclone) activity. Talling *et al.* (2007) showed that the distal debris flow deposit that they identified was underlain by a turbidite deposit. This was assumed to have originated from the same event, showing the turbidites are able to move large volumes of sediment over large distances and high velocities.

(2) **Wave-induced suspension:** Some density currents are initiated through wave-induced suspension of sand or fine-grained particles, especially in the vicinity of shelf-break canyons (Inman, Aubrey and Pawka, 1976).

15.13 Sediment Budgets for Modern Margins

Unfortunately, few modern margins have properly calibrated sediment budgets that allow a full understanding of the sediment cascade. Those that do have adequate sediment budgets are constrained by the comparatively short period of observation, which may mean that high-magnitude–low-frequency events, such as large submarine slides or density currents, have been omitted. Those areas that do have properly calibrated sediment budgets for the continental shelf and slope show considerable variation in observed pattern (Table 15.1). Note that even for these well-calibrated examples there are considerable margins for error (unknown sinks), but nonetheless it does appear that 25–45% of sediment that is released from terrigenous sources is currently stored on the continental shelf, with up to a further *c.* 30% being stored on the continental slope. These values of storage on the shelf probably should be considered an upper boundary as low-frequency–high-magnitude events that may be a key component of sediment movement have probably not been captured by this dataset. A number of other areas have partially calibrated sediment budgets. Examples include:

- the Columbia River (USA), which discharges *c.* $10^{10}\,\mathrm{kg\,yr^{-1}}$ of sediment, of which approximately 65% accumulates on the continental shelf – with the highest concentrations being located in the mid-shelf region (Nittrouer *et al.*, 1979);

- the Changjiang River (China), which discharges *c.* $10^{11}\,\mathrm{kg\,yr^{-1}}$ of sediment, of which almost all accumulates on the continental shelf – about 40% of this total is sedimented in the region of the river mouth (DeMaster *et al.*, 1985);

- the Huanghe River (China), which discharges *c.* $10^{12}\,\mathrm{kg\,yr^{-1}}$ of sediment, of which about 85% accumulates on the shelf – with about 15% travelling to the deep ocean (Alexander, Demaster and Nittrouer, 1991);

- the Amazon River (Brazil), which discharges *c.* $10^{12}\,\mathrm{kg\,a^{-1}}$ of sediment, of which almost all accumulates on the shelf – although some movement along shelf is noted (Kuehl, Demaster and Nittrouer, 1986).

Table 15.1 A summary of key sediment cascades for continental shelf and slope systems that are adequately constrained

Name	Suspended sediment supply $10^6\,m^3\,yr^{-1}$	Inner shelf %	Middle–outer shelf %	Stored on continental slope %	Sink unknown %	Deep ocean	Along-shelf transport	Notes and references
Eel Canyon, California (USA), Pacific	19	*c.* 10	20	32	38	—	—	Sommerfield *et al.*, 2007
Santa Cruz Shelf, (USA) Pacific Ocean	2.5	44		8	31	—	17	Eittreim *et al.*, 2002
West Hokkaido forearc shelf, (Japan), Pacific Ocean	1.9	25		13–25	—	50–62	—	Noda and TuZino, 2007
Poverty Bay (New Zealand), Pacific Ocean	15	40		1	—	59	—	Slope and shelf rate estimated from highest recorded values Orpin *et al.*, 2006

Wheatcroft and Sommerfield (2005) suggested that there are four key factors that determine the rate of sediment accumulation on the continental shelf.

(1) **Timing of sediment delivery:** larger rivers may have a shorter sediment delivery distance in the ocean than do smaller systems. The explanation is that small rivers have short delivery times, meaning that they tend to discharge during periods when the sea is still being affected by the energetic storm that induced the high flow. Larger rivers thus tend to discharge high sediment loads into calmer water, allowing more proximal sedimentation. This may be of greater importance in the Pacific Rim, where very short, steep rivers are affected by highly energetic storms (tropical cyclones).

(2) **Wave climate:** there is considerable variation in wave climate around the world. There is strong evidence that more energetic wave climates favour more distant sediment deposition.

(3) **Shelf geometry:** it is intuitive that wider shelves will allow greater sediment deposition. However, it is likely to be a little more complex than this as the presence of canyons into even a wide shelf may increase rates of sediment loss to the continental slope and beyond. Additionally, even small increases in continental shelf gradient may increase the rate of sediment movement on the shelf.

(4) **Pathways of sediment transport:** different sediment mechanisms may have very different rates of movement associated with them. Thus, for example, where cross-shelf transport rates are high the rate of sediment loss may be reduced substantially.

15.14 Conclusion

The comparative poverty of our understanding of the sediment cascade in the continental shelf and slope environment is remarkable given the size and economic importance of this area. As Talling *et al.* (2007), there may well be potential mechanisms for sediment movement that are not included in a contemporary description of the sediment cascade for this environment. In addition, our poor understanding of magnitude–frequency relationships means that even those components that we do understand are difficult to analyse in terms of a long-term sediment budget. Nonetheless, it is clear that at present, even with the large increases in sediment production that have accompanied human development in river basins, the majority of shelves trap the largest proportion of incoming sediments, with comparatively small amounts reaching the deep ocean floor. However, it is also likely that our failure to adequately understand the number and magnitude of large-scale submarine landslides, which transport vast amounts of sediment from the shelf and upper slope to the deep ocean, means that the sediment budget of the deep marine environment probably underestimates the true current rate of sediment accumulation.

References

Alexander, C.R., Demaster, D.J. and Nittrouer, C.A. (1991) Sediment accumulation in a modern epicontinental-shelf setting: the Yellow Sea. *Marine Geology*, **98**, 31–72.

Curray, J.R. and Moore, D.G. (1974) Sedimentary and tectonic processes in the Bengal Deep-sea Fan and Geosyncline, in *Continental Margins* (eds C.A. Burk and C.L. Drake), Springer Verlag, New York, pp. 617–627.

DeMaster, D.J., McKee, B.A., Nittouer, C.A. *et al.* (1985) Rates of sediment accumulation and particle reworking based on radiochemical measurements from continental shelf deposits in the East China Sea. *Continental Shelf Research*, **4**, 143–158.

Eittreim, S.L., Xu, J.P., Noble, M. and Edwards, B.D. (2002) Towards a sediment budget for the Santa Shelf. *Marine Geology*, **181** (1–3), 235–248.

Emery, K.O. and Milliman, J.D. (1970) Quaternary Sediments of the Atlantic Continental Shelf of the United States. *Quaternaria*, **12**, 3–18.

Harris, P.T., Baker, E.K., Cole, A.R. and Short, S.A. (1993) A preliminary study of sedimentation in the tidally dominated Fly River Delta, Gulf Of Papua. *Continental Shelf Research*, **13** (4), 441–472.

Huhnerbach, V. and Masson, D.G. (2004) Landslides in the North Atlantic and its adjacent seas: an analysis of their morphology, setting and behaviour. *Marine Geology*, **213** (1–4), 343–362.

Inman, D.L., Aubrey, D.G. and Pawka, S.S. (1976) Application of nearshore processes to the Nile Delta. Proceedings of the Seminar on the Nile Delta Sedimentology, UNESCO/ASRT/UNDP, Alexandria, pp. 205–255.

Imran, J. and Syvitski, J.P.M. (2000) Impact of extreme river events on coastal oceanography. *Oceanography*, **13** (3), 85–92.

Johnson, H.D. and Baldwin, C.T. (1986) Shallow siliciclastic seas, in *Sedimentary Environment and Facies* (ed. H.G. Reading), Blackwell Scientific Publications, Oxford, pp. 229–282.

Kao, S.J., Jan, S., Hsu, S.C. *et al.* (2008) Sediment Budget in the Taiwan Strait with High Fluvial Sediment Inputs from Mountainous Rivers: New Observations and Synthesis. *Terrestrial Atmospheric and Oceanic Sciences*, **19** (5), 525–546.

Kuehl, S.A., Demaster, D.J. and Nittrouer, C.A. (1986) Nature of sediment accumulation on the Amazon continental shelf. *Continental Shelf Research*, **6**, 209–225.

Haflidason, H., Sejrup, H.P., Nygard, A. *et al.* (2004) The Storegga slide: architecture, geometry and slide development. *Marine Geology*, **213**, 201–234. doi: 10.1016/j.margeo.2004.10.007.

Khain, V.E., Levin, L.E. and Tuliani, L.I. (1971) The volume of sedimentary layers and probable reserves of carbons in the depression of the world ocean. *Reports of the Russian Academy of Sciences*, **200**, 5.

Lisitzin, A.P. (1996) *Oceanic Sedimentation: Lithology and Geochemistry*, American Geophysical Union, Danvers, Massachusetts, p. 400.

Masson, D.G. (1994) Late Quaternary turbidity current pathways to the Madeira Abyssal Plain and some constraints on turbidity current mechanisms. *Basin Research*, **6**, 17–33.

Masson, D.G., Harbitz, C.B., Wynn, R.B. *et al.* (2006) Submarine landslides: processes, triggers and hazard prediction. *Philosophical Transactions of the Royal Society A–Mathematical Physical and Engineering Sciences*, **364** (1845), 2009–2039.

McCool, W.W. and Parsons, J.D. (2004) Sedimentation from buoyant fine-grained suspensions. *Continental Shelf Research*, **24**, 1129–1142.

Morton, R.A. (1981) Formation of storm deposits by wind-forced currents in the Gulf of Mexico and the North Sea, in *Holocene Marine Sedimentation in the North Sea Basin* (eds S-.D. Nio, R.T.E. Shüttenhelm and Tj.C.E. Van Weering), Blackwell Science, Boston, MA, pp. 385–396.

Mulder, T. and Syvitski, J.P.M. (1995) Turbidity currents generated at river mouths during exceptional discharges to the world oceans. *Journal of Geology*, **103**, 285–299.

Nittrouer, C.A. and Wright, L.D. (1994) Transport of particles across continental shelves. *Reviews of Geophysics*, **32** (1), 85–113.

Nittrouer, C.A., Sternberg, R.W., Carpenter, R. and Bennett, J.T. (1979) The use of Pb-210 geochronology as a sedimentological tool: application to the Washington continental shelf. *Marine Geology*, **31**, 297–316.

Noda, A. and TuZino, T. (2007) Characteristics of sediments and their dispersal systems along the shelf and slope of an active forearc margin, eastern Hokkaido, northern Japan. *Sedimentary Geology*, **201** (3–4), 341–364. doi: 10.1016/j.sedgeo.2007.07.002.

Orpin, A.R., Alexander, C., Carter, L. *et al.* (2006) Temporal and spatial complexity in post-glacial sedimentation on the tectonically active, Poverty Bay continental margin of New Zealand. *Continental Shelf Research*, **26** (17–18), 2205–2224. doi: 10.1016/j.csr.2006.07.029.

Parsons, J.D., Friedrichs, C.T., Traykovski, P. *et al.* (2007) The mechanics of marine sediment gravity flows, in *Continental Margin Sedimentation: Transport to Sequence Stratigraphy* (eds C.A. Nittrouer, J. Austin, M. Field *et al.*), IAS Special Publication 37. Blackwell Publishing, Oxford, pp. 275–338.

Ramsey, L.A., Hovius, N., Lague, D. and Liu, C.-S. (2006) Topographic characteristics of the submarine Taiwan orogen. *Journal of Geophysical Research*, **111**, F02009. doi: 10.1029/2005JF000314.

Sommerfield, C.K. and Wheatcroft, R.A. (2007) Late Holocene sediment accumulation on the northern California shelf: Oceanic, fluvial, and anthropogenic influences. *Geological Society of America Bulletin*, **119** (9–10), 1120–1134.

Sommerfield, C.K., Ogston, A.S., Mullenbach, B.L. *et al.* (2007) Oceanic dispersal and accumulation of river sediment, in *Continental Margin Sedimentation: Transport to Sequence Stratigraphy* (eds C. Nittrouer, J. Austin, M. Field *et al.*), Ias special Publication 37. Blackwell Publishing, Oxford, pp. 157–212.

Swift, D.J.P. and Field, M.E. (1981) Evolution of a classic sand ridge field: Maryland sector, North American inner shelf. *Sedimentology*, **28**, 461–482. doi: 10.1111/j.1365-3091.1981.tb01695.x.

Talling, P.J., Wynn, R.B., Masson, D.G. *et al.* (2007) Onset of submarine debris flow deposition far from original giant landslide. *Nature*, **450**, 541–544.

Teague, W.J., Jarosz, E., Wang, D.W. and Mitchell, D.A. (2007) Observed Oceanic Response over the Upper Continental Slope and Outer Shelf during Hurricane Ivan. *Journal of Physical Oceanography*, **37**, 2181–2206.

Walling, D.E. and Webb, B.W. (1996) Erosion and sediment yield: a global overview, in *Erosion and Sediment Yield: Global and Regional Perspectives* (eds D.E. Walling and B.W. Webb), IAHS Publication **236**, International Association of Hydrological Sciences, Wallingford, pp. 3–19.

Wheatcroft, R.A. and Sommerfield, C.K. (2005) River sediment flux and shelf sediment accumulation rates on the Pacific Northwest margin. *Continental Shelf Research*, **25** (3), 311–332. doi: 10.1016/j.csr.2004.10.001.

Wright, L.D. and Friedrichs, C.T. (2006) Gravity-driven sediment transport on continental shelves: A status report. *Continental Shelf Research*, **26** (17–18), 2092–2107.

Wynn, R.B., Weaver, P.P.E., Masson, D.G. and Stow, D.A.V. (2002) Turbidite depositional architecture across three inter-connected deep-water basins on the Northwest African Margin. *Sedimentology*, **49**, 669–695.

Yokoyama, Y., Lambeck, K., De Deckker, P. *et al.* (2000) Timing of the Last Glacial Maximum from observed sea-level minima. *Nature*, **406**, 713–716.

Index

Sediment Cascades: An Integrated Approach Edited by Timothy Burt and Robert Allison

Surrey
Fairlands Youth Club, Guildford
Haslemere Methodist Youth Club

Tyne & Wear
Beth Jacobs Girls' Club, Gateshead
Blaydon & District Swimming & Life-Saving Club
1509 (Blaydon) Squadron, Air Training Corps
Caprian Amateur Dramatic & Operatic Society, Gateshead
Christchurch Youth Club, Felling
Fellside Junior Tennis Club
Gateshead Amateur Swimming Club
Gateshead Battalion Boys' Brigade
Gateshead East Primary Atheletics Assocation
Gateshead Sea Cadet Corps
Gateshead 2nd Scout Group
Gateshead 8th Scout Group
Gateshead 9th Scout Group
Gateshead 12th Scout Group
Gateshead 15th Scout Group
Gateshead 19th Scout Group
Gateshead 22nd Scout Group
Gateshead 26th Scout Group
Gateshead 33rd Scout Group
Gateshead Trampoline Club
Hertfordshire House Community Assocation, Gateshead
Quykham Venture Scout Unit, Newcastle
Redheugh Boys Club
St Theresa's Club for Handicapped Children, Gateshead
Smailes Lane Youth Club, Rowlands Gill
South Shields YMCA
Whickham Badminton Club
Whickham Choral Union
Whickham Tennis Club
1st Windy Nook Girls' Brigade Company
Windy Nook Methodist Junior and Youth Clubs
Winlaton Centre Badminton Club
Winlaton Centre Netball Club

West Midlands
All Saints Youth Club, Sedgley
Quarry Bank Community Association, Brierley Hill
Tipton Green Methodist Boys' Club

West Yorkshire
Canterbury Avenue Youth & Community Assocation
Earlsheaton High School, Kirklees
Netherton Youth Centre, Huddersfield
Phoenix Youth Centre, Huddersfield
R. M. Grylls Youth Club, Liversedge

Slough District Scout Council
Solihull Young National Trust
Somerset and South Avon Federation of Young Farmers' Clubs
Somerset Council for Voluntary Youth Services
Stafford District Voluntary Services
Suffolk Association of Youth
Suffolk Council for Voluntary Youth Services
Surrey Association of Youth Clubs and Surrey Physically Handicapped and Able Bodied
Sussex Youth Association

Thamesdown Voluntary Service Council, Wiltshire

Welwyn Hatfield Young Volunteers
West End Co-ordinated Voluntary Services for Homeless Single People, London
Worcestershire Federation of Young Farmers' Clubs
Worcestershire Girl Guides Association

Youth Action, Tameside
Youth in Action, Sutton

5. Other Local Youth Organisations, Clubs and Units

Avon
Three Lamps Methodist Youth Club, Bristol

Berkshire
Caversham Park Village Association
Wokingham Methodist Youth Club

Cambridgeshire
St Neots Centre, Huntingdon
Save the Children, Cambridge Project

Cheshire
Halton Lodge Youth Club, Runcorn

Devon
Whipton Youth Club, Exeter

Dorset
Harewood Centre, Bournemouth

East Sussex
Elm Court Centre, Seaford

Essex
Druid Venture Scout Group, Southend

Gloucestershire
Bisley Youth Club
Dursley Shell Group
Fairford Youth Club
Lonsdale Methodist Youth Club, Gloucester
Northway Club, Tewkesbury
Roxburgh House Youth Club
Sitmoc—Young People's Fellowship, Cheltenham
Soudley Youth Club
Stonehouse Community Centre
Winchcombe Youth Club
Wotton-under-Edge Youth Club

Greater Manchester
Firswood & District Community Association

Hampshire
2nd Aldershot Venture Scout Unit
Chandlers Ford Methodist Youth Club, Eastleigh
Romsey Methodist Youth Club
Weston Park Centre, Southampton

Hertfordshire
Sea Rangers Assocation, Potters Bar

Humberside
Snaith Boys' Club, Goole
Snaith Girls' Friendly Society, Goole

Kent
Culverstone Community Association, Meopham
Faversham County Youth Club
Maidstone YMCA
St Thomas' Youth Club, Canterbury
Showfields Youth Club, Tunbridge Wells

Lincolnshire
Grantham Play Association

London Borough of Enfield
Enfield Churches' (Noah's Ark) Open Youth Centre
33rd Enfield Guides
10th Enfield Sea Scouts
Enfield Youth Theatre
4th New Southgate Guide Company and Brownie Pack

London Borough of Greenwich
Greenwich Voluntary Workers' Bureau

London Borough of Islington
Springboard, Islington

London Borough of Merton
7th Morden Scouts
8th Morden Scouts

London Borough of Newham
Carpenters' and Dockland Centre
Church Army Hartley Centre
Eastlea Centre, Canning Town
Fairbairn House Boys Club, Plaistow
Kensington Youth Centre
Little Eye Club, Manor Park
Shipman Youth Centre

London Borough of Southwark
Southwark Community Development Project for the Mentally Handicappe

London Borough of Wandsworth
Balham Job & Training Workshop
Springfield Methodist Youth Club

Merseyside
Great Georges Project, Liverpool
Worcester Youth Centre, Bootle

North Yorkshire
Pocklington Youth Centre, York

Northamptonshire
Abington Community Association, Northampton

Northumberland
Buffalo Youth Club, Blyth
Cramlington Detached Youth Project
Cramlington Village Community Centre
Seaton Delaval Youth Club, Whitley Bay
South Beach Youth Club

Somerset
Lighthouse Methodist Youth Club, Burnham-on-Sea

South Yorkshire
Barnsley Sea Cadet Corps
Beckett Centre, Barnsley
Bolton-on-Dearne Church Lads' & Church Girls' Brigade
Elsecar Parish Church Open Youth Club
Willowgarth Youth Club
Young People's Interest Group, Barnsley

S. James the Great Youth Group, Pudsey
St John's Great Horton Youth Groups, Bradford

Wiltshire
Quest—Queen's Drive Methodist Church, Swindon

Others
Redbrook Youth Club, Gwent

6. Local Advisory and Co-ordinating Committees

Bradford Youth and Community Consultative Committee
Kirklees Area Youth Committees
Liverpool Youth Organisations Committee
Market Drayton Youth Leisure Committee
Tameside Youth and Community Services Advisory Committee
Washington Youth Organisations Council
Whitchurch Area Youth Committee, Shropshire
Woking District Youth and Community Committee

7. Youth Councils and Forums

Airedale Youth Council
Alnwick and District Youth Council
Amble District Youth Council

Barnet London Borough Council of Youth
Basildon Youth Council
Berwick Youth Council
Bournemouth Youth and Community Council
Burntwood Youth Council, Walsall
Bury St Edmunds Youth Forum

Chelmsford District Youth Council
Churchill Area Youth Council, Avon

Dudley Region Action Group

Enfield Youth Council

Formby Council of Youth

Gateshead Metropolitan Borough Youth Council

Hillingdon Youth Assembly

Leeds Youth Council

Morpeth Youth Council

North East Derbyshire Youth Council
North Trent Youth Members' Council

Ryde Area Youth Forum

Spen Valley and Mirfield Youth Council
Stoke-on-Trent Youth Members' Council

Sutton Young People's Council

Wansbeck Youth Council
Weston Area Youth Council

8. Information, Advice and Counselling Services

Advice Services Alliance

British Association for Counselling
Bromley Youth Project

Just Ask Advisory & Counselling Service, London

London Youth Advisory Centre

National Association of Young People's Counselling & Advisory Services

Off the Record, Bristol
Off the Record, Leeds
Open Door, Hornsey
Outlet, Oxford

Reading Youth Counsellors

Service Six Counselling, Wellingborough

Thirty Three, Cambridge

Under 21, Leytonstone

Walk-In Centre, London
Wandsworth Youth Information Counselling Project
Wycombe Youth Information Service

Young Persons' Advisory Service, Liverpool

9. Religious Bodies and Youth Organisations

Anglican Youth People's Association

Baha'i Faith
Bradford Diocese
British Council of Churches

Catholic Youth Service Council
Christian Movement for Peace
Church of England Board of Education

Derby Diocesan Council of Education

Methodist Association of Youth Clubs
Methodist Assocation of Youth Clubs—Wessex Region

Peterborough Diocesan Youth Committee

Quaker Workcamps

Southwark Diocesan Youth and Children's Council

Toc H

United Reformed Church

Young Christian Students

10. **Ethnic Minority Organisations and Race Relations Bodies**

Afro-Asian Society Council of Immigrants Welfare Service
Association for Jewish Youth
Associazioni Cristiane Lavoratori Italiani

Bangladesh Welfare Association
Black Youth and Community Workers' Association
Britain Turkish Cyprus Committee

Commission for Racial Equality

Hindu Cultural Society

Khayal

Manchester Council for Community Relations

National Association for Asian Youth
North Kirklees Community Relations Council

Overseas Chinese Education Centre

Pakistan Association, Liverpool
Preston Muslim Society

Runnymede Trust

Slough Community Relations Council

Ukrainian Community Centre
Union of Maccabi Associations

Warwick District Community Relations Council
Working Group on Training for Ethnic Minorities in Youth and Community Work

11. **Professional Bodies, Associations and Trade Unions**

Assistant Masters and Mistresses Association
Association of District Secretaries

Bournemouth & Christchurch Full-time Workers

Cleveland Social Services Workers' Group
Community and Youth Work Training Agencies' Staff Group
Community and Youth Service Assocation*
Community and Youth Service Assocation—Humberside Branch
Community and Youth Service Assocation—Norfolk Branch
Community and Youth Service Assocation—South-West Branch
Community and Youth Service Assocation—West Yorkshire Branch

Dudley Leaders' Council

Halesowen Leaders' Council
Headmasters' Conference

Leicestershire Youth Lobby
London Neighbourhood Youth Workers' Group

Managerial, Professional & Staff Liaison Group

*renamed The Community and Youth Workers' Union from April 1982.

National Association of Careers & Guidance Teachers
National Association of Head Teachers
National Association of Schoolmasters/Union of Women Teachers—Swindon Branch
National Association of Teachers in Further and Higher Education
National Association of Youth and Community Education Officers
National Association of Youth and Community Education Officers—Liverpool Branch
National Union of Mineworkers
National Union of Teachers
National Working Party of Young Volunteer Organisers

Principal Youth and Community Officers

Secondary Heads Association
Society of Education Officers
Southwark Project Workers' Group
Standing Conference of Principal Youth and Community Officers—London and Home Counties

Trades Union Congress

Warwickshire Leaders' Council

Youth Development Association

12. **Training Institutions, Advisory Councils and Other Bodies**

Brunel University, Regional Training Consultative Unit

Consultative Group on Youth and Community Work Training

Development Training Advisory Group

East Anglian Regional Advisory Council for Further Education

Federation of Community Work Training Groups

In-Service Training and Education Panel

Joint National Committee on Training for Playleadership

Manchester Polytechnic
Matlock College of Higher Education

National Group for Trainers
North-Western Regional Advisory Council for Further Education
Northern Council for Further Education

South-West Regional Council for Further Education
Southern Regional Council for Further Education
Sunderland Polytechnic

University of London Goldsmiths' College

West Midlands Advisory Council for Further Education
Westhill College

Yorkshire and Humberside Council for Further Education

13. **Other Educational Bodies**

Advisory Council for Adult and Continuing Education
Association for Liberal Education

British Council

Central Bureau for Educational Visits and Exchanges

Further Education Curriculum Review and Development Unit

14. **Sports Bodies**

British Schools Lawn Tennis Association

Central Council for Physical Recreation

National Playing Fields Association

Sports Council

15. **Other Organisations**

Equal Opportunities Commission

Library Assocation

National Organisation of Labour Students

Pilkington Brothers Ltd

Social Affairs Unit

Tone Vale Hospital, Taunton

16. **Individuals**

Dr. J. Adair

Mr. P. R. Barritt
Mr. P. Blake

Mrs. B. Crabtree

Mr. B. Davies
Mr. J. L. Deacon
Mr. D. J. G. Delmer
Mr. M. W. Dybeck

Prof. S. J. Egglestone

Mr. W. H. Farrar
Mr. W. Faulkner
Mr. W. Fortescue

Mr. R. S. George
Mr. D. Goodall
Mr. R. C. Goodey
Ms. C. T. Green
Mr. S. Griffiths
Mr. R. Gutfreund

Mr. F. W. G. Hardy

Revd. J. O. James

Mr. W. H. Keen
Mr. P. D. Killeen

Mr. R. B. Larkins
Mr. A. E. Lawrance and others
Mr. R. A. Lund
Mr. K. R. R. Lunn

Mr. D. McCarrick
Manchester Polytechnic Students
Mr. D. Marsland
Ms. J. Mays

Ms. D. J. Owens

Mr. W. Parvin & others
Mr. G. Peglar
Mr. L. Richards

Mr. B. Savell
Mr. C. Smith
Mr. J. R. Smith
Mr. S. W. Smith
Ms. A. Stoker

Mr. J. L. Thompson

Mr. A. Wadsworth
Mr. R. J. Webb
Mr. E. Wilkinson

Anonymous respondents

Various radio tapes

APPENDIX C: INTERIM STATEMENT*

1 The varied nature of the provision made by local authorities and voluntary bodies for young people in their leisure time, and the difficulty of simply defining its objectives and principles, make it vulnerable to two kinds of danger—on the one hand, a laissez-fair attitude disguised as 'voluntaryism' and on the other hand a tendency to seek simplistic solutions. In carrying out the first part of our terms of reference ('To report on present provision, both statutory and voluntary') we propose to avoid these dangers if we can by taking as our starting point the situation in which young people find themselves today, then analysing their specific needs (bearing in mind the widely differing circumstances, traditions and beliefs of different communities), and assessing the extent to which these are met (or not met). It would obviously not be sufficient to describe the provision made for young people—we shall also assess the effectiveness of its many varieties and put forward suggestions for future development, including the financial and legislative implications.

2 Recent events in inner-city areas may seem to give our work a certain topicality. But it would be a mistake to view the Youth Service simply in terms of a rescue operation. The Youth Service exists for all young people to use at all times wherever they are, and its effectiveness in helping with the problems of particular communities must depend upon a correct adjustment to the needs of youth in general. We are surrounded by a 'cloud of witnesses'. Warnings have for years been uttered, by community leaders, the churches, social workers, and youth workers, about the degree to which young people, especially in inner cities but not only there, are alienated from society. In extreme cases their frustration and despair leads to a rage directed against the outward forms of social control; but the basic alienation is what matters most. If the causes of this are not sought out and dealt with, the national calamity which may result seems likely to transcend the immediate effects of particular outbreaks of violence.

3 Unemployment has often been cited as one of the influences which create this alienation; and indeed it is a very potent cause, but not the only one. What we see lacking in the lives of many of the younger generation, particularly in our inner cities, is not so much employment as the whole range of experience which young people need in order to recognise the rights and responsibilities of adulthood. By 'experience' we here mean circumstances which often pass unnoticed when present, but whose absence may later be felt in a calamitous way—the experience, for example, of being valued for ones own sake; of belonging to a peer group; of having to submit to exacting routines; of taking responsibility collectively and being answerable to and for other people; and (not least) of being totally absorbed in a pleasurable activity. The full quality and range of such experience, marking the passage from the dependency of childhood to the independence and interdependence of adulthood, come from many sources. Employment certainly plays its part because (as is widely and rightly recognised) a young person's first job encapsulates many types of experience in one, including a vital measure of financial independence and the ability to satisfy felt needs. Every young person must have some way of finding this independence.

* Issued by the Review Group, with the agreement of the Secretary of State, in October 1981.

Alienation results when young people are deprived not just of employment, but of an appropriate mix of all these types of experience—secure family life, sustaining neighbourhood links, achievement in education, a social milieu in which they can take pride, financial independence, status in general.

4 All this we would expect to argue more fully in our final report. It may seem as though in advancing the above observations at this time we have anticipated the full and proper consideration of the mass of evidence which is now reaching us from local authorities, voluntary bodies, young people themselves and many organisations at national and local levels. Even a first perusal, however, suggests that though there are differences in language and also in priorities, there is a substantial degree of consensus on the needs of young people today. As we have said above, there is no mystery about the links between the deprivation of essential experience and the alienation and the despairing violence which result from this. How to prevent this is another matter and one to which we shall have to give much more consideration.

5 Meantime, it seems useful to us to advance certain positions. In the *first* place, since the forms of deprivation are so many and various it seems clear that no one service or agency can deal with them all. An attack on a broad front will be necessary. One of the traditional roles of the Youth Service has been to supplement, or in extreme cases provide substitutes for, the providers of the formative types of experience referred to above. But today the Youth Service must stand alongside the employment service, the further education and training services, schools, the social services, the probation services etc. All forms of social care must be brought into play. *Secondly*, it seems to us that, while this pluralism is not only unavoidable but indeed of benefit in providing flexibility of response to an infinite variety of needs, nevertheless some form of focus is needed both at national and local level. We shall have to study the evidence in much more detail before we can recommend what form this focus should take; but somewhere, somehow the essential management functions of evaluating, making policy, assigning roles and monitoring performance must be carried out. *Thirdly*, and turning to the Youth Service itself, we believe that it follows from what has been said above about the experiential needs of young people that the Youth Service must continue to offer the widest possible range of choice. The precise 'mix' of experience required differs from one young individual to another. It is important that young people themselves should be free to choose the form of experience which is suitable for them. No one can or should make that choice for them. It is equally important that they should know what is available and have a say in determining what it should be. Many varieties of provision have been devised within the Youth Service in order to make that power of choice an effective one. We feel sure that the public generally does not appreciate how wide and varied that provision is; and we shall aim to document it in our report. It includes, for example, clubs in all their many forms, leisure-time facilities, counselling services, adventure pursuits, enterprise groups, various forms of community service, and informal contact with adults other than family. All these varieties of provision are valuable in one way or another. Only one approach is and must be wrong: that is, to pick on one single variety—for example community service—and elevate it to the status of a universal medicine

which can and should be administered to all. That would be the very negation of what the Youth Service should be doing.

6 *Finally*, we must affirm a belief in the need for a strong and effective presence on the part of the Youth Service, both nationally and especially in areas of acute social deprivation. There are many areas where this presence is manifest and effective, and there are many areas where it is not. It is not always a question of the total amount of resources in money and buildings. It is often a question of choosing the right policy, of using resources to the best advantage, of taking steps to mobilise and encourage voluntary efforts of many kinds—in short, of arranging for the effective exercise locally of the management functions referred to above, both within the Youth Service and in relation to all forms of social care.

7 It should not be supposed that even a greatly enhanced Youth Service can alone supply all experiential needs of young people. In particular, work (however differently it may be defined in the future) is such a powerful provider of educative experience that it is difficult to see what could take its place with a great majority of young people. We welcome the emphasis placed on vocational preparation in the broad context of personal development in the MSC's 'New Training Initiative', on which we propose to offer comments shortly. But work cannot be all in all, any more than the Youth Service can be. We affirm our belief that now, more perhaps than at any time, it is necessary to take seriously the informal and voluntary terms of social education of the young in order to preserve their confidence in themselves and in their society.

APPENDIX D: THE Q-SEARCH REPORT—A BRIEF NOTE

At an early stage in its work the Review Group decided that it needed access to the opinions of young people to a more comprehensive extent than could be afforded by the submissions from young people's organisations and individual young people and the contacts made in meetings, visits and radio programmes. After investigation of possibilities and with professional advice, the Group decided to commission Q-Search of London to carry out a survey which would report in an objective and organised way on the views of young people in general. Given the limited time available, it was decided to confine the survey to young people between the ages of 14 and 19 which would cover the main span of the years of adolescence. For these young people, the survey would investigate the experience and activity followed at times when they had a choice and their reasons for choice of and evaluation of these experiences and activities. Information was wanted on what factors of a list of factors including enjoyment, relevance to development and adulthood, opportunties for responsibility, financial considerations and accessibility, most influenced young people when choosing activities; on their knowledge of the role and scope of the Youth Service; and their perception of their own problems. The sample of young people investigated was required to cover the social class spectrum, allow for at least proportional representation of the two largest ethnic minorities (the actual balance used is given below) and give a fair representation of the age-balance within the overall age-range, a fair balance of the two sexes and fair representation of different environments and areas of the country. While the overall sample would not in a survey of this type be large enough to permit contrastive analysis of the views of each group, it was hoped to have indications of the differing attitudes of the two sexes and the upper and lower halves of the age-group.

Method

The method proposed and carried out by Q-Search consisted of two stages: a 'qualitative stage' in which groups of young people gave their opinions in discussions on the matter being surveyed, and in which interviews with pairs and individuals were also held; and a 'quantitative stage' in which individual young people completed questionnaires, compiled after analysis of the qualitative stage and in discussion with the Review Group. The work on the two stages was carried out in August 1981 and January 1982 respectively, and the final report was completed in June 1982.

Sample

Qualitative stage

Each *group* contained 6 or 7 young people, and all groups were single-sexed. Group discussions lasted 3½ hours.

12 groups chosen to give equal representation to 6 regions of the country, 3 environmental areas (inner urban, outer urban and rural), 2 age-groups (14–15, 17–18), the two sexes, and two socio-economic groups (ABC1, C2DE)

Interviews were with pairs of young people for the 14/15 age group and with individuals for the rest. Each interview lasted 1½ hours.

12 interviews with young people from ethnic minorities chosen to give equal representation to two minority groups (Asian and West Indian), 3 regions of the country, 2 age-groups as above, the two sexes, and two socio-economic groups as above.
8 interviews with Caucasians chosen to give equal representation to 4 regions of the country, 3 environmental areas as above, 4 age-groups (14/15, 15/16, 16/17, 18/19), the two sexes and two socio-economic groups as above.

Quantitative stage

635 completed questionnaires were received.
The balance aimed at was:

A racial balance of 70% Caucasian, 15% Asian and 15% West Indian.
Equal representation of age-groups 14–16 and 17–19, the two sexes, and socio-economic groups as for qualitative stage and a fair geographical spread.

Amongst those who had passed minimum school-leaving age, equal representation of those in education, in employment and unemployed (including those on YOP schemes).

Printed in England for Her Majesty's Stationery Office by Commercial Colour Press, London E.7.
Dd.201686 C20 10/82